Modern Filter Theory and Design

Modern Filter Theory and Design

Editors

Gabor C. Temes
Professor, Electrical Sciences and Engineering Department
University of California, Los Angeles

and

Sanjit K. Mitra
Professor, Department of Electrical Engineering
University of California, Davis

A Wiley-Interscience Publication

JOHN WILEY & SONS

New York London Sydney Toronto

Library of Congress Cataloging in Publication Data

Temes, Gabor C 1929–
Modern filter theory and design.

"A Wiley-Interscience publication."
Includes bibliographical references.
1. Electric filters. 2. Electric networks.
I. Mitra, Sanjit Kumar, joint author. II. Title.

TK7872.F5T46 621.3815'32 73-7905
ISBN 0-471-85130-2

Printed in the United States of America

10 9 8 7 6 5 4 3 2 1

This book is dedicated to
Sydney Darlington
a pioneer of modern filter theory

Preface

1. The Aim of the Book. A filter is, in the generalized sense of the word, defined as an electric circuit which supplies a prescribed response to a given excitation. This response is usually *different* from the excitation in some specified ways, and is defined in terms of its behavior in either the frequency or the time domains. Thus, a filter performs a function which is typically much more sophisticated than the tasks of simple amplification or logical operations most often encountered in electronics.

Since the first introduction of filters in the early part of this century, there have been almost continuous advances in filter theory and technology. With the development of new computer components and design techniques, many new varieties of filtering methods have appeared and many are used in the practice today. Quite truly, "Filters have so permeated electronic technology that it is hard to conceive a modern world without them."*

Useful varieties of filters include lumped LC filters, crystal and ceramic filters, mechanical filters, microwave filters, lumped and distributed active RC filters, switched filters, *N*-path filters, and digital filters. Even though all these circuits have essentially the same objectives, enough differences exist to warrant a detailed study of each separate method.

Because of the variety of techniques and the mathematical sophistication accompanying them, filter theory and design was for many years regarded as a highly specialized subject. Its knowledge was scattered, available mostly in journal articles and handbook chapters, rather than comprehensive volumes.** More recently, this material has apparently reached the critical mass and coherence necessary for publication in the form of books; as a result, a number of such works appeared, authored or edited by experts such as E. Christian, Ph. Geffe, J. Herrero and G. Willoner, L. Huelsman, D. Humpherys, J. Skwirzynski, S. Stefanescu, A. Zverev, and others. These books performed a highly useful function in collecting the material available in the literature and presenting it in a coherent manner. However, they

* A. I. Zverev, *IEEE Spectrum*, **3**, 129–131 (March, 1966).

** There were a few exceptions, which included Cauer's important book, *Synthesis of Linear Communication Networks*, first published in 1941.

were primarily aimed at the industrial specialist, and typically covered only a few aspects of the total field.

The authors of this book regard filter design as the *mathematically based synthesis of practical electric circuits.* Therefore we believe that the fundamentals of modern filter theory and design should be learned at universities in comprehensive courses outlining the most important features of all filter categories. Courses in selective topics of filter design having indeed been taught for several years at leading universities in the United States and abroad.

The *main purpose* of this book is then to serve as a textbook for a new course in circuit theory and design. This new course should follow the courses now widely given in circuit analysis and in network synthesis. The level of the book is such that senior undergraduate or first year graduate students should be able to follow it easily. We also hope to provide the industrial designer with a summary of circuit design theory. The book does *not* aspire to serve as a handbook but rather as a text giving the foundations needed to understand the theory and the motivation behind the more application-oriented papers and volumes. The material contained in this book has also been successfully used as a text for intensive short courses in circuit design.

It is significant to note that the authors of this work have a dual background, having worked both in the industry as leading circuit design specialists, and at universities as engineering educators. Thus they are well qualified to write a textbook in a discipline which was born and reached maturity in the industry, and has now entered the classrooms of engineering schools.

2. Organization of the Book. Chapters 1–3 cover the basic theory and design of linear filters built from discrete passive components. Chapters 4–5 apply this theory to the special case of resonator filters. Chapter 6 discusses computer-aided optimization methods. Chapters 7–9 describe the theory and design of filters containing linear active and distributed elements. Finally, Chapters 10–12 treat circuits with switches and digital filters. Thus, all presently important areas of filter theory are at least briefly discussed. Much of the material included has never been published in a book; some has not been published in any form. None, to our knowledge, is obsolete as of 1973.

3. Prerequisites. The preparation assumed on the part of the reader consists of (1) an undergraduate level engineering mathematics course which discussed complex analysis, Laplace and Fourier transformation; (2) the usual circuit analysis course using such text as Desoer–Kuh, *Basic Circuit Theory*, or Wing, *Circuit Theory*, or Van Valkenburg, *Network Analysis*, or any similar text; and (3) a basic course in network synthesis, using such textbook as Balabanian, *Network Synthesis*, or Van Valkenburg, *Introduction to Modern Network Synthesis* or any of the numerous equivalent works.

4. Teaching the Book. The content of the book can be taught in 3 quarters or 2 semesters, in 4 weekly lecture hours; or, it can be covered somewhat abridged in one semester, using 6 weekly lecture hours. For the one semester, 6 lecture hours per week presentation, Chapters 4, 5, 6, and 10 may be discussed only lightly or dropped altogether.

5. Acknowledgments. The editors are grateful to all the authors, who contributed high-quality text under the severe constraints of stylistic and tutorial uniformity and within a rigid time schedule. They are also much indebted to Professors R. Gadenz, H. J. Orchard, and G. Szentirmai for critical readings of the completed manuscript. Finally, the support of the National Science Foundation under Grants GK 35904X and GK 14736 is gratefully acknowledged.

The nonalphabetical ordering of the editors' names has been decided by a tossed coin.

Los Angeles, California — *Gabor C. Temes*
Davis, California — *Sanjit K. Mitra*

Contents

9

ACTIVE DISTRIBUTED $\overline{RC}$ NETWORKS
by Ralph W. Wyndrum, Jr. **375**

10

APPLICATION, CHARACTERIZATION, AND DESIGN OF SWITCHED FILTERS
by M. R. Aaron **415**

11

N-PATH FILTERS
by L. E. Franks **465**

12

DIGITAL FILTERS
by Roger M. Golden **505**

INDEX 559

1

Introduction

S. K. Mitra
University of California
Davis, California

H. J. Orchard and G. C. Temes
University of California
Los Angeles, California

The theory of filters owes its origin to Wagner and Campbell, who in 1915 advanced the concept of passive electric wave filters. Since then, the theory and practice of filter design has advanced considerably with a consequent broadening of the term "filter." Presently, a filter is more or less defined as a network which is required to have a prescribed response for a given excitation. The response requirement may be given either in terms of time or frequency; in the latter case, usually selectivity is implied.

The theory of passive lumped filters, more commonly known as classical filter theory, has received the most attention during the last fifty years. The initial highly successful image-parameter theory has been gradually replaced by the more efficient insertion-loss theory of Cauer, Darlington, and Piloty. This process was accelerated by the increasing availability of high-speed digital computers, which has made it possible to implement economically filter-design techniques based on the latter, more complex, approach. In many applications, filters with sharply selective attenuation characteristics are needed. This necessitates the use of crystal and ceramic elements. Due to the complex models of these devices, a special theory had to be developed for design purposes. In the microwave range, the network elements are of a distributed nature and again require special design methods. However, some of the basic techniques used here evolved from the classical filter theory.

Probably the most significant recent impact on the progress of filter design and theory came from modern advances in micro-electronics. These led to the rapid development of inductorless active filters with lumped and distributed elements, of N-path filters and of digital filters. The development of digital transmission systems also required filters to work with switches—the theory of resonant transfer filters was thus born. In all of these new approaches to filter design, the computer has played an increasingly important role, and often new techniques have been developed either to take advantage of the high speed of modern-day computers or to get around the inherent limitations of digital computers.

The purpose of this book is to present the basic theory and some of the more important design techniques used in each of the above approaches to filter design.

1.1 BASIC CONCEPTS

In this book we are concerned mainly with single-input, single-output type filters, shown symbolically in Figure 1-1. The filter is assumed to be

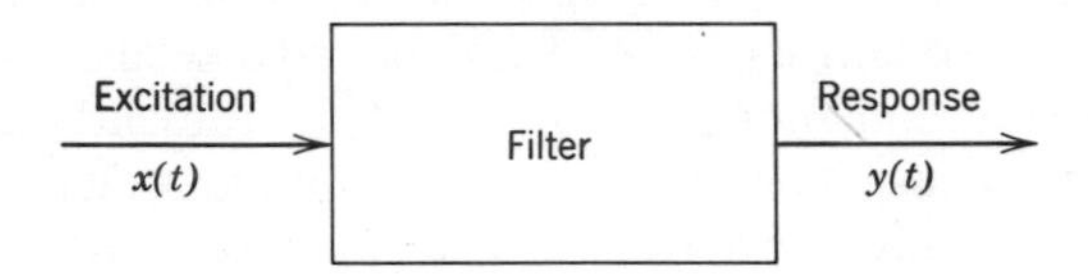

FIGURE 1-1 Single-input, single-output filter.

composed of only linear elements, so that the output variable (the response) will be linearly related to the input variable (the excitation). This relation characterizing the filter can be expressed either in the time domain or in the frequency domain.

Continuous-Signal Filters

In the filters described in Chapters 3 through 11 the input variable, $x(t)$, and the output variable, $y(t)$, are continuous functions of time which are related through a convolution integral

$$y(t) = \int_0^\infty h(t, \tau)x(\tau)\, d\tau. \tag{1.1}$$

When the filter is a time-invariant network, which is the more common case, (1.1) simplifies to

$$y(t) = \int_0^\infty h(t - \tau)x(\tau)\, d\tau. \tag{1.2}$$

By setting $x(t)$ equal to the unit impulse function $\delta(t)$ in (1.2), the integral can be evaluated directly to give

$$y(t) = h(t). \tag{1.3}$$

Thus $h(t)$ is seen to be the response of the filter to a unit impulse applied at $t = 0$. It is called the *impulse response* and, via the convolution integral, it completely characterizes the response of the filter (assumed to be initially at rest) to any excitation.

This time-domain relationship between input and output variables can be converted to the frequency domain by taking the Laplace transform of (1.2). This gives

$$Y(s) = H(s)X(s), \tag{1.4}$$

where $Y(s)$, $X(s)$, and $H(s)$ are the Laplace transforms respectively of $y(t)$, $x(t)$, and $h(t)$. $H(s)$ is a network function, defined in the usual way as the ratio of the Laplace transform of the response to the Laplace transform of the excitation.

If the filter is composed of lumped elements, then $H(s)$ will be a rational function and its poles will be the natural frequencies of the system. For the filter to be stable, these natural frequencies must lie in the left half of the complex s-plane. If the filter contains distributed elements, then $H(s)$ will be a more complicated type of function, but its singularities will still play a role analogous to that of the poles of a rational function, and, for stability, they also must be restricted to the left half of the s-plane.

The zeros of $H(s)$ are the complex frequencies for which the filter produces no output. These zeros may, in principle, lie anywhere in the complex s-plane, although frequently the selection of a specific circuit arrangement for the filter will impose constraints upon their location. In concluding these few comments on continuous-signal filters, it is necessary to explain an apparent inconsistency in the notations used by practical filter designers. When the external behavior of a filter is being considered in the context of the system to which it belongs, it will probably be described most conveniently by a conventional output/input type of network function. However, in the detailed numerical and experimental work necessary in designing and testing the filter, it usually proves simpler to describe the performance with an input/output type transfer function.

The reader will therefore find both types of transfer functions being used in the later chapters of this book, and he must be careful not to confuse one usage with the other. To make matters slightly worse, the standard symbol for the input/output function used with reactance two-port networks has, for over 30 years, been $H(s)$! The authors regret this potential source of confusion and hope that all usages are adequately defined where they first occur.

Digital Filters

For the digital filters described in Chapter 12, the input and output variables are sequences of numbers, each having a finite number of digits and usually residing in the register of a digital computer. The output sequence of numbers is generated, in order, one number at a time, from the input sequence by the action of a digital computer. The latter constitutes the digital filter. It may consist of a special-purpose computer with a wired-in program, or it may be a general-purpose computer with a program stored in the addressable memory.

As the input and output variables are just sequences of numbers, they can be represented quite adequately merely by attaching an integer subscript to the input and output symbols, e.g., x_n and y_n. Most commonly, however, the input sequence is obtained by sampling, at regularly spaced points in time, a continuous signal $x(t)$ which may, for example, be a voltage, or a force, or a temperature, etc.. A quantized version of each sample is then represented by one number in the input sequence.

To indicate the dependence of the input sequence upon the sampled function $x(t)$ and upon the sampling interval T, it is customary to write it as $x(nT)$. In a real-time digital filter one new member of the output sequence is calculated in each sampling interval T. The input and output sequences are thus generated at the same rate and so, to retain a consistent notation, the output sequence is written $y(nT)$. If required, a continuous output signal $y(t)$ can then be reconstructed from the sequence $y(nT)$. The name of the device which produces $y(t)$ as a voltage is digital-to-analog converter.

Although a digital filter is extremely flexible and can be made to produce almost any type of relationship between input and output, it is usually arranged to act as a linear system and then $y(nT)$ is related to $x(nT)$ by a discrete version of (1.2), namely

$$y(nT) = \sum_{m=0}^{n} h(nT - mT)x(mT). \tag{1.5}$$

By setting $x(mT)$ equal to the special signal

$$\begin{aligned} x(0) &= 1 \\ x(mT) &= 0, \qquad m \neq 0, \end{aligned} \tag{1.6}$$

which, for a digital filter, plays a role analogous to that of the unit impulse, (1.5) gives

$$y(nT) = h(nT). \tag{1.7}$$

The sequence $h(nT)$ which characterizes the behavior of the digital filter can thus be regarded as a kind of "impulse response."

In place of the Laplace transform, the theory of digital filters uses the z-transform [1] which, for a sequence $h(nT)$, is defined as

$$H(z) = \sum_{m=0}^{\infty} h(mT)z^{-m}. \tag{1.8}$$

Taking the z-transform of (1.5) gives [1]

$$Y(z) = H(z)X(z), \tag{1.9}$$

where $Y(z)$, $X(z)$, and $H(z)$ are the z-transforms respectively of $y(nT)$, $x(nT)$, and $h(nT)$. Equation (1.9) is the exact analog of (1.4). $H(z)$ can be regarded as a transfer function (of the complex variable z) which, just like the $H(s)$ for continuous-signal filters, is the ratio of the transform of the response divided by the transform of the excitation.

Normally $H(z)$ is a rational function of z, and the location of its poles determines whether the filter is stable. If $H(z)$ is of degree k it can be expanded into partial fractions

$$H(z) = \sum_{r=1}^{k} \frac{A_r}{1 - z^{-1}a_r}. \tag{1.10}$$

By adding the binomial expansions of each of the k fractions one obtains an explicit expression for the nth term of the sequence $h(nT)$, as the coefficient of z^{-n}:

$$h(nT) = \sum_{r=1}^{k} A_r a_r^n. \tag{1.11}$$

If this sequence $h(nT)$ is not to diverge as $n \to \infty$ then clearly $|a_r|$ must be less than or equal to unity for all r. This is equivalent to saying that the poles of $H(z)$ must lie within or on the unit circle.

It should be realized that at the input and output of a digital filter proper, there is no signal such as exists in the electrical filters described in Chapters 2 through 11, only a collection of numbers. One cannot, therefore, talk of the frequency response of the digital filter—it does not exist.

Nevertheless, if one considers the complete system, starting with the continuous signal $x(t)$ which is being sampled to provide the input sequence to the digital filter, and finishing with the continuous signal $y(t)$ which is being reconstructed from the output sequence, then this system *does* have both a frequency response and an impulse response. Much of the effort in designing a digital filter comes through relating the $H(z)$ of the digital filter to the transfer function of this overall continuous system, and of selecting $H(z)$ so that the continuous system has some desired characteristic.

Loss, Phase, and Delay

The performance of a filter in the frequency domain is usually specified in terms of the real and imaginary parts of the *logarithm* of the transfer function, rather than directly in terms of the transfer function itself. This follows the practice that was found most convenient in the design of transmission systems, consisting of a large number of cascaded subsystems. In such a system, the overall transfer function is the *product* of those of the subsystems; the logarithmic transfer function, on the other hand, is simply the *sum* of the elementary logarithmic functions.

The *loss* α, in nepers, and the *phase* β, in radians, are defined, quite generally, by

$$\alpha + j\beta \triangleq \ln\left\{\frac{\text{Laplace transform of excitation}}{\text{Laplace transform of response}}\right\}\Bigg|_{s=j\omega}. \tag{1.12}$$

By this definition the position of the zero point on the loss scale depends, rather fortuitously, on the specific choice of input and output variables; normally it is adjusted by adding an appropriate constant to α, so that the zero value of α corresponds to some simple physical condition, e.g., maximum transfer of power, the output level at zero frequency, etc..

In practical design work the theoretical units of nepers and radians are almost always converted to decibels and degrees, respectively, via the relations

$$1 \text{ neper} = (20 \log_{10} e) \text{ decibels} \cong 8.686 \text{ decibels}$$

$$1 \text{ radian} = \frac{180}{\pi} \text{ degrees} \cong 57.296 \text{ degrees.}$$

Associated with the phase β (in radians) is the *envelope delay* or *group delay* (these terms will be used interchangeably),

$$T_g \triangleq \frac{d\beta}{d\omega} \text{ seconds.} \tag{1.13}$$

Closely related to the envelope delay is the *phase delay* β/ω, but this quantity has no useful meaning except in lowpass systems, is not readily measureable, and hence is rarely employed.

1.2 THE NATURE OF A FILTER SPECIFICATION

In this section we discuss the characteristics which a filter is usually called upon to provide and the manner in which they are specified to the designer by the systems engineer who needs the filter.

The purpose of most linear filters is to separate a wanted signal from a mixture of the wanted signal and one or more unwanted signals. In general, the energy spectra of the wanted and unwanted signals will overlap, and a complete separation by means of a linear time-invariant network is then impossible. In these circumstances the best that can be done is to shape the loss and phase characteristics so that the ratio of wanted-to-unwanted signals at the output of the filter is maximized. The derivation of this best shaping for the loss and phase must take into account the detailed behavior of the spectra of all the signals involved, and it is a relatively complicated process. A typical example of this situation arises in pulse transmission where the received pulse is contaminated with noise.

Here we are concerned with the simpler and more common condition usually found, for example, in frequency-division multiplex systems in which the spectra do not overlap. In principle, complete separation of the wanted signal is then possible. In practice, the degree of separation that is achieved and the quality of transmission of the wanted signal are limited, partly by economic considerations, and partly by the quality of available components.

Passbands

The frequency band occupied by the wanted signal is referred to as the *passband.* In this band the ideal requirement for the filter is to provide constant loss and constant envelope delay so that the wanted signal will be transmitted with no distortion.

With a finite, lumped network one cannot achieve *exactly* constant loss over a finite band of frequencies, so it is customary to specify instead some acceptable upper and lower limits between which the loss can vary. These limits are chosen to accommodate not only the variations of loss that would exist even with an ideal network, but also the effects of component tolerances and of component dissipation. The latter effects are normally greater at the edges of the passband, and for this reason the limits there are often set further apart. Figure 1-2 illustrates an example. The permissible response must pass between the shaded barriers (broken line).

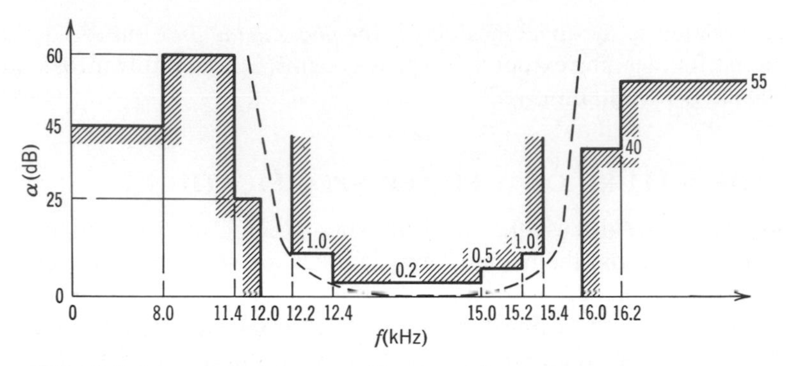

FIGURE 1-2 Typical loss specification. The broken line represents a permissible loss response.

Envelope Delay

For the same reason that one cannot get exactly constant loss over a finite band, it is also impossible to get exactly constant envelope delay. If it is important that the envelope delay of the transmitted signal be controlled, then upper and lower limits for the delay will be prescribed over the passband in a fashion similar to those for the loss, with the limits frequently being set further apart at the edges of the band (see Figure 1-3). Here, the broadening of the limits at the band edges is to help reduce the size of the filter, rather than to accommodate the effects of component imperfections, since the degree of the transfer function increases rapidly as one increases the fraction of the passband over which a tight tolerance on delay is demanded.

The minimum value of the delay in the passband can never be made zero, and in a tightly specified filter may be quite substantial. There are no very

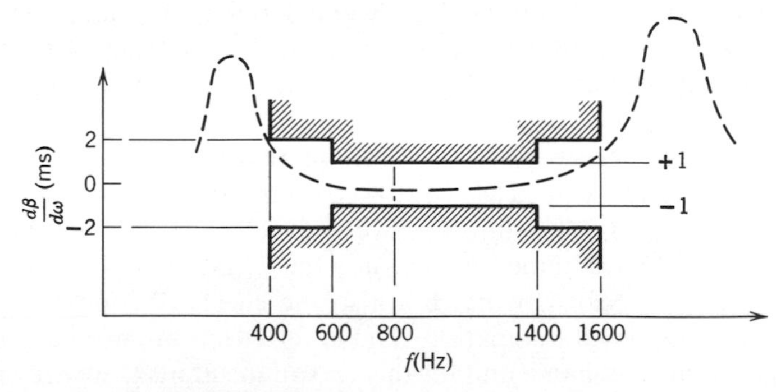

FIGURE 1-3 Typical group-delay specification. The broken line represents a permissible group-delay response.

simple rules for predicting with worthwhile precision just what flat delay one should expect, and consequently there is rarely any attempt to place limits on it. To do so might seriously handicap the designer. Normally one just has to accept whatever flat delay turns up as a result of meeting the specification on the constancy of the delay. The specification is thus concerned only with *relative* delay.

Often there is no need to control the envelope delay of the transmitted signal within the filter, either because it is of no interest, as in the case of a voice signal, or else because it will be corrected, together with the delay distortion due to other pieces of tandem-connected equipment, at some remote point in the system. In this event no specification is laid down for the delay, and it is assumed that a minimum phase network will be supplied.

Stopbands

The frequency bands occupied by the unwanted signals are referred to as the *stopbands*. In these bands the common form of specification merely requires the loss, relative to the lower limit set for the passband, to be equal to or greater than some minimum amount.

As far as the user is concerned, it is normally of no interest whether the actual filter loss in the stopband exceeds the amount specified by 1 dB or 20 dB, and in this sense stopband responses have only lower limits, in contrast to passbands where there are both upper and lower limits. Sometimes a filter designer is forced to provide much more than the specified minimum loss over parts of the stopband in order just to meet the specification at other parts. In general, however, he tries as far as possible to give no more stopband loss than is absolutely necessary, partly because more loss usually means a more expensive filter, and partly because the quality of the passband response goes down as the stopband loss goes up.

The minimum loss specified over the stopbands is set by the requirements of the system in which the filter is to operate. Sometimes the system designer is unable, for one reason or another, to be certain exactly how much loss is really needed at each frequency, and then it is customary to set the specification at the same constant loss for all frequencies. The value of this constant loss is chosen as an amount that is thought to be quite safe.

On other occasions the system designer can predict very accurately what minimum loss-frequency curve is needed, and the stopband specification given to the filter designer then takes the form of either a polygon or step-function approximation to this calculated curve. Such a specification usually gives the filter designer more confidence in its genuineness even though it may involve more effort to meet it.

Transition Bands

The loss response of a filter is a continuous function of frequency and is thus unable to have jump discontinuities. For this reason there must always be some interval in the frequency spectrum, separating the edge of the passband from the edge of the stopband, in which the loss can rise from the low value in the passband to that required in the stopband. These intervals are described as *transition bands.*

The bandwidth allocated to these transition bands is one of the main factors determining the size of filter needed to meet the specification. Moreover, as one decreases the width of the transition bands, not only does the complexity of the filter increase, but also it becomes more difficult to meet the specification set for the passband. Narrower transition bands mean that the loss has to change more rapidly near the passband edge, and this causes the passband response to be more sensitive to both dissipation and component tolerances.

Classes of Filters

In general a filter may have several, distinct passbands, and special design methods for such networks have occasionally been discussed in the literature. But usually a filter specification calls for only one passband, and all practical design techniques start with this assumption.

The most general single-passband filter is the *bandpass* filter in which the two edges of the passband are at finite nonzero frequencies and where there is a stopband on each side of the passband. The stopband below the passband extends down to zero frequency, and the one above extends to infinite frequency. The two stopbands may have quite different specifications and the filter is then referred to as an *asymmetric* bandpass filter.

Special types of bandpass filter specifications may possess some symmetry in their stopbands. If the latter are symmetrical on a logarithmic frequency scale around the passband, then the filter is called a *geometrically symmetric* filter. Alternatively, the stopband specification may be symmetrical on an arithmetic frequency scale around the passband, and the filter is then described as an *arithmetically symmetric* filter. Obviously the arithmetic symmetry cannot extend beyond twice the midband frequency, but by then the specification is usually unimportant anyway. Geometrically symmetric filters are specially easy to design, via reactance transformations, as will be shown later in this chapter, while the design of arithmetically symmetric filters, which are important in data systems and frequency-modulated systems, is generally quite difficult.

Two important limiting cases of the bandpass filter occur. The first is where the lower stopband disappears and the lower edge of the passband is set to zero frequency. The asymmetric bandpass filter then degenerates into a *lowpass* filter. The second is where the upper stopband disappears and the upper edge of the passband tends to infinity. This circuit is referred to as a *highpass* filter.

To every highpass filter there corresponds a lowpass filter, and vice versa. Each can be found from the other by replacing the variable s by ω_0^2/s, where ω_0 is some suitable constant having the dimensions of angular frequency. The replacement may be made at any stage in the design up to and including the actual networks. One can, for example, transform a transfer function for a lowpass filter into the transfer function for a highpass filter by means of this reciprocal transformation on the variable. Alternatively, applying the transformation to the element immittances turns inductors into capacitors, and vice versa, and this *reactance* transformation changes a lowpass filter into a highpass filter.

This relationship makes it unnecessary to have separate design procedures for both lowpass and highpass filters and the latter are normally designed by first transforming their specification to that of an equivalent lowpass filter. The lowpass filter is designed, and at the very last stage the highpass network is obtained by transforming the lowpass network. The reader is referred to Section 1.5 for the details of this process.

In addition to these three main classes of single-passband filters, namely lowpass, highpass, and bandpass, there is occasionally some demand for a special kind of double-passband filter called a *bandstop* filter. This has a single stopband separating two passbands, one extending down to zero frequency and the other extending up to infinite frequency. General asymmetric bandstop filters can be designed, but there are difficulties both in handling the two different passbands in the approximation stage and also in their realization as a ladder circuit. Consequently, bandstop filters are usually designed by a simple reactance transformation from a lowpass filter. The characteristics of the resulting bandstop filter are then symmetrical on a logarithmic frequency scale.

1.3 THE FILTER DESIGN PROCESS

The various steps which an engineer usually follows in designing a filter are laid out in the flow chart of Figure 1-4. Box 1 contains the specification of the performance which the filter must provide. It will describe the required loss and phase (or delay) behavior, as outlined in the previous section, together with any other desirable features, such as the behavior of the impedances at the ports, signal level, size, weight, cost, etc., which the system engineer considers important.

The next step, shown in Box 2, involves choosing the type of filter which can most easily be made to meet the specification. The choice will be

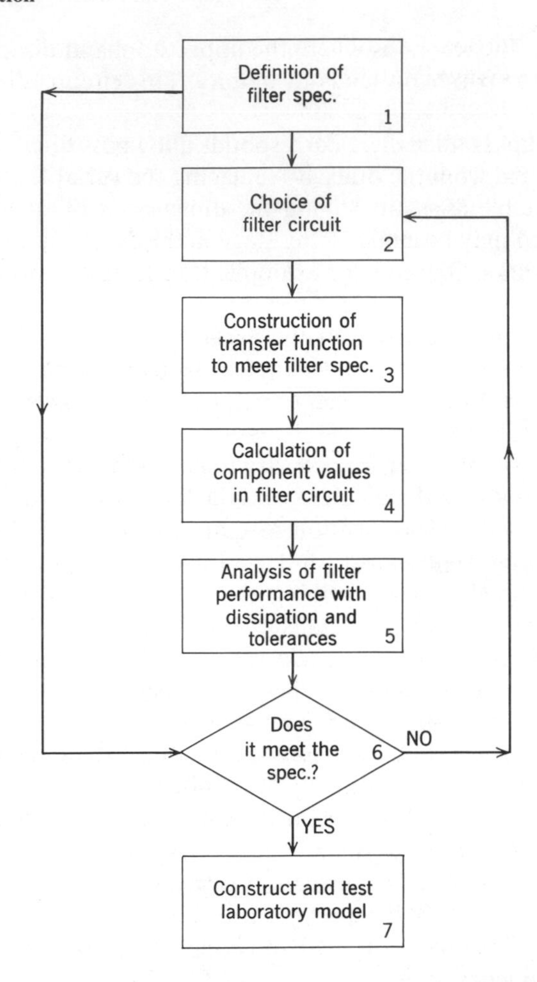

FIGURE 1-4 The flow-chart of filter design operations.

influenced by such factors as the frequency range of operation, degree of selectivity, impedance level, size, and cost. The filter designer has a wide variety of filter types to choose from, as the later chapters in this book will show, and considerable experience is necessary in making a wise selection for a specific job.

Once the class of filter has been decided the detailed design work can start. The first stage is indicated in Box 3. Each different class of filter tends to be approached in a different way, but in all cases there is some *approximation process* which is carried out in order to construct, or at least to specify, a suitable transfer function. The latter must simultaneously meet the specification set in Box 1 and be appropriate to the class of filter chosen in Box 2. Details of many standard techniques for carrying out this approximation process are described in Chapter 2.

The step from a knowledge of the transfer function to a knowledge of the component values (or their equivalents in the case of distributed networks) is shown in Box 4. Frequently, though fortunately not always, this requires a large amount of tedious numerical work, factoring polynomials, making continued-fraction expansions and similar operations. Nowadays this work is almost always done on a digital computer.

When the filter has been defined in terms of its component values the next step, shown in Box 5, is to analyze the performance of the filter. This is done firstly with the nominal component values in order to confirm the correctness of all the steps carried out in Boxes 3 and 4; and then with errors in the components in order to see whether the original specification will still be satisfied by the filter when made with practical components. The errors will be due partly to tolerances on the values (e.g., a random distribution of 2% errors in inductances and capacitances), and partly to dissipation and other parasitic effects such as lead inductance, inductor self-capacitance, etc. Although, in theory, many of these parasites can be included in the original design, and occasionally are, their effect is often so small that it is not worth while complicating the design to account for them.

At this stage the computed performance is compared with the original specification, Box 6. If it does not meet the specification, then the designer must repeat the steps 2 through 5 with different choices for the various parameters as indicated by the way in which the first design failed. Eventually, unless the specification cannot be met by any real network, the designer will find a network whose computed response is satisfactory. He then passes to the practical step of Box 7.

The mathematical models which we have for electrical filters are normally so good that the construction of a laboratory model no longer serves the purpose it used to have before the digital computer was available, namely to discover by measurement whether the design was satisfactory. Instead, the laboratory model is used more as a first prototype for production and as a sample for trials by the system engineer. Investigations of cross talk, from input to output and between the filter and other equipment, of overloading, and of nonlinear distortion are also more conveniently carried out with a laboratory sample than by calculation.

1.4 TRANSFER FUNCTIONS FOR REACTANCE FILTERS

Historically, lumped linear time-invariant reactance two-port circuits were the first filter networks used. Even today, their attractive properties (low cost, low sensitivity to component tolerances, simplicity, etc.) make them the most widely used filters. Furthermore, their theory forms the basis of much of the design theory of other, more general, filter classes. For this reason, in the first part of the book (in the remainder of Chapter 1, and in Chapters 2 and 3), the properties, network functions and design techniques of reactance filters will be discussed.

Transducer Function, Characteristic Function

Consider the doubly loaded reactance two-port shown in Figure 1-5. The impedances $Z_1(s)$ and $Z_2(s)$ or, equivalently, the reflection coefficients

$$\rho_1(s) = \frac{Z_1(s) - R_1}{Z_1(s) + R_1}$$

$$\rho_2(s) = \frac{Z_2(s) - R_2}{Z_2(s) + R_2} \tag{1.14}$$

describe the matching of the filter to its terminations R_1, R_2. The power transmission through the filter can be conveniently described by the *transducer power ratio*, defined as the ratio of the maximum power P_{max} available from the generator to the power P_2 actually absorbed by the load* R_2:

$$\frac{P_{max}}{P_2} = \frac{E^2/4R_1}{|V_2|^2/R_2} = \left|\frac{1}{2}\sqrt{\frac{R_2}{R_1}}\frac{E}{V_2}\right|^2. \tag{1.15}$$

For a passive two-port, no power gain is possible and hence $P_{max}/P_2 \geq 1$. For such networks therefore it is meaningful to define an auxiliary quantity, the *characteristic function* $K(s)$ by the equation

$$\frac{P_{max}}{P_2} = 1 + |K(s)|^2, \qquad s = j\omega. \tag{1.16}$$

Equation (1.16) remains valid even for active networks, if P_{max} denotes an arbitrarily specified maximum output power rather than the available gener-

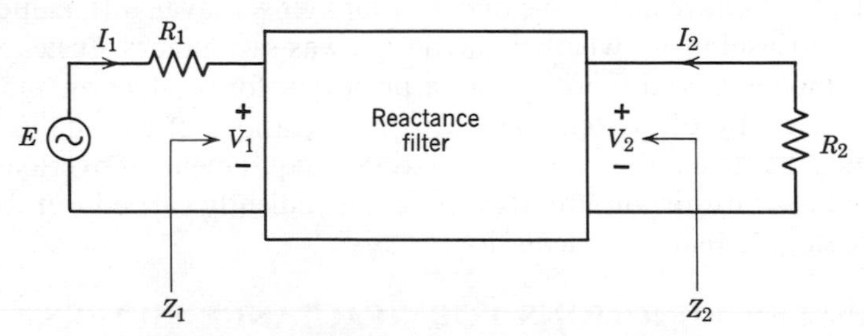

FIGURE 1-5 Doubly terminated reactance filter.

* Here we assume that the generator voltage E is real.

ator power. (Some power limits always exist in practice, due to the limited dynamic range of all circuits, and can, as a rule, easily be established *a priori.*) The advantages of introducing the characteristic function will become apparent soon.

If, inspired by (1.15), we also define the *transducer function* $H(s)$ by

$$H(s) = \frac{1}{2}\sqrt{\frac{R_2}{R_1}}\frac{E}{V_2(s)} \tag{1.17}$$

then, from (1.15)–(1.17), the relation

$$|H(s)|^2 = 1 + |K(s)|^2, \qquad s = j\omega \tag{1.18}$$

is obtained. As explained earlier, for the lumped linear time-invariant circuits considered in this section all network functions are rational in s with real coefficients, and hence (1.18) may be rewritten in the form

$$H(s)H(-s) = 1 + K(s)K(-s), \qquad s = j\omega. \tag{1.19}$$

Equation (1.19) is easier to use in actual computations than (1.18).

For the special, but most important case of *reactance two-ports,* no power is lost in the two-port. Then, (1.14) and (1.17) give, after some calculations,

$$|\rho_1(j\omega)|^2 = |\rho_2(j\omega)|^2 = 1 - \frac{P_2}{P_{max}}. \tag{1.20}$$

Hence, using (1.18),

$$|\rho_1(j\omega)|^2 = |\rho_2(j\omega)|^2 = \frac{|K(j\omega)|^2}{|H(j\omega)|^2}. \tag{1.21}$$

Accordingly, for reactance two-ports, one can also define $K(s)$ by the relation*

$$K(s) = \rho_1(s)H(s). \tag{1.22}$$

Since $K(s)$, $H(s)$, and $\rho_1(s)$ are all rational functions, they can be written in the form

$$H(s) = \frac{e(s)}{p(s)}$$

$$K(s) = \frac{f(s)}{p(s)}$$

$$\rho_1(s) = \frac{f(s)}{e(s)} \tag{1.23}$$

* The alternative definition $K(s) = \rho_2(s)H(s)$ is also possible.

where $e(s), f(s)$, and $p(s)$ are all polynomials in s. Note that H and K have the same denominator $p(s)$, due to (1.19); the form of $\rho_1(s)$ is then obtained from (1.22). In terms of e, f, and p, (1.19) becomes

$$e(s)e(-s) = p(s)p(-s) + f(s)f(-s) \tag{1.24}$$

where we use analytic continuation to remove the condition $s = j\omega$, i.e., to extend validity over the whole s-plane. Equation (1.24) (or any of its various alternative forms) is often called the *Feldtkeller equation.*

The approximation problem is usually solved by finding any two of the three polynomials e, f, and p; the third one can be calculated using (1.24). The synthesis then proceeds from e, f, and p. Hence, the properties of these polynomials which are necessary or sufficient for the realization of the two-port are of great interest. It will be shown in Chapter 3 that the following properties hold:

1. $e(s)$ is a strictly Hurwitz polynomial, i.e., its zeros are in the inside of the left half-plane;
2. $p(s)$ is a pure even or pure odd polynomial;
3. the degree of $e(s)$ is greater than or equal to that of either $f(s)$ or $p(s)$.

The zeros of $e(s)$ are the *natural modes* of the filter; the zeros of $p(s)$ are the *loss poles* (also called *transmission zeros*), while the zeros of $f(s)$ are often called *reflection zeros* or *zero-loss frequencies.*

Usually, the approximation problem is solved in terms of the rational functions given in (1.23). The polynomials constituting the numerators and denominators of these functions must satisfy the three constraints given above, as well as (1.24), the Feldtkeller equation.

Loss, Phase, and Delay for Reactance Two-Ports

The loss α, phase β, and the group delay T_g, defined previously in Section 1.1, can all be obtained for reactance two-ports directly from $H(j\omega)$. The zero-loss point then corresponds to maximum transfer of power:

$$\alpha(\omega) = 10 \log_{10}\left(\frac{P_{\max}}{P_2}\right) = 20 \log_{10}|H(j\omega)|. \tag{1.25}$$

The phase, as seen from (1.12), is defined as the difference between the angles of E and of $V_2(j\omega)$, which is by (1.17) simply the angle of $H(j\omega)$. Hence,

$$\beta(\omega) = \tan^{-1}\left[\frac{H_o(s)/j}{H_e(s)}\right]_{s=j\omega} = \frac{1}{j}\tanh^{-1}\left[\frac{H_o(j\omega)}{H_e(j\omega)}\right]. \tag{1.26}$$

Here, the subscripts e and o indicate the even and odd parts of $H(s)$, respectively:

$$H_e(s) = [H(s) + H(-s)]/2$$
$$H_o(s) = [H(s) - H(-s)]/2. \tag{1.27}$$

From $\beta(\omega)$, the group delay T_g can be obtained using its definition, (1.13):

$$T_g = \frac{d\beta(\omega)}{d\omega} = \left[\frac{d}{ds}\tanh^{-1}\frac{H_o(s)}{H_e(s)}\right]_{s=j\omega} = \left\{Ev\left[\frac{H'(s)}{H(s)}\right]\right\}_{s=j\omega}. \tag{1.28}$$

Here, $Ev[f(s)]$ denotes the even part of $f(s)$ and $H'(s) \equiv dH(s)/ds$. Since $H(s) = e(s)/p(s)$, T_g can also be written in the form

$$T_g(\omega) = \left\{Ev\left[\frac{e'(s)}{e(s)} - \frac{p'(s)}{p(s)}\right]\right\}_{s=j\omega} = \left\{Ev\left[\frac{e'(s)}{e(s)}\right]\right\}_{s=j\omega}. \tag{1.29}$$

The last equality holds only for reactance two-ports, where $p(s)$ is a purely even or odd polynomial and hence $p'(s)/p(s)$ is odd. Evidently, by simple analytic continuation,

$$T_g(s) = Ev\left[\frac{H'(s)}{H(s)}\right] = Ev\left[\frac{e'(s)}{e(s)}\right] \tag{1.30}$$

can be defined as a function of s which for $s = j\omega$ coincides with the group delay.

An important special case is that of the *allpass functions*, described by $e(s) = [e_a(s)]^2$, $p(s) = e_a(s)e_a(-s)$

$$H(s) = \frac{e(s)}{p(s)} = \frac{[e_a(s)]^2}{e_a(s)e_a(-s)} = \frac{e_a(s)}{e_a(-s)}. \tag{1.31}$$

Here, the left-half-plane loss poles cancel half of the doubled natural modes. These functions have the property that $|H(j\omega)| \equiv 1$ and thus $\alpha(\omega) \equiv 0$; hence their name. For these functions, with $e(s) \equiv [e_a(s)]^2$,

$$T_g(s) = 2Ev\left[\frac{e_a'(s)}{e_a(s)}\right]. \tag{1.32}$$

For modulated signals, $T_g(s)$ describes the amount of delay suffered by the modulating time function which forms the envelope of the signal. For unmodulated (baseband, video) signals, the variations of the phase delay $T_{ph}(\omega)$ define the relative delay of the component sine-waves of the signal. Even then, however, the group delay is often used, since it is usually a more sensitive indicator of deviations from the ideal linear-phase behavior than the phase delay.*

The main advantages of dealing with $T_g(\omega)$ rather than $\beta(\omega)$ are, however, its simple mathematical form, which avoids manipulation of transcendental functions and hence makes both theoretical and numerical analysis and design relatively simple, and the fact that it is easily measured.

Impulse Response, Step Response, Time Response

If the circuit of Figure 1-5 is excited by a general signal $e(t)$ rather than by a sine-wave,† the Laplace transform of the output signal $v_2(t)$ can be obtained from (1.17):

$$V_2(s) = \frac{1}{2}\sqrt{\frac{R_2}{R_1}}\frac{E(s)}{H(s)} \tag{1.33}$$

where $E(s) = \mathscr{L}[e(t)]$ is the Laplace transform of $e(t)$. If we also denote

$$h(t) = \mathscr{L}^{-1}\left[\frac{1}{2}\sqrt{\frac{R_2}{R_1}}\frac{1}{H(s)}\right] = \frac{1}{2}\sqrt{\frac{R_2}{R_1}}\mathscr{L}^{-1}\left[\frac{p(s)}{e(s)}\right] \tag{1.34}$$

then, by (1.2),

$$v_2(t) = h(t)*e(t) = \int_0^\infty h(t-\tau)e(\tau)\,d\tau, \tag{1.35}$$

where the asterisk $*$ denotes the convolution operation defined by the integral. Specifically, if $e(t)$ is the unit impulse $\delta(t)$, then

$$v_2(t) = h(t)*\delta(t) = h(t). \tag{1.36}$$

Hence, the time function $h(t)$ defined in (1.34) is the *impulse response* of the circuit of Figure 1-5. If, on the other hand, $e(t)$ is chosen as the unit step function

$$e(t) = u(t) = \begin{cases} 1, & t \geq 0 \\ 0, & t < 0 \end{cases} \tag{1.37}$$

then $E(s) = 1/s$ and hence

$$v_2(t) = g(t) = \mathscr{L}^{-1}\left[\frac{1}{2}\sqrt{\frac{R_2}{R_1}}\frac{p(s)}{se(s)}\right]. \tag{1.38}$$

The function $g(t)$ is called the *step response* of the filter.

* Very often, for linear-phase lowpass filters, the deviation from the exact linear phase is a nearly periodic function of ω with a rapidly increasing amplitude. Then the phase- and group-delay errors will be weighted replicas of the phase error; T_{ph} with a $1/\omega$ weighting factor, T_g without a weighting factor. Hence, T_{ph} will *attenuate* the sensitive parts of the phase error versus ω curve.

† $e(t)$ must not be confused with $e(s)$, the numerator polynomial of $H(s)$.

For filters excited by signals other than steady-state sine-waves, the behavior of the resulting output $v_2(t)$ is of great interest. For general input, $v_2(t)$ is given by (1.35); for impulse or step input by $h(t)$ or $g(t)$, respectively. Usually it is adequate to design a network for an acceptable $h(t)$ or $g(t)$, although sometimes $v_2(t)$ is specified for a given $e(t)$. The process of satisfying such requirements is called *time domain approximation.*

Theoretically, it is of course possible to take the Laplace transform of $e(t)$ and the required response $v_2(t)$; then (1.33) can form the basis of finding $H(s)$ by approximating in the complex s-plane or on the $j\omega$-axis. Unfortunately, however, approximation errors in the s-plane and in the t-domain are only loosely connected. For example, a small sine-wave-like error in the frequency response $H(j\omega)$ may result in unwanted sharp peaks (echoes) of considerable amplitude in the impulse response $h(t)$. Hence, time domain approximation should preferably be done directly in terms of the time variable t, rather than indirectly in terms of the frequency variable ω, even though this will almost always entail much more work. We will not discuss time domain approximation in this book; the reader is referred to the comprehensive work by Su[2] for information. Equations (1.34)–(1.38) enable the designer, however, to check the time responses of a filter designed from frequency domain considerations.

1.5 SCALING AND REACTANCE TRANSFORMATIONS

Scaling

The numbers describing electrical quantities in filter theory are often exceedingly large (e.g., 10^8 Hz) or small (e.g., 10^{-11} F) when expressed in their conventional SI units. This causes inconveniences in manual calculations and can lead to computer over- or underflow in computer-aided circuit analysis or design. Also, theoretical as well as numerical results can be simplified and made more general if certain key quantities are arbitrarily assigned the value 1. For example, if in the circuit of Figure 1-5 R_1 and R_2 are equal, then using the common value of R_1 and R_2 as impedance unit will considerably simplify all theoretical equations describing the operation of the circuit. This arbitrary assignment of units is called *scaling* or *normalization.*

It is obvious that not all units in the "normalized" circuit may be chosen at will. For example, if the units of impedance, radian frequency, and capacitance are selected as Z_0, ω_0, and C_0, respectively, then before normalization the equation

$$|Z| = 1/\omega C \tag{1.39}$$

holds for a capacitor. After normalization, it is natural to require that the normalized quantities z, Ω, and c be also related by

$$|z| = 1/\Omega c. \tag{1.40}$$

But, since by definition the normalized values are given by

$$\begin{aligned} z &= Z/Z_0 \\ \Omega &= \omega/\omega_0 \\ c &= C/C_0, \end{aligned} \tag{1.41}$$

(1.40) requires that the units Z_0, ω_0, and C_0 satisfy

$$Z_0 = 1/\omega_0 C_0. \tag{1.42}$$

Similar considerations lead to the relation

$$Z_0 = \omega_0 L_0 \tag{1.43}$$

where L_0 is the unit inductance.

In addition, the phase angle of the sine-wave time functions in the circuit is given by

$$\beta = \omega t. \tag{1.44}$$

Introducing the unit time t_0 and the unit frequency ω_0, the phase angle in the normalized circuit is

$$B = (\omega/\omega_0)(t/t_0). \tag{1.45}$$

For physical reasons, we want to keep $\beta = B$. (For example, a normalized phase inverter should still act as a phase inverter!) This requires, therefore, that the units ω_0, t_0 be related by

$$\omega_0 t_0 = 1,\ t_0 = 1/\omega_0. \tag{1.46}$$

From the above, it should be clear that only *two* variables can be scaled independently. In practice, these are usually chosen to be impedance and frequency. At the start of the design, some convenient values Z_0 and ω_0 are selected as reference values for impedance and angular frequency, respectively,* and the normalized filter is designed. Afterwards, the final real network element values and responses may be obtained from the normalized ones using the units Z_0, ω_0 and

$$L_0 = Z_0/\omega_0,\ C_0 = 1/(\omega_0 Z_0) \tag{1.47}$$

$$t_0 = 1/\omega_0. \tag{1.48}$$

* Z_0 is usually the (positive real) value of one of the terminating resistances, while ω_0 the angular frequency at some convenient point, e.g., at a passband edge.

One big advantage afforded by normalization is that it makes the compilation of "filter catalogs" possible. These are tables of the element values or critical frequencies (zeros and poles) of filters, usually normalized to $R_1 = R_2 = 1$ and $\omega_0 = 1$ (for filters) or $t_0 = 1$ (for delay networks). Equations (1.48) then enable the designer to adapt the tabulated circuit for his specific requirements. A thorough (and, at the time of its publication, up-to-date) tabulation of these catalogs is given in reference [3].

Reactance Transformations

A useful and simple way for generalizing results obtained for lowpass filters to more complicated filter types is provided by the *reactance transformations.* These enable the designer to transform a lowpass filter (often called *prototype* filter) into a bandpass, highpass, bandstop, or multiband filter satisfying prescribed loss requirements, using only elementary manipulations.

The essence of reactance transformation is to replace the frequency variable ω_{LP} of the lowpass prototype filter by a realizable reactance function $X(\omega)$

$$\omega_{LP} = X(\omega) = A\omega \frac{(\omega_2^2 - \omega^2)(\omega_4^2 - \omega^2)\cdots}{(\omega_1^2 - \omega^2)(\omega_3^2 - \omega^2)\cdots} \tag{1.49}$$

where $0 \le \omega_1 < \omega_2 < \omega_3 \cdots$ and A is a positive real constant. This transformation replaces the reactance of an inductor L_{LP} in the prototype by the new reactance

$$\omega_{LP} L_{LP} = L_{LP} A\omega \frac{(\omega_2^2 - \omega^2)(\omega_4^2 - \omega^2)\cdots}{(\omega_1^2 - \omega^2)(\omega_3^2 - \omega^2)\cdots} \tag{1.50}$$

while that of a capacitor C_{LP} becomes

$$\frac{-1}{\omega_{LP} C_{LP}} = -\frac{1}{C_{LP} A\omega} \frac{(\omega_1^2 - \omega^2)(\omega_3^2 - \omega^2)\cdots}{(\omega_2^2 - \omega^2)(\omega_4^2 - \omega^2)\cdots}. \tag{1.51}$$

These new reactances are always realizable and can, if desired, be considerably more complicated than their prototypes.

The transformation of the frequency axis, produced by the mapping of (1.49), is illustrated in Figure 1-6. As the figure shows, the transformation is, in general, one to many; hence, the passband $-\omega_p \le \omega_{LP} \le \omega_p$ of the prototype becomes a set of passbands centered around $\omega = 0, \pm\omega_2, \pm\omega_4$, ... for the final filter. (The passbands are indicated by heavy line segments in Figure 1-6.)

Since only the independent variable ω_{LP} of the prototype loss response $\alpha_{LP}(\omega_{LP})$ has been changed, the new loss response $\alpha(\omega)$ can be found by assigning the loss value $\alpha_{LP}(\omega_{LP})$ to all ω-values corresponding to ω_{LP}.

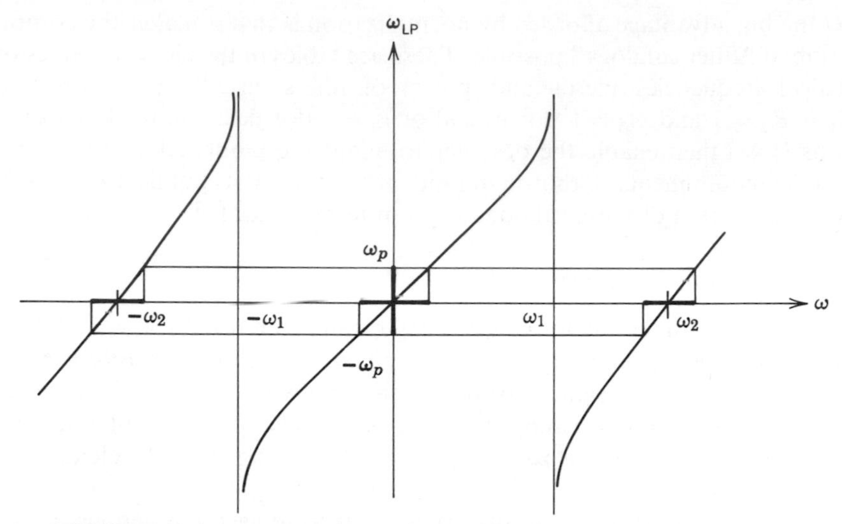

FIGURE 1-6 The mapping of the frequency axis in a reactance transformation.

Although this technique is applicable to the design of multiband filters, its main usefulness lies in simpler transformations. The four simplest cases will be illustrated next.

1. $\omega_{LP} = A\omega$:

A comparison with (1.41) shows directly that this "transformation" amounts only to a scaling of the frequency axis.

2. $\omega_{LP} = -A/\omega$:

Here an inductive prototype reactance is transformed into a capacitive reactance according to

$$\begin{aligned} \omega_{LP} L_{LP} &\rightarrow -AL_{LP}/\omega = -1/\omega C \\ C &= 1/AL_{LP}, \end{aligned} \tag{1.52}$$

while a capacitive reactance becomes an inductive one:

$$\begin{aligned} -1/\omega_{LP} C_{LP} &\rightarrow \omega/AC_{LP} = \omega L \\ L &= 1/AC_{LP}. \end{aligned} \tag{1.53}$$

Hence, inductors and capacitors change roles in the prototype and final circuits.* The loss response of the prototype is transformed, as indicated in

* Frequency-independent elements (resistors, sources, gyrators, etc.) remain, of course, unchanged.

Figure 1-7, into that of a *highpass* filter. For prescribed highpass loss characteristics, the required selectivity of the lowpass prototype is

$$k \triangleq \omega_p/\omega_s = \omega_{s\text{HP}}/\omega_{p\text{HP}} \tag{1.54}$$

where the subscript HP refers to the highpass filter. The passband and stopband losses α_p and α_s of the lowpass filter can be directly obtained from the α_p and α_s of the highpass. Therefore, the lowpass prototype circuit may be obtained. Thereafter the highpass filter can be derived by using (1.52)–(1.53).

3. $\omega_{\text{LP}} = A\dfrac{\omega_2^2 - \omega^2}{-\omega}$:

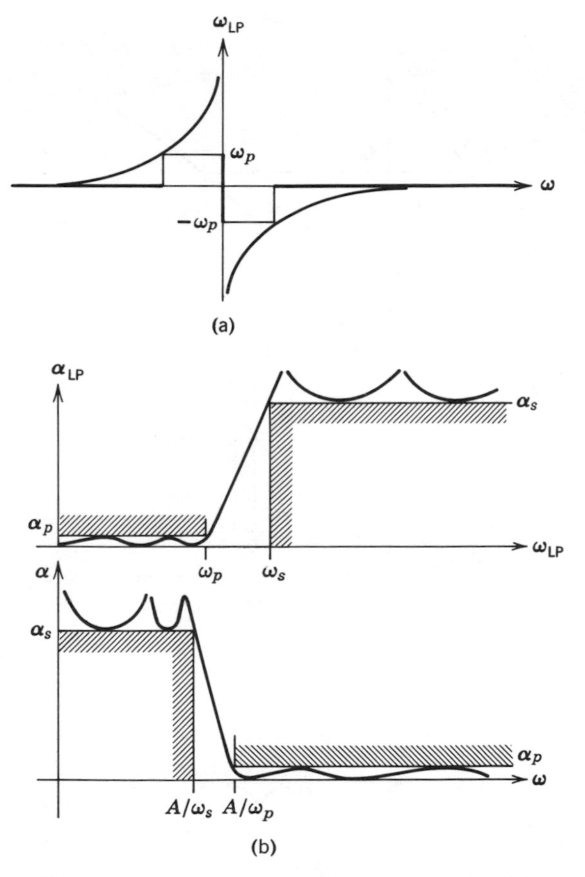

FIGURE 1-7 Lowpass-to-highpass transformation.

Here, an inductor is replaced by a series-tuned circuit:

$$\omega_{LP} L_{LP} \to A L_{LP} \omega - \frac{A L_{LP} \omega_2^2}{\omega} = \omega L - \frac{1}{\omega C}$$
$$L = A L_{LP}$$
$$C = \frac{1}{L \omega_2^2} \tag{1.55}$$

while a capacitor is transformed into a parallel-tuned circuit:

$$\omega_{LP} C_{LP} - A C_{LP} \omega - \frac{A C_{LP} \omega_2^2}{\omega} = \omega C - \frac{1}{\omega L}$$
$$C = A C_{LP}$$
$$L = \frac{1}{C \omega_2^2} \,. \tag{1.56}$$

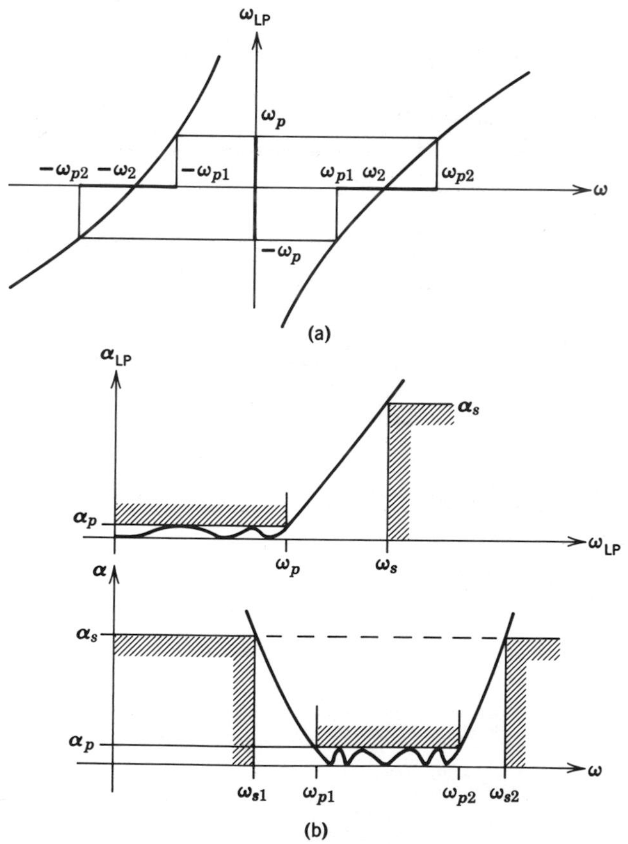

FIGURE 1-8 Lowpass-to-bandpass transformation.

The transformation of the ω_{LP} axis and of α_{LP} is illustrated in Figure 1-8. Evidently now a *bandpass filter* characteristic is obtained. The passband limit ω_{LP} of the lowpass filter is mapped to ω_{p2} and $-\omega_{p1}$, while $-\omega_{LP}$ to ω_{p1} and $-\omega_{p2}$ (Figure 1-8(a)). Hence, ω_{p2} and $-\omega_{p1}$ are solutions of the quadratic equation

$$\omega^2 - \frac{\omega_p}{A}\omega - \omega_2^2 = 0, \tag{1.57}$$

and therefore

$$\begin{aligned} -\omega_{p1}\omega_{p2} &= -\omega_2^2 \\ \omega_{p2} - \omega_{p1} &= \omega_p/A. \end{aligned} \tag{1.58}$$

Similar equations connect the stopband limit ω_s of the prototype to ω_{s1} and ω_{s2}, the stopband limits of the bandpass filter:

$$\begin{aligned} \omega_{s1}\omega_{s2} &= \omega_2^2 \\ \omega_{s2} - \omega_{s1} &= \omega_s/A. \end{aligned} \tag{1.59}$$

It should be noted that ω_{p1} and ω_{p2}, as well as ω_{s1} and ω_{s2}, display geometric symmetry about ω_2. This property holds for any pair of frequencies corresponding to a single ω_{LP}, and hence *only geometrically symmetrical bandpass loss characteristics are obtainable using this transformation.* This also means, of course, that the minimum stopband loss α_s of the bandpass filter is the same in both stopbands.

For a given geometrically symmetrical bandpass loss response, the design procedure can now be based upon the preceding discussion. The steps are the following:

(i) The selectivity parameter of the prototype is found from

$$k = \frac{\omega_p}{\omega_s} = \frac{A(\omega_{p2} - \omega_{p1})}{A(\omega_{s2} - \omega_{s1})} = \frac{\omega_{p2} - \omega_{p1}}{\omega_{s2} - \omega_{s1}}. \tag{1.60}$$

(ii) The loss limits α_p and α_s of the prototype are obtained from the (unchanged) loss limits α_p and α_s of the bandpass response.

(iii) The passband frequency limit ω_p is chosen in a convenient way (e.g., as $\omega_p = 1$) and the lowpass filter designed.

(iv) A and ω_2^2 are obtained from

$$A = \frac{\omega_p}{\omega_{p2} - \omega_{p1}} \quad \text{or} \quad A = \frac{\omega_s}{\omega_{s2} - \omega_{s1}}$$

$$\omega_2^2 = \omega_{p1}\omega_{p2} = \omega_{s1}\omega_{s2} \tag{1.61}$$

and then (1.55) and (1.56) used to transform the reactive elements of the lowpass prototype into those of the final bandpass filter.

Often, the condition $\omega_{p1}\,\omega_{p2} = \omega_{s1}\,\omega_{s2}$ is not valid for the initial specifications. Then, one or more of these limits must be modified before the above outlined steps can commence.

4. $\omega_{\mathrm{LP}} = A\dfrac{\omega}{\omega_1^2 - \omega^2}$:

This transformation leads to a *bandstop* response (Figure 1-9). The analysis is entirely analogous to that given for the bandpass case above, and is therefore omitted. Here,

$$\omega_{\mathrm{LP}} L_{\mathrm{LP}} \to \frac{L_{\mathrm{LP}} A}{\dfrac{\omega_1^2}{\omega} - \omega} \tag{1.62}$$

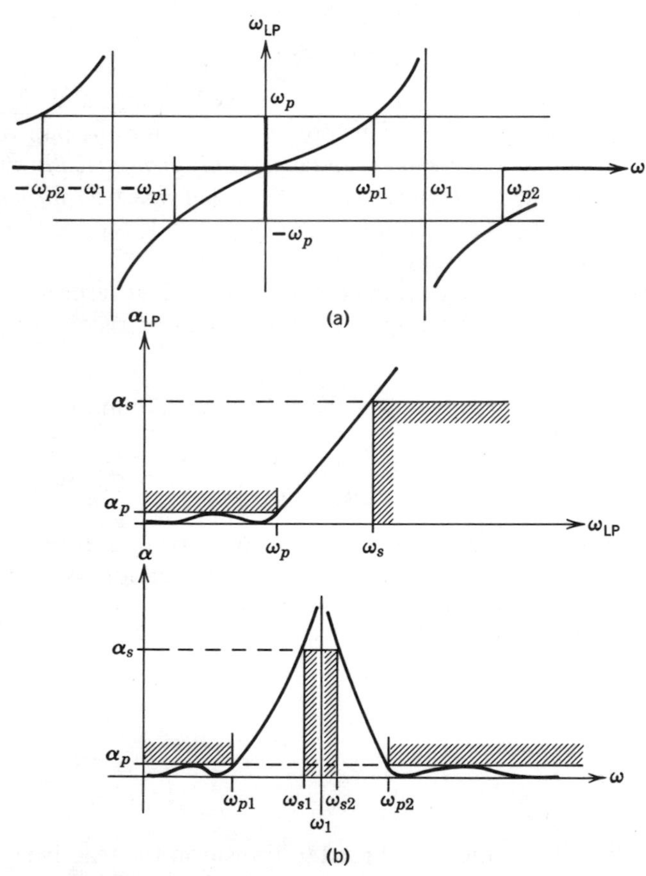

FIGURE 1-9 Lowpass-to-bandstop transformation.

which represents the parallel combination of a capacitor $1/(L_{LP}A)$ and an inductor $L_{LP}A/\omega_1^2$. The other design relations are

$$\omega_{p1}\omega_{p2} = \omega_{s1}\omega_{s2} = \omega_1^2$$
$$A = \omega_p(\omega_{p2} - \omega_{p1}) = \omega_s(\omega_{s2} - \omega_{s1}). \tag{1.63}$$

The design process for bandstop filters is based upon these relations. It is very similar to the procedure for bandpass filters; its derivation is left for the reader as an exercise.

More complicated reactance transformations are seldom needed and hence will not be discussed here.

It is important to note that the reactance transformations are not directly useful for transforming linear-phase lowpass filters into linear-phase highpass, bandpass, or bandstop circuits. This is so because the linear-lowpass phase $\beta_{LP} = \omega_{LP}T$ is replaced in the transformation by

$$AT\omega\frac{(\omega_2^2 - \omega^2)(\omega_4^2 - \omega^2)\cdots}{(\omega_1^2 - \omega^2)(\omega_3^2 - \omega^2)\cdots}$$

which (except for the trivial $\omega_{LP} = A\omega$ case) is no longer linear in ω.

In addition to the reactance transformations, there exist more general mappings, called *Moebius* transformations, which provide additional degrees of freedom for the designer. These require more complicated and laborious design techniques, and are hence seldom used. The interested reader is referred to reference [4] for a detailed discussion of these transformations.

Somewhat different transformations are useful for active RC filters. These will be discussed in Chapter 8.

1.6 SUMMARY

Following a brief discussion of the various filter types and the general aspects of filter specifications and design, the network functions useful for describing the transmission properties of doubly terminated reactance two-ports were given. Some elementary methods were also presented for the scaling of circuits, and for transforming lowpass filters into other filter types: highpass, bandpass, and bandstop filters.

The discussions of this chapter should enable the reader to understand the broad implications of Boxes 1 and 2 in the flowchart of Figure 1-4. The next chapter will provide detailed information on how the operations indicated in Box 3 are performed. Chapter 3 will cover the calculation of the element values (Box 4) for passive reactance filters; the chapters following thereafter will describe the properties and realization of other filter types. The remaining tasks, represented by Boxes 5–7 in the flowchart, involve more or less routine circuit analysis and the building of the filter; these will not be discussed in this book.

PROBLEMS

1.1 Prove the validity of (1.20) for reactance two-ports. (Hint: Use (1.14)–(1.15); note that for reactance two-ports the input power P_1 and the output power P_2 are equal:

$$P_1 = |I_1|^2 \operatorname{Re}(Z_1) = P_2 = |V_2|^2/R_2.)$$

1.2 (a) Derive (1.62) and (1.63) of the text.

(b) Find a step-by-step procedure, similar to the one given in Section 1.5 for bandpass filters, for the design of bandstop filters with geometrically symmetric frequency response.

1.3 Verify the relation (1.9) from (1.5) and (1.8). (Hint: Substitute the power-series expressions for $H(z)$, $X(z)$, and $Y(z)$ into (1.9); compare the coefficients of like powers of z.)

1.4 Calculate and plot the loss, phase, group delay, and phase delay from the transducer function

$$H(s) = 2\frac{s^3 + 2s^2 + 2s + 1}{s^2 + 2}$$

for $0 \le \omega \le 1$.

1.5 Find and plot the impulse and step response of the filter with the $H(s)$ of Problem 1.4.

1.6 What can you predict about the impulse response of a filter for which $e(s)$ has $j\omega$-axis zeros? What if some zeros lie in the right half of the s-plane?

1.7 Let $F(s)$ be an arbitrary network function. Show that if each resistance R_i and inductance L_i in the circuit is multiplied by a factor k, while each capacitance C_i is divided by k, then $F(s)$ is changed into

$kF(s)$, if $F(s)$ is measured in Ω,
$F(s)/k$, if $F(s)$ is measured in $1/\Omega$,
$F(s)$, if $F(s)$ is dimensionless.

1.8 Let the impulse response of a filter be $h(t) = e^{-2t}u(t)$, where $u(t)$ is the unit step function:

$$u(t) = \begin{cases} 1, & \text{for } t \ge 0 \\ 0, & \text{for } t < 0. \end{cases}$$

What is the response of the filter to the excitations
(a) $u(t)$?
(b) $tu(t) - (t - 1)u(t - 1)$?

REFERENCES

[1] E. I. Jury, *Theory and Application of the* z-*Transform Method*, John Wiley, New York, 1964.
[2] K. L. Su, *Time Domain Synthesis of Linear Networks*, Prentice-Hall, Englewood Cliffs, New Jersey, 1971.
[3] H. J. Orchard and G. C. Temes, "Filter design using transformed variables," *IEEE Trans. Circuit Theory*, **CT-15**, No. 4, 385–408 (Dec., 1968).
[4] E. Christian and E. Eisenmann, *Filter Design Tables and Graphs*, John Wiley, New York, 1966.

2

Approximation Theory

G. C. Temes
University of California
Los Angeles, California

The first step in filter synthesis is to find mathematical functions which, on the one hand, satisfy the specifications of the filter and, on the other, can be exactly realized by a practical circuit. The latter requirement restricts the mathematical properties of these functions. For example, in the usual case of lumped linear time-invariant circuits, whether active or passive, all network functions can be obtained from the Kirchhoff equations wherein the coefficients are built from the elementary impedance functions (R, sL, $1/sC$) through multiplication, division, addition and subtraction. As a consequence, all these network functions must be *rational* functions of s, the complex frequency; i.e., they must be of the form $P(s)/Q(s)$, where $P(s)$ and $Q(s)$ are polynomials in s. Hence, in this case any requirements specified for the frequency response of the filter must be satisfied by rational functions of s. Similar considerations restrict the time response of these networks to exponential, polynomial and trigonometric functions of the time t or their combinations.

The situation becomes more general for networks containing distributed as well as lumped elements. There, hyperbolic and other transcendental functions may also enter into the expressions. Time-variable elements extend the possible forms of network functions too.

A discussion of distributed and time-variable circuits, however, will be postponed until Chapters 7, 9 and 10. In this chapter, we will only investigate methods useful in approximating the most commonly specified filter characteristics using network functions appropriate for lumped linear time-invariant circuits.

The design requirements for a lumped linear filter may specify its behavior in the frequency, or in the time domain (sometimes in both). The case of time-domain design will not be discussed in this work, for lack of space. The interested reader is referred to the recent excellent book by K. Su,* for a thorough treatment.

With these restrictions, the topic discussed in this chapter will be to find the rational functions $H(s)$ and $K(s)$ such that they satisfy the realizability requirements, mentioned in Chapter 1, and simultaneously meet the design requirements.

As outlined in Chapter 1, the properties usually specified for filters include the loss response, phase or delay characteristics, time response, the behavior of the input and output impedances, etc. Furthermore, sometimes several of these properties are prescribed simultaneously. As may be expected, these requirements are often conflicting, and some compromise is necessary. Even in the case of the simplest specifications, the idealized response usually calls for unrealizable properties (e.g., zero attenuation throughout a passband or infinite attenuation over a stopband), and hence an *error* between ideal and actual responses is unavoidable. This error should, of course, be kept as small as possible. The size of the error over a range of time or frequency can be measured by various criteria. If the behavior of the network function (e.g., loss) is of interest in the vicinity of only one value of its independent variable (e.g., frequency), then a reasonable requirement may be a *maximally flat* error. The error is called maximally flat at a point if its value as well as the highest possible number of its derivatives vanishes there. Such a behavior of the error $e(x)$ at x_0 is illustrated in Figure 2-1. It is evident from the figure, and can also be verified by considering the Taylor-series expansion of $e(x)$ around x_0, that this condition guarantees small error values in the immediate neighborhood of x_0. It is equally clear that it guarantees nothing whatever far away from x_0.

In many applications, the error value must be kept small over a whole range of the x-axis, between limits x_1 and x_2. One obvious way to keep the error small over the range is to keep its maximum absolute value small. If the maximum absolute error in the range is as small as possible, we call the network function a *minimax* or *Chebyshev approximation* of the specified

* K. Su, *Time-Domain Synthesis of Linear Networks*, Prentice-Hall, Englewood Cliffs, New Jersey, 1971.

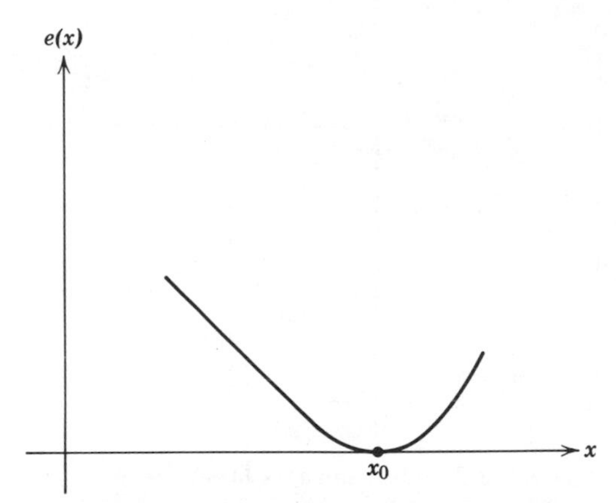

FIGURE 2-1 Maximally flat error function.

ideal performance. Very often, the minimax error has the *equal-ripple* behavior indicated in Figure 2-2.

Other methods for defining "small" errors over a range of independent variables are also possible and practical. Discussion of these is postponed until Chapter 6.

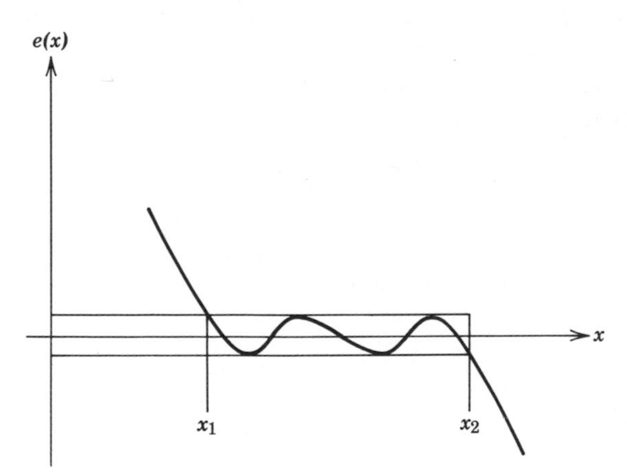

FIGURE 2-2 Equal-ripple error function.

2.1 IDEAL LOWPASS LOSS FUNCTIONS

Lowpass Filter Specifications

In many applications, only the loss response of the filter is of interest. In others, the phase or delay response is also specified, but is corrected, using constant-loss "allpass" networks, only *after* the desired loss characteristics

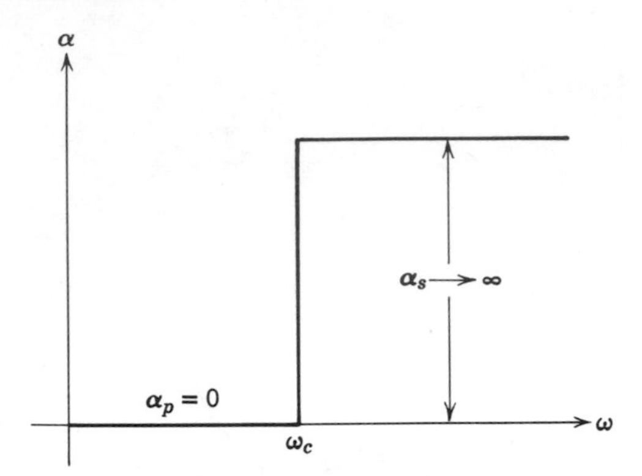

FIGURE 2-3 Ideal lowpass loss-response.

have been achieved. Hence, solving the approximation problem for some simple idealized loss responses is of great interest.

The simplest idealized loss function is that illustrated in Figure 2-3. It has zero loss up to a frequency ω_c, and infinite loss for all higher frequencies. This response cannot be satisfied with a finite circuit (See Problem 2.2). Hence, the specification is usually rephrased in a more practical form, in which the loss is required to be less than a prescribed upper limit α_p in the range $0 \leq \omega \leq \omega_p$, and then take on values higher than another specified (lower) limit α_s for $\omega_s \leq \omega < \infty$ (Figure 2-4).Here, as elsewhere in the book, the loss is the so-called *transducer loss*, which is defined (in decibels) by

$$\alpha = 10 \log_{10}\left[\frac{P_{\max}}{P_2}\right] = 20 \log_{10}|H(j\omega)|. \tag{2.1}$$

$P_{\max}$ is chosen here as the available generator power for passive reactance filters; it can be chosen arbitrarily for active circuits.

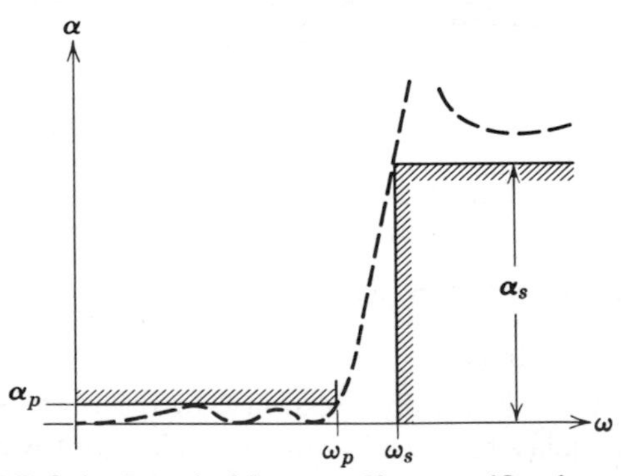

FIGURE 2-4 Practical lowpass filter specification and response.

For some passive filters the generator and load resistances are not equal, and hence, without using transformers, zero transducer loss cannot be achieved for zero frequency. Then a "flat loss," α_0, is unavoidable (Figure 2-5). However, this merely represents a constant multiplier $10^{\alpha_0/20}$ in $H(s)$, as can be seen from (2.1).

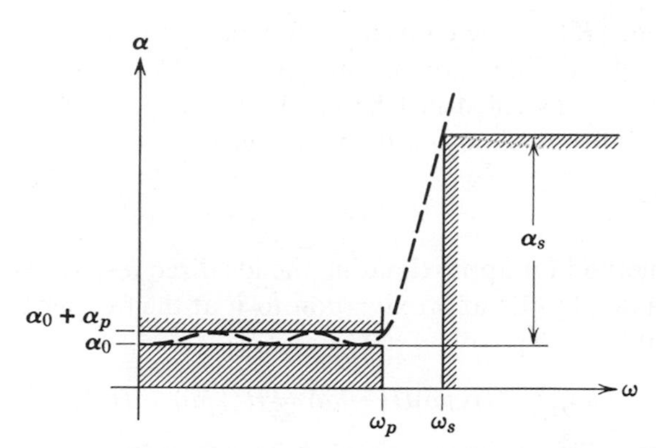

FIGURE 2-5 Filter response with flat loss.

While the specifications for lowpass filters are given in terms of the loss response $\alpha(\omega)$, the solution of the approximation problem is almost always carried out in terms of the characteristic function $K(\omega)$. The reasons for this are the following:

1. $K(\omega)$ is a polynomial or a rational function of ω; $\alpha(\omega)$ is not. Approximation with polynomials or rational functions is relatively convenient.
2. By (2.1) and (1.16),

$$\alpha(\omega) = 10 \log_{10}[1 + |K(j\omega)|^2]. \tag{2.2}$$

From this relation, it is easy to see that
if $\alpha = 0$, then $|K|^2 = 0$
if $\alpha \to \infty$, then $|K|^2 \to \infty$

$$\text{if } \frac{d\alpha(\omega)}{d\omega} = \, > \, < 0, \text{ then } \frac{d|K|^2}{d\omega} = \, > \, < 0$$

$$\text{if } \alpha' = \alpha'' = \cdots = \alpha^{(n)} = 0, \text{ then } [|K|^2]' = [|K|^2]'' = \cdots = [|K|^2]^{(n)} = 0$$

where the prime denotes differentiation with respect to ω. From these relations, it can be seen that the frequency-dependent properties of $|K|^2$ duplicate in many ways those of α; in fact, the curves of $\alpha(\omega)$ and $|K(j\omega)|^2$ differ only in their vertical scales.

3. For filters, the loss must be low in some frequency bands and high in some others. This can be best achieved if all zeros of $K(s)$ are on the $j\omega$-axis in the low-loss bands (*passbands*) and also all poles of $K(s)$ are on the $j\omega$-axis in the high-loss bands (*stopbands*). Hence, placing these zeros and poles is a one-dimensional, rather than a two-dimensional, problem—an obvious simplification.

In conclusion, $|K|^2$ is the convenient function to use when only the loss characteristics of the filter are of interest. It will also be noted that the realizability conditions listed in Chapter 1 are the least stringent for $K(s)$; this allows flexibility in choosing its parameters.

Butterworth Filters

The simplest method for approximating the idealized response of Figure 2-3 is to find a maximally flat approximation to it at the frequency origin. It is easy to see that*

$$|H(j\omega)|^2 = H(j\omega)H(-j\omega) = H_e^2(j\omega) - H_0^2(j\omega) \tag{2.3}$$

is an *even* function of ω; hence, so is $\alpha(\omega)$. Therefore, the discussion may be carried out in terms of ω^2 as independent variable. Then the maximally flat condition requires

$$\alpha(\omega^2) = \frac{d\alpha(\omega^2)}{d(\omega^2)} = \frac{d^2\alpha(\omega^2)}{d(\omega^2)^2} = \cdots = \frac{d^{(n-1)}\alpha(\omega^2)}{d(\omega^2)^{n-1}} = 0, \tag{2.4}$$

all relations being valid for $\omega^2 = 0$. Equations (2.4) may, due to the relations between α and $|K|^2$, also be rephrased in terms of $|K|^2$ in the form

$$|K|^2 = [|K|^2]' = [|K|^2]'' = \cdots = [|K|^2]^{(n-1)} = 0 \tag{2.5}$$

where again $\omega^2 = 0$ and the indicated differentiations are with respect to ω^2. Equations (2.5) impose n conditions upon $|K|^2$, which therefore must have at least n free parameters to satisfy them. Let $|K|^2$, a rational function in ω^2, be in the form

$$|K|^2 = C^2 \frac{\omega^{2n} + a_{n-1}\omega^{2(n-1)} + \cdots + a_1\omega^2 + a_0}{Q_m(\omega^2)} \tag{2.6}$$

where Q_m is an mth-degree polynomial in ω^2 with nonzero constant term. Then it is easy to verify that by choosing the n coefficients zero:

$$a_0 = a_1 = \cdots = a_{n-1} = 0,$$

* Here, as in the rest of the book, $f_e(x)$ (or $f_0(x)$) denotes the even (odd) part of $f(x)$, i.e., $f_e(x) = [f(x) + f(-x)]/2$, $f_0(x) = [f(x) - f(-x)]/2$.

the resulting function

$$|K|^2 = \frac{C^2\omega^{2n}}{Q_m(\omega^2)} \tag{2.7}$$

satisfies (2.5). Since $C^2 > 0$, the condition $|K|^2 \geq 0$ requires $Q_m(\omega^2) \geq 0$. The simplest case is $Q_m(\omega^2) \equiv 1$. Then

$$|K|^2 = C^2\omega^{2n} \tag{2.8}$$

and hence

$$K(s) = \pm Cs^n \tag{2.9}$$

can be chosen. The filter described by (2.9) is usually called* a *Butterworth filter*.

The discussions of Chapter 3 will make it obvious that, in order to proceed with the actual realization of the filter, the approximation process must supply the following information:

1. The degree n of the filter; n is the degree of $e(s)$. It is thus the number of natural modes or the *total* number of loss poles (including those at zero and infinite frequency).
2. the characteristic function $K(s)$.
3. the transducer function $H(s)$.
4. the loss poles (contained also in H and K).

In order to gain all this information for the Butterworth filter, we combine (2.2) and (2.9):

$$\alpha(\omega) = 10 \log_{10}[1 + C^2\omega^{2n}]. \tag{2.10}$$

Here, C and n must be chosen so as to satisfy the usual conditions

$$\begin{aligned} \alpha(\omega) &\leq \alpha_p \quad \text{for} \quad |\omega| \leq \omega_p \\ \alpha(\omega) &\geq \alpha_s \quad \text{for} \quad |\omega| \geq \omega_s \end{aligned} \tag{2.11}$$

(see Figure 2-4). Since the number of elements required in the eventual realization of the filter increases with increasing degrees, n should be kept as low as possible. It must, however, be an integer. Figure 2-6 illustrates the effect of n on the response given by (2.10), for $C = 1$. Clearly, higher n gives lower loss in the passband and higher loss in the stopband. If the requirements of (2.11) are just barely satisfied (with the equality sign), then

$$\begin{aligned} \alpha_p &= 10 \log_{10} [1 + C^2\omega_p^{2n}] \\ \alpha_s &= 10 \log_{10}[1 + C^2\omega_s^{2n}] \end{aligned} \tag{2.12}$$

*After its discoverer.

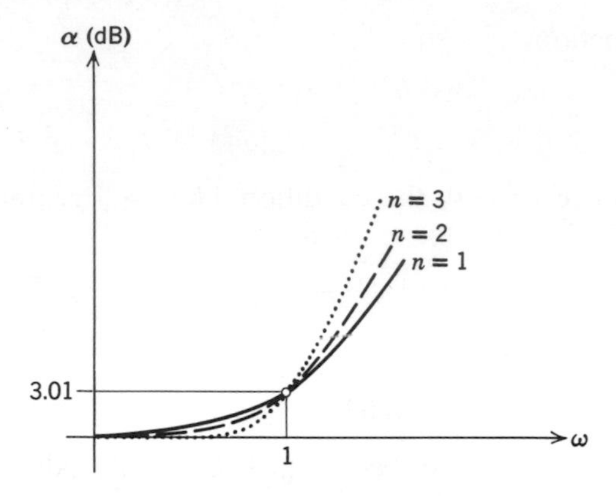

FIGURE 2-6 Butterworth loss responses.

or

$$C^2\omega_p^{2n} = 10^{\alpha_p/10} - 1$$
$$C^2\omega_s^{2n} = 10^{\alpha_s/10} - 1$$

Dividing these two equations and solving for n gives

$$n \geq \frac{\log_{10} k_1}{\log_{10} k}, \tag{2.13}$$

where the abbreviated notations

$$k_1 = \left[\frac{10^{\alpha_p/10} - 1}{10^{\alpha_s/10} - 1}\right]^{1/2} \tag{2.14}$$

and

$$k = \frac{\omega_p}{\omega_s} \tag{2.15}$$

have been used.* The $\geq$ sign was needed in (2.13) because a noninteger n must be rounded *up* to the nearest higher integer. After this increase of n, only one of the relations of (2.11) can be satisfied with the equal sign. If the second one is chosen† then

$$C = \frac{\sqrt{10^{\alpha_s/10} - 1}}{\omega_s^n} \tag{2.16}$$

results for the constant factor.

* Often k is called *selectivity parameter*, while k_1 is the *discrimination parameter*.

† This is usually advisable, since the "safety margin" resulting from an inequality relation is normally needed at the more critical passband limit ω_p.

Next, the Feldtkeller equation is used to calculate $H(s)$. From (1.24) and (2.9) we have, with $f(s) = Cs^n$ and $p(s) \equiv 1$,

$$e(s)e(-s) = 1 + C^2(-1)^n s^{2n} \tag{2.17}$$

and hence, the zeros of $e(s)e(-s)$ appear as the roots of the equation

$$s^{2n} = C^{-2}(-1)^{n-1} = \frac{e^{j\pi(n-1+2k)}}{C^2}, \qquad k = 1, 2, \ldots. \tag{2.18}$$

They are therefore

$$s_k = C^{-1/n} \exp j\left(\frac{\pi}{2} + \pi\frac{2k-1}{2n}\right), \qquad k = 1, 2, \ldots, 2n. \tag{2.19}$$

Since $e(s)$ is a strictly Hurwitz polynomial, only the left-half-plane roots s_1, $s_2, \ldots, s_n$ qualify as its zeros. These roots lie at equal angles on a half circle of radius $C^{-1/n}$ in the left half of the s-plane. They are illustrated, for $n = 5$, in Figure 2-7.

Since the leading coefficient of $e(s)$, by (2.17), is $\pm C$, we have

$$H(s) \equiv e(s) = \pm C \prod_{k=1}^{n} (s - s_k)$$

$$K(s) \equiv f(s) = \pm Cs^n \tag{2.20}$$

with n, C, and s_k given by (2.13), (2.16), and (2.19), respectively.

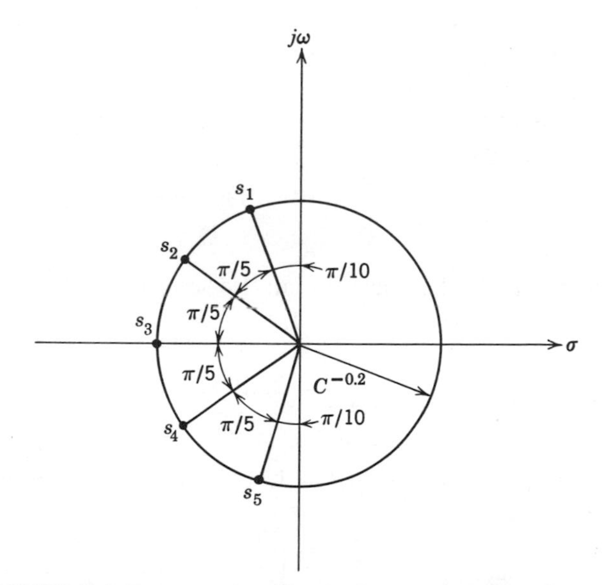

FIGURE 2-7 Natural modes for Butterworth filter ($n = 5$).

The root factors and coefficients of $e(s)$ can now easily be computed [27].

Example 2-1. Find n, $H(s)$, and $K(s)$ for a Butterworth filter with the specifications $\alpha \leq 0.1$ dB for $f \leq 3$ MHz, $\alpha \geq 60$ dB for $f \geq 12$ MHz. In order to obtain an integer n, satisfy

(i) the passband specification,
(ii) the stopband specification,

without safety margin.

Solution: From (2.14) and (2.15),

$$k_1 = \sqrt{\frac{10^{0.1/10} - 1}{10^{60/10} - 1}} = 0.15262 \times 10^{-3}, \qquad k = \frac{3}{12} = 0.25.$$

Then (2.13) gives $n \geq 6.33888$. Therefore, $n = 7$ should be chosen. To simplify the following calculations, frequency normalization will be introduced.

(a) If the passband specifications are exactly met, the coefficient C is given by (2.12) as

$$C = \frac{\sqrt{10^{\alpha_p/10} - 1}}{\Omega_p^n} = \frac{0.152620418}{\Omega_p^7}$$

where Ω_p is the normalized passband limit. For convenience, the unit radian frequency ω_0 is chosen such that $C = 1$, i.e.

$$\omega_0 = \frac{2\pi(3 \times 10^6)}{(0.152620418)^{1/7}} = 24.656299 \times 10^6.$$

(b) If the stopband specifications are met without any margin, the coefficient C is given by (2.16) as

$$C = \frac{\sqrt{10^{\alpha_s/10} - 1}}{\Omega_s^n} = \frac{999.9995}{\Omega_s^7},$$

where Ω_s is the normalized stopband limit. For convenience, the unit radian frequency ω_0 is chosen such that $C = 1$, i.e.,

$$\omega_0 = \frac{2\pi(12 \times 10^6)}{(999.9995)^{1/7}} = 28.105397 \times 10^6.$$

(Note that *any* ω_0 may be satisfactory, but of course ω_0 must be a *unique* value.) The normalized $s_k (k = 1, \ldots, 7)$ are given now in both cases by (2.19), using $C = 1$:

$$s_k = \exp j\left(\frac{\pi}{2} + \pi\frac{2k - 1}{14}\right), \qquad k = 1, \ldots, 7.$$

Then, from (2.20):

$$K(s) = s^7$$

$$\begin{aligned} H(s) &= \prod_{k=1}^{7} (s - s_k) \\ &= (s+1)(s^2 + 0.4450s + 1)(s^2 + 1.2470s + 1)(s^2 + 1.8019s + 1) \\ &= s^7 + 4.4940s^6 + 10.0978s^5 + 14.5920s^4 \\ &\quad + 14.5920s^3 + 10.0978s^2 + 4.4940s + 1 \end{aligned}$$

where s and s_k are also normalized. The value of ω_0 is different in cases (a) and (b), as mentioned above. The denormalized (physical) expression for $H(s)$ may be obtained by replacing s by s/ω_0 in the above expressions.

Chebyshev Filters

The ideal lowpass characteristic may also be approximated in the vicinity of the frequency origin in an equal-ripple manner (Figure 2-8). As in the case of Butterworth filters, we initially restrict the discussion to the simple case when both $K(s)$ and $H(s)$ are polynomials, rather than general rational functions. Instead of deriving the proper $K(s)$ in a logical manner (which is straightforward but tedious [1]), the function will be directly given, and its properties verified afterward. Postulating

$$\begin{aligned} |K|^2 &= k_p^2 \cos^2 nu(\omega) \\ u(\omega) &= \cos^{-1} (\omega/\omega_p), \end{aligned} \tag{2.21}$$

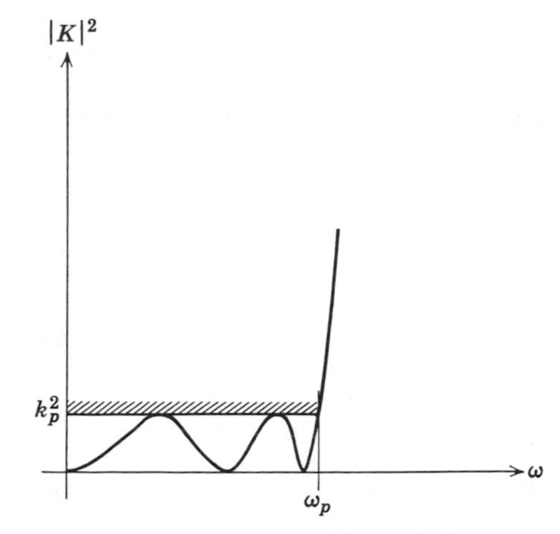

FIGURE 2-8 Equal-ripple passband filter response.

we see that $|K|^2$ is a polynomial in $(\omega/\omega_p)^2 = \cos^2 u$, since

$$\cos(nu) = \mathrm{Re}[(e^{ju})^n] = \mathrm{Re}[(\cos u + j \sin u)^n]$$
$$= \cos^n u - \binom{n}{2}(\cos^{n-2} u)(1 - \cos^2 u)$$
$$+ \binom{n}{4}(\cos^{n-4} u)(1 - \cos^2 u)^2 - + \cdots. \qquad 2.22)$$

Also, for $|\omega| < \omega_p$, u is real and hence $|K|^2 \leq k_p^2$. In fact, as ω grows from 0 to ω_p, u can be considered to grow from $-\pi/2$ to 0, and hence $|K|^2$ oscillates n times between 0 and k_p^2. For values of $\omega > \omega_p$, on the other hand, $|K|^2$ is a monotone increasing function of ω^2 (see Problem 2.5). It has the optimality property that of all nth-order polynomials $P_n(\omega^2)$ in ω^2, restricted to values $0 \leq P_n(\omega^2) \leq k_p^2$ for $\omega^2 \leq \omega_p^2$, it increases fastest* for $\omega^2 \geq \omega_p^2$.

The polynomial

$$T_n(x) = \cos(n \cos^{-1} x) \qquad (2.23)$$

entering the expression for $|K|^2$ is called the nth-order *Chebyshev polynomial.* Hence, the filters realized from (2.21) are commonly known as *Chebyshev filters.*

For values $\omega > \omega_p$, the relation

$$\omega = \omega_p \cos u = \omega_p \cosh(ju) \qquad (2.24)$$

may be used to show that now

$$u = -j \cosh^{-1}(\omega/\omega_p) \qquad (2.25)$$

is pure imaginary, and therefore

$$|K|^2 = k_p^2 \cosh^2 nju = k_p^2 \cosh^2(n \cosh^{-1} \omega/\omega_p) \qquad (2.26)$$

is real and greater than k_p^2. (Equation (2.26) also shows again the monotonic rise of $|K|^2$ with ω for $\omega \geq \omega_p$.) For given α_p, α_s, ω_p, and ω_s, (2.2), (2.21), and (2.26) give

$$|K|^2_{\omega=\omega_p} = 10^{\alpha_p/10} - 1 \leq k_p^2$$
$$|K|^2_{\omega=\omega_s} = 10^{\alpha_s/10} - 1 \geq k_p^2 \cosh^2(n \cosh^{-1} \omega_s/\omega_p) \qquad (2.27)$$

so that the necessary degree satisfies

$$n \geq \frac{\cosh^{-1}(1/k_1)}{\cosh^{-1}(1/k)} \qquad (2.28)$$

where the abbreviated notations of (2.14) and (2.15) were again used.

*See Problems 2.7 and 2.8 for the proof of special cases of this statement.

To extend (2.21) to complex variables, we replace ω by s/j and regard $u = v + jw$ and $s = \sigma + j\omega$ as complex quantities. Then,

$$|K(s)|^2 = k_p^2 \cos^2 nu(s) = k_p^2 T_n^2(s/j\omega_p)$$
$$u(s) = \cos^{-1}(s/j\omega_p). \tag{2.29}$$

The computation of the natural modes proceeds from the Feldtkeller equation

$$k_p^2 \cos^2[n(v_k + jw_k)] + 1 = 0, \qquad k = 1, 2, \ldots, n. \tag{2.30}$$

Here, $u_k = v_k + jw_k$ is the kth natural mode, in the complex u-plane. Using the relation

$$\cos(x + jy) = \cos x \cosh y - j \sin x \sinh y, \tag{2.31}$$

it can easily be shown (Problem 2.6) that the s-plane natural modes satisfy

$$s_k = \sigma_k + j\omega_k = j\omega_p \cos(v_k + jw_k)$$
$$\sigma_k = \omega_p \sin v_k \sinh w_k$$
$$\omega_k = \omega_p \cos v_k \cosh w_k. \tag{2.32}$$

It is instructive to derive the locus on which the s_k lie. By (2.32),

$$\frac{\sigma_k^2}{\omega_p^2 \sinh^2 w_k} + \frac{\omega_k^2}{\omega_p^2 \cosh^2 w_k} = 1; \tag{2.33}$$

hence, the natural modes lie on an *ellipse*. The major axis of the ellipse is on the $j\omega$-axis and its size is $\omega_p \cosh w_k$; the minor axis is on the σ-axis and is of size $\omega_p \sinh w_k$. The natural modes are illustrated, for $n = 5$, in Figure 2-9.

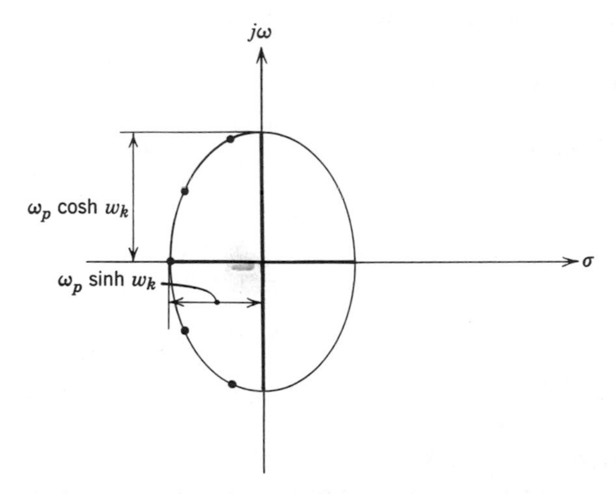

FIGURE 2-9 Natural modes for fifth-degree Chebyshev filter.

The zeros of $H(s) \equiv e(s)$ are now known; its constant factor may be found from

$$\lim_{\omega \to \infty} |H|^2 = \lim_{\omega \to \infty} |K|^2 = \lim_{\omega \to \infty} k_p^2 \left(\frac{\omega}{\omega_p}\right)^{2n} \tag{2.34}$$

where the Feldtkeller equation and (2.21) and (2.22) were utilized. Hence,

$$H(s) \equiv e(s) = \pm k_p \prod_{k=1}^{n} \left(\frac{s}{\omega_p} - \sin v_k \sinh w_k - j \cos v_k \cosh w_k\right). \tag{2.35}$$

The reflection zeros are easily shown to be at $\omega_p \cos\{[(2k-1)/n](\pi/2)\}$, $k = 1, 2, \ldots, n$, using (2.21).

The Chebyshev polynomials needed in $K(s)$ are easily generated if we note that (for $\omega_p \equiv 1$)

$$\begin{aligned} T_{m+1}(\omega) &= \cos(mu + u) = \cos mu \cos u - \sin mu \sin u \\ T_{m-1}(\omega) &= \cos mu \cos u + \sin mu \sin u \\ T_{m+1}(\omega) + T_{m-1}(\omega) &= 2\omega T_m(\omega), \end{aligned} \tag{2.36}$$

and hence, with

$$\begin{aligned} T_0 &= 1 \\ T_1 &= \omega \\ T_{m+1} &= 2\omega T_m - T_{m-1}, \end{aligned} \tag{2.37}$$

all $T_n(\omega)$ can readily be found.

Example 2-2. Find n, $K(s)$, and $H(s)$ of a Chebyshev filter which satisfies the same specifications as the Butterworth filter of the previous example.

Solution: From (2.28), now

$$n \geq \frac{\cosh^{-1} 6552.2212}{\cosh^{-1} 4} = \frac{9.48071}{2.06344} = 4.59462.$$

Therefore, $n = 5$ is sufficient now.

Normalizing the frequency such that the passband limit radian frequency Ω_p is equal to 1, i.e., by choosing $\omega_0 = 2\pi(3 \times 10^6)$, (2.37) gives:

$$T_5(\omega) = 16\omega^5 - 20\omega^3 + 5\omega.$$

Here $k_p = \sqrt{10^{0.1/10} - 1} = 0.152620418$, and therefore by (2.29) we can choose

$$\begin{aligned} K(s) &= 0.152620418(16s^5 + 20s^3 + 5s) \\ &= 2.44192669s^5 + 3.05240836s^3 + 0.76310209s. \end{aligned}$$

From (2.30)–(2.32), the natural modes are obtained as

$$s_k = \sigma_k + j\omega_k, \qquad k = 1, \ldots, 5$$

where

$$\sigma_k = \sin v_k \sinh w_k$$
$$\omega_k = \cos v_k \cosh w_k$$

and

$$v_k = \pm\frac{2k-1}{n}\frac{\pi}{2}$$

$$w_k = \pm\frac{1}{n}\sinh^{-1}\left(\frac{1}{k_p}\right)$$

$$n = 5, \qquad k_p = 0.152620418.$$

Then (2.35) gives

$$\begin{aligned} H(s) = 0.152620418&(s + 0.538914)(s + 0.166534 - j1.080372) \\ &\cdot (s + 0.166534 + j1.080372)(s + 0.435991 - j0.667707) \\ &\cdot (s + 0.435991 + j0.667707). \end{aligned}$$

Filters with Finite Loss Poles

For both Butterworth and Chebyshev filters, $K(s)$ and $H(s)$ were polynomials. Hence, all their poles, which are also the loss poles of the circuit, occur for $\omega \to \infty$. More selectivity may be obtained by moving these poles to finite ω-values. In fact, the $\alpha(\omega)$ curve may be visualized as a flexible steel band which may be "nailed down" to the $j\omega$-axis by reflection zeros and pushed up to great heights using loss poles (Figure 2-10). Certainly the best selectivity may be obtained* by distributing the loss zeros and poles in some fashion throughout the pass- and stopbands, respectively.

The simplest way to introduce finite loss poles and also to maintain control over the passband is to reinstate the polynomial $Q_m(\omega^2)$ into the denominator of a Butterworth characteristic function, as in (2.7). Choosing $m \leq n/2$ and

$$Q_m(\omega^2) = \prod_{i=1}^{m} (\omega^2 - \omega_i^2)^2$$

$$K(s) = \pm C \frac{s^n}{\prod_{i=1}^{m} (s^2 + \omega_i^2)} \tag{2.38}$$

*These qualitative considerations may also be supported by quantitative proof [2].

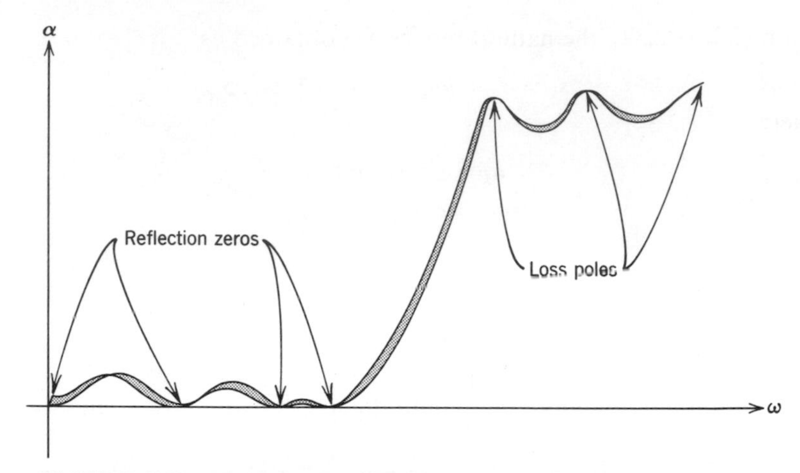

FIGURE 2-10 The role of zeros and poles in shaping the loss-response.

it is easy to verify that $|K|^2$ and hence α are still maximally flat at the frequency origin. Now, however, there are also $2m$ loss poles at the $\pm\omega_i$, and only $(n - 2m)$ loss poles at infinite frequency.

The stopband behavior of the loss will depend strongly, and the passband behavior somewhat, on the location of the loss poles ω_i. Hence, finding proper values for the ω_i is important. The simplest situation is that depicted in Figure 2-11, where the stopband loss oscillates between α_s and ∞ in an "equal-ripple" fashion. This *Chebyshev stopband* behavior may be obtained in a straightforward manner from the Chebyshev passband characteristic function of (2.21), in the following steps:

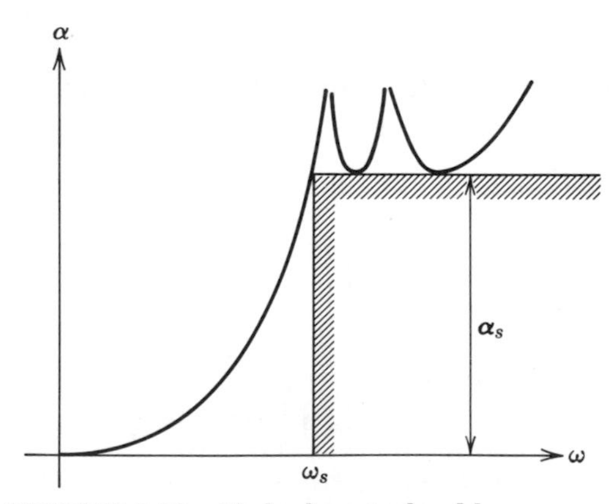

FIGURE 2-11 Chebyshev stopband loss-response.

1. Replace ω/ω_p by ω_s/ω.
2. Replace $|K|^2$ by $|K|^{-2}$.

These steps will result in the transformation of $|K|^2$ illustrated in Figure 2-12. The resulting equations are

$$|K|^2 = [k_p^2 \cos^2 nu]^{-1}$$
$$u(\omega) = \cos^{-1}(\omega_s/\omega) \tag{2.39}$$

with the loss poles at

$$\omega_i = \frac{\omega_s}{\cos \frac{2i-1}{n}\frac{\pi}{2}}, \qquad i = 1, 2, \ldots, n. \tag{2.40}$$

For given α_p, α_s, ω_p, and ω_s, the necessary degree is still given by (2.28). Because of the form of $|K|^2$ as given by (2.39), this filter is often called an "inverse Chebyshev" filter.

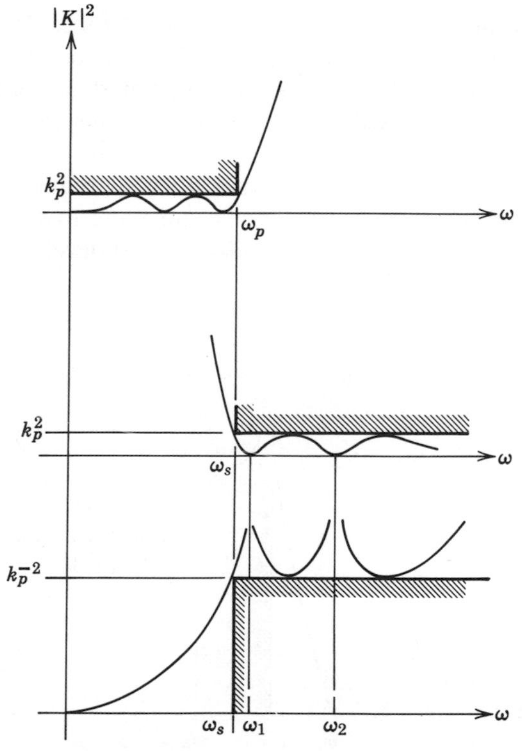

FIGURE 2-12 Transformation of Chebyshev passband response into Chebyshev stopband response.

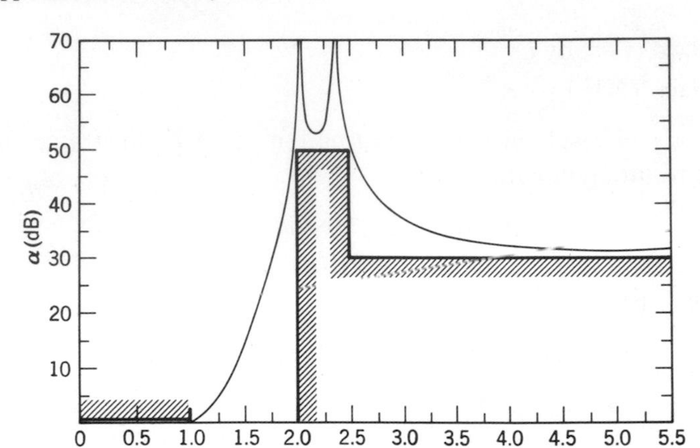

FIGURE 2-13 Filter characteristic with frequency-dependent stopband loss specifications.

If the required minimum stopband loss α_s is not constant, but varies with ω, then the loss-pole distribution given by (2.40) may be very inefficient, and the loss poles should be found so as to accommodate the actual requirements (see Figure 2-13). A design technique for this case is described in reference [3].

Next, we shall generalize the Chebyshev filter to obtain filters which have equal-ripple behavior in their passbands and finite loss poles ω_i in their stopbands (Figure 2-14). Although a more logical development is possible

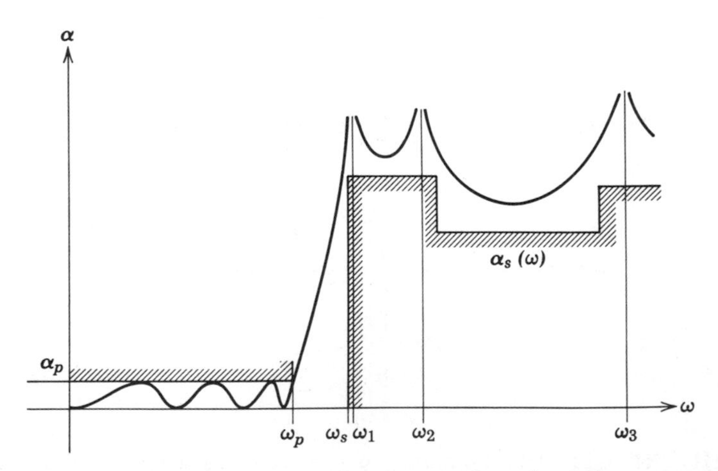

FIGURE 2-14 Equal-ripple passband, general stopband filter response.

[4], a heuristic approach will be used here. As a first step, the new complex independent variable

$$z = \sqrt{1 + \omega_p^2/s^2}, \qquad \mathrm{Re}(z) \geq 0 \tag{2.41}$$

is introduced. This transforms the passband ($s = j\omega$, $|\omega| \leq \omega_p$) onto the imaginary axis (jy-axis) of the z-plane; it maps the stopband ($s = j\omega$, $|\omega| > \omega_p$) on the $0 < x \leq 1$ portion of the real x-axis. For example, the loss response shown in Figure 2-14 becomes that shown in the three-dimensional diagram of Figure 2-15. Next, we define the polynomial

$$E(z^2) + zF(z^2) = \prod_{i=1}^{n} (z + z_i) \tag{2.42}$$

where the z_i are the transformed loss poles, $z_i = \sqrt{1 - \omega_p^2/\omega_i^2}$. E and F are even polynomials, and hence, E is the even, zF the odd part of the overall expression. It is important to keep in mind that (2.41) transforms the loss poles located at both $+j\omega_i$ and $-j\omega_i$ in the s-plane to $+\sqrt{1 - \omega_p^2/\omega_i^2}$, and hence all finite $j\omega$-axis poles appear *doubled* in the z-plane, and also that poles at $\omega \to \infty$ do enter $E + zF$ via factors of the form $(z + 1)$. Now, examine the behavior of the function

$$|K|^2 = k_p^2 \frac{E^2}{E^2 - z^2F^2} = \frac{k_p^2}{1 - \left(\dfrac{zF}{E}\right)^2} \tag{2.43}$$

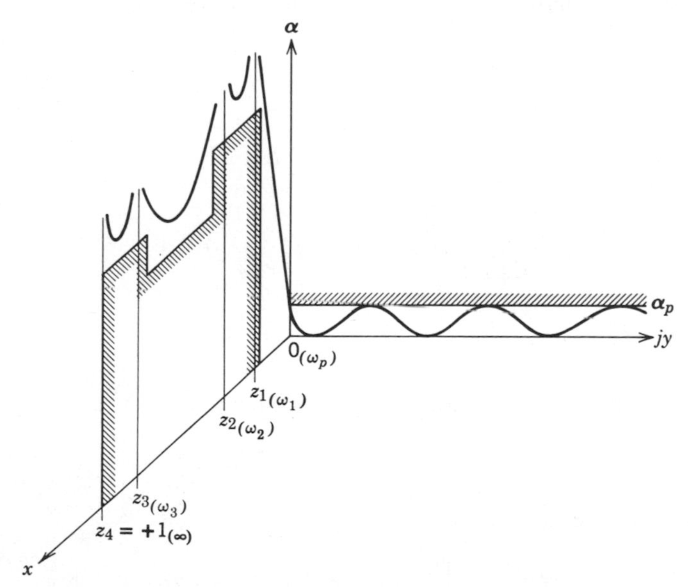

FIGURE 2-15 Transformed loss-response in the z-plane.

in the z-plane. By (2.42), this z-plane function has poles at the z_i in the stopband. In the transformed passband, i.e., on the jy-axis, zF/E behaves as a reactance since zF and E are the odd and even parts, respectively, of the Hurwitz polynomial $E + zF$ (see (2.41) and (2.42)). Hence, yF/E is a monotone increasing function of y, with interlaced zeros and poles. Consequently, $1 - (zF/E)^2 = 1 + (yF/E)^2$ varies between 1 (when $yF/E = 0$) and ∞ (when $yF/E \to \infty$); and therefore $|K|^2$ oscillates between 0 and k_p^2. In conclusion, then, $|K|^2$ has the required poles and the required Chebyshev passband behavior. Since it is a function of z^2 only, it is a simple matter to substitute $z^2 = 1 + \omega_p^2/s^2$ into (2.43), and thus to obtain the s-plane characteristic function. Hence, for prescribed ω_i, $|K|^2$ can be found in straightforward steps. In order to find the ω_i for a prescribed frequency-dependent loss $\alpha_s(\omega)$, let $|K|^2$ be decomposed:

$$|K|^2 = k_p^2 \frac{E^2}{E^2 - z^2F^2} = \frac{k_p^2}{2}\left[1 + \frac{1}{2}\frac{E + zF}{E - zF} + \frac{1}{2}\frac{E - zF}{E + zF}\right]. \tag{2.44}$$

Here, by (2.2),

$$k_p^2 = 10^{\alpha_p/10} - 1 \tag{2.45}$$

and hence, $k_p^2 < 1$ for the usual case of $\alpha_p < 3$ dB. On the other hand, we note that normally in the transformed stopband $\alpha \geq 30$ dB so that $|K|^2 \gg 1$. Hence, either the second *or* the third term on the extreme right of (2.44) must be much greater than 1. (They cannot be both large, since they are reciprocals of each other.) In the stopband ($0 < z \leq 1$), $E - zF$ has the n zeros z_i, while $E + zF$ has none. Hence,

$$\frac{1}{2}\frac{E + zF}{E - zF} \gg 1 \gg \frac{1}{2}\frac{E - zF}{E + zF} \tag{2.46}$$

and

$$|K|^2 \cong \frac{k_p^2}{4}\frac{E + zF}{E - zF} = \frac{k_p^2}{4}\prod_{i=1}^{n}\frac{z_i + z}{z_i - z}. \tag{2.47}$$

Thus, in the stopband

$$\alpha(\omega) \cong 10\log_{10}|K|^2 = 10\log_{10}(10^{\alpha_p/10} - 1) - 6.02 + \sum_{i=1}^{n} 10\log_{10}\frac{z_i + z}{z_i - z}. \tag{2.48}$$

Introducing the new variable

$$\gamma = \frac{1}{2}\ln(1 - \omega_p^2/\omega^2) = \ln z, \tag{2.49}$$

the terms under the summation become

$$10 \log_{10} \frac{e^{\gamma_i} + e^{\gamma}}{e^{\gamma_i} - e^{\gamma}} = 10 \log_{10} \frac{e^{\gamma_i - \gamma} + 1}{e^{\gamma_i - \gamma} - 1} = 10 \log_{10} \coth \frac{\gamma_i - \gamma}{2}. \quad (2.50)$$

Separating the double poles z_i corresponding to finite $\pm\omega_i$, therefore,

$$\alpha(\omega) \cong 10 \log_{10}(10^{\alpha_p/10} - 1) - 6.02 + \sum_{i=1}^{m} 20 \log_{10} \coth \frac{|\gamma - \gamma_i|}{2}$$
$$+ 10(n - 2m)\log_{10} \coth \frac{|\gamma|}{2}. \quad (2.51)$$

Hence, $\alpha(\omega)$ is the weighted sum of several "template functions," identical in shape but centered at different values $\gamma = \gamma_i$. The template function

$$\alpha_t = 20 \log_{10} \coth \frac{|\gamma - \gamma_i|}{2} \quad (2.52)$$

is shown in Figure 2-16. The parameters γ_i can be found using graphical methods, or, preferably, by way of an iterative algorithm readily programmed for a computer [3].

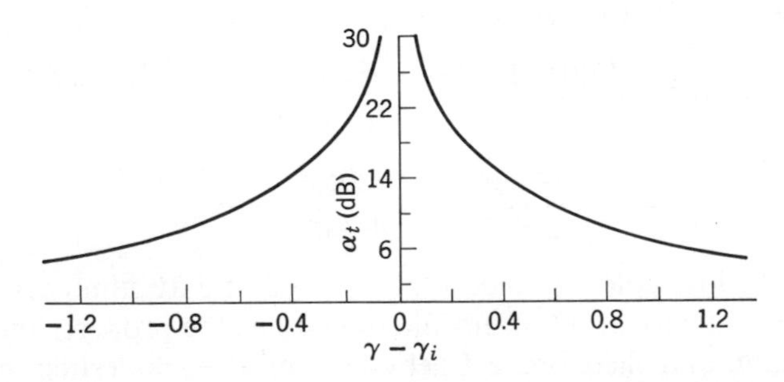

FIGURE 2-16 Template curve for equal-ripple passband filters.

Example 2-3. Find $|K(s)|^2$ for a Chebyshev passband lowpass filter with $n = 3$, one loss pole at $\omega_{1,2} = \pm 2$ and one at infinity. In the passband $(|\omega| \leq 1)$ the maximum permissible loss is 0.3 dB.

Solution: Equation (2.41), for $\omega_p = 1$, gives

$$z = \sqrt{1 + \frac{1}{s^2}}, \qquad \text{Re}(z) \geq 0.$$

Therefore,

$$s_{1,2} = \pm j2 \Rightarrow z_1 = z_2 = \frac{\sqrt{3}}{2},$$

$$s_3 \to \infty \Rightarrow z_3 = 1.$$

By (2.42):

$$E + zF = \left(z + \frac{\sqrt{3}}{2}\right)^2 (z+1) = z^3 + (1+\sqrt{3})z^2 + \left(\frac{3}{4} + \sqrt{3}\right)z + \frac{3}{4}$$

and

$$E - zF = \left(-z + \frac{\sqrt{3}}{2}\right)^2 (-z+1) = -z^3 + (1+\sqrt{3})z^2 - \left(\frac{3}{4} + \sqrt{3}\right)z + \frac{3}{4}.$$

Equation (2.43) then gives:

$$|K(s)|^2 = \frac{k_p^2[(1+\sqrt{3})z^2 + \frac{3}{4}]^2}{(\frac{3}{4} - z^2)^2(1 - z^2)}$$

where k_p^2 is given by (2.45) for $\alpha_p = 0.3$ dB, i.e.,

$$k_p^2 = 0.071519305.$$

Transforming to the s-plane by using (2.41):

$$|K(s)|^2 = \frac{16(0.0715193)(-s^2)(3.48205s^2 + 2.73205)^2}{(s^2+4)^2}$$

$$= \frac{-1.144309s^2(3.48205s^2 + 2.73205)^2}{(s^2+4)^2}.$$

Evidently, the loss poles are indeed at $\omega = \pm 2$ and at infinity, as specified.

An important special set of this filter class has the property that $\alpha_s(\omega) \equiv \alpha_{s0} =$ constant, and therefore, a Chebyshev response is desired in both the passband and stopband (Figure 2-17). These circuits can be treated exactly and lead to expressions containing elliptic functions; for this reason, they are called elliptic filters. Their analysis, although not too complicated, is quite lengthy and will hence be omitted. The interested reader is referred to the excellent tutorial exposition of reference [6].

It should be noted that it is practical to regard the elliptic filter as just another Chebyshev passband filter, and to use the template method to approximate the specified $\alpha_s = \alpha_{s0}$ in the stopband. The computational effort and accuracy will be comparable to that obtainable by using the conventional techniques of elliptic filter design [6, 7].

Finally, it should be noted that for given α_p, α_s, ω_p, and ω_s, the elliptic

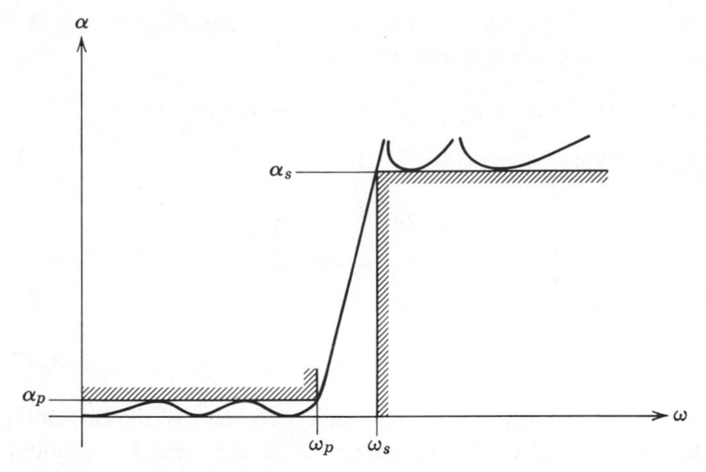

FIGURE 2-17 Elliptic filter response.

function requires the lowest degree of all rational characteristic functions. This is due to its Chebyshev properties in both frequency bands. Hence, if only the loss response is of interest, the elliptic function may be regarded as the most efficient and economical approximation to the ideal response shown in Figure 2-3.

2.2 LINEAR PHASE FUNCTIONS

In many practical applications, the waveform of a nonsinusoidal signal passing through the filters should be preserved without significant distortion. This usually requires that the phase response of the filter in its passband be linear, or, equivalently, that both phase-and group-delay be constant in the passband. As was the case with the loss response, this flat delay may be approximated using various criteria, to be discussed next.

Maximally Flat Group-Delay Polynomials

To obtain a maximally flat approximation around $\omega = 0$ to a constant group delay (and hence to a linear phase), (1.26) is used with a specified $\beta_s(\omega) = \omega T_0$:

$$j\beta_s(\omega) = j\omega T_0 \overset{!}{=} \tanh^{-1}[H_0(s)/H_e(s)]_{s=j\omega}$$

$$\tanh sT_0 = \frac{e^{2sT_0} - 1}{e^{2sT_0} + 1} \overset{!}{=} [e_0(s)/e_e(s)]^{\pm 1}$$

$$e^{2sT_0} \overset{!}{=} \pm \frac{e(s)}{e(-s)}, \qquad s = j\omega. \tag{2.53}$$

Here, the + sign is valid for even $p(s)$, the − sign for odd $p(s)$. Assuming, for definiteness, an even $p(s)$ and writing

$$e(s) = a_0 + a_1 s + a_2 s^2 + \cdots + a_{n-1} s^{n-1} + s^n, \tag{2.54}$$

(2.53) gives, by Taylor expansion,

$$e^{2sT_0} e(-s) \stackrel{!}{=} e(s)$$

$$\left[1 + (2sT_0) + \frac{(2sT_0)^2}{2!} + \frac{(2sT_0)^3}{3!} + \cdots\right][a_0 - a_1 s + a_2 s^2 - + \cdots]$$

$$\stackrel{!}{=} a_0 + a_1 s + a_2 s^2 + \cdots + a_n s^n. \tag{2.55}$$

For maximally flat approximation, the coefficients of the two polynomials on the two sides of (2.55) should agree. Since the right hand side does not contain terms in s^{n+1}, s^{n+2}, ..., the coefficients of these terms on the left-hand side should also vanish up to and including the term s^{2n}. Therefore, from the coefficients of s^{n+1}, s^{n+2}, ..., s^{2n}

$$\frac{(2T_0)^{n+1}}{(n+1)!} a_0 - \frac{(2T_0)^n}{n!} a_1 + - \cdots + (-1)^{n-1} \frac{(2T_0)^2}{2!} a_{n-1} + (-1)^n \frac{(2T_0)}{1!} = 0$$

$$\frac{(2T_0)^{n+2}}{(n+2)!} a_0 - \frac{(2T_0)^{n+1}}{(n+1)!} a_1 + - \cdots + (-1)^{n-1} \frac{(2T_0)^3}{3!} a_{n-1} + (-1)^n \frac{(2T_0)^2}{2!} = 0$$

$$\cdot \quad \cdot \quad \cdot \quad \cdot \quad \cdot \quad \cdot \quad \cdot \quad \cdot \quad \cdot \quad \cdot \quad \cdot \quad \cdot \quad \cdot \quad \cdot \quad \cdot$$

$$\frac{(2T_0)^{n+k}}{(n+k)!} a_0 - \frac{(2T_0)^{n+k-1}}{(n+k-1)!} a_1 + - \cdots + (-1)^{n-1} \frac{(2T_0)^{k+1}}{(k+1)!} a_{n-1} + (-1)^n \frac{(2T_0)^k}{k!} = 0$$

$$\cdot \quad \cdot \quad \cdot \quad \cdot \quad \cdot \quad \cdot \quad \cdot \quad \cdot \quad \cdot \quad \cdot \quad \cdot \quad \cdot \quad \cdot \quad \cdot \quad \cdot$$

$$\frac{(2T_0)^{2n}}{(2n)!} a_0 - \frac{(2T_0)^{2n-1}}{(2n-1)!} a_1 + - \cdots + (-1)^{n-1} \frac{(2T_0)^{n+1}}{(n+1)!} a_{n-1} + (-1)^n \frac{(2T_0)^n}{n!} = 0. \tag{2.56}$$

It will next be shown that these n equations are satisfied by the n coefficients*

$$a_i = \frac{(2n-i)!}{i!\,(n-i)!} (2T_0)^{i-n}, \qquad i = 0, 1, 2, \ldots, n-1. \tag{2.57}$$

Substituting a_i as given by (2.57) into (2.56), the kth equation becomes

$$\sum_{i=0}^{n} (-1)^i \frac{(2n-i)!}{i!\,(n-i)!} \frac{(2T_0)^k}{(n+k-i)!} = 0. \tag{2.58}$$

* Note that (2.57) is also valid for $a_n \equiv 1$.

Omitting the unimportant $(2T_0)^k$ factor and introducing $j \equiv n - i$,

$$\sum_{j=0}^{n}(-1)^j \frac{(n+j)!}{j!\,(n-j)!}\frac{1}{(k+j)!} =$$

$$\frac{1}{n!}\sum_{j=0}^{n}(-1)^j\binom{n}{j}\frac{(n+j)!}{(k+j)!} = 0, \qquad k = 1, 2, \ldots, n. \tag{2.59}$$

It is easy to see that this expression is the same as

$$\frac{1}{n!}\frac{d^{n-k}}{dx^{n-k}}[x^n(1-x)^n], \qquad k = 1, 2, \ldots, n \tag{2.60}$$

for $x = 1$. Since this latter function is surely zero for $x = 1$, due to the presence of a $(1 - x)$ factor in each term, the proof is complete.

Problem 2.20 contains a verification of the essential fact that the a_i given by (2.57) also provides a matching for the coefficients of $s^0, s^1, s^2, s^3, \ldots, s^n$ in (2.55).

The coefficients a_i given in (2.57) have been tabulated for $T_0 = 1$ [27]. Next, the behavior of $e(s)$ for large n will be examined.

For $n \to \infty$,

$$\frac{a_i}{a_0} = \left[\frac{(2n-i)!}{(2n)!\,2^{-i}}\right]\left[\frac{n!}{(n-i)!}\right]\frac{T_0^i}{i!} = \frac{n(n-1)(n-2)\cdots(n-i+1)}{n\left(n-\frac{1}{2}\right)\left(n-\frac{2}{2}\right)\cdots\left(n-\frac{i-1}{2}\right)}\frac{T_0^i}{i!} \to \frac{T_0^i}{i!}$$

and hence

$$e(s) = \sum_{i=0}^{n} a_i s^i \to \sum_{i=0}^{n} a_0 \frac{(sT_0)^i}{i!} \to a_0 e^{sT_0}. \tag{2.61}$$

Thus, for a constant $p(s)$, $H(j\omega) \propto e^{j\omega T_0}$ which is the required characteristic of the ideal system.

For large but finite n,

$$e(j\omega)e(-j\omega) = [e_e(j\omega)]^2 - [e_0(j\omega)]^2 \to a_0^2 e^{(\omega T_0)^2/(2n-1)} \tag{2.62}$$

as can be shown by Taylor expansion for large n and for $\omega T_0 \ll n$. Hence, choosing $p(s) \equiv a_0$, the limiting loss response (in decibels) is the Gaussian function

$$\alpha(\omega) = 10 \log_{10} e^{(\omega T_0)^2/(2n-1)} = (10 \log_{10} e)\frac{(\omega T_0)^2}{2n-1} \tag{2.63}$$

and the 3-dB frequency is $\omega_{3\text{ dB}} = T_0^{-1}\sqrt{(2n-1)\ln 2}$.

The actual group-delay performance can also be found using series expansion. The result is the expected maximally flat approximation to the constant group delay T_0:

$$T_G = T_0\left[1 - \left(\frac{\omega^n}{a_0}\right)^2 + - \cdots\right]$$

$$T_G \cong T_0\left[1 - \left(\frac{2^n n!}{(2n)!}\right)^2 (\omega T_0)^{2n}\right]. \tag{2.64}$$

Equations (2.63) and (2.64) hold already quite well for values of n as low as 3 and can hence be used to establish the required degree n for a prescribed T_0 and a specified loss or delay distortion at some ω_p.

By (2.62)–(2.64), for $n > 3$ and $\omega T_0 < n$, the approximation

$$H(j\omega) \cong \exp\left[\frac{(\omega T_0)^2}{2(2n-1)} + j\omega T_0\right] \tag{2.65}$$

holds. Hence, if in the circuit of Figure 1-5, we choose $R_1 = R_2 = 1$ and $E(t) = 2\,\delta(t)$, the corresponding impulse response of the system will be given by:

$$v_2(t) = \mathscr{F}^{-1}\left\{\exp\left[-j\omega T_0 - \frac{T_0^2\omega^2}{2(2n-1)}\right]\right\}$$

$$= \frac{\sqrt{(2n-1)/\pi}}{\sqrt{2T_0}} \exp\left[-\frac{2n-1}{2T_0}(t - T_0)^2\right] \tag{2.66}$$

where the well-known relations of Fourier transformation theory [9]

$$\sqrt{\frac{\alpha}{\pi}}\, e^{-\alpha t^2} \leftrightarrow e^{\omega^2/4\alpha}$$

$$f(t - T_0) \leftrightarrow \mathscr{F}^{-1}[f(t)]e^{-j\omega T_0} \tag{2.67}$$

have been used. Hence, the amplitude and time responses are both approximately Gaussian.

The maximally flat approximation to a constant group delay around $\omega = 0$, described above, was originally derived by W. E. Thomson [10]. The resulting $e(s)$ polynomials, with coefficients given in (2.57), are related to the Bessel polynomials and are themselves known as Lommel polynomials. For these reasons, the circuits derived from these $e(s)$ are usually called Bessel and/or Thomson filters.

Example 2-4. Find n and $e(s)$ for a Bessel/Thomson filter which approximates $T_0 = 2\ \mu\text{sec}$ with a maximum error of 20% in the 0–0.4-MHz frequency range. What is the loss at 0.4 MHz?

Solution: From (2.64), here

$$\left(\frac{2^n n!}{(2n)!}\right)^2 (2\pi 0.8)^{2n} \leq 0.2.$$

By substituting $n = 1, 2, \ldots,$ it is found that $n = 8$ is the necessary degree. Therefore, the coefficients a_i, obtained from (2.57), are:

$$a_0 = \frac{2027025}{256} 10^{48} = 7918.0664 \times 10^{48}$$

$$a_1 = \frac{2027025}{128} 10^{42} = 15836.1328 \times 10^{42}$$

$$a_2 = \frac{945945}{64} 10^{36} = 14780.3906 \times 10^{36}$$

$$a_3 = \frac{270270}{32} 10^{30} = 8445.9375 \times 10^{30}$$

$$a_4 = \frac{51975}{16} 10^{24} = 3248.4375 \times 10^{24}$$

$$a_5 = \frac{6930}{8} 10^{18} = 866.2500 \times 10^{18}$$

$$a_6 = \frac{630}{4} 10^{12} = 157.5000 \times 10^{12}$$

$$a_7 = \frac{36}{2} 10^{6} = 18.0000 \times 10^{6}$$

$$a_8 = 1$$

and, by (2.54),

$$e(s) = \sum_{i=0}^{8} a_i s^i.$$

The loss at 0.4 MHz is obtained from (2.63):

$$\alpha[2\pi(0.4)10^6] = 4.342945 \frac{(2\pi(0.4)2)^2}{15}$$

$$= 7.315311 \text{ dB}.$$

The reader is urged to repeat these calculations in normalized units (choosing, e.g., $T_0 = 1$) to see how the large powers of ten vanish in normalized calculations.

Equal-Ripple Group-Delay Approximation

In the preceding section we have seen that a significant improvement resulted when the maximally flat (Butterworth) loss response was replaced by the equal-ripple (Chebyshev) approximation. Such an improvement is to be expected also in the case of delay approximation [12]. Hence, the approximation of a constant group delay in an equal-ripple manner will be discussed next.

In order to obtain the equations which a polynomial $e(s)$ must satisfy if it is to yield an equal-ripple approximation of a constant group delay T_0, the properties of the group-delay function $T_g(s)$ must be analyzed. By (1.29), for a reactance two-port

$$T_g(s) = \mathrm{Ev}\left[\frac{e'(s)}{e(s)}\right] = \mathrm{Ev}\left[\frac{d}{ds}\ln e(s)\right]. \tag{2.68}$$

Therefore, if $e(s) = k\prod_{i=1}^{n}(s - s_i)$,

$$T_g(s) = \mathrm{Ev}\left[\sum_{i=1}^{n}(s - s_i)^{-1}\right] = \sum_{i=1}^{n}\left[\frac{1/2}{s - s_i} - \frac{1/2}{s + s_i}\right]$$
$$= \sum_{i=1}^{n}\frac{s_i}{s^2 - s_i^2}. \tag{2.69}$$

Hence, for any set of conjugate complex s_i with negative real parts, (2.69) represents a group-delay function realizable by a reactance two-port. Note the $\pm 1/2$ residues and also that, for $s \to \infty$, $T_g(s) \to (\sum_{i=1}^{n} s_i)/s^2$.

Next, the desired response (Figure 2-18) will be examined. Permitting a

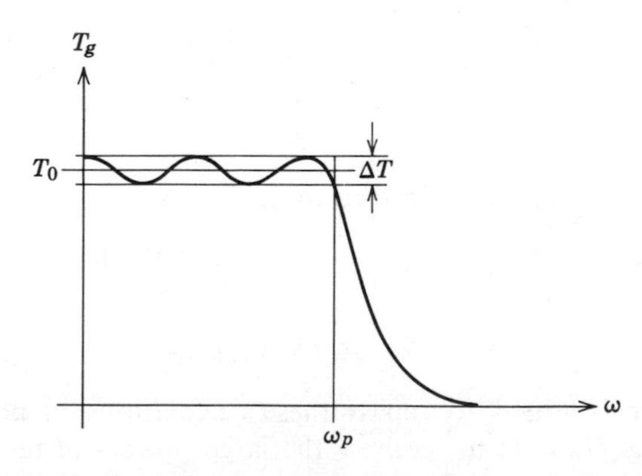

FIGURE 2-18 Equal-ripple group-delay response.

peak-to-peak ripple ΔT about a mean delay T_0 for $|\omega| \leq \omega_p$, $T_g(\omega)$ may evidently be written in the form

$$T_g(\omega) = T_0 - \frac{\Delta T}{2} R(\omega). \tag{2.70}$$

Here, $R(\omega)$ must be an even rational function of ω which also has to satisfy the following conditions:

1. $R(\omega)$ oscillates between $+1$ and -1 for $|\omega| \leq \omega_p$.
2. $\lim_{\omega\to\infty} R(\omega) = 2T_0/\Delta T$, to assure $\lim_{\omega\to\infty} T_g(\omega) = 0$. For large ω, $R(\omega) \to 2T_0/\Delta T + \text{const.}/\omega^2$.
3. Creating the complex function* $R(s)$ by analytic continuation, this must have conjugate complex poles s_i in the left half s-plane, as well as poles at their negatives $-s_i$. The residues at the s_i must be $-1/\Delta T$; and at $-s_i$, $+1/\Delta T$.

All these conditions follow directly from (2.69) and (2.70), as well as from Figure 2-18.

To find a function $R(s)$ satisfying conditions 1–3, we again introduce the mapping of (2.41) and consider the behavior of the function

$$R(z) = \frac{1}{2}\left[\frac{E + zF}{E - zF} + \frac{E - zF}{E + zF}\right] \tag{2.71}$$

where $E + zF = \prod_{i=1}^{n}(z + z_i)$ as before, and where all the z_i are in the right half of the z-plane. On the imaginary jy-axis, the two terms of $R(z)$ behave as the allpass functions $(1/2)e^{\pm j\alpha(y)}$ and hence $R(jy) = \cos\alpha(y)$ oscillates† between ± 1. Recalling that by (2.41) the jy-axis corresponds to $|\omega| \leq \omega_p$, it is seen that $R(z)$ as a function of ω satisfies condition 1.

Since $z \to 1$ as $\omega \to \infty$, condition 2 requires

$$R(1) = \frac{1}{2}\left[\prod_{i=1}^{n} \frac{1 + z_i}{-1 + z_i} + \prod_{i=1}^{n} \frac{1 - z_i}{-1 - z_i}\right] = \frac{2T_0}{\Delta T}. \tag{2.72}$$

To satisfy condition 3, the partial fraction expansion of T_g, calculated from (2.70), must coincide with that given in (2.69). Combining (2.70) and (2.71),

$$T_g(z) = T_0 - \frac{\Delta T}{4}\left[\prod_{i=1}^{n} \frac{z_i + z}{z_i - z} + \prod_{i=1}^{n} \frac{z_i - z}{z_i + z}\right]$$

$$\lim_{z\to z_k} T_g(z) = -\lim_{z\to z_k} \frac{\Delta T}{4}\left[2z_k \prod_{\substack{i=1\\ i\neq k}}^{n} \frac{z_i + z_k}{z_i - z_k}\right] \frac{1}{z_k - z} \tag{2.73}$$

* Obtained by replacing ω by s/j in $R(\omega)$.

† This same conclusion may be drawn by noting that $R(z)$ and the $|K|^2$ of (2.44) are related, and that they satisfy $R(z) = 2k_p^{-2}|K|^2 - 1$.

while a transformation of the $T_g(s)$ of (2.69) into the z-plane gives

$$T_g(z) = \sum_{i=1}^{n} \frac{\pm \dfrac{\omega_p}{\sqrt{z_i^2 - 1}}}{\dfrac{\omega_p^2}{z^2 - 1} - \dfrac{\omega_p^2}{z_i^2 - 1}} = \sum_{i=1}^{n} \frac{\pm\sqrt{z_i^2 - 1}(z^2 - 1)}{\omega_p(z_i^2 - z^2)}$$

$$\lim_{z \to z_k} T_g(z) = \lim_{z \to z_k} \frac{\pm(z_k^2 - 1)^{3/2}}{\omega_p 2z_k(z_k - z)}. \tag{2.74}$$

Here the $\pm$ signs must be selected such that each $s_i = \pm\omega_p(z_i^2 - 1)^{-1/2}$ and hence each $(z_i^2 - 1)^{1/2}$ has a negative real part. Equating the z-plane residues

$$-\frac{\Delta T}{2} z_k \prod_{\substack{i=1 \\ i \neq k}}^{n} \frac{z_i + z_k}{z_i - z_k} = \frac{\pm(z_k^2 - 1)^{3/2}}{2\omega_p z_k}$$

or:

$$\frac{\pm(z_k^2 - 1)^{3/2}}{z_k^2} + \omega_p \, \Delta T \prod_{\substack{i=1 \\ i \neq k}}^{n} \frac{z_i + z_k}{z_i - z_k} = 0, \qquad k = 1, 2, \ldots, n. \tag{2.75}$$

For a given ω_p and ΔT, the n complex z_k may be found from the n complex simultaneous nonlinear equations (2.75). If a good initial approximation is available for the z_k, then iterative techniques (such as Newton's method) will rapidly converge to the solution. The mean delay T_0 can then be found from (2.72). Very good initial approximation is provided by the nearly equal-ripple linear-phase functions discussed by Darlington [11], which, however, requires *a priori* knowledge of T_0. Having obtained the z_i, then $s_i = \omega_p(z_i^2 - 1)^{-1/2}$, $\mathrm{Re}(s_i) < 0$ can be calculated.

In most practical situations T_0 is exactly prescribed and there are limits set on ΔT (or ω_p). Then (2.72) and (2.75) must be solved simultaneously for the z_i and ΔT (or ω_p).

The natural modes obtained by solving (2.75) (with prescribed ΔT) for the s_i have been thoroughly tabulated [12].

Example 2-5. Find s_1, s_2, and T_0 for $\Delta T = .04$ s and $n = 2$ for an equal-ripple delay approximation. Assume $\omega_p = 1$ rad/sec.

Solution: From (2.72) and (2.75),

$$\frac{1}{2}\left[\frac{1 + z_1}{-1 + z_1} \frac{1 + z_2}{-1 + z_2} + \frac{1 - z_1}{-1 - z_1} \frac{1 - z_2}{-1 - z_2}\right] = 50T_0$$

$$\frac{\pm(z_1^2 - 1)^{3/2}}{z_1^2} + 0.04 \frac{z_2 + z_1}{z_2 - z_1} = 0$$

$$\frac{\pm(z_2^2 - 1)^{3/2}}{z_2^2} + 0.04 \frac{z_1 + z_2}{z_1 - z_2} = 0.$$

Solving for z_1 and z_2, and using (2.41) to transform into the s-plane, we find

$$s_{1,2} = -1.236807 \pm j0.918276$$

and $T_0 = 1.062$.

Linear-Phase Filters with Chebyshev Stopbands

One common disadvantage of the linear-phase filters discussed thus far in this section is that they cannot provide appreciable loss selectivity even if $H(s)$ has a very high degree. This can be seen, e.g., by comparing the Gaussian loss response $\alpha = 3(\omega/\omega_{3\text{ dB}})^2$ which the Bessel filter approaches for large n with the loss response of a high-order Chebyshev or elliptic filter. If a sharply selective loss characteristic is desired, the introduction of finite loss poles ω_i can be useful. This is equivalent to incorporating a denominator $\prod_{i=1}^{m}(s^2 + \omega_i^2)$ into $H(s)$. Since this denominator is pure even, in the passband (where $|\omega| < \omega_i$) its value is a positive real number and thus does not affect the phase or delay in any way. A straightforward procedure for choosing the ω_i is obtained by specifying a Chebyshev stopband [13, 14]. Consider the mapping of the complex frequency variable

$$w = \sqrt{1 + s^2/\omega_s^2}, \qquad \text{Re}(w) \geq 0. \tag{2.76}$$

This is very similar to the z-transformation introduced in (2.41), but it maps the $\omega_s \leq \omega \leq \infty$ portion of the $j\omega$-axis (i.e., the *stopband*) on the imaginary w-axis. Now defining

$$E(w^2) + wF(w^2) = \prod_{i=1}^{n} (w + w_i) \tag{2.77}$$

and choosing

$$|H|^2 = 10^{\alpha_s/10} \frac{E^2 - w^2F^2}{E^2} \tag{2.78}$$

we conclude that $E + wF$ is a Hurwitz polynomial, and thus, for $w = jv$, wF/E is a reactance function in v. Therefore $-(wF/E)^2$ oscillates between $+0$ and $+\infty$ taking on only positive values, and the loss

$$\alpha(w) = 10 \log_{10}|H|^2 = \alpha_s + 10 \log_{10}\left[1 - \left(w\frac{F}{E}\right)^2\right] \tag{2.79}$$

oscillates between α_s and $+\infty$ *no matter what the values of the* w_i *are.*

Comparison of (2.77) and (2.78) shows that the w_i are the transformed natural modes. Hence, choosing the w_i as the transforms of the zeros of any of the linear-phase $e(s)$ polynomials (Bessel [10], Darlington [11], and Ulbrich-Piloty [12]) results in an $H(s)$ function giving linear phase in the passband and a Chebyshev stopband performance for $\omega \geq \omega_s$.

It should be noted that (2.78) does not guarantee the usually desirable condition $\alpha = 0$ for $\omega = 0$. This can be achieved if α_s is chosen to satisfy

$$1 = 10^{\alpha_s/10}\left[1 - \left(\frac{wF(w^2)}{E(w^2)}\right)^2_{w=1}\right] \tag{2.80}$$

since $w = 1$ corresponds, by (2.76), to $\omega = 0$. In addition, a loss pole at infinite frequency, i.e., $\alpha \to \infty$ for $\omega \to \infty$ can also be obtained merely by choosing n odd, since $\omega \to \infty$ corresponds to $w \to \infty$ and since for odd n the polynomial $E^2 - w^2F^2$ is of higher order ($2n$) than E^2 (which is of order $2n - 2$). For even n, the $\alpha \to \infty$ for $\omega \to \infty$ condition can also be achieved, but it requires an iterative procedure [13].

For a desirable linear-phase polynomial $e(s)$ and for given ω_s, the procedure is then to transform the roots s_i of $e(s)$ via (2.76) into

$$w_i = \sqrt{1 + s_i^2/\omega_s^2}, \qquad \mathrm{Re}(w_i) \geq 0. \tag{2.81}$$

Next, the even part $E(w^2)$ of the polynomial

$$E + wF = \prod_{i=1}^{n} (w + w_i)$$

is found.* The minimum stopband loss α_s obtainable can then be derived from (2.80). Retransforming $E(w^2)$ using (2.76) into the s-domain, the required transducer function is

$$H(s) = 10^{\alpha_s/20}\left[\frac{E(w)|_{w=0}}{e(s)|_{s=j\omega_s}}\right]\frac{e(s)}{E(s)}. \tag{2.82}$$

Here the factor in the square bracket is needed to insure that $H(j\omega_s) = 10^{\alpha_s/20}$. (This property is, by virtue of (2.78), automatically present in the w-plane, but must be re-established in the s-plane.)

Example 2-6. Find $H(s)$ from the second-order Thomson polynomial for $T_0 = 1$:

$$e(s) = s^2 + 3s + 3 = \left(s + \frac{3 + j\sqrt{3}}{2}\right)\left(s + \frac{3 - j\sqrt{3}}{2}\right).$$

Use $\omega_s = 3$. How much is α_s?

* Since $E + wF$ is a Hurwitz polynomial, all roots of $E(w^2)$ lie on the jv-axis, so that $E(w^2)$ is of the form $C\prod_{k=1}^{m}(w^2 + v_k^2)$. In the s-plane this expression becomes $C\prod_{k=1}^{m}(\omega_k^2 - \omega^2)/\omega_s^2$.

Solution: If we specify $\omega_s = 3$, then

$$w_1 = \left[1 + \left(\frac{3 + j\sqrt{3}}{2}\right)^2 /9\right]^{1/2} = \sqrt{\frac{7 + \sqrt{52}}{12}} + j\sqrt{\frac{-7 + \sqrt{52}}{12}}$$

$$w_2 = w_1^* \text{ (conjugate of } w_1\text{)}$$

$$E = w^2 + |w_1|^2 = w^2 + \sqrt{13}/3$$

$$F = [(7 + \sqrt{52})/3]^{1/2}.$$

Hence, by (2.80)

$$10^{\alpha_s/10} = \left[1 - \frac{F^2}{E^2}\right]^{-1}_{w=1} = 22 + 6\sqrt{13} \cong 43.62$$

$$\alpha_s \cong 16.4 \text{ dB}.$$

By (2.82), therefore,

$$H(s) = \sqrt{43.62} \frac{\sqrt{13}/3}{|-\omega_s^2 + 3j\omega_s + 3|_{\omega_s=3}} \frac{s^2 + 3s + 3}{1 + s^2/9 + \sqrt{13}/3}$$

$$H(s) = \sqrt{43.62} \frac{s^2 + 3s + 3}{s^2 + 19.81}.$$

The loss poles are at $\pm j\sqrt{19.81} \cong \pm j4.45$. $H(0) = 1$, as required.

The Chebyshev stopband improves the selectivity of a linear-phase filter considerably. Even so, however, its selectivity is inadequate for many applications. Although several other attempts have also been made to combine linear phase with ideal lowpass loss response [15, 16], none has fully succeeded. Hence, in most practical cases the "brute force" approach of designing a filter from its loss specifications and then phase equalizing it by way of cascading allpass networks appears to be the best procedure.

Finally, it should again be noted that the usually desirable conditions $H(0) = 1$ and $H(\infty) \to \infty$ are not automatically satisfied for the $|H|^2$ obtained from (2.78). This, however, does not create a problem if an active realization is used.

2.3 BANDPASS FILTER FUNCTIONS

Bandpass Filters with Unsymmetrical Loss Specifications

The reactance and frequency transformations of the previous chapter can be used to transform Butterworth, Chebyshev, and inverse Chebyshev filters as well as elliptic type filters, into bandpass filters. Even the most general frequency transformations, however, restrict strongly the bandpass characteristics which can be achieved this way. Fortunately, the techniques used to obtain general stopband lowpass filter characteristics with maximally flat or equal-ripple passbands can easily be extended for bandpass filters. The squared characteristic function for the former case (Figure 2-19) is given by

$$|K|^2 = C^2 \frac{(\omega^2 - \omega_0^2)^{2n}}{\prod_{i=1}^{m}(\omega^2 - \omega_i^2)^2}. \tag{2.83}$$

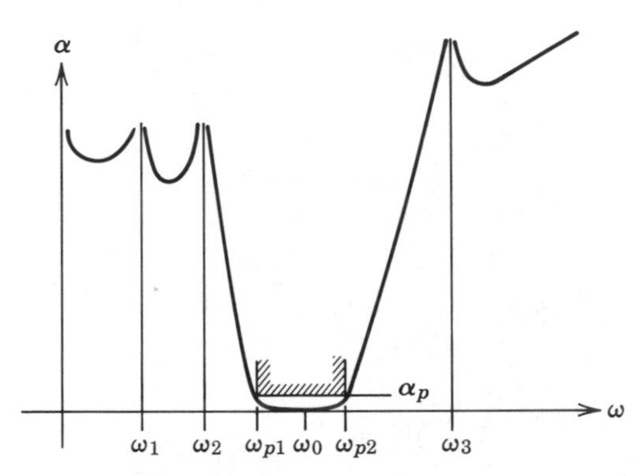

FIGURE 2-19 Maximally flat passband bandpass filter response.

It is directly obvious that this $|K|^2$ leads to a maximally flat loss response at ω_0 and to loss poles at the ω_i. If at ω_{p1} and ω_{p2} the passband loss should be α_p, then

$$C^2 = (10^{\alpha_p/10} - 1)\frac{\prod_{i=1}^{m}(\omega_p^2 - \omega_i^2)^2}{(\omega_p^2 - \omega_0^2)^{2n}} \tag{2.84}$$

where ω_p can be either ω_{p1} or ω_{p2}.

The rest of the design procedure becomes identical with that given for lowpass filters.

The generalization of the z-mapping used for Chebyshev passband, general stopband lowpass filters to bandpass filter characteristics is also fairly straightforward. The transformation*

$$z = \sqrt{\frac{s^2 + \omega_{p2}^2}{s^2 + \omega_{p1}^2}}, \qquad \text{Re}(z) \geq 0 \tag{2.85}$$

* The alert reader will note that the bandpass functions and relations reduce to their counterparts valid for the lowpass case when $\omega_0 \to 0$ for maximally flat passband, or $\omega_{p1} \to 0$ for equal-ripple passband.

maps the passband ($s = j\omega$, $\omega_{p1} \leq |\omega| \leq \omega_{p2}$) onto the imaginary jy-axis of the z-plane; the lower stopband ($s = j\omega$, $|\omega| < \omega_{p1}$) moves to the $\omega_{p2}/\omega_{p1} \leq z < \infty$ range, the upper stopband ($s = j\omega$, $|\omega| > \omega_{p2}$) to the $0 < z \leq 1$ range of the positive real z-axis (Figure 2-20).

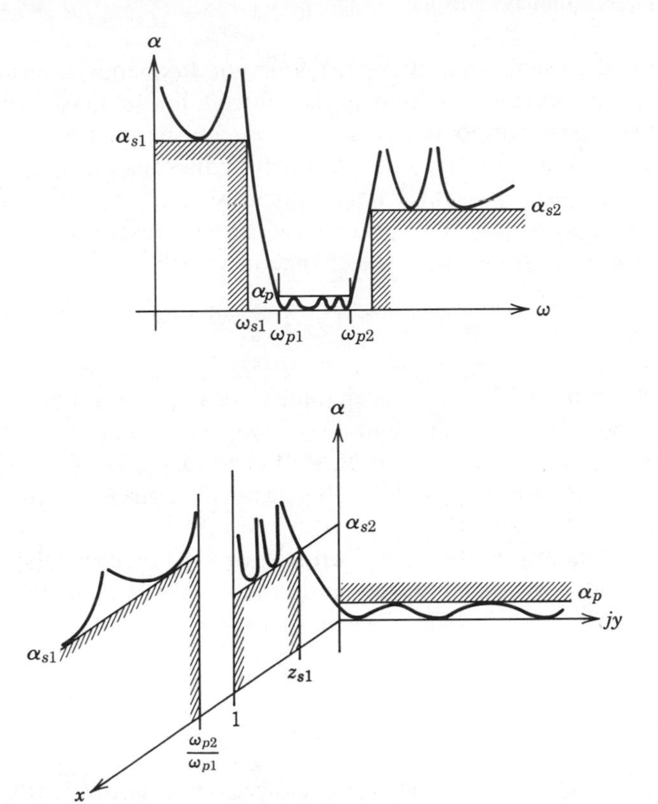

FIGURE 2-20 Equal-ripple passband bandpass filter response in *s*-and *z*-planes.

The derivations given for the lowpass case now become applicable, except that for a rational $K(s)$ the overall degree n must be *even*. Modifying slightly the definition of γ as given in (2.49), to

$$\gamma = \ln z = \frac{1}{2} \ln\left(\frac{\omega^2 - \omega_{p2}^2}{\omega^2 - \omega_{p1}^2}\right), \tag{2.86}$$

the template method also becomes applicable.

A filter-function class peculiar to bandpass characteristics is that of parametric filters. These circuits possess reflection zeros (zeros of K) on the real axis (σ-axis) of the s-plane. For this reason, they require special approximation techniques; they will be discussed in Chapter 3.

Linear-Phase Bandpass Filters

As mentioned earlier, none of the reactance or frequency transformations discussed in the previous section will transform a linear-phase lowpass filter into a linear-phase bandpass circuit. For narrow bandpass characteristics (i.e., when the bandwidth $\omega_{p2} - \omega_{p1}$ is much smaller than $(\omega_{p1} + \omega_{p2})/2$), an *approximately* linear passband phase response may be obtained in the manner discussed below. If a linear-phase lowpass response is available in the $-\hat{\omega}_p \leq \hat{\omega} \leq \hat{\omega}_p$ range, then the change of variable

$$\omega = \frac{\omega_{p2} - \omega_{p1}}{2\hat{\omega}_p}\hat{\omega} + \frac{\omega_{p2} + \omega_{p1}}{2} \tag{2.87}$$

will linearly translate the lowpass characteristics up along the $j\omega$-axis. The new response will be centered at $(\omega_{p2} + \omega_{p1})/2$ rather than 0, and the lowpass bandlimits $-\hat{\omega}_p$, $\hat{\omega}_p$ will be shifted to ω_{p1} and ω_{p2}, respectively. Due to the linear nature of the shift, a linear lowpass phase response remains linear after the transformation.

The only difficulty is that a realizable lowpass function $\hat{H}(\hat{s})$, with real coefficients and hence real and conjugate complex zeros/poles will no longer retain these essential properties after the change of s-variable, corresponding to (2.87)

$$s = \frac{\omega_{p2} - \omega_{p1}}{2\hat{\omega}_p}\hat{s} + j\frac{\omega_{p2} + \omega_{p1}}{2}. \tag{2.88}$$

The change will introduce complex coefficients and, equivalently, complex zeros and poles unaccompanied by their conjugates (Figure 2-21(a) and (b)).

To re-establish realizability, the conjugates of all zeros and poles can arbitrarily be annexed to the function (Figure 2-21(c)). The complete, new expression therefore becomes

$$H(s) = \hat{H}\left[\frac{2\hat{\omega}_p}{\omega_{p2} - \omega_{p1}}\left(s - j\frac{\omega_{p2} + \omega_{p1}}{2}\right)\right]\hat{H}\left[\frac{2\hat{\omega}_p}{\omega_{p2} - \omega_{p1}}\left(s + j\frac{\omega_{p2} + \omega_{p1}}{2}\right)\right] \tag{2.89}$$

where $\hat{H}(\hat{s})$ is the original lowpass response.

The appended lower-half-plane zeros and poles do, of course, affect the passband response somewhat and tend to spoil the phase linearity. They are, however, much farther away (at a distance of approximately $\omega_{p1} + \omega_{p2}$) and

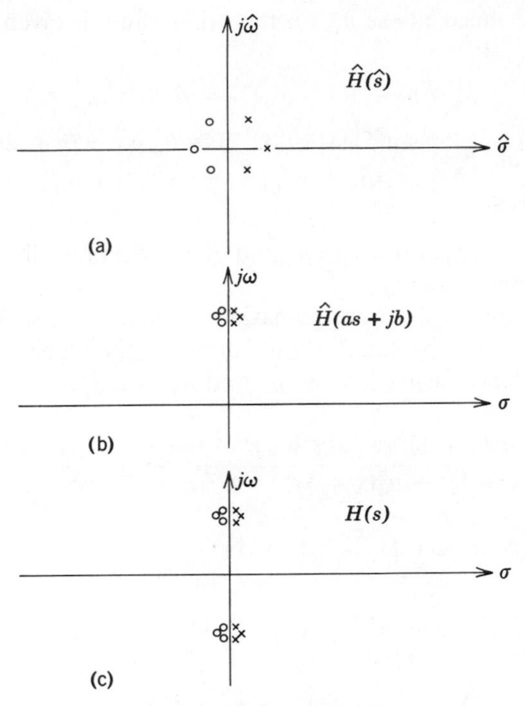

FIGURE 2-21 Lowpass-to-bandpass transformation for narrowband filters.

hence for narrowband filters with poles and zeros close to the $j\omega$-axis* their effect is small.

For bandpass circuits which do not satisfy these conditions, exact maximally flat and equal-ripple delay may be obtained by appropriate modifications of the design procedures discussed in Section 2.2.

To find a polynomial $e(s)$ representing *maximally flat bandpass delay* at $s = j\omega_0$, recall that the phase angle β of

$$e(s) = s^n + a_{n-1} s^{n-1} + \cdots + a_1 s + a_0 \tag{2.90}$$

for $s = j\omega$ satisfies

$$\tan \beta = \frac{a_1 \omega - a_3 \omega^3 + a_5 \omega_5 - + \cdots}{a_0 - a_2 \omega^2 + a_4 \omega^4 - + \cdots}. \tag{2.91}$$

* In this connection, note that the $\hat{\sigma}$-coordinates of the lowpass zeros and poles are also multiplied by $(\omega_{p2} - \omega_{p1})/2\hat{\omega}_p$.

The required linear phase β_s, on the other hand, is given by

$$\beta_s = \beta_0 + (\omega - \omega_0)T_0 = \beta_0 + xT_0$$

$$\tan\beta_s = \frac{\sin(\beta_0 + xT_0)}{\cos(\beta_0 + xT_0)} = \frac{\tan\beta_0 \cos xT_0 + \sin xT_0}{\cos xT_0 - \tan\beta_0 \sin xT_0} \tag{2.92}$$

where $x \equiv \omega - \omega_0$, and where β_0 and T_0 are the prescribed phase angle and delay, respectively, at ω_0.

For a maximally flat approximation around $\omega = \omega_0$, i.e., $x = 0$, the expressions for $\tan\beta$ and $\tan\beta_s$ should agree, to the highest possible power of x (*not* ω!). Hence, using $b \triangleq \tan\beta_0$ and $\omega_0 = 1$,

$$\frac{a_1(x+1) - a_3(x+1)^3 + a_5(x+1)^5 - + \cdots}{a_0 - a_2(x+1)^2 + a_4(x+1)^4 - + \cdots} \overset{!}{=} \frac{b\cos xT_0 + \sin xT_0}{\cos xT_0 - b\sin xT_0}$$

$$\begin{aligned}
&\left[a_0 - a_2(x+1)^2 + a_4(x+1)^4 - + \cdots\right] \\
&\cdot \left[b\sum_{i=0}^{\infty}(-1)^i\frac{(xT_0)^{2i}}{(2i)!} + \sum_{i=0}^{\infty}(-1)^i\frac{(xT_0)^{2i+1}}{(2i+1)!}\right] \overset{!}{=} \\
&\left[a_1(x+1) - a_3(x+1)^3 + a_5(x+1)^5 - + \cdots\right] \\
&\cdot \left[\sum_{i=0}^{\infty}(-1)^i\frac{(xT_0)^{2i}}{(2i)!} - b\sum_{i=0}^{\infty}(-1)^i\frac{(xT_0)^{2i+1}}{(2i+1)!}\right]
\end{aligned} \tag{2.93}$$

is obtained. Equating the coefficients of $x^0, x^1, x^2, x^3, \ldots$ in (2.93), the linear system of equations

$$\begin{aligned}
&ba_0 - a_1 - ba_2 + a_3 + ba_4 - a_5 - + \cdots = 0 \\
&T_0 a_0 - (1 - bT_0)a_1 - (2b + T_0)a_2 + (3 - bT_0)a_3 + (4b + T_0)a_4 - (5 - bT_0)a_5 - + \cdots = 0 \\
&-b\frac{T_0^2}{2}a_0 - \left(-bT_0 - \frac{T_0^2}{2}\right)a_1 - \left(b + 2T_0 + b\frac{T_0^2}{2}\right)a_2 + \left(3 - 3bT_0 - \frac{T_0^2}{2}\right)a_3 + - \cdots = 0 \\
&\cdots\cdots\cdots\cdots\cdots\cdots\cdots
\end{aligned} \tag{2.94}$$

is obtained. It can be solved for $a_0, a_1, \ldots, a_{n-1}$.

Combining the first and third, second and fourth, etc. equations, the expressions simplify somewhat. The resulting coefficients have been tabulated [26].

Equal-ripple bandpass delay can be obtained on the basis of the same considerations as those used in the lowpass case. Equations (2.68)–(2.73) remain valid, but in order to obtain the equal-ripple delay between ω_{p1} and ω_{p2} (rather than 0 and ω_p), the z-variable must be defined by (2.85) rather than (2.41). (For a rational $T_g(s)$, n must again be even.) Therefore, transformation of the group delay function $T_g(s)$ of (2.69) into the z-plane yields

$$T_g(z) = \sum_{i=1}^{n} \frac{\pm\sqrt{\dfrac{\omega_{p2}^2 - z_i^2\omega_{p1}^2}{z_i^2 - 1}}}{\dfrac{\omega_{p2}^2 - z^2\omega_{p1}^2}{z^2 - 1} - \dfrac{\omega_{p2}^2 - z_i^2\omega_{p1}^2}{z_i^2 - 1}}$$

$$= \sum_{i=1}^{n} \frac{\pm\sqrt{(\omega_{p2}^2 - z_i^2\omega_{p1}^2)(z_i^2 - 1)}\,(z^2 - 1)}{(\omega_{p2}^2 - \omega_{p1}^2)(z_i^2 - z^2)}$$

$$\lim_{z \to z_k} T_g(z) = \frac{\pm(\omega_{p2}^2 - z_k^2\omega_{p1}^2)^{1/2}(z_k^2 - 1)^{3/2}}{(\omega_{p2}^2 - \omega_{p1}^2)2z_k(z_k - z)} \tag{2.95}$$

instead of (2.74).* Here, the $\pm$ signs are to be chosen such that all $s_i = \pm[(\omega_{p2}^2 - z_i^2\omega_{p1}^2)/(z_i^2 - 1)]^{1/2}$ have negative real parts.

Equating the residues at $z = z_k$ in (2.73) and (2.95) gives

$$-\frac{\Delta T}{2} z_k \prod_{\substack{i=1 \\ i \neq k}}^{n} \frac{z_i + z_k}{z_i - z_k} = \pm\frac{(\omega_{p2}^2 - z_k^2\omega_{p1}^2)^{1/2}(z_k^2 - 1)^{3/2}}{2z_k(\omega_{p2}^2 - \omega_{p1}^2)}$$

$$\pm\frac{(z_k^2 - 1)^{3/2}}{z_k^2} + \Delta T \frac{\omega_{p2}^2 - \omega_{p1}^2}{(\omega_{p2}^2 - z_k^2\omega_{p1}^2)^{1/2}} \prod_{\substack{i=1 \\ i \neq k}}^{n} \frac{z_i + z_k}{z_i - z_k} = 0, \qquad k = 1, 2, \ldots, n. \tag{2.96}$$

Equation (2.96) again represents n nonlinear simultaneous equation in the n unknown z_i. These must be solved iteratively, from a good initial approximation. The latter may be obtained from the lowpass equal-ripple delay with approximately the same ΔT and with a degree $\hat{n} = n/2$, by applying the shifting process described in connection with (2.87)–(2.89).

Having obtained a flat-delay polynomial $e(s)$, its maximally flat or equal-ripple delay will remain unchanged if it is divided by a second polynomial $p(s)$ containing only $j\omega$-axis zeros or, more generally, only zeros with quadrantal s-plane symmetry. Hence, such a pure even or odd $p(s)$ can be used to provide additional selectivity for the bandpass loss response.

*Note that for $\omega_{p1} \to 0$, $\omega_{p2} \to \omega_p$, (2.95) and (2.74) coincide, as do (2.96) and (2.75).

To obtain *Chebyshev stopband, linear-phase passband bandpass filter* response, the transformation of (2.76) should be generalized for bandpass filters. Consider the mapping [14]

$$w = \sqrt{\frac{s^2 + \omega_{s2}^2}{-s^2 - \omega_{s1}^2}}, \qquad \mathrm{Re}(w) \geq 0. \tag{2.97}$$

This transforms the passband ($\omega_{s1} < \omega < \omega_{s2}$) to the positive real w-axis; it transforms both the lower stopband ($|\omega| \leq \omega_{s1}$) and the upper stopband ($|\omega| \geq \omega_{s2}$) to the imaginary jv-axis, to $\omega_{s2}/\omega_{s1} \leq |v| \leq \infty$ and to $0 \leq |v| \leq 1$, respectively. The gap between 1 and ω_{s2}/ω_{s1} is occupied by the image of the real s-axis (σ-axis).

Now if the roots of a linear phase $e(s)$ are mapped using (2.97) to values $w_1, w_2, \ldots, w_n$, then $E + wF$ can be created as before in (2.77), and the $|H|^2$ of (2.78) leads to a response which, in the s-plane, retains the phase linearity of $e(s)$ and which at the same time exhibits a Chebyshev-stopband loss behavior. For $H(s)$ to be rational, the degree n must be even.

To achieve the desired loss discrimination between pass- and stopbands, the loss $\alpha(\omega)$ must be zero at some point in the passband. In addition, for realizability in the form of a passive ladder network, there must be loss poles at $\omega = 0$ ($|v| = \omega_{s2}/\omega_{s1}$) and at $\omega \to \infty$ ($|v| = 1$), but there should be no loss pole on the real s-axis ($1 < |v| < \omega_{s2}/\omega_{s1}$). These conditions are not automatically valid for the $|H|^2$ obtained from (2.97), (2.77), and (2.78). If they must be satisfied in a given problem, then iteration can be used to modify $|H|^2$ appropriately.

2.4 SUMMARY

The purpose of this chapter was to give a brief description of filter approximation methods. These methods enable the designer of lumped linear time-invariant filters, active or passive, to obtain realizable rational functions of the complex variable s which satisfy requirements on the loss response or the phase or delay response. In addition, special techniques were described which permit *simultaneous* control of phase and loss. Many of the results initially derived for lowpass loss responses can be generalized to highpass, bandpass, or bandstop filters by using appropriate transformations of the elements or network functions of the lowpass filter.

Finally, procedures were given for solving the approximation problem directly for the important filter class of bandpass networks. These procedures may be used for circuits which cannot be derived from lowpass filters by either frequency or reactance transformations.

The following chapters will show how the functions obtained by using the approximation methods introduced in this chapter form the basis of filter realization. Also, extensions of these methods will be given, for parametric filters (in Chapter 3), general requirements (Chapter 6), distributed networks (Chapters 7 and 9), as well as switched and digital filters (Chapters 10–12). These extensions, however, are based to a considerable extent on the more classical theory just presented.

PROBLEMS

2.1 Show that if

$$F(s) = \frac{a_m s^m + a_{m-1} s^{m-1} + \cdots + a_1 s + a_0}{b_n s^n + b_{n-1} s^{n-1} + \cdots + b_1 s + b_0}$$

where $s = j\omega$ and all a_i, b_j, ω are real, then

$$F(s)F(-s) = |F(s)|^2.$$

2.2 Prove that no finite lumped linear circuit can realize the loss response illustrated in Figure 2-3.
(Hint: Assume a rational form for $H(s)$ with a numerator of finite degree n. Pick $n+1$ points with $|\omega| < \omega_c$ and attempt to find the numerator coefficients so as to satisfy the required response.)

2.3 Calculate $K(s)$ and $H(s)$ for a Butterworth filter satisfying

$$\alpha \leq 0.2 \text{ dB} \quad \text{for} \quad f \leq 4 \text{ kHz}$$
$$\alpha \geq 40 \text{ dB} \quad \text{for} \quad f \geq 10 \text{ kHz}.$$

If the 40 dB loss is exactly achieved at 10 kHz, what is the actual loss at 4 kHz? Why?

2.4 What is the asymptotic rate of rise for $\alpha(\omega)$ for a Butterworth filter, if α is in decibels and ω is measured on a logarithmic scale, in decades?

2.5 Show that

$$|K|^2 = k_p^2 \cos^2(n \cos^{-1} \omega/\omega_p)$$

is a monotone increasing function of ω for $\omega \geq \omega_p$.
(Hint: Investigate the degree of $d|K|^2/d(\omega^2)$. How many zeros does it have for $\omega^2 < \omega_p^2$?).

2.6 Prove (2.32) for the natural modes of a Chebyshev filter.
(Hint: Start with the Feldtkeller equation and make use of the relation (2.31) for $\cos(x + jy)$.)

2.7 Show that

$$|K|^2 = \cos^2(3 \cos^{-1} \omega)$$

has a greater slope at $\omega^2 = 1$ than any other cubic polynomial $P_3(\omega^2)$ going through the (1, 1) point and satisfying $0 \leq P_3(\omega^2) \leq 1$ for $0 \leq \omega^2 \leq 1$. (Hint: Assume the opposite. Then the curves shown below are valid. How many intersections occur?)

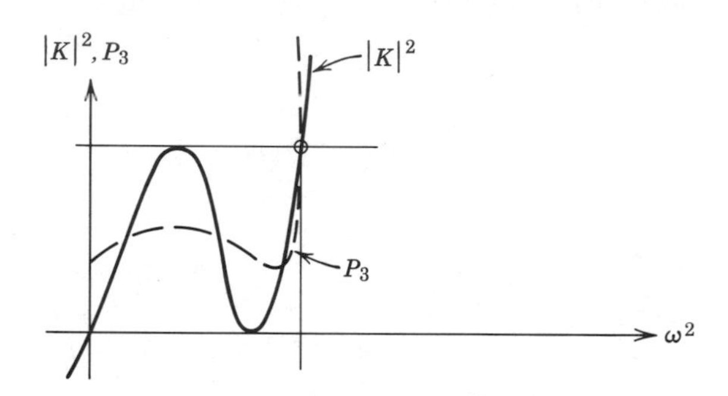

2.8 Show that $|K|^2 = \cos^2(n \cos^{-1} \omega)$ reaches a higher value at an arbitrary $\omega^2 > \omega_p^2$ than any other nth-order polynomial $P_n(\omega^2)$ satisfying $0 \leq P_n(\omega^2) \leq 1$ for $0 \leq \omega^2 \leq \omega_p^2$.
(Hint: Generalize the proof by contradiction derived for Problem 2.7.)

2.9 Prove that if both $\omega_p \to 0$ and $k_p \to 0$ in a Chebyshev filter such that $\omega_p/k_p^{1/n} \to$ constant, the Chebyshev filter becomes a Butterworth filter.
(Hint: Examine the location of the natural modes.)

2.10 Show that the reflection zeros of a Chebyshev filter are at $\omega_p \cos[(2k - 1)/n] \times (\pi/2)$, $k = 1, 2, \ldots, n$.

2.11 Show that for the same degree n, passband loss α_p and limit ω_p, the asymptotic loss response ($\omega \gg \omega_p$) of a Chebyshev filter is $(n - 1)\,6.02$ dB more than that of a Butterworth filter.

2.12 Find $K(s)$ and $H(s)$ of a Chebyshev filter for $\alpha_p = 0.3$ dB, $\alpha_s = 52$ dB, $\omega_p = 125$ kHz, $\omega_s = 200$ kHz.

2.13 For fixed n, ω_p, and ω_s, how much does α_p need to be increased to gain 10 dB in α_s? Does the increase depend on whether the filter is Butterworth, Chebyshev, or elliptic? Why or why not?

2.14 Find $K(s)$ for an inverse Chebyshev filter with $\alpha_s = 40$ dB, $\alpha_p = 3$ dB, $\omega_p = 25$ MHz, $\omega_s = 40$ MHz.

2.15 Find the natural modes of a Butterworth filter with $n = 5$, $\alpha_p = 1$ dB, $\omega_p = 1$ kHz. Transform them into the z-plane using (2.41). Repeat with a Chebyshev filter having identical specifications.

2.16 Construct $|K|^2$ for a Chebyshev passband behavior in the z- and s-planes, with $n = 5$, $\alpha_p = 0.2$ dB, $\omega_p = 1$, $\omega_1 = 1.1$, $\omega_2 = 2$.

2.17 Compare the necessary degrees for $\alpha_p = 0.2$ dB, $\alpha_s = 60$ dB, $\omega_p = 1.5$ MHz, $\omega_s = 1.7$ MHz for Butterworth and Chebyshev filters.

2.18 Prove that for given α_p, α_s, ω_p, and ω_s, an elliptic filter meeting the requirements exactly is always of lower degree than the corresponding Butterworth filter.
(Hint: Plot $|K|^2$ for the two filters, e.g., with $n = 3$. Count the intersections of the curves and draw the appropriate conclusions.)

2.19 Develop explicit formulas for the natural modes of an inverse Chebyshev filter with given n, ω_s, and α_s.

2.20 Derive the relations between the a_i of (2.55) by matching the coefficients of s^0, $s^1, s^2, \ldots, s^n$. Show, by substituting into the first few relations, that the a_i given in (2.57) satisfies them.

2.21 Calculate and plot the phase and group delay for the first-, second-, and third-order Thomson-Lommel polynomials.

2.22 Show that the Lommel polynomials satisfy the recursion relation

$$L_n(s) = (2n - 1)L_{n-1}(s) + s^2 L_{n-2}(s).$$

(Hint: Use (2.57).)

2.23 Show that the parameters of the equal-ripple group delay function satisfy

$$-\sum_{i=1}^{n} s_i^{-1} = T_0 - (-1)^n \, \Delta T/2.$$

2.24 (a) Rewrite (2.21)–(2.34) and (2.53)–(2.67) choosing convenient normalized units.
(b) Rewrite (2.7)–(2.20) using $\omega_0 = C^{-1/n}$ as the frequency unit.

2.25 (a) What is the loss response of a filter derived from a lowpass prototype via

$$\omega_{\mathrm{LP}} = A \frac{(\omega^2 - \omega_2^2)(\omega^2 - \omega_4^2)}{\omega(\omega^2 - \omega_3^2)}\;?$$

(b) How will the reactances of the prototype filter be transformed?

2.26 (a) The passband width of a geometrically symmetric bandpass filter is 1350 Hz. The stopband limits are $f_{s1} = 575$ Hz and $f_{s2} = 2000$ Hz. Find the passband limits.

(b) Assuming $R_1 = R_2 = 10^4\ \Omega$, design the bandpass filter from the normalized lowpass prototype shown below.
The prototype was designed for $R_1 = R_2 = 1$, $\omega_p \omega_s = 1$.

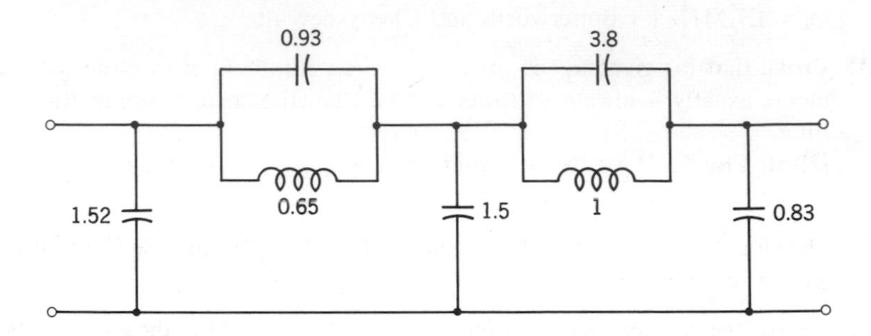

2.27 Using the narrowband relation (2.89), transform the fifth-degree Thomson lowpass delay function into one giving flat delay between 999 Hz and 1001 Hz. Let $T = 1$ and $\hat{\omega} = 1$ for the lowpass function.

REFERENCES

[1] E. A. Guillemin, *Synthesis of Passive Networks*, John Wiley, New York, 1957, Chapter 14.

[2] A. Papoulis, "On the approximation problem in filter design," *IRE Conv. Rec.*, **5**, Pt. 2, 175–185 (1957).

[3] B. R. Smith and G. C. Temes, "An iterative approximation procedure for automatic filter synthesis," *IEEE Trans. Circuit Theory*, **12**, No. 1, 107–112 (March, 1965).

[4] F. I. Kuo and W. Magnuson, Eds., *Computer-Oriented Circuit Design*, Prentice-Hall, Englewood Cliffs, New Jersey, 1969, Chapter 6.

[5] G. Szentirmai, "Nomographs for designing elliptic-function filters," *PIRE*, **48**, No. 1, 113–114 (January, 1960).

[6] A. J. Grossman, "Synthesis of Tchebycheff parameter symmetrical filters," *PIRE*, **45**, No. 4, 454–473 (April, 1957).

[7] H. J. Orchard, "Computation of elliptic functions of rational fractions of a quarterperiod," *IRE Trans. Circuit Theory*, **CT-5**, No. 4, 352–355 (Dec. 1958).

[8] H. J. Orchard and G. C. Temes, "Filter design using transformed variables," *IEEE Trans. Circuit Theory*, **CT-15**, No. 4, 385–408 (Dec., 1968).

[9] A. Papoulis, *The Fourier Integral and its Applications*, McGraw-Hill, New York, 1962.

[10] W. E. Thomson, "Delay networks having maximally flat frequency characteristics," *Proc. IEE* (*London*), **96**, 487–490 (Nov., 1949).

[11] S. Darlington, "Network synthesis using Tchebyshev polynomial series," *Bell Syst. Techn. J.*, **31**, 613–665 (July, 1952).

[12] E. Ulbrich and H. Piloty, " Über den Entwurf von Allpässen, Tiefpässen und Bandpässen mit einer in Tschebyscheffschen Sinne approximiert konstanten Gruppenlaufzeit," *Arch. Elek. Übertr.*, **14**, No. 10, 451–457 (Oct., 1960).
[13] R. Unbehauen, " Low-pass filters with predetermined phase or delay and Chebyshev stopband attenuation," *IEEE Trans. Circuit Theory*, **CT-15**, No. 4, 337–341 (Dec., 1968).
[14] G. C. Temes and M. Gyi, " Design of filters with arbitrary passband and Chebyshev stopband attenuation," *IEEE Internat. Conv. Rec.*, **15**, Pt. 5, 2–12 (1967).
[15] B. J. Bennett, " Synthesis of electric filters with arbitrary phase characteristics," *IRE Internat. Conv. Rec.*, Pt. 5, 19–26 (1953).
[16] Y. Peless and T. Murakami, "Analysis and synthesis of transitional Butterworth-Thomson filters and bandpass amplifiers," *RCA Rev.*, **18**, 60–94 (March, 1957).
[17] H. Ruston, " Synthesis of RLC Networks by Discrete Tschebyscheff Approximations in the Time Domain," *Tech. Rept. No. 107, Electr. Def. Group, Dept. of Electr. Engn.*, U. of Michigan (April, 1960).
[18] E. A. Guillemin, op. cit., Chapter 15.
[19] See, e.g., D. C. Handscomb, Ed., *Methods of Numerical Approximation*, Pergamon Press, Oxford, 1966, Sec. III.15.
[20] R. Fischl, " Optimal Chebyshev approximation at discrete points in the time domain," *Proc. Seventh Midwest Symp. Circuit Theory*, 33–43 (May, 1964).
[21] J. Jess and H. W. Schüssler, "A class of pulse-forming networks," *IEEE Trans. Circuit Theory*, **CT-12**, No. 2, 296–299 (June, 1965); also, " On the design of pulse-forming networks," ibid., No. 3, 393–400 (Sept., 1965).
[22] D. A. Spaulding, " Synthesis of pulse-shaping networks in the time-domain," *Bell Syst. Techn. J.*, **48**, 2425–2444 (Sept., 1969).
[23] W. E. Thomson, " The synthesis of a network to have a sine-squared impulse response," *Proc. IEE* (*London*), **99**, Pt. 111, 373–376 (Nov., 1952); *Discussion*, **100**, 110 (1953).
[24] H. E. Kallmann, " Transversal filters," *PIRE*, **28**, No. 7, 302–310 (July, 1940).
[25] E. Christian and E. Eisenmann, *Filter Design Tables and Graphs*, John Wiley, New York, 1966.
[26] H. J. Orchard and G. C. Temes, " Maximally flat approximation techniques," *Proc. IEEE*, **56**, No. 1, 65–66 (Jan., 1968).
[27] N. Balabanian, *Network Synthesis*, Prentice-Hall, Englewood Cliffs, New Jersey, 1958, Chapter 9.

3

Passive Filters with Lumped Elements

H. J. Orchard and G. C. Temes
University of California
Los Angeles, California

In this chapter we shall discuss the principles of the insertion-loss theory as applied to the design of passive filters with lumped components. Much of this basic material, however, is also directly applicable to the design of filters constructed with other components such as, e.g., mechanical resonators, microwave resonators, and active devices, and the reader is well advised to obtain some familiarity with the ideas described here before reading the later chapters that deal with these more specialized networks.

3.1 GENERAL TWO-PORT REACTANCE NETWORKS

We assume that the filter is a two-port network of lumped inductors and capacitors, driven at port 1 by a generator of internal impedance R_1 and emf E, and loaded at port 2 by a resistance R_2, as shown in Figure 3-1(a). The maximum power available from the resistive generator is $E^2/4R_1$ and would be delivered to R_2 if the reactance network were replaced by an ideal transformer which exactly matched R_1 to R_2, as shown in Figure 3-1(b). The voltage across R_2 would then be $V_2' = (E^2R_2/4R_1)^{1/2}$, the largest possible value of V_2.

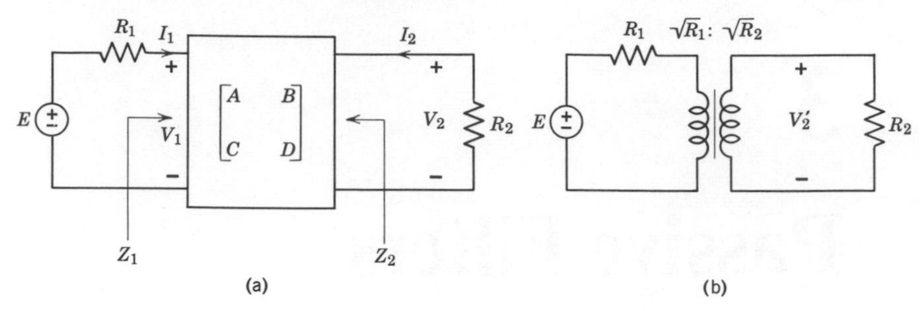

FIGURE 3-1 (a) Reactance twoport network with source and load. (b) Circuit for defining reference voltage V'_2.

Taking this as our reference we can then define the *transducer function* $H(s)$ as

$$H(s) = \frac{V'_2}{V_2} = \frac{E}{2V_2}\sqrt{\frac{R_2}{R_1}}. \tag{3.1}$$

It follows from the definition that $|H(j\omega)|^2$ is the ratio of the maximum power available from the generator to the actual power in R_2, and hence that $|H(j\omega)| \geq 1$ for all real ω.

The *loss* α in nepers and the *phase* β in radians are related to H by

$$\alpha + j\beta = \ln H(j\omega).$$

In practical applications α and β are usually expressed in decibels and degrees, respectively. From the phase in radians we derive the *envelope delay*, $T \triangleq d\beta/d\omega$, in seconds.

From the transducer function, which describes the external behavior of the reactance network when operated between resistive terminations, we need to be able to derive a set of parameters which describe the reactance two-port network by itself, i.e., unterminated. This is conveniently achieved through the *chain matrix* which relates the input voltage and current of the network to its output voltage and current by

$$\begin{bmatrix} V_1 \\ I_1 \end{bmatrix} = \begin{bmatrix} A & B \\ C & D \end{bmatrix} \cdot \begin{bmatrix} V_2 \\ -I_2 \end{bmatrix} \tag{3.2}$$

where V_1, I_1, V_2, and I_2 are as shown in Figure 3-1(a).

The four elements of this matrix are all rational functions of the complex frequency variable s with real coefficients, and they share a common denominator polynomial whose roots are the frequencies of infinite loss. For a reactance network, A and D are even functions of s while B and C are odd functions. Moreover, for a reciprocal network the determinant of the matrix is unity, i.e., $AD - BC = 1$. If required, the impedance and admittance matrices of the network can be obtained from the chain matrix elements, without further calculation, through the equations:

$$\mathbf{Z} = \frac{1}{C}\begin{bmatrix} A & 1 \\ 1 & D \end{bmatrix}, \qquad \mathbf{Y} = \frac{1}{B}\begin{bmatrix} D & -1 \\ -1 & A \end{bmatrix}. \tag{3.3}$$

This shows that all four functions A/C, D/C, B/D, and B/A are the driving-point impedances of a reactance network.

By introducing into (3.2) the relations

$$E = V_1 + I_1 R_1$$
$$0 = V_2 + I_2 R_2$$

which obviously hold for the circuit in Figure 3-1(a), one can quickly show that

$$H(s) = \frac{(AR_2 + DR_1) + (B + CR_1 R_2)}{2\sqrt{R_1 R_2}}. \tag{3.4}$$

The even and odd parts of $H(s)$ are evidently

$$H_e(s) = \frac{AR_2 + DR_1}{2\sqrt{R_1 R_2}}$$

$$H_o(s) = \frac{B + CR_1 R_2}{2\sqrt{R_1 R_2}}, \tag{3.5}$$

but without some further information it is not possible to separate A from D or B from C, as we attempt to find A, B, C, and D from a given $H(s)$.

In order to solve for the matrix elements, it would be helpful to have available an auxiliary function, similar to $H(s)$ but in which the signs of C and D are reversed, namely:

$$K(s) \triangleq \frac{(AR_2 - DR_1) + (B - CR_1 R_2)}{2\sqrt{R_1 R_2}}. \tag{3.6}$$

The even and odd parts of $K(s)$ would then be:

$$K_e(s) = \frac{AR_2 - DR_1}{2\sqrt{R_1 R_2}}$$

$$K_o(s) = \frac{B - CR_1 R_2}{2\sqrt{R_1 R_2}}, \tag{3.7}$$

and given these functions one could immediately solve (3.5) and (3.7) for the chain matrix to get:

$$\begin{bmatrix} A & B \\ C & D \end{bmatrix} = \frac{1}{\sqrt{R_1 R_2}} \begin{bmatrix} (H_e + K_e)R_1 & (H_o + K_o)R_1 R_2 \\ (H_o - K_o) & (H_e - K_e)R_2 \end{bmatrix}. \tag{3.8}$$

The problem of finding the chain matrix from a given transducer function would thus be reduced to that of constructing $K(s)$ from $H(s)$.

The relationship between these two functions can be found through an examination of the expressions for $H(s)H(-s)$ and $K(s)K(-s)$. Since reversing the sign of s reverses the sign of the odd part of a function, but does not affect its even part, it follows that:

$$H(s)H(-s) = \frac{(AR_2 + DR_1)^2 - (B + CR_1 R_2)^2}{4R_1 R_2}$$

and

$$K(s)K(-s) = \frac{(AR_2 - DR_1)^2 - (B - CR_1 R_2)^2}{4R_1 R_2}.$$

Expanding the squared terms in these expressions and then subtracting gives:

$$H(s)H(-s) - K(s)K(-s) = AD - BC = 1.$$

We are thus led to the fundamental equation* in the design process

$$H(s)H(-s) = 1 + K(s)K(-s). \tag{3.9}$$

This equation, as we shall now show, allows us to find $K(s)$ from a prescribed $H(s)$, or alternatively, as is more frequently the case, to find $H(s)$ from a prescribed $K(s)$.

Both $H(s)$ and $K(s)$ are rational functions of s with real coefficients, possessing the same degree. They both have the same poles as the elements of the chain matrix; these poles are the frequencies of infinite loss through the network. As the chain matrix elements are either even or odd functions of s, it follows that these poles must be the roots of either an even or an odd polynomial. Each complex pole must therefore be accompanied in H and K by both its conjugate and its negative, the four roots forming altogether a quad arranged symmetrically around the origin.

Exceptions to this occur when the poles are on the j-axis, so that the conjugate and the negative coincide, and also when the poles are real and so appear in positive-negative pairs. Poles may also occur at the origin and at

* Here, the reader will observe that the $K(s)$, defined in (3.6), is identical with the characteristic function which was defined from a different standpoint by (1.22).

infinity, although the latter exist by virtue of the excess of the degree of the numerator of H and K over the degree of the denominator. Figure 3-2 shows all these various possibilities. Any of these poles may be of any multiplicity.

The zeros of H are the natural frequencies of the resistance-terminated reactance network. As the latter is a passive and therefore stable system, these zeros must lie inside the left half of the s-plane. Apart from this there

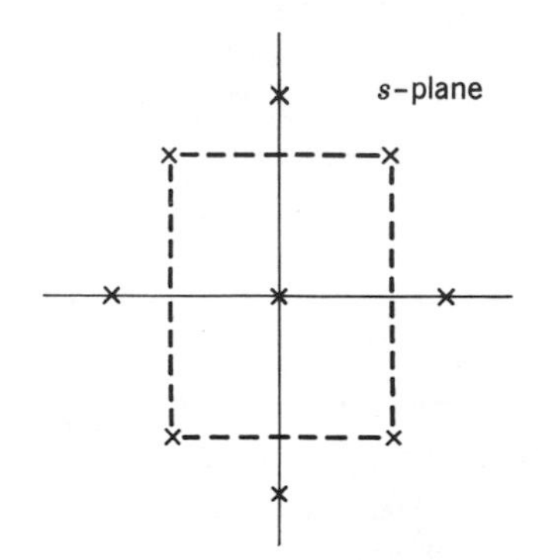

FIGURE 3-2 Possible locations for the poles of the transducer function $H(s)$ in the finite part of the s-plane.

are no other restrictions on the zeros, and the numerator polynomial of H is thus a quite general, real, strictly Hurwitz polynomial. We denote it by $e(s)$. There are no restrictions of this kind on the zeros of K, and the corresponding polynomial, which plays a companion role to that of $e(s)$, is called $f(s)$.

With the substitution of

$$H(s) = \frac{e(s)}{p(s)} \quad \text{and} \quad K(s) = \frac{f(s)}{p(s)} \tag{3.10}$$

into (3.9), we get*

$$e(s)e(-s) = p(s)p(-s) + f(s)f(-s). \tag{3.11}$$

If we are given $K(s)$ and wish to find $H(s)$, we proceed as follows. Since $f(s)$ and $p(s)$ are assumed known, we can form the right hand side of (3.11) as a single polynomial in s^2 and then find its roots. In general this will have to be done numerically, although in special cases† one may derive explicit formulas for the roots. The roots of the right hand side are naturally also the roots of the left hand side and therefore can be distributed between $e(s)$ and $e(-s)$. This distribution is quite unique since all the left-half-plane roots

* Note that since $|H(j\omega)| \geq 1$, the degree of $e(s)$ is at least as high as that of $p(s)$, and hence, by (3.11), as that of $f(s)$.

† Such as for the Butterworth and Chebyshev polynomial approximations discussed in Chapter 2. These can be regarded as limiting cases of the elliptic-function filter, for which explicit formulas also exist.

must be allocated to $e(s)$ and all their negatives to $e(-s)$. Recombination of the factors containing the roots of $e(s)$ yields the complete polynomial.

Conversely, if we are given $H(s)$ and wish to find $K(s)$, we can rewrite (3.11) as

$$f(s)f(-s) = e(s)e(-s) - p(s)p(-s) \tag{3.12}$$

and then proceed to form and factor the right hand side, just as before, in order to get the roots of the left hand side. Knowing these roots we must then decide which to allocate to $f(s)$ and which to $f(-s)$. But as long as conjugate pairs of complex roots are kept together in the same polynomial, and the negative of every root in $f(s)$ is placed in $f(-s)$, we are otherwise free to allocate the roots just as we please.

To each different arrangement of left-half- and right-half-plane roots in $f(s)$, there will correspond a different chain matrix and therefore a different network. All these various networks will have the same transducer function $H(s)$, but they will present different impedances seen into the ports. These impedances, denoted by $Z_1(s)$ and $Z_2(s)$ in Figure 3-1(a), are conveniently specified through the corresponding *reflection coefficients*

$$\rho_1(s) = \frac{Z_1(s) - R_1}{Z_1(s) + R_1} \qquad \text{and} \qquad \rho_2(s) = \frac{Z_2(s) - R_2}{Z_2(s) + R_2}. \tag{3.13}$$

In terms of the chain matrix elements and the terminating resistances, one can easily derive for $\rho_1(s)$ and $\rho_2(s)$ the expressions

$$\rho_1(s) = \frac{(AR_2 - DR_1) + (B - CR_1R_2)}{(AR_2 + DR_1) + (B + CR_1R_2)}$$
$$\rho_2(s) = \frac{(-AR_2 + DR_1) + (B - CR_1R_2)}{(AR_2 + DR_1) + (B + CR_1R_2)}, \tag{3.14}$$

from which it follows that

$$\rho_1(s) = \frac{K(s)}{H(s)} = \frac{f(s)}{e(s)}$$
$$\rho_2(s) = -\frac{K(-s)}{H(s)} = \mp\frac{f(-s)}{e(s)}; \qquad \begin{array}{l} -\text{ for } p(s) \text{ even} \\ +\text{ for } p(s) \text{ odd.} \end{array} \tag{3.15}$$

These formulas show that $\rho_1(s)$ and $K(s)$ have exactly the same zeros. It is frequently helpful, when trying to decide the allocation of roots to $f(s)$ in the construction of the characteristic function, to think instead of the effect that the choice will have upon the reflection coefficients. We note, as a consequence of (3.15), that the zeros of $\rho_2(s)$ are the *negatives* of the zeros of $\rho_1(s)$. Whatever the choice of roots for $f(s)$, however, we cannot change the modulus of ρ_1 at real frequencies, for by dividing (3.9) throughout by $H(s)H(-s)$ we get

$$1 = \frac{1}{H(s)H(-s)} + \frac{K(s)K(-s)}{H(s)H(-s)},$$

which for real frequencies becomes

$$1 = |H(j\omega)|^{-2} + |\rho_1(j\omega)|^2. \tag{3.16}$$

Equation (3.16) is the famous Feldtkeller energy equation for reactance two-port networks; the left-hand side can be thought of as the maximum available power from the generator, normalized to unity, while the right-hand side shows how this power is split between the fraction that is transmitted through the network to the resistive termination at port 2 and the fraction that is reflected back to the generator. For a given $H(s)$ we thus see that $|\rho_1(j\omega)|$ is fixed.

The different networks corresponding to the different possible allocations of roots to $f(s)$ will have not only different impedances Z_1, Z_2 seen at the ports, but also different internal circuit arrangements. One half of these networks will simply be duals of the other half, but within any group that has a common circuit diagram we find, from one network to another, different element values. There are no simple rules to guide one in selecting a distribution of roots in $f(s)$ for the best (i.e., most uniform) distribution of element values, and the only practical solution in any specific case seems to be to design all possible networks and select the best by inspection. When $H(s)$ is of high degree there will be a great many networks and the selection may have to be organized by computer.

Another consideration in choosing the distribution of roots may be a wish to control the sensitivities of the transfer function to tolerances on the terminating resistors R_1 and R_2. A simple calculation shows that if one wishes to achieve minimum sensitivity of $|H|$ to one of the terminations, then the reflection coefficient at the port concerned should have all its zeros in the left half of the s-plane. This will cause this reflection coefficient to be a minimum-phase function, thereby having its zeros and poles in the same half plane, i.e., as close as possible. The other reflection coefficient will then be, by (3.15), a *maximum* phase-shift function, and this leads to maximum sensitivity of $|H|$ to the termination at that port.

Alternating the zeros of both reflection coefficients between the two half planes leads, of course, to a compromise, with $|H|$ moderately sensitive to variations of both terminating resistors.

Example 3-1. Given $K(s) = s^3$, find $H(s)$ and the chain matrix. (Assume $R_1 = R_2 = 1$.)

Solution: By (3.11):

$$e(s)e(-s) = 1 \cdot 1 + s^3(-s^3) = 1 - s^6$$
$$= (1 - s^2)(1 + s^2 + s^4) = (1 + s)(1 - s)(1 + s + s^2)(1 - s + s^2).$$

Therefore, the Hurwitz factor is

$$H(s) = e(s) = (1 + s)(1 + s + s^2) = 1 + 2s + 2s^2 + s^3.$$

The chain matrix is then given by (3.8):

$$H_e(s) = 1 + 2s^2$$
$$H_o(s) = 2s + s^3$$
$$K_e(s) = 0$$
$$K_o(s) = s^3$$

and

$$\begin{bmatrix} A & B \\ C & D \end{bmatrix} = \begin{bmatrix} 1 + 2s^2 & 2s + 2s^3 \\ 2s & 1 + 2s^2 \end{bmatrix}.$$

3.2 FILTER CIRCUITS

For most purposes, the preferred filter circuit is a doubly terminated reactance two-port. Its advantages include:

1. An ability to produce any physically realizable loss response;
2. A possibility of ladder network realization;
3. A near minimum number of components for most filter specifications;
4. A maximum power transfer (zero flat loss) in the passband resulting in low passband sensitivity to element variations.

To illustrate this last point, consider the circuit of Figure 3-3(a). Let the nominal loss response be that shown in Figure 3-3(b) and consider the change in the loss response at $\omega = \omega_1$ with variations in the arbitrarily selected element C_7. Figure 3-3(c) indicates the variation of $\alpha(\omega_1)$ with changes of C_7 from its nominal value C_7^0; obviously, the sensitivity $\partial\alpha/\partial C_7$ at ω_1 is zero since $\alpha(\omega_1) = 0$ and $\alpha \geq 0$. A similar argument for $\omega = \omega_2$, illustrated in Figure 3-3(d), indicates that if the loss is small (but not zero), the sensitivity is ordinarily low (although not exactly zero, either). It is this inherently low sensitivity in the passband which makes it possible to build very large filter circuits with practical components*.

* Note that the sensitivity argument outlined above is not necessarily restricted to reactance two-port networks; e.g., it remains valid for a constant-resistance network or even an active network provided it operates to give maximum transfer of power.

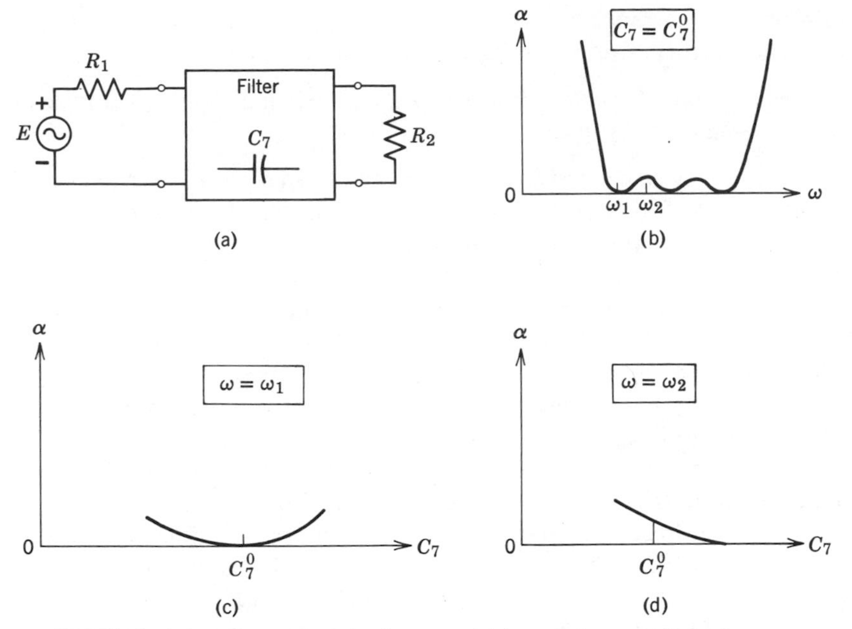

FIGURE 3-3 Illustrating the low sensitivity of the transducer loss at passband frequencies to errors in a typical component.

Needless to say, the maximum power-flow property afforded by resistively terminated reactance two-ports is desirable for other reasons as well, such as noise immunity, less gain needed in the system, etc. For all these reasons, it is usually bad strategy to design reactance filters with high flat loss or with only one resistive termination.

Reactance Ladders

Historically the first (and still the most widely used) configuration, the ladder network is illustrated in Figure 3-4. In addition to the low passband

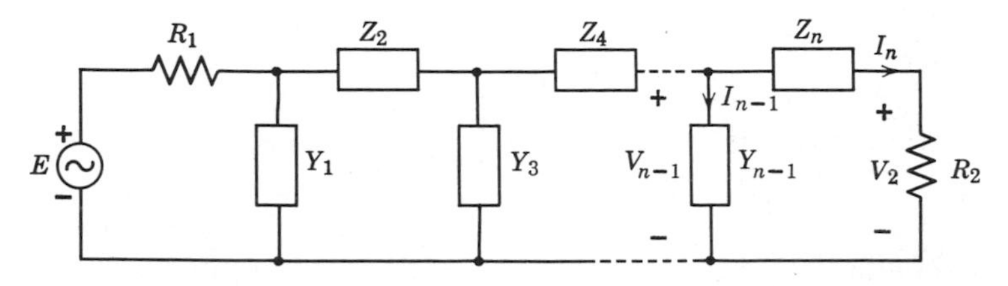

FIGURE 3-4 Ladder network.

sensitivity discussed above, ladder circuits also possess advantageous properties in their stopbands, based on the following fact:

A ladder network can have a loss pole if and only if a shunt admittance or a series impedance has a pole.

This property is instinctively obvious; for a formal proof, refer to Problem 3.2 at the end of the chapter. As a consequence, each branch is responsible for creating a loss pole, and, vice versa, each finite nonzero loss pole can be identified with a branch. (Poles at zero and infinite frequency are often created by two or more branches.) This makes the tuning of the ladder filter relatively simple. Equally important, due to this property, the loss poles of ladders are fairly insensitive to element variations, as compared to the loss poles in circuits which depend on a bridge-type balance of several branch impedances to obtain a loss pole.

Evidently, a reactance ladder must have all its loss poles on the $j\omega$-axis. Hence, it cannot realize H or K functions with poles that are on the real s-axis or occur in conjugate complex quads (Figure 3-2). In fact, not even all transfer functions with only $j\omega$-axis poles can be realized with a ladder without incurring negative element values. Hence, ladders are not perfectly general (canonical) reactance two-ports. Nevertheless, they are usable in most practical cases and their somewhat limited applicability is amply compensated for by their other advantages.

Lattices

Assume that $K(s)$ is purely odd. Then, by (3.7),

$$AR_2 = DR_1$$

and hence, by (3.14), $\rho_1(s) = \rho_2(s)$ and consequently $Z_1/R_1 = Z_2/R_2$. Also, by (3.3), $z_{11}/R_1 = z_{22}/R_2$. Thus, if $R_1 = R_2$ (which can be achieved by cascading an ideal transformer with the two-port), the ports of the network may be interchanged without any change in its electrical behavior. Such a circuit is called electrically *symmetric*.

It is easily shown that for this restricted but important class of two-ports, there always exists a realization in the form of a symmetric *lattice* network (Figure 3-5(a); for a convenient shorthand notation, see Figure 3-5(b)). For, by inspection, the open-circuit impedance parameters of the circuit are $z_{11} = z_{22} = (Z_b + Z_a)/2$ and $z_{12} = (Z_b - Z_a)/2$. Hence, $Z_a = z_{11} - z_{12}$ and $Z_b = z_{11} + z_{12}$ results. Let now N be an arbitrary symmetric two-port (Figure 3-6(a)). The two impedances obtained by combining $\pm 1:1$ ideal transformers with N in the ways shown in Figure 3-6(b) and (c) are easily

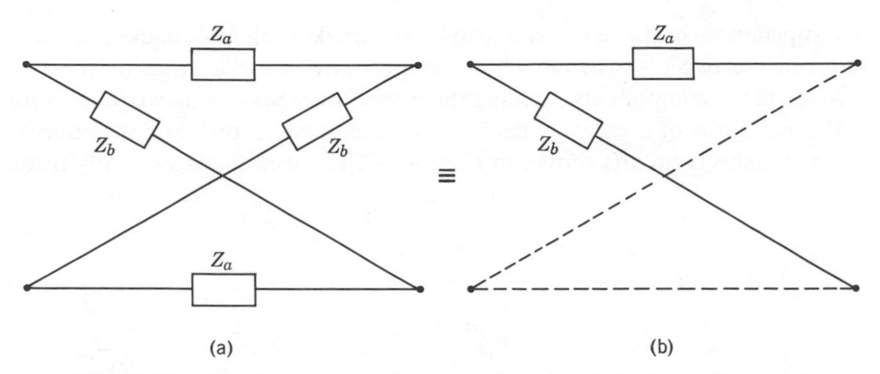

FIGURE 3-5 Symmetric lattice network and its abbreviated diagrammatic representation.

seen to equal $2(z_{11} \mp z_{12})$, i.e., $2Z_a$ and $2Z_b$. Hence, if N is realizable, so are all branch impedances of its lattice equivalent.

One major disadvantage of the symmetric lattice is the large number of components required; this number equals twice the degree of the polynomial $e(s)$, i.e., it is twice the theoretical minimum needed. Another disadvantage is that loss poles are obtained through a bridge balance of all branches, i.e., by making $Z_a = Z_b$ at each pole frequency. Finally, for satisfactory performance,

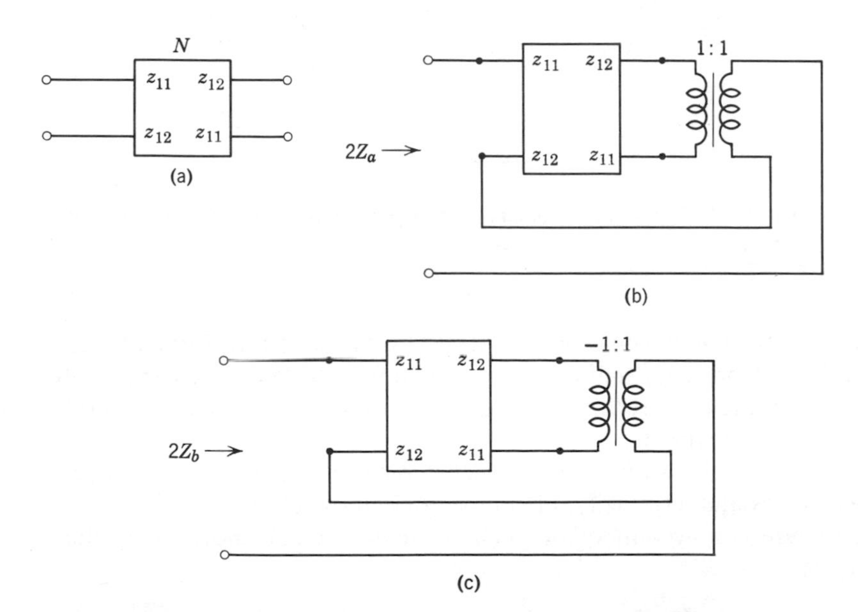

FIGURE 3-6 Showing how the arm impedances of a lattice network are related to a given symmetrical network to which it is equivalent.

the impedances in the two series arms must track at all frequencies, as must those in the two cross arms. These properties call for a large number of low-tolerance components, making the lattice an expensive network to build and tune. Some of these drawbacks can be eliminated by using the equivalent unbalanced circuits shown in Figure 3-7. These circuits essentially trade

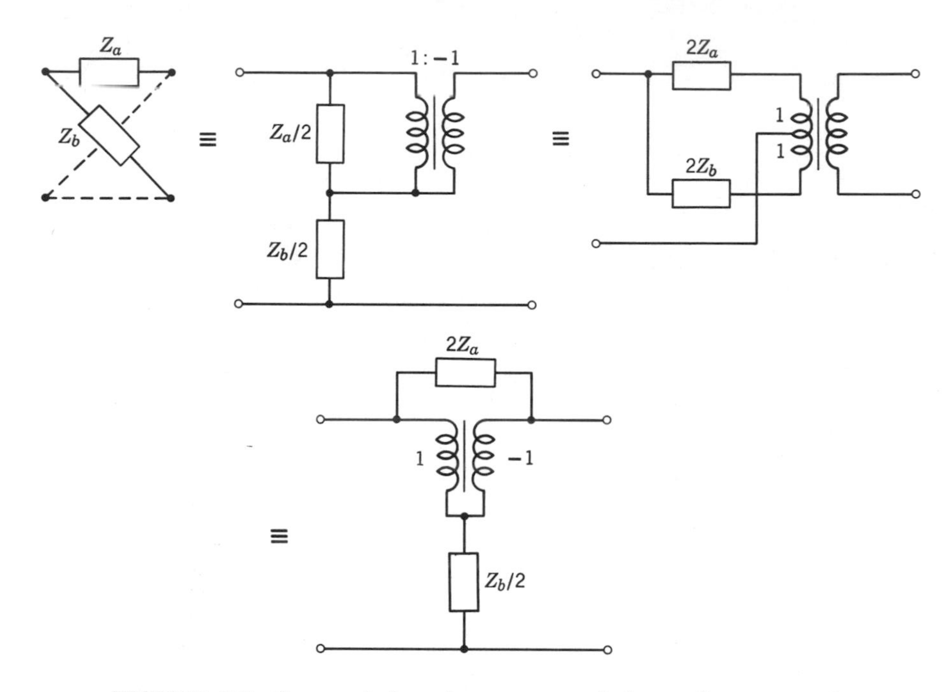

FIGURE 3-7 Some unbalanced two-port equivalents of a symmetric lattice.

half of the components of the lattice for an ideal transformer*. They are called (incorrectly) half-lattices or economy bridges. Even these circuits, however, are more sensitive in their stopbands than a comparable ladder—if one can be realized.

Lattices and half-lattices are used mostly to realize filters with complex loss poles and, especially, filters using electromechanical resonators. Their structure can accommodate such resonators much more easily than the ladder.

* The large number of elements in the symmetric lattice arises because the network is not only symmetric with respect to the ports, but also balanced with respect to ground.

Cauer's Canonical Two-Port

A two-port realization which uses the minimum number of components and is always realizable was given by Cauer. The circuit is shown (for a compact impedance matrix [3]) in Figure 3-8(a). Each series-connected two-port within the circuit corresponds to a term in the partial-fraction expansion of the impedance parameters and can be simply calculated from it. The reader is referred, e.g., to reference [3] for the details. Figure 3-8(b) shows the dual of this circuit (for a compact admittance matrix).

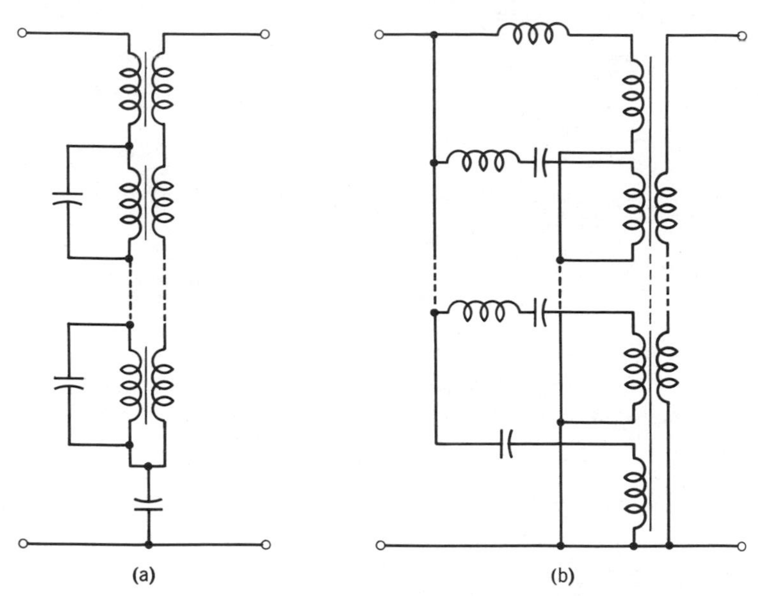

FIGURE 3-8 The canonical two-port reactance networks due to Cauer.

These networks depend on an intricate balance of all element values for their loss poles. Also, they use a precise multiwinding transformer or a multitude of closely coupled inductors. For these reasons, they are almost never used in practical applications.

Darlington's Canonical Tandem Realization

It was shown by Darlington and Piloty that a cascade connection of the sections illustrated in Figure 3-9 can produce any realizable reactance two-port characteristics. Each section corresponds to a loss-pole pair or quad. Sections A and B are, of course, standard components of a ladder network realizing $j\omega$-axis poles. Section C is similar to a Brune-section; for the realization of $j\omega$-axis poles the mutual inductance is positive, while for

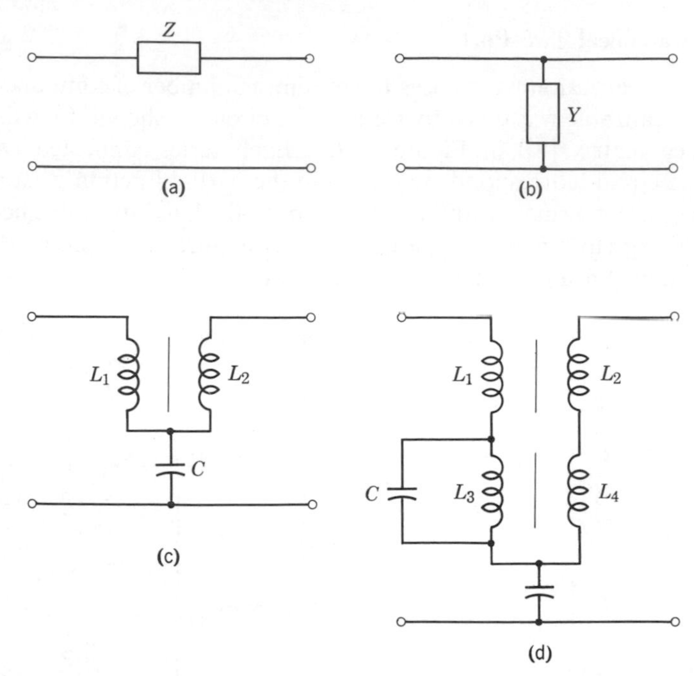

FIGURE 3-9 The canonical tandem section networks due to Darlington and Piloty. All inductors have unit coupling coefficients.

real (i.e., σ-axis) poles it is negative. The D-type section is needed for the realization of a complex pole quad. Evidently, it is a simple (fourth degree) version of the Cauer two-port (Figure 3-8(a)).

The design of Darlington's two-port can be carried out directly, using special algorithms (see Reference [4]). It is more convenient, however, to design it by first finding an equivalent ladder circuit with negative or complex elements and then transforming that into the final circuit.

If all loss poles are on the $j\omega$-axis and at least one of them is at zero or infinite frequency, then a ladder realization may be possible. Specifically, no D sections will be needed and all C sections will have positive couplings. Hence, in their equivalent ladder representations (Figure 3-10), the center branch has only positive elements. This may make it possible to combine the positive and negative elements of cascaded sections so as to eliminate the components with negative values. Consider, e.g., a fifth-degree filter with loss poles at $\pm j\omega_1$, $\pm j\omega_2$, and ∞. The Darlington realization will be that shown in Figure 3-11(a); using the equivalences of Figure 3-10, we obtain the circuit of Figure 3-11(b). Obviously, it depends on the specific element values whether or not we have achieved our goal of realizing a ladder network with only positive elements.

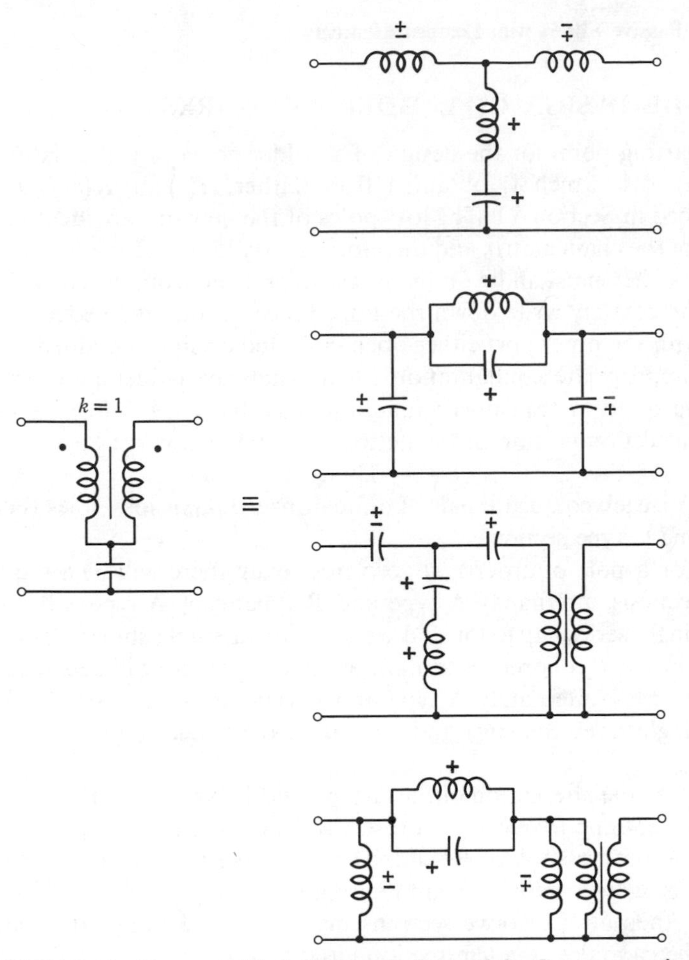

FIGURE 3-10 Four possible ladder equivalents of the C-type canonical tandem section for the case where the loss pole lies on the *j*-axis. Note that one element is always negative.

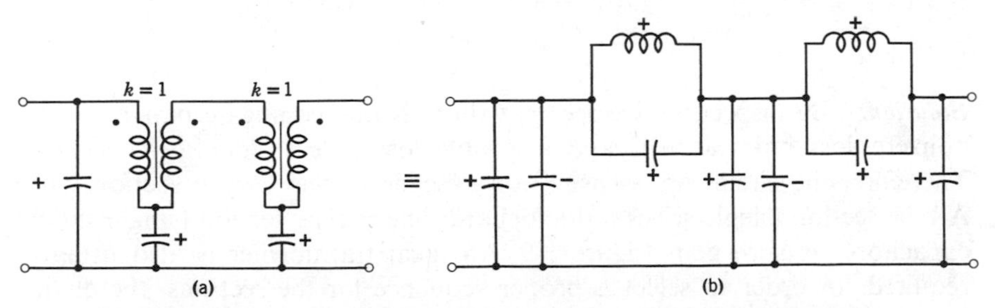

FIGURE 3-11 Illustrating how a ladder network is derived from a canonical tandem section network.

3.3 THE DESIGN OF LADDER NETWORKS

The starting point for the design of a ladder network is the chain matrix of the network which is obtained from either $H(s)$ or $K(s)$ via the steps described in Section 3.1. The loss poles of the network are identical with the poles of the chain matrix and therefore clearly known; for a reactance ladder network they must all lie on the j-axis. Moreover, from the chain matrix one can immediately write down the impedance or admittance seen into either port with the other port either open circuited or short circuited, using (3.3).

An appropriate configuration for a reactance ladder network that can provide a given transducer function may be derived from a canonical-tandem-section version of the network. In this network:

1. For each conjugate pair of finite nonzero j-axis loss poles there will be one C-type section.
2. For a pole of order r at zero frequency there will be a sequence of r sections, alternately A type and B type; each A type will consist of a single series capacitor and each B type a single shunt inductor.
3. For a pole of order m at infinite frequency there will be a sequence of m sections, alternately A type and B type; each A type will consist of a single series inductor and each B type a single shunt capacitor.

As long as the alternation of the A-type and B-type sections for the poles at zero and infinite frequency is preserved, and each sequence starts in such a way (with either an A or B section) as to make the impedance behavior at the ports agree at zero and infinite frequency with that dictated by the chain matrix, then all the above sections may be placed in any desired order to form the canonical tandem section circuit.

Example 3-2. Design a ladder network from a given characteristic function

$$K(s) = \frac{-76s^4 + 11s^3 - 33s^2 + 12s - 4}{4\sqrt{3}\,(s^2 + 4)}.$$

Solution: By inspection, we see that there is one conjugate pair of finite nonzero loss poles at $\pm j2$ and a double loss pole at infinite frequency. Therefore, the circuit will consist of a cascade of one C-type section, one A-type section (single series inductor) and one B-type section (single shunt capacitor), as shown in Figure 3-9. An ideal transformer is also usually required. In order to select a proper sequence for the sections, the chain matrix must be derived. This matrix will also be required for the calculation of the element values.

By (3.10) and (3.11)

$$\begin{aligned} e(s)e(-s) &= 48(s^2+4)^2 + (-76s^4 + 11s^3 - 33s^2 + 12s - 4) \\ &\quad \cdot (-76s^4 - 11s^3 - 33s^2 - 12s - 4) \\ &= 5776s^8 + 4895s^6 + 1481s^4 + 504s^2 + 784 \\ &= (76s^4 + 125s^3 + 135s^2 + 84s + 28) \\ &\quad \cdot (76s^4 - 125s^3 + 135s^2 - 84s + 28). \end{aligned}$$

Therefore,

$$H(s) = \frac{76s^4 + 125s^3 + 135s^2 + 84s + 28}{4\sqrt{3}\,(s^2+4)}$$

so that

$$H_e = \frac{76s^4 + 135s^2 + 28}{4\sqrt{3}\,(s^2+4)}$$

$$H_o = \frac{125s^3 + 84s}{4\sqrt{3}\,(s^2+4)}$$

$$K_e = \frac{-76s^4 - 33s^2 - 4}{4\sqrt{3}\,(s^2+4)}$$

$$K_o = \frac{11s^3 + 12s}{4\sqrt{3}\,(s^2+4)}.$$

Assuming $R_1 = R_2 = 1$, (3.8) then gives the chain matrix

$$\begin{bmatrix} A & B \\ C & D \end{bmatrix} = \begin{bmatrix} \dfrac{102s^2 + 24}{4\sqrt{3}(s^2+4)} & \dfrac{136s^3 + 96s}{4\sqrt{3}(s^2+4)} \\ \dfrac{114s^3 + 72s}{4\sqrt{3}(s^2+4)} & \dfrac{152s^4 + 168s^2 + 32}{4\sqrt{3}(s^2+4)} \end{bmatrix}.$$

Examination of the behavior of this matrix around infinite frequency then shows that a suitable configuration is that given in Figure 3-12(a).

The C-type sections can then be replaced by one of the four ladder equivalences given in Figure 3-10. When the loss pole lies above the filter passband, a "lowpass" type ladder equivalent should be used, and when the loss pole lies below the passband, a "highpass" type is to be used. The negative elements appearing in these ladder equivalents must now be combined with similar positive elements either in other C-type equivalents, or else in the A- or B-type sections. In filters of any complexity the number of different possible ladder arrangements which can be derived is extremely

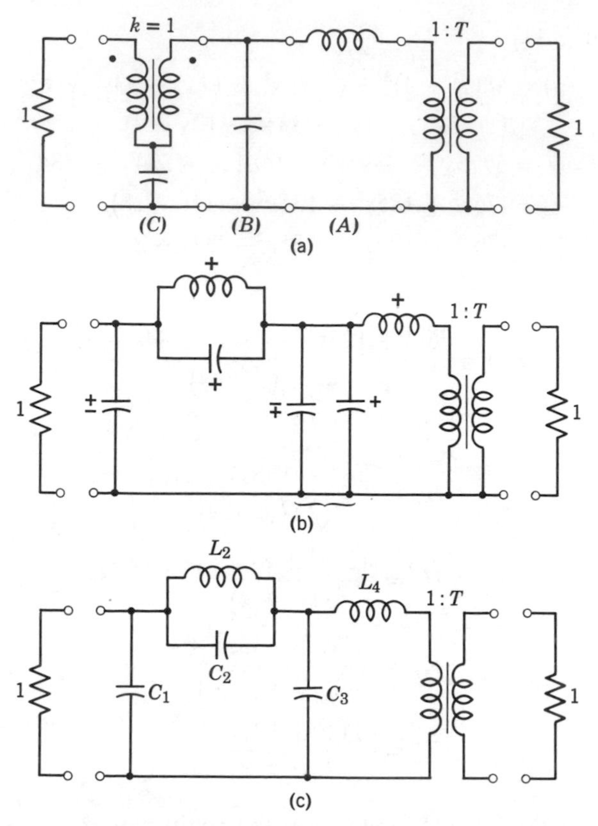

FIGURE 3-12 Illustrating the steps in the development of a fourth-degree lowpass filter network.

high, and the particular choice that is best for a specific practical application must be left to the discretion of the designer.

Continuing with the example, since the filter evidently passes zero frequency, a lowpass-type ladder equivalent of the C-type section should be chosen (see Figure 3-10). The circuit configuration then becomes that shown in Figure 3-12(b). Combining the shunt capacitance of the C-type section with the positive capacitance of the B-type section gives the final configuration of Figure 3-12(c). At this stage the signs of the shunt capacitances in the C-type section are unknown and one must just hope that C_1 and C_3 in Figure 3-12(c) will prove to be positive.

As mentioned in the example, in principle, an ideal transformer is also required in the network, most conveniently arranged to appear at one of the ports. But in practice the ideal transformer is almost always eliminated, either by introducing impedance transformations within the filter, which cancel the effect of the ideal transformer, or else by arranging to change the terminating impedance at the port where the transformer appears.

When a suitable ladder configuration has been found, there remains the job of calculating the component values. At this stage the ladder consists of a sequence of series and shunt branches which fall into one of two classes:

1. Parallel-tuned circuits in series branches or series-tuned circuits in shunt branches resonating at one of the j-axis loss poles. These resonances are known.
2. Single inductors or capacitors in series or in shunt. These components contribute to the (possibly multiple) loss poles at zero and/or infinite frequency.

The calculation of the component values is now made by taking one or more of the four input impedances (open circuit or short circuit at port 1 or port 2) and developing them into the particular ladder form chosen for the network. Usually the impedance of lower degree at either port is used, partly because it is simpler, and partly because it is better conditioned (see Section 3.5). This means that one branch at the end of the ladder contributes nothing to such an impedance at the other end, and therefore will not be found from its development. This missing branch, together with the ideal transformer turns ratio, is subsequently found from a development of the impedance at the second port.

The ladder development consists of a sequence of partial or total removals of the poles of the impedance or admittance at each point as one progresses down the ladder. The tuned circuits in category 1 above always correspond to full removals. Those components in category 2 that immediately precede such a full removal of a tuned circuit, represent partial removals of poles at zero or infinite frequency, which are made to prepare for the full removal of the tuned circuit. The remaining components in category 2 represent full removals of poles. The process is well known [3, 4] and will only be illustrated by continuing the previous example. It was shown earlier that the network given in Figure 3-12 is the required ladder, derived from the canonical-tandem sections, and that the parallel-tuned circuit resonates at the given loss pole $s = \pm j2$. Initially, the network is to operate between unit terminating resistances. Starting from port 1 we pick the admittance of lower degree, which in this case is the admittance with port 2 open circuited; here it is:

$$Y_1 = \frac{C}{A} = \frac{19s^3 + 12s}{17s^2 + 4}.$$

The inductance L_4 and the ideal transformer thus play no part in determining Y_1, and we have the situation shown in Figure 3-13(a). The first capacitance C_1 is found from the fact that, at the resonance of L_2 and C_2, its

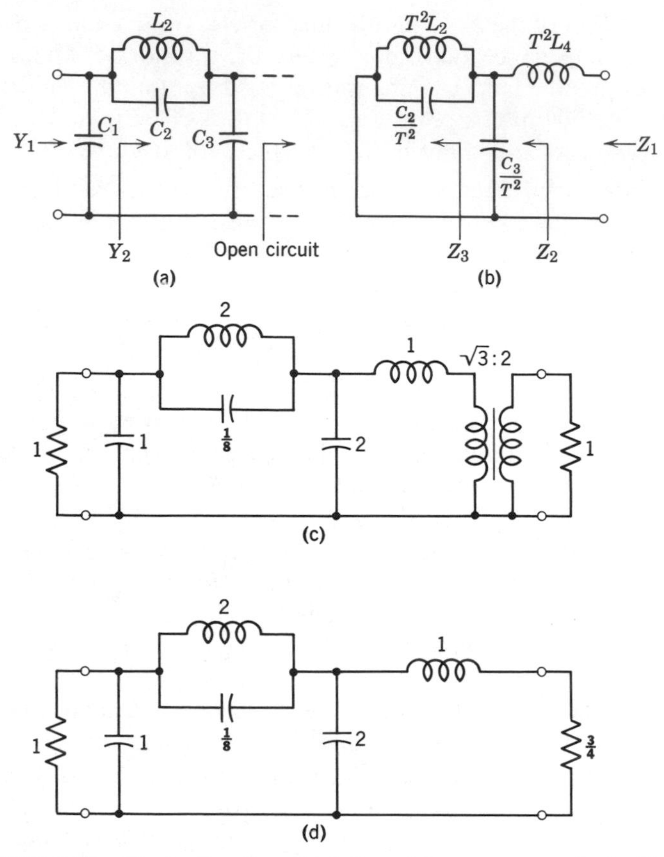

FIGURE 3-13 Illustrating the calculation of the components in the filter of Figure 3-12(c) from the impedances at port 1 and port 2.

admittance equals Y_1. Thus at $s = j2$

$$j2C_1 = Y_1(j2) = j2$$

and so $C_1 = 1F$. The admittance of C_1 is subtracted from Y_1 to give

$$Y_2 = Y_1 - C_1 s = \frac{19s^3 + 12s}{17s^2 + 4} - s$$

$$= \frac{2s(s^2 + 4)}{17s^2 + 4}.$$

$1/Y_2$ has a pole at $s = \pm j2$, and the impedance of the $L_2 C_2$ combination is given by the terms in the partial-fraction expansion of $1/Y_2$ around these poles:

$$\frac{1}{Y_2} = \frac{17s^2 + 4}{2s(s^2 + 4)} = \frac{8s}{s^2 + 4} + \frac{1}{2s}.$$

So $L_2 = 8/4 = 2H$ and $C_2 = (1/8)F$. The remainder,

$$\frac{1}{Y_2} - \frac{8s}{s^2 + 4} = \frac{1}{2s}$$

gives the impedance of C_3, whence $C_3 = 2F$. This completes the development from port 1 and gives all the components except for L_4 and the turns ratio T of the ideal transformer.

Next we turn to port 2 where the impedance of lower degree is the impedance with port 1 short circuited:

$$Z_1 = \frac{B}{A} = \frac{68s^3 + 48s}{51s^2 + 12}.$$

At this stage the simplest procedure is first to develop this impedance on the assumption that the ideal transformer has been removed and, therefore, that the remainder of the network to the left of the transformer has been scaled in impedance by a factor T^2, i.e., the inductances are all multiplied by T^2 and the capacitances divided by T^2. Then, by comparing the values of those elements which are found from both port 1 and port 2, one can calculate the value of T.

Proceeding on this basis we get the situation shown in Figure 3-13(b). The impedance of T^2L_4 provides the pole of Z_1 at infinite frequency so that:

$$T^2L_4 = \lim_{s\to\infty} Z_1/s = 68/51 = 4/3.$$

The impedance of T^2L_4 is next subtracted from Z_1 to give:

$$Z_2 = \frac{68s^3 + 48s}{51s^2 + 12} - \frac{4s}{3} = \frac{32s}{51s^2 + 12}.$$

At $s = \pm j2$ the parallel-tuned circuit formed by T^2L_2 and C_2/T^2 has infinite impedance and the admittance of C_3/T^2 constitutes the whole of $1/Z_2$. Thus

$$\frac{j2C_3}{T^2} = \frac{1}{Z_2(j2)} = j3,$$

whence $C_3/T^2 = 3/2$. Subtracting the admittance of C_3/T^2 from $1/Z_2$ gives

$$\frac{1}{Z_3} = \frac{1}{Z_2} - \frac{sC_3}{T^2} = \frac{51s^2 + 12}{32s} - \frac{3s}{2} = \frac{3s^2 + 12}{32s}.$$

Z_3 is the impedance of T^2L_2 in parallel with C_2/T^2 so that

$$T^2L_2 = 32/12 = 8/3$$
$$C_2/T^2 = 3/32.$$

Since $C_2 = (1/8)F$, as obtained in the course of the development from port 1, it follows that $T^2 = 4/3$. Knowing T^2 one can then calculate L_4 as $(4/3)/T^2 = 1$. The original circuit of Figure 3-12 is now completely designed and the component values are shown on Figure 3-13(c). Rather than attempt to make the ideal transformer, onc would, in practice, normally replace the combination of transformer and 1 ohm terminating resistance by a 3/4 ohm termination, as shown in Figure 3-13(d).

An Improved Realization Algorithm

The operations described above for the ladder expansion are unnecessarily complicated and diverse. In addition, they lead to some loss of numerical accuracy, since a subtraction of nearly equal numbers must usually be carried out in every stage. All these disadvantages may be avoided by using a simple and general algorithm especially developed* for the realization of

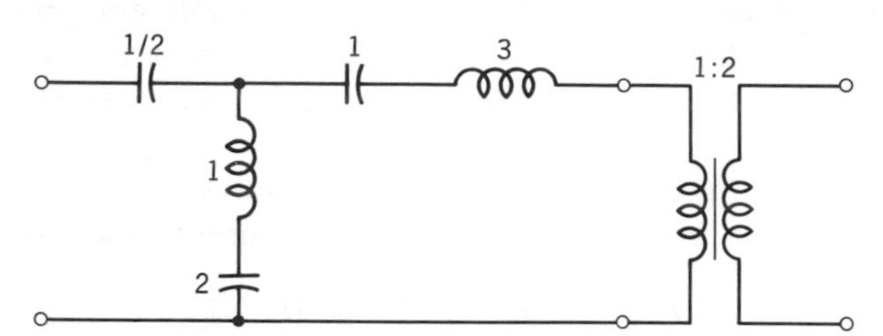

FIGURE 3-14 Example circuit for the general ladder realization algorithm.

ladders [5]. The use of the algorithm will be illustrated with the simple example shown in Figure 3-14. The first step of the design is demonstrated in Figure 3-15(a). From the figure,

$$Z_1 = \frac{1}{sC_1} + \frac{1}{\dfrac{s/L_2}{s^2 + \omega_2^2} + \dfrac{1}{Z_3}}$$

where $\omega_2^2 = 1/L_2 C_2$. Let $Z_1 = U(s^2)/sV(s^2)$ and $Z_3 = \bar{U}(s^2)/s\bar{V}(s^2)$, where U, V, $\bar{U}$, and $\bar{V}$ are polynomials in s^2. Then, after reduction to a single fraction, the above becomes

* *Editors' note:* This algorithm was discovered by H. J. Orchard.

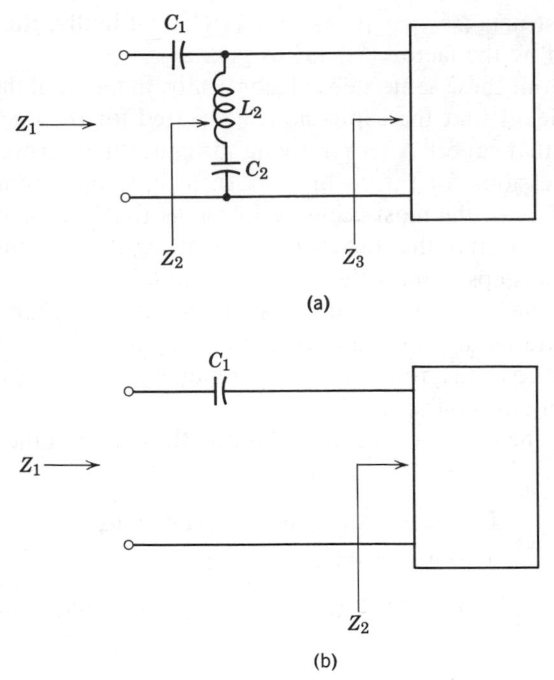

FIGURE 3-15 Details in the ladder realization of the circuit of Figure 3-14.

$$\frac{U}{sV} = \frac{(1/C_1)[(1/L_2)\bar{U} + (s^2 + \omega_2^2)\bar{V}] + (s^2 + \omega_2^2)\bar{U}}{s[(1/L_2)\bar{U} + (s^2 + \omega_2^2)\bar{V}]}. \tag{3.17}$$

By equating separately the numerators and denominators, and with an obvious substitution, we get the following two equations:

$$U = (1/C_1)V + (s^2 + \omega_2^2)\bar{U} \tag{3.18}$$

$$V = (1/L_2)\bar{U} + (s^2 + \omega_2^2)\bar{V}. \tag{3.19}$$

In (3.18), U, V, and ω_2^2 are known; the unknowns are C_1 and the coefficients of the polynomial $\bar{U}$. Similarly, in (3.19), V, $\bar{U}$, and ω_2^2 are assumed to be known ($\bar{U}$ through the solution of (3.18)), while the unknowns are L_2 and the coefficients of $\bar{V}$. The two equations are thus identical in form and can be solved, one after the other, by using a common algorithm.

If a ladder circuit is designed by the simple methods described earlier in this section, this is equivalent to solving (3.18) (and then (3.19)) in the following way. One sets $s^2 = -\omega_2^2$ and obtains $1/C_1 = U(-\omega_2^2)/V(-\omega_2^2)$.

Then one constructs $(s^2 + \omega_2^2)\bar{U}$ as $U - (1/C_1)V$. Finally, this last polynomial is divided by the factor $s^2 + \omega_2^2$ to give $\bar{U}$.

By carrying out these same steps algebraically, in terms of the coefficients, it becomes evident that the expressions generated for the coefficients of $\bar{U}$ contain terms that cancel. After removing the cancelling terms, the resulting algebraic expressions for the coefficients then represent, from a computational point of view, the most accurate formulas that can be derived.

A direct evaluation of the coefficients of $\bar{U}$, using these formulas, involves more arithmetic steps than in the simple method, but it gives higher accuracy because in the latter case the terms which should cancel are calculated in separate operations and thus do not match exactly. The formulas for the coefficients are easy to program for a computer, and the extra computing time is normally insignificant.

To illustrate the nature of the formulas and the extent of the cancellations let

$$U = u_0 + u_1 s^2 + u_2 s^4 + u_3 s^6 + u_4 s^8$$
$$V = v_0 + v_1 s^2 + v_2 s^4 + v_3 s^6 + v_4 s^8$$

and, for convenience, $a_1 = 1/C_1$ and $x = -\omega_2^2$. The equation to be solved is then

$$U = a_1 V + (s^2 - x)\bar{U}. \tag{3.20}$$

Setting $s^2 = x$ gives immediately $a_1 = U(x)/V(x)$. Then

$$U - a_1 V = (u_0 - a_1 v_0) + (u_1 - a_1 v_1)s^2 + (u_2 - a_1 v_2)s^4 + (u_3 - a_1 v_3)s^6 + (u_4 - a_1 v_4)s^8.$$

Dividing this polynomial in the usual way by $s^2 - x$ gives, by (3.20), the coefficients of $\bar{U}$ as:

$$\begin{aligned}
\bar{u}_3 &= (u_4 - a_1 v_4)\\
\bar{u}_2 &= (u_3 - a_1 v_3) + (u_4 - a_1 v_4)x\\
\bar{u}_1 &= (u_2 - a_1 v_2) + (u_3 - a_1 v_3)x + (u_4 - a_1 v_4)x^2\\
\bar{u}_0 &= (u_1 - a_1 v_1) + (u_2 - a_1 v_2)x + (u_3 - a_1 v_3)x^2 + (u_4 - a_1 v_4)x^3.
\end{aligned}$$

Consider the expression for $\bar{u}_1$, for example. Replacing a_1 by $U(x)/V(x)$ and multiplying throughout by $V(x)$ one gets

$$\begin{aligned}
\bar{u}_1 V(x) &= u_2 V(x) - v_2 U(x) + xu_3 V(x) - xv_3 U(x) + x^2 u_4 V(x) - x^2 v_4 U(x)\\
&= V(x)(u_2 + u_3 x + u_4 x^2) - U(x)(v_2 + v_3 x + v_4 x^2)\\
&= (v_0 + v_1 x + \underline{v_2 x^2 + v_3 x^3 + v_4 x^4})(u_2 + u_3 x + u_4 x^2)\\
&\quad - (u_0 + u_1 x + \underline{u_2 x^2 + u_3 x^3 + u_4 x^4})(v_2 + v_3 x + v_4 x^2)\\
&= (v_0 + v_1 x)(u_2 + u_3 x + u_4 x^2) - (u_0 + u_1 x)(v_2 + v_3 x + v_4 x^2).
\end{aligned}$$

The underlined parts of $V(x)$ and $U(x)$, when multiplied by the associated factors, obviously cancel. Similar cancellations occur in the expressions for the other coefficients. The results for all the coefficients (after cancellations) are given in Table 3-1. The general rule for solving this problem for any U and V should be obvious from an examination of the pattern of the entries.

TABLE 3-1

$$\Delta \triangleq V(x) = v_0 + v_1 x + v_2 x^2 + v_3 x^3 + v_4 x^4$$
$$a_1 \Delta = u_0 + u_1 x + u_2 x^2 + u_3 x^3 + u_4 x^4 = U(x)$$
$$\bar{u}_0 \Delta = v_0(u_1 + u_2 x + u_3 x^2 + u_4 x^3) - u_0(v_1 + v_2 x + v_3 x^2 + v_4 x^3)$$
$$\bar{u}_1 \Delta = (v_0 + v_1 x)(u_2 + u_3 x + u_4 x^2) - (u_0 + u_1 x)(v_2 + v_3 x + v_4 x^2)$$
$$\bar{u}_2 \Delta = (v_0 + v_1 x + v_2 x^2)(u_3 + u_4 x) - (u_0 + u_1 x + u_2 x^2)(v_3 + v_4 x)$$
$$\bar{u}_3 \Delta = (v_0 + v_1 x + v_2 x^2 + v_3 x^3)u_4 - (u_0 + u_1 x + u_2 x^2 + u_3 x^3)v_4$$

For a full removal of a pole of Z_1 at $s = 0$, as shown in Figure 3-15(b), with $Z_1 = U/sV$, one gets $Z_2 = s\bar{U}/V$, i.e., the denominator polynomial V remains unchanged. The equations relating the polynomials reduce to

$$U = (1/C_1)V + s^2\bar{U}. \tag{3.21}$$

Application of the preceding equations (Table 3-1) with $x = 0$ gives C_1 and $\bar{U}$.

The analysis of the circuit of Figure 3-16(a) can be performed in a similar fashion (see Problem 3.10) to give a pair of equations of the general form of

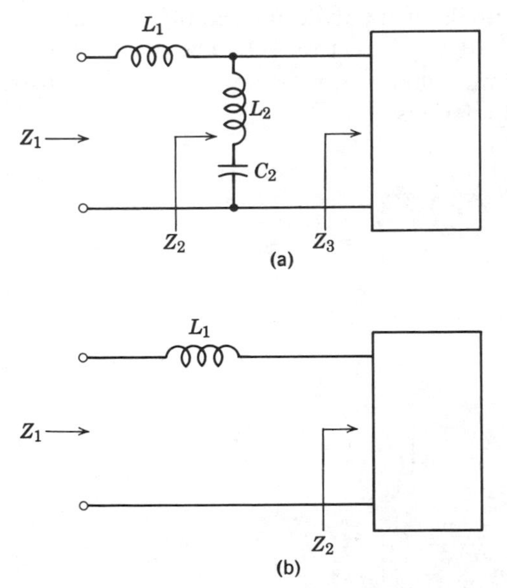

FIGURE 3-16 Further illustration of ladder realization.

(3.18) and (3.19). The full removal of a pole of Z_1 at infinite frequency (Figure 3-16(b)) is again merely a special case in which only one polynomial equation need be solved.

It is easily shown that the circuits of Figure 3-17(a) and Figure 3-17(b) are duals of those in Figure 3-15(a) and Figure 3-16(a), respectively, and that their design relations are therefore mathematically identical (Problem 3.11).

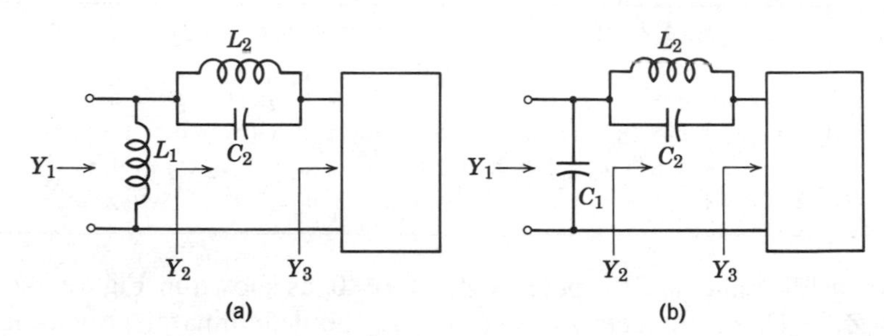

FIGURE 3-17 Circuits dual to those in Figures 3-15 and 3-16.

It can now be concluded that all relations involved in the ladder development are in the form

$$P(s^2) = a_1 Q(s^2) + (ys^2 - x)R(s^2), \tag{3.22}$$

where P and Q are known polynomials and (y, x) can be any one of the four sets $(1, -\omega_2^2)$, $(1, 0)$, $(\omega_2^{-2}, -1)$, $(0, -1)$. The polynomial $Q(s^2)$ may include a factor s^2. Then the unknown constant a_1 and the polynomial R may be found from the relations:

$$\Delta = \sum_{i=0}^{n} q_i x^i y^{n-i}$$

$$a_1 \Delta = \sum_{i=0}^{n} p_i x^i y^{n-i}$$

$$r_k \Delta = \left[\sum_{i=0}^{k} q_i x^i y^{k-i}\right]\left[\sum_{i=k+1}^{n} p_i x^{i-k-1} y^{n-i}\right]$$

$$- \left[\sum_{i=0}^{k} p_i x^i y^{k-i}\right]\left[\sum_{i=k+1}^{n} q_i x^{i-k-1} y^{n-i}\right], \; k = 0, 1, \ldots, (n-1), \tag{3.23}$$

where

$$P(s^2) = \sum_{i=0}^{n} p_i s^{2i}$$

$$Q(s^2) = \sum_{i=0}^{n} q_i s^{2i} \qquad \text{(Note: } q_n \text{ may be zero)}$$

$$R(s^2) = \sum_{i=0}^{n-1} r_i s^{2i}, \tag{3.24}$$

(see Problem 3.12). For example, for $n = 3$, the equations are

$$\begin{aligned}
\Delta &= q_0 y^3 + q_1 xy^2 + q_2 x^2 y + q_3 x^3 \\
a_1 \Delta &= p_0 y^3 + p_1 xy^2 + p_2 x^2 y + p_3 x^3 \\
r_0 \Delta &= q_0(p_1 y^2 + p_2 xy + p_3 x^2) - p_0(q_1 y^2 + q_2 xy + q_3 x^2) \\
r_1 \Delta &= (q_0 y + q_1 x)(p_2 y + p_3 x) - (p_0 y + p_1 x)(q_2 y + q_3 x) \\
r_2 \Delta &= (q_0 y^2 + q_1 xy + q_2 x^2)p_3 - (p_0 y^2 + p_1 xy + p_2 x^2)q_3.
\end{aligned} \tag{3.25}$$

In the special case when $x = 0$ and $y = 1$, corresponding to a full removal of a pole at $s = 0$, these formulas reduce to

$$\begin{aligned}
\Delta &= q_0 \\
a_1 &= p_0/q_0 \\
r_0 &= (q_0 p_1 - p_0 q_1)/q_0 \\
r_1 &= (q_0 p_2 - p_0 q_2)/q_0 \\
r_2 &= (q_0 p_3 - p_0 q_3)/q_0.
\end{aligned} \tag{3.26}$$

Here, the expressions for the r_k are identical with those obtained in the simplest approach and the algorithm offers no advantage—one would use it only to achieve uniformity in the programming.

Minimum-Inductor Filters

Inductors using magnetic core materials are normally bulky, heavy, and expensive* compared to the capacitors used in the filter. Hence, it is desirable to minimize their number in the circuit. Realization of a conjugate pair of finite nonzero $j\omega$-axis loss poles, of course, requires an inductor; a loss pole at zero (or infinite) frequency can, however, be obtained using only a series (or shunt) capacitor. Hence, it appears that the minimum possible number of inductors equals the number of finite nonzero pole pairs. Each pair of real poles (realized with a C-type section) also requires one inductor,

* Typically, five times as costly as capacitors.

and a complex quad of poles two inductors (if a tapped inductor is regarded as a single element). This minimum number is indeed obtainable for lowpass and highpass filters; e.g., Figures 3-18(a) and (b) illustrate a lowpass and highpass circuit, respectively, with all inductors used for the realization of finite nonzero loss poles*.

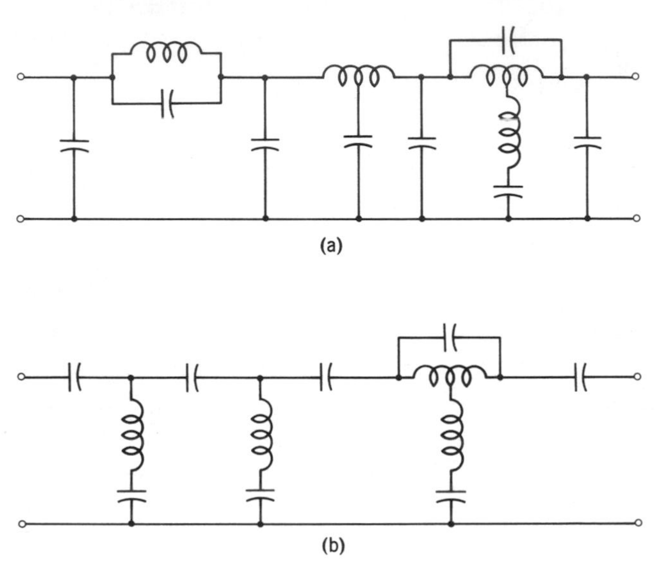

FIGURE 3-18 Examples of minimum-inductor lowpass and highpass filters.

For bandpass filters the situation is more involved. Experience shows that in order to obtain positive element values during the ladder development, loss poles in the upper stopband must be realized using lowpass-type sections (Figures 3-19(a) and (b)) while poles in the lower stopband require highpass-type sections (Figures 3-19(c) and (d)). The minimum inductor requirement then suggests cascading sections of the types shown in Figures 3-19(a) and (c) to realize a bandpass characteristic. Such a circuit is illustrated in Figure 3-20 for the case when there are two poles in each stopband, and, in addition, one pole at zero and one at infinite frequency. This circuit does use the least possible number of inductors for this set of loss poles, and, in this sense of the word, is highly efficient. It has, however, a subtle limitation which makes it usable only through special design techniques.

* The sections containing coupled inductors represent practical realizations for type-*D* sections, corresponding to complex loss poles.

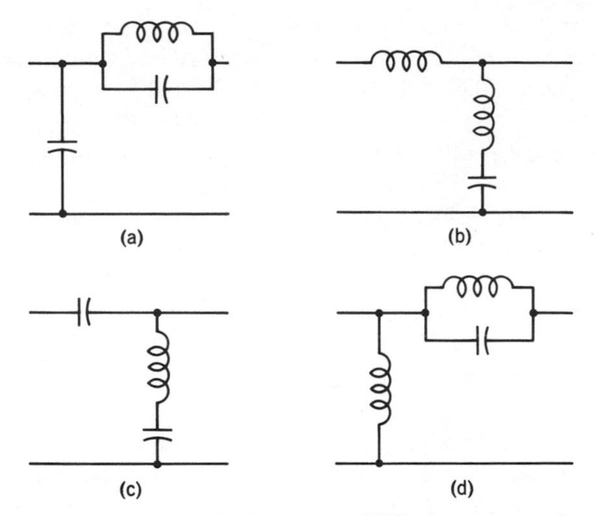

FIGURE 3-19 Basic lowpass and highpass type sections.

To understand the restriction on the minimum-inductor circuit, consider the behavior of the reflection coefficients

$$\rho_1(s) = \frac{f(s)}{e(s)} = \frac{Z_1(s) - R_1}{Z_1(s) + R_1}$$

$$\rho_2(s) = \mp \frac{f(-s)}{e(s)} = \frac{Z_2(s) - R_2}{Z_2(s) + R_2}. \tag{3.27}$$

At zero frequency, both Z_1 and Z_2 become infinite; at infinite frequency both impedances tend to zero. Consequently,

$$\begin{aligned}\rho_1(0) &= \rho_2(0) = +1\\ \rho_1(\infty) &= \rho_2(\infty) = -1.\end{aligned} \tag{3.28}$$

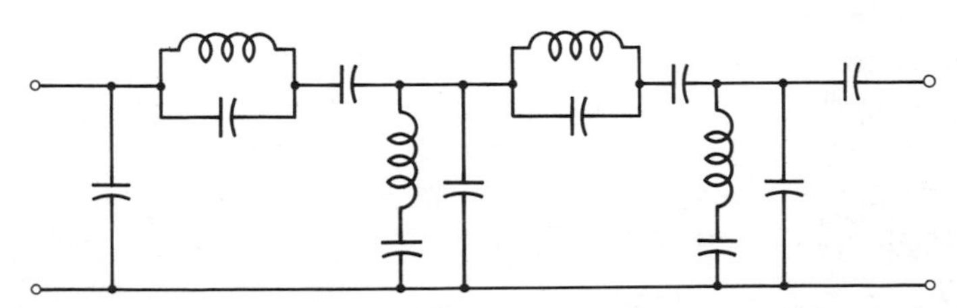

FIGURE 3-20 Minimum-inductor bandpass filter which requires two reflection zeros on the real *s*-axis.

Following the behavior of $\rho_1(s)$ and $\rho_2(s)$ along the *positive* real s-axis (where $e(s)$ can have no zeros), we must therefore find a zero of *both* $f(s)$ and $f(-s)$. In conclusion then, the circuit of Figure 3-20 (and, by straightforward generalization, any circuit built exclusively from the sections of Figures 3-19(a) and (c)) has at least one reflection zero each along both the positive and negative *real* axes. Since all popular and efficient transfer functions, as described in Chapter 2, have their reflection zeros in or near the passband, the circuit of Figure 3-20 cannot be used directly to realize these functions. In fact, the real reflection zeros have little effect on the response, and their value can be regarded as an almost free design parameter. For this reason, this truly minimum-inductor filter is often referred to as a *parametric filter*. The design of this circuit requires special approximating methods [7, 5] and is beyond the scope of this book.

Consider now the bandpass filter shown in Figure 3-21. This circuit is similar to the parametric filter of Figure 3-20, but it has two added inductors, corresponding to one additional pole at both zero and infinite frequency. It is easy to see that for this circuit $\rho_1(0) = \rho_1(\infty) = -1$,

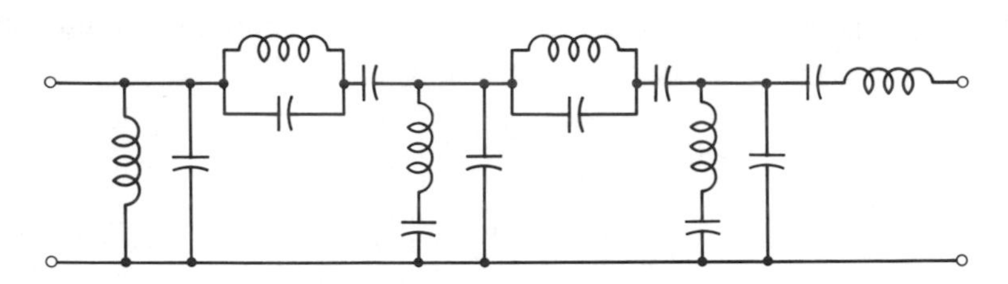

FIGURE 3-21 Minimum-inductor bandpass filter which can have all its reflection zeros on the *j*-axis inside the passband.

$\rho_2(0) = \rho_2(\infty) = +1$, and hence no real reflection zeros need occur. Hence, bandpass circuits with at least two poles of loss at both zero and infinite frequency *can* realize conventional transfer functions, with all reflection zeros in or near the passband. Similar conclusions can be reached if the numbers of loss poles at zero and infinite frequencies are 1 and 3, or 3 and 1 as verified in Problem 3.14.* Hence, a "conventional" minimum-inductor circuit has typically $n/2$ inductors, where n is the degree of the filter, i.e., the total number of loss poles, including those at zero and at infinite frequency. The corresponding number for parametric filters is $(n/2) - 1$.

* At least one loss pole is usually desirable at both zero and infinite frequency, to make the practical sections of Figures 3-18 to 3-19 applicable.

3.4 LATTICE NETWORK DESIGN

It was pointed out in Section 3.2 that symmetrical lattices and/or their transformer-coupled equivalents are important design vehicles, especially for resonator filters. The design of these circuits consists essentially of finding the lattice impedances Z_a and Z_b from a prescribed $H(s)$ and $K(s)$. Since the realization of a reactive impedance is a well-explored and quite easy task (see Reference [3]), the design is trivial once Z_a and Z_b have been found.

The calculation of Z_a and Z_b in terms of $H(s)$ and $K(s)$ can be based on results obtained earlier. As shown in Section 3.1 ((3.3)–(3.6)), $H(s)$ and $K(s)$ and the impedance parameters z_{ij} may be expressed in terms of the chain matrix elements A, B, C, and D, as well as the terminations R_1 and R_2 of the filter. In addition, it was shown in Section 3.2 that for a symmetric lattice the lattice impedances Z_a, Z_b, and the z_{ij} are related by

$$Z_a = z_{11} - z_{12} = \frac{A-1}{C} \qquad Z_b = z_{11} + z_{12} = \frac{A+1}{C}$$

whence the chain matrix is

$$\mathbf{T} = \frac{1}{Z_b - Z_a}\begin{bmatrix} Z_b + Z_a & 2Z_aZ_b \\ 2 & Z_b + Z_a \end{bmatrix}. \tag{3.29}$$

Substituting these relations into (3.4)–(3.6) and using $R_1 = R_2 = R$, we obtain

$$H(s) = \frac{(Z_b + R)(Z_a + R)}{R(Z_b - Z_a)}$$

$$K(s) = \frac{Z_aZ_b - R^2}{R(Z_b - Z_a)}. \tag{3.30}$$

Also, (3.8) and (3.29) may be used to express Z_a and Z_b directly in terms of $H(s)$ and $K(s)$. The result is

$$Z_a = R\frac{H_e + K_e - 1}{H_o - K_o} = R\frac{H_e - 1}{H_o - K_o}$$

$$Z_b = R\frac{H_e + K_e + 1}{H_o - K_o} = R\frac{H_e + 1}{H_o - K_o} \tag{3.31}$$

since $K_e \equiv 0$ for a symmetric two-port.

Equations (3.31) might be regarded as the solution to our problem. It can easily be verified, however, that the degrees of both the numerators and denominators of Z_a and Z_b, as given by (3.31), are too high, and that cancellations of common factors must be carried out (see Problem 3.15). Since in practical design these factors may not easily be detected and, due to numerical inaccuracies, may not even remain common, it is much preferable to devise a design approach which avoids these cancelling common factors. Such a technique will be described next.

Writing the rational functions $H(s)$, $K(s)$, $Z_a(s)$, and $Z_b(s)$ in terms of their numerator and denominator polynomials,

$$H(s) = \frac{e(s)}{p(s)}, \qquad K(s) = \frac{f(s)}{p(s)} \tag{3.32}$$

$$Z_a(s) = R\frac{n_a(s)}{d_a(s)}, \qquad Z_b(s) = R\frac{n_b(s)}{d_b(s)} \tag{3.33}$$

and substituting these relations into (3.30), one finds

$$H(s) = \frac{(n_a + d_a)(n_b + d_b)}{(n_b d_a - n_a d_b)} = \frac{e(s)}{p(s)} \tag{3.34}$$

$$K(s) + 1 = \frac{(n_a + d_a)(n_b - d_b)}{(n_b d_a - n_a d_b)} = \frac{f(s) + p(s)}{p(s)}$$

$$K(s) - 1 = \frac{(n_a - d_a)(n_b + d_b)}{(n_b d_a - n_a d_b)} = \frac{f(s) - p(s)}{p(s)}. \tag{3.35}$$

From (3.34) and (3.35) one may deduce that:

1. Since each of the five polynomials, $p(s) = (n_b d_a - n_a d_b)$, n_a, d_a, n_b, and d_b must be purely even or purely odd, the possible parity choices are those shown in Table 3-2.

TABLE 3-2

$p(s)$	n_a	d_a	n_b	d_b
Even	Even	Odd	Odd	Even
	Odd	Even	Even	Odd
Odd	Even	Odd	Even	Odd
	Odd	Even	Odd	Even

2. The natural modes, which are the zeros of $e(s) = (n_a + d_a)(n_b + d_b)$, fall into two groups. Those in the first group which belong to $(n_a + d_a)$ are also zeros of $f(s) + p(s)$. Those in the second group which belong to $(n_b + d_b)$ are zeros of $f(s) - p(s)$.

On the basis of these observations the following design procedure suggests itself.

1. Find all the natural modes.
2. Substitute the natural modes into $f(s) + p(s)$ and/or $f(s) - p(s)$. From those modes, which are zeros of $f(s) + p(s)$, form the polynomial $(n_a + d_a)$; from those that are zeros of $f(s) - p(s)$, form $(n_b + d_b)$.
3. The four polynomials n_a, d_a, n_b, and d_b are then chosen as the odd or even parts of $(n_a + d_a)$ or $(n_b + d_b)$ according to one or other of the possibilities enumerated in Table 3-2.

As mentioned before, the rest of the design, involving the realization of Z_a and Z_b, is straightforward.

It should be pointed out that an alternative design procedure based on the factoring of $f(s) + p(s)$ or $f(s) - p(s)$ is also possible, and it is essentially equivalent to the one presented here. One advantage of the described process is that it is easily performed in terms of a transformed frequency variable (see Section 3.5).

Rewriting $H(s)$ in the forms

$$
\begin{aligned}
H(s) &= \frac{(Z_b + R)(Z_a + R)}{R(Z_b - Z_a)} = -\frac{(Z_a + R)(Z_b + R)}{R(Z_a - Z_b)} \\
&= -\frac{(R^2/Z_b + R)(R^2/Z_a + R)}{R(R^2/Z_b - R^2/Z_a)} = \frac{(R^2/Z_a + R)(R^2/Z_b + R)}{R(R^2/Z_a - R^2/Z_b)}
\end{aligned}
\tag{3.36}
$$

shows that, apart from a reversal of polarity, the lattices which have impedances Z_a, Z_b; *or* Z_b, Z_a; *or* R^2/Z_a, R^2/Z_b; *or* R^2/Z_b, R^2/Z_a are equivalent in their transfer properties.

Example 3-3. Design a lattice filter from a given $K(s) = s^3$, using both design approaches described in this section.

Solution: (a) From a previous example (Section 3.1),

$$K(s) = s^3 \text{ gives } H(s) = s^3 + 2s^2 + 2s + 1,$$

and therefore:

$$H_e(s) = 2s^2 + 1, H_o(s) = s^3 + 2s, K_o(s) = s^3.$$

Assuming $R_1 = R_2 = 1$, (3.31) gives:

$$Z_a = \frac{2s^2}{2s}; \qquad Z_b = \frac{2s^2 + 2}{2s}.$$

After cancellation of common factors, Z_a and Z_b become:

$$Z_a = s; \qquad Z_b = s + \frac{1}{s}.$$

The corresponding lattice is shown in Figure 3-22.

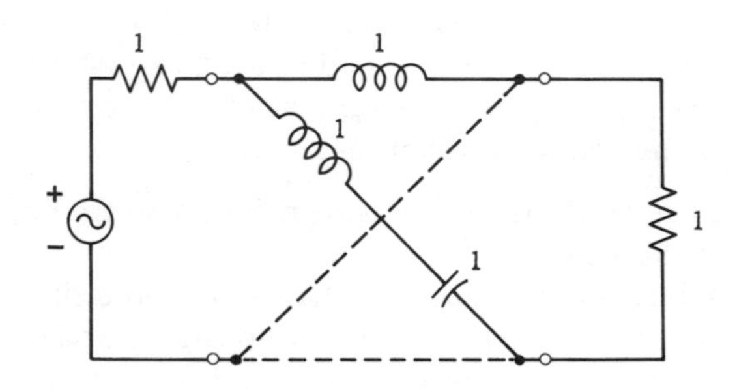

FIGURE 3-22 Circuit for the design example of a lattice filter.

(b) Using the design steps 1 to 3:

1. Since $H(s) = (s + 1)(s^2 + s + 1)$, the natural modes are -1, $-0.5 \pm j0.866$.
2. Here, $f(s) = s^3$, $p(s) = 1$, $f(s) + p(s) = s^3 + 1$ and $f(s) - p(s) = s^3 - 1$. Clearly, $s_1 = -1$ is a zero of $f(s) + p(s)$, and therefore $n_a + d_a = s + 1$. Similarly, the natural modes $s_{2,3} = -0.5 \pm j0.866$ are zeros of $f(s) - p(s)$, and hence $n_b + d_b = s^2 + s + 1$.
3. Using Table 3-2 for an even $p(s)$, and choosing the second possibility offered by the table, the following results:

$$n_a(s) = s \quad ; \qquad d_a(s) = 1$$
$$n_b(s) = s^2 + 1; \qquad d_b(s) = s.$$

Finally, by (3.33):

$$Z_a(s) = s; \quad Z_b(s) = \frac{s^2 + 1}{s} = s + \frac{1}{s}.$$

No cancellation is required to get the lattice impedances using this approach.

3.5 NUMERICAL ILL-CONDITIONING IN FILTER DESIGN

The previous discussions in this chapter may have suggested that, if the design steps are carried out exactly as described, the values of the filter elements will be obtained without difficulty. This, unfortunately, is not the case. Although the design process presents no problems for simple lowpass filters of degree up to about five, it is found that, in more complicated filters, a ladder development of the impedance at one port will give element values which differ somewhat from those obtained through a corresponding development at the other port.

The discrepancy between these two sets of element values increases rapidly with the degree of the filter until, for still quite moderately sized networks, there is almost no correlation between them. The natural conclusion is that, probably, neither set of element values is correct, and that somewhere in the design process accuracy is being lost.

At one time it was thought that the loss of accuracy occurred through repeated calculations involving subtractions between nearly equal numbers as, e.g., in evaluating $\sqrt{a+\epsilon}-\sqrt{a}$ when ϵ is very small compared with a. The solution in this simple case is merely to compute $\epsilon/(\sqrt{a+\epsilon}+\sqrt{a})$, and it was hoped that similar artifices would be found to solve the accuracy problem in filter design.

In fact, although some of the inaccuracy certainly arises from such causes, it turns out that by far the greatest part is actually due to *ill-conditioning* in one form or another. (Ill-conditioning is a general-purpose mathematical name to describe what, in engineering terms, would usually be referred to as high sensitivity.)

The most important ill-conditioning in the filter design occurs when the information about the filter is stored in the coefficients of the polynomials forming the chain matrix (or, equally, the **Z** or **Y** matrix). Very small changes in these coefficients result in large changes in the element values of the ladder network; in other words, the element values are extremely sensitive to errors in the coefficients.

This sensitivity is so high that, for filters of only moderate complexity, the mere act of truncating the exact coefficients to ten decimal digits may be sufficient to disturb some of the element values quite violently, or even occasionally to make them negative. In the latter case the chain matrix is no longer physically realizable! After the coefficients have been so truncated, it is of no help to carry out the subsequent arithmetic steps with higher (or even with infinite) precision; the damage done by the truncation is irreparable.

Example 3-4. To illustrate this sensitivity of the element values to tolerances on the coefficients, consider the simple seventh-degree lowpass filter shown in Figure 3-23, whose element values are given in row 1 of Table 3-3.

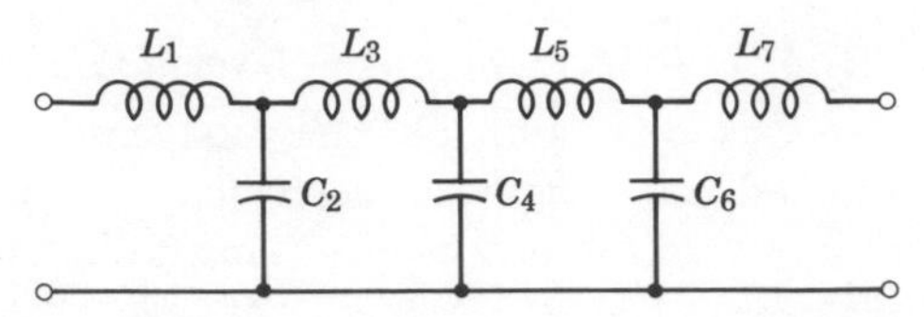

FIGURE 3-23 Circuit referred to in the example to illustrate ill-conditioning.

The input impedance at port 1 with port 2 short circuited is

$$Z(s) = \frac{32s^7 + 64s^5 + 38s^3 + 6s}{32s^6 + 48s^4 + 18s^2 + 1}.$$

If the coefficient of s^5 in $Z(s)$ is changed by 1% from 64 to 64.64, and the corresponding ladder network found by development of this modified impedance, the element values change to those given in row 2 of Table 3-3. It is seen that, even in this very simple case, the 1% error in the coefficient of s^5 is magnified up to a 19% error in L_7.

TABLE 3-3

Element values	L_1	C_2	L_3	C_4	L_5	C_6	L_7
Initial values	1	2	2	2	2	2	1
1% change in s-variable coefficient	1	1.923	1.745	1.775	2.062	2.184	1.193
1% change in z-variable coefficient	1.006	1.988	1.994	2.025	1.994	1.988	1.006

The high sensitivity of the elements to the coefficients of the chain-matrix polynomials is the product of two effects. Firstly, the polynomials themselves are very ill-conditioned [8]. In general, an ill-conditioned polynomial is one whose zeros are very sensitive to errors in the coefficients. The geometrical arrangement of the zeros and their relationship to the origin determine the extent of the ill-conditioning. Here the ill-conditioning of the chain-matrix polynomials is due to the way in which the roots are distributed along the j-axis, and it is made worse by the fact that the zeros tend to cluster together near the passband edges*. The effect in narrow-band bandpass filters is further exaggerated by the closeness of the passband edges.

Secondly, the zeros of the various polynomials are required to interlace because they belong to positive-real functions, and the information about the element values resides primarily in the *distances between the zeros* of the polynomials. Small movements of the zeros can thus cause large changes in the elements. This effect evidently just enhances the basic ill-conditioning of the polynomials.

A further factor contributing to the overall loss of accuracy in the design is the ill-conditioning of the polynomials $e(s)$ and $f(s)$. This is nowhere as severe as that of the polynomials appearing in the chain matrix, but it is sufficient to limit slightly the accuracy with which one can locate the zeros (by whatever means one uses), and this in turn causes small errors in the chain matrix.

The simplest way of overcoming the ill-conditioning is the obvious one of using multiple-precision arithmetic in all the calculations. Since, however, the necessary number of digits is very roughly equal to the degree of the filter, this leads to extravagant amounts of multiple precision and to lengthy computer times for large filters. More efficient and sophisticated are the following two methods:

1. The product method [9]. In this technique, all polynomials are stored and manipulated in factored form (or "product" form) thus: $P(s) = k\prod_i(s - s_i)$. Hence, the information is stored in the *zeros*, rather than the coefficients. This improves the conditioning of the design because it eliminates the ill-conditioning of the polynomials. However, it can do nothing about the close packing of the zeros near passband edges and the ill-conditioning from this source is unaffected.
2. The transformed variable method [5]. This technique replaces s^2 everywhere by a bilinear expression in z^2, where z is the transformed complex frequency. The mapping is chosen so that, in the z-plane, the critical frequencies are more widely separated, and hence all the polynomials concerned become very much better conditioned. The net result is that the sensitivity of the ladder elements to the coefficients of the z-variable polynomials is even less than their sensitivity to the zeros of the s-variable polynomials.

An objective comparison of the two methods involves several considerations. The product method is conceptually simple and easy to program†. It

* Even such seemingly innocent polynomials as $P(s) = (s-1)(s-2)\cdots(s-19)(s-20)$, which show little clustering, may be highly ill-conditioned, as pointed out by Wilkinson [8].

† It uses the same Newton-type subroutine over and over again.

requires, however, a factoring of a linear combination of factored polynomials in every step of the design process, and hence it leads to even more lengthy computations than would multiple precision. The transformed-variable method, to be demonstrated later, is somewhat harder to understand and to program, but it is very fast to execute and the improvement in accuracy is greater than that obtainable with the product method. There is no reason, of course, why the two methods could not be combined, carrying out all calculations in terms of factored polynomials of z rather than s.

A bilinear transformation is used in preference to any other mathematical form because it is the only simple function that gives a one-to-one correspondence between the two variables in both directions of the transformation. Moreover, it preserves the degree of any rational function whose variable is so changed. The transformation here has to be expressed in terms of the *squares* of the two variables so that it can simultaneously correct for the ill-conditioning at both positive and negative frequencies.

The bilinear mapping of s^2 which leads to optimum accuracy varies somewhat from filter to filter. It was found, however, that the transformation already described in Chapter 2 in connection with the equal-ripple passband approximation is a serviceable compromise:

$$z^2 = \frac{s^2 + \omega_2^2}{s^2 + \omega_1^2}, \quad \mathrm{Re}(z) \geq 0, \quad \text{for bandpass filters;}$$
$$z^2 = 1 + \omega_2^2/s^2, \quad \mathrm{Re}(z) \geq 0, \quad \text{for lowpass filters,} \tag{3.37}$$

where ω_1 and ω_2 are the lower and upper passband limits, respectively. Using this transformation makes the design of equal-ripple passband filters especially easy. In addition, since it transforms the two passband limits to zero and to infinity in the z-plane, it can be expected to separate the critical frequencies (tightly clustered in the s-plane around these two points) in a nearly optimum fashion for a wide variety of filters. This separation was, in fact, observed in countless applications.

The improvement in the degree of ill-conditioning due to using the z-variable may be demonstrated with the simple example given above. Expressed in terms of the z-variable, the short-circuit input impedance of the seventh-degree lowpass filter of Figure 3-23 is:

$$Z = \frac{6z^6 + 20z^4 + 6z^2}{\sqrt{z^2 - 1}\,(z^6 + 15z^4 + 15z^2 + 1)}$$

where $\omega_2 = 1$ is assumed.

If the coefficient of z^4 in the numerator is increased by 1% from 20 to 20.2, the element values now change to those given in row 3 of Table 3-3. In contrast to the figures in row 2 these show an error from the original values in row 1 of never more than 1.25%.

As the transformation from s to z is given in terms of the *squares* of the two variables, the relationship between s and z involves a square root. Thus, in the lowpass case for example, a polynomial such as $e(s)$ which can be written in the form $U(s^2) + sV(s^2)$, where U and V are even polynomials in s, will transform into:

$$\frac{\bar{U}(z^2) + \sqrt{z^2 - 1}\,\bar{V}(z^2)}{(z^2 - 1)^{n/2}}, \qquad \text{for } n \text{ even}$$

or

$$\frac{\sqrt{z^2 - 1}\,\bar{U}(z^2) + \bar{V}(z^2)}{\sqrt{z^2 - 1}\,(z^2 - 1)^{(n-1)/2}}, \qquad \text{for } n \text{ odd.}$$

Here $\bar{U}$ and $\bar{V}$ are the corresponding even polynomials in z. The surd factor $\sqrt{z^2 - 1}$ is never completely eliminated and remains as a multiplier to distinguish the transformed versions of the odd and even parts of the s-variable polynomial. Only in purely even polynomials such as $e(s)e(-s)$ will the surd be absent.

In carrying out a design with the z-variable, the change from s to z must be made right at the very beginning, and it is important that no s-variable representations be used, even implicitly, at any intermediate stage. The biggest nuisance in the whole z-variable design is the need to handle the surd multiplier in the proper way when manipulating polynomials. A careless disregard for this may demolish the accuracy very easily.

With a little care one can reconstruct all the steps of the s-variable design process, including the ladder development, in terms of the z-variable. There are a few places where different approaches are possible, and the choice of the right one for conserving accuracy is not always obvious. It is not possible here to discuss these subtleties, and for further details the reader is referred to the literature [5].

To illustrate the general ideas of the process, Table 3-4 gives, side-by-side for comparison, the corresponding steps in the s and z variables for the case of a third-degree Butterworth lowpass filter. Here, $\omega_2 = 1$ and so $s = (z^2 - 1)^{-1/2}$. It is hoped that this table will be completely self-explanatory.

TABLE 3-4

Formula	s-variable design	z-variable design
$K(s)$	$= s^3$	$= \dfrac{1}{(z^2-1)\sqrt{z^2-1}}$
$H(s)H(-s)$ $= 1 + K(s)K(-s)$	$= 1 - s^6$	$= 1 - \dfrac{1}{(z^2-1)^3}$
		$= \dfrac{z^6 - 3z^4 + 3z^2 - 2}{(z^2-1)^3}$
	$= (1-s^2)(1+s^2+s^4)$	$= \dfrac{(z^2-2)(z^4-z^2+1)}{(z^2-1)^3}$
$H(s)$[Hurwitz factor of $H(s)H(-s)$]	$= (1+s)(1+s+s^2)$	$= \dfrac{(\sqrt{z^2-1}+1)(z^2+\sqrt{z^2-1})}{(z^2-1)\sqrt{z^2-1})}$
	$= 1 + 2s + 2s^2 + s^3$	$= \dfrac{2z^2 - 1 + (z^2-1)\sqrt{z^2-1}}{(z^2-1)\sqrt{z^2-1}}$
$R_1 = R_2 = 1$, hence $\mathbf{T} = \begin{bmatrix} H_e + K_e & H_o + K_o \\ H_o - K_o & H_e - K_e \end{bmatrix}$	$= \begin{bmatrix} 1 + 2s^2 & 2s + 2s^3 \\ 2s & 1 + 2s^2 \end{bmatrix}$	$= \dfrac{1}{(z^2-1)\sqrt{z^2-1}} \begin{bmatrix} (z^2+1)\sqrt{z^2-1} & 2z^2 \\ 2z^2 - 2 & (z^2+1)\sqrt{z^2-1}) \end{bmatrix}$

$\frac{1}{y_{11}} = \frac{B}{D}$	$= \frac{2s + 2s^3}{1 + 2s^2}$	$= \frac{2z^2}{(z^2 + 1)\sqrt{z^2 - 1}}$
$L_1 = \lim_{s \to \infty} \frac{1}{sy_{11}}$	$= \lim_{s \to \infty} \frac{2 + 2s^2}{1 + 2s^2} = 1$	$= \lim_{z \to 1} \frac{2z^2}{z^2 + 1} = 1$
$Z_a = \frac{1}{y_{11}} - 1 \cdot s$	$= \frac{2s + 2s^3}{1 + 2s^2} - 1 \cdot s$ $= \frac{s}{1 + 2s^2}$	$= \frac{2z^2}{(z^2 + 1)\sqrt{z^2 - 1}} - \frac{1}{\sqrt{z^2 - 1}} = \frac{\sqrt{z^2 - 1}}{z^2 + 1}$
$\frac{1}{Z_a}$	$= \frac{1 + 2s^2}{s} = 2s + \frac{1}{s}$ $= C_2 s + \frac{1}{L_3 s}$	$= \frac{z^2 + 1}{\sqrt{z^2 - 1}} = \frac{2}{\sqrt{z^2 - 1}} + \sqrt{z^2 - 1}$ $= \frac{C_2}{\sqrt{z^2 - 1}} + \frac{\sqrt{z^2 - 1}}{L_3}$

$L_1 = 1$ $L_3 = 1$

$y_{11} \longrightarrow$ $C_2 = 2$

Z_a

PROBLEMS

3.1 Derive (3.14) for ρ_1 and ρ_2. (Hint: Use (3.2) and the relations following (3.3).)

3.2 Prove that a ladder can have a loss pole only at a pole of a series impedance or a shunt admittance. (Hint: Refer to Figure 3-4. At a pole of loss, $V_2 = 0$, and hence $I_n = 0$. If $Z_n < \infty$, therefore, $V_{n-1} = 0$. Thus, if $Y_{n-1} < \infty$, $I_{n-1} = 0$. Work your way back until you obtain $E = 0$, i.e., a contradiction.)

3.3 Show that if $K(s)$ is purely even, then the elements of the chain matrix satisfy $B = CR_1 R_2$, and hence $\rho_1(s) = -\rho_2(s)$ and $Z_1/R_1 = R_2/Z_2$. Such a network is called *antimetric*. What is the effect of interchanging the ports and replacing the antimetric circuit by its dual? (Hint: Show first that for dual circuits, $(z_{ij})_a = R_1 R_2 (y_{ij})_b$.)

3.4 Prove that the input impedances of the circuits of Figure 3.6(b) and (c) are $2(z_{11} - z_{12})$ and $2(z_{11} + z_{12})$, respectively.
(Hint: Connect a 1 A current source at the port and calculate the resulting voltage across the input terminals.)

3.5 Prove that the two-ports of Figure 3-7 are electrically equivalent to the symmetric lattice shown in Figure 3-5.

3.6 Prove the network equivalences of Figure 3-10.
(Hint: Use Y-Δ transformation, and make use of the fact that $k = 1$.)

3.7 Show that the transducer function can be calculated from the impedance parameters according to

$$H(s) = \frac{z_{11} R_2 + z_{22} R_1 + z_{11} z_{22} - z_{12}^2 + R_1 R_2}{2\sqrt{R_1 R_2}\, z_{12}}$$

(Hint: Use (3.3) and (3.4).)

3.8 Realize the circuit of Figure 3-14 from the impedance parameters

$$z_{11} = \frac{2s^2 + 5}{2s}, \qquad z_{22} = \frac{16s^2 + 6}{s}.$$

The loss poles are at 0, $\pm j/\sqrt{2}$, and ∞. Assume $R_1 = R_2 = 1\ \Omega$.

3.9 Prove (3.21) for the circuit of Figure 3-15(b).

3.10 Show that for the circuit of Figure 3-16(a), either the relations $Z_1 = sU/V$, $Z_3 = s\bar{U}/\bar{V}$, $U = L_1 V + (s^2/\omega_2^2 + 1)\bar{U}$, $V = C_2 s^2 \bar{U} + (s^2/\omega_2^2 + 1)\bar{V}$, *or* the relations $Z_1 = U/sV$, $Z_3 = \bar{U}/s\bar{V}$, $U = L_1 s^2 V + (s^2/\omega_2^2 + 1)\bar{U}$, $V = C_2 \bar{U} + (s^2/\omega_2^2 + 1)\bar{V}$ hold.

3.11 Derive the design relations for the circuits shown in Figure 3-17. Prove that they are in the same mathematical forms as the relations valid for the circuits of Figures 3-15(a) and 3-16(a).

3.12 Prove (3.23) from (3.22).
(Hint: Match coefficients of s^0, s^2, s^4, ..., s^{2n} and solve the resulting equations by Cramer's rule, as described in the text for the simpler case when $y = 1$.)

3.13 Illustrate the limitations of parametric filters by comparing the location of the critical frequencies of the elementary filters shown.

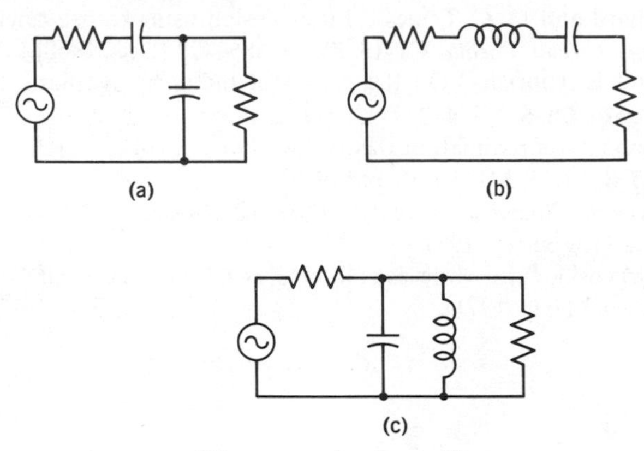

Elementary bandpass filters.

3.14 Draw simple bandpass filters with:
(a) One loss pole at zero and three loss poles at infinite frequency;
(b) Three loss poles at zero, and one loss pole at infinite frequency.
Show that neither needs to have real reflection zeros.

3.15 Carry out the design of a symmetric lattice network from $K(s) = s^5$:
(a) From (3.31), cancelling all possible common factors in Z_a and Z_b;
(b) Using the design steps 1 to 3 of Section 3.4.
Design a ladder and a "half-lattice" circuit (Figure 3-7) from K. Compare the number of elements in the three circuits.

3.16 To illustrate the effects of narrow-band zero clustering, consider the polynomial with zeros $s_1 = j0.9999$ and $s_2 = j1.0001$.
(a) Calculate the coefficients of the polynomial;
(b) Show that by storing only seven digits of each coefficient and refactoring the truncated polynomial thus obtained, the erroneous zeros $s_1 = s_2 = j1$ result.

REFERENCES*

[1] H. W. Bode, "A general theory of electric wave filters," *J. Math. Phys.*, **13**, 275–362 (1934).

* We have not included here Darlington's classical paper (*J. Math. Phys.*, **18**, 257–353 (1939)) in which much of the material discussed in this chapter was first published. The conciseness of Darlington's paper makes it difficult reading for the average reader of this book.

[2] H. W. Bode, *Network Analysis and Feedback Amplifier Design*, D. Van Nostrand, New York, 1945, Chapter X.
[3] N. Balabanian, *Network Synthesis*, Prentice-Hall, Englewood Cliffs, New Jersey, 1958.
[4] E. A. Guillemin, *Synthesis of Passive Networks*, John Wiley, New York, 1957, Section 7.4.
[5] H. J. Orchard and G. C. Temes, " Filter design using transformed variables," *IEEE Trans. Circuit Theory*, **CT-15**, No. 4, 385–408 (Dec., 1968).
[6] R. Saal and E. Ulbrich, " On the design of filters by synthesis," *IRE Trans. Circuit Theory*, **CT-5**, No. 4, 284–327 (Dec., 1958).
[7] H. Watanabe, "Approximation theory for filter-networks," *IRE Trans. Circuit Theory*, **CT-8**, No. 3, 341–356 (Sept., 1961).
[8] J. H. Wilkinson, *Rounding Errors in Algebraic Processes*, Prentice-Hall, Englewood Cliffs, New Jersey, 1963.
[9] J. K. Skwirzynski, " On synthesis of filters," *IEEE Trans. Circuit Theory*, **CT-18**, No. 1, 152–163 (Jan., 1971).

4

Crystal and Ceramic Filters

George Szentirmai
Cornell University
Ithaca, New York

Piezoelectric crystals are used in electric filters for their extraordinarily stable resonant frequencies and very small dissipative losses. These properties become essential if and when the bandwidth or the transition range from the passband to the stopband of a filter becomes very narrow.

On the other hand, the equivalent electrical circuit of a piezoelectric resonator contains at least three elements with certain relationships between them. The difficulty of designing filters with such elements is mainly that of accommodating this restricted combination of elements in the generally unrestricted filter structures.

Among filter designers, the use of the older image-parameter design technique persisted considerably longer for the design of crystal filters than for the design of purely electrical filters, precisely because one has a much more direct control over the structure through this theory. Furthermore, the restrictions caused by the element value relationships of the crystal equivalent circuit can be converted into restrictions on the achievable performance more easily.

In the last decade however, the insertion-loss technique made considerable inroads, and today we have a sizeable arsenal of techniques and methods for crystal filter design through this modern method.

This chapter will henceforth consider only the insertion-loss design method; about the older and still used image-parameter method the reader should consult the literature [1–5].

4.1 PIEZOELECTRIC CRYSTALS AND THEIR EQUIVALENT CIRCUITS

Naturally we are unable to consider the theory of piezoelectric (and ceramic) resonators in detail here. Suffice it to say that these are three-dimensional mechanically oscillating bodies, and as such they have usually very many modes of oscillations and are capable of oscillating at the fundamental and many of the higher harmonic frequencies of each of these modes. These oscillations are excited through the piezoelectric properties of the material and by the arrangement and shape of the electrodes and the crystal itself, one has the capability of selecting certain modes and harmonics of oscillations over others.

At the electrical terminals, the observable equivalent electric circuit will contain an infinite number of (lossy) series resonant circuits, each representing a particular harmonic of a particular mode, in parallel (Figure 4-1). Note

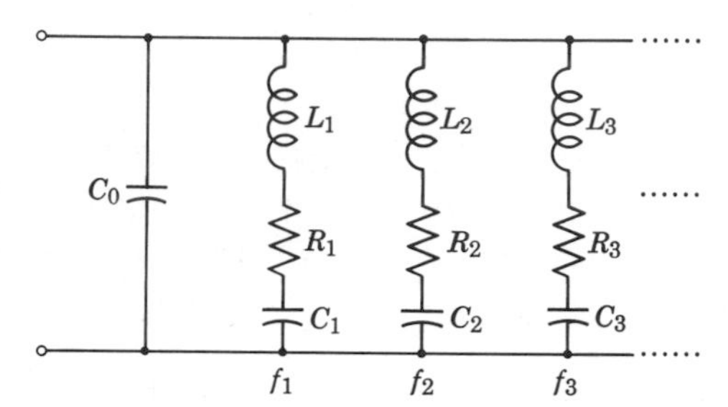

FIGURE 4-1 Electrical equivalent circuit of piezoelectric resonator.

that the resonant frequencies are not necessarily harmonically related even if they belong to the same mode. The capacitance C_0 represents the static capacitance of the electrode arrangement.

If we wish to operate the crystal in the close neighborhood of one, say f_1, of its resonant frequencies (where f_1 need *not* be the lowest possible resonant frequency of the crystal), then we may neglect all other branches and use the simplified equivalent circuit of Figure 4-2. We must, however, keep in mind that this is valid only in a narrow frequency band around f_1; just how narrow the band is depends on a great number of factors. This fact causes little difficulty in the design and construction of very narrow bandpass filters, but it complicates the situation in the design of other filter types. We will consider this problem again later in this chapter.

Considering now the simplified equivalent circuit of Figure 4-2, the series-resonant circuit has an extremely high Q ranging anywhere from about 10,000 to over a million for natural quartz. For ceramic resonators the value

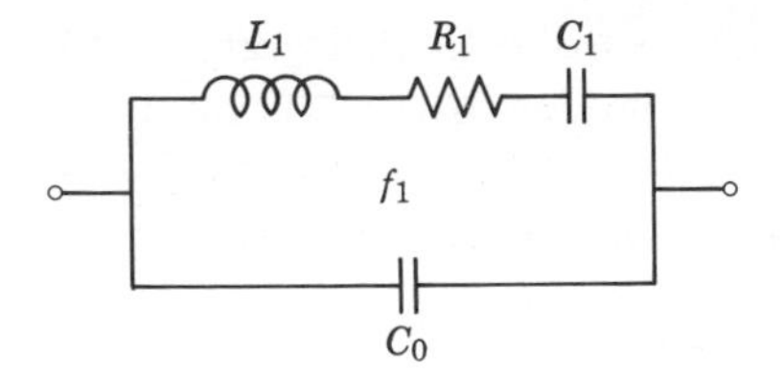

FIGURE 4-2 Simplified electrical equivalent circuit.

of Q is of the order of a few thousand. Consequently, under normal circumstances, the series resistance R_1 can be neglected. The ratio of capacitors

$$r = \frac{C_0}{C_1} \tag{4.1}$$

is dependent on the electromechanical coupling and the order of resonance that is utilized. For fundamental resonance of quartz crystals, the minimum value obtainable for this ratio is about 125. For ceramic material, it is somewhat lower, and for some recently discovered materials it may go as low as 20 to 30. Naturally, it can always be increased by connecting an additional shunt capacitor across the crystal terminals.

Henceforth, we will assume that the piezoelectric crystals are represented by their equivalent circuits of the form of Figure 4-2 with $R_1 = 0$ and a capacitance ratio

$$r = \frac{C_0}{C_1} \geq r_{\min}, \tag{4.2}$$

where $r_{\min}$ is a parameter we are given at the start of our design. We note also that in production the individual electrical parameters of the equivalent circuit cannot be controlled to better than a few percent tolerance, but the resonant frequency f_1 can. It is therefore customary to specify a crystal by the parameter triplet (f_1, L_1, and C_0) rather than (L_1, C_1, and C_0).

Finally, we note that the general impedance level of the equivalent circuit is rather high for the most commonly used quartz crystal resonators. Typically, the order of magnitude of L_1 is several henries. Hence crystals must be used in high impedance circuits, or alternatively, transformers are employed to bring the impedance level down to tolerable values.

4.2 VERY NARROW BAND LADDER FILTERS

The simplest way to adapt the equivalent circuit of a crystal for bandpass filters is to find a filter configuration already containing branches of the proper form.

As ladders are the most often used structures due to their very good sensitivity properties, one would first look for a ladder structure containing either pure capacitive branches, or branches realizable by crystals or crystal-capacitor combinations. Such a structure is shown in Figure 4-3 and can be

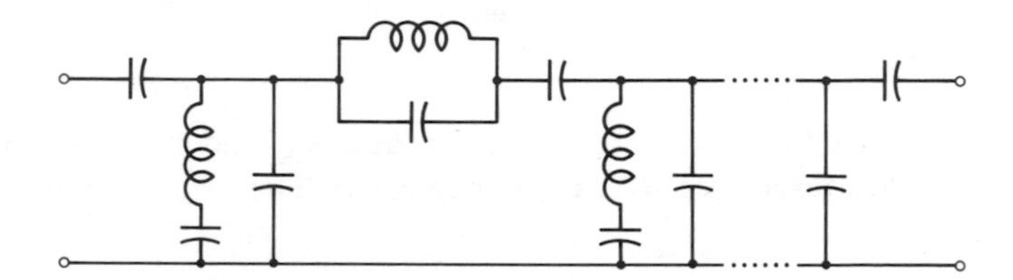

FIGURE 4-3 Crystal-capacitor realizable ladder filter.

obtained by the insertion loss design techniques described in Chapter 3. This filter must necessarily be of the symmetrical-parametric type with a single transmission zero at both zero and infinite frequencies. The branches containing inductive elements are either already in the right form or can be converted into such a form by elementary operations. The two operations used almost exclusively for such conversions are the (capacitive) impedance transformation (Figure 4-4) and the equivalence of Figure 4-5. The method is best illustrated by some simple examples.

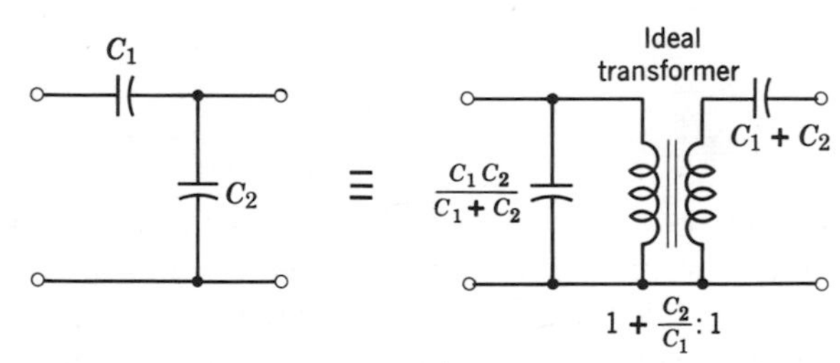

FIGURE 4-4 Impedance (Norton) transformation.

Example 4-1. Let us design a bandpass filter with passband from 800 kHz to 800.4 kHz. The passband performance is to be of the equal-ripple type with .25 dB ripple. The minimum required stopband loss is 50 dB below 799.75 kHz and above 800.65 kHz.

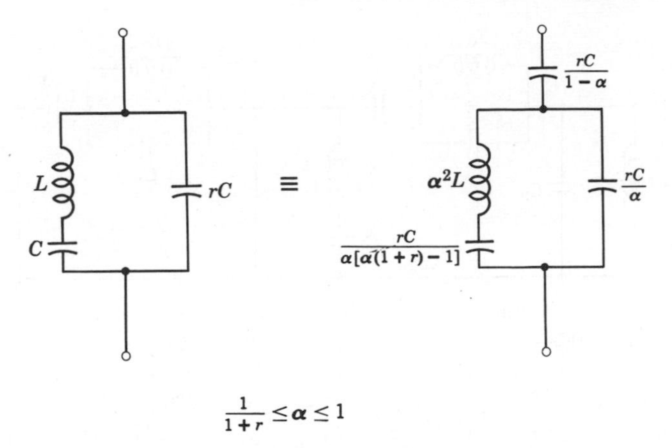

$$\frac{1}{1+r} \leq \alpha \leq 1$$

FIGURE 4-5 Two-terminal equivalence.

Solution: The specified response can be achieved with two finite transmission zeros below the band and two above the band at

$$f_{10} = 799.06 \text{ kHz}$$
$$f_{20} = 799.71 \text{ kHz}$$

and at

$$f_{30} = 800.69 \text{ kHz}$$
$$f_{40} = 801.34 \text{ kHz}$$

respectively. The sequence of transmission zeros in the realization has some minor influence on the element value distribution; the best one of the four possible alternating* sequences is in this case $f_{20}, f_{40}, f_{10}, f_{30}$, leading to the configuration and element values shown in Figure 4-6.

The first step here is to check the capacitance ratios. These are shown on the figure and are upper limits in this case, since some of the capacitances will be needed for impedance transformations, and hence the final r values will be less than those indicated. Incidentally, the capacitance ratio for the series branches can be calculated easily if they are converted into the equivalent parallel Foster form. A quick check of the equivalence of Figure 4-5 shows the correctness of the formula used in Figure 4-6.

The numerical values of the capacitance ratios indicate that we will have no trouble in realizing this structure using crystals and capacitors only.

* That is, a transmission zero in the lower stopband is followed by one in the upper stopband and vice versa. This is a preferable arrangement since it leads to better (larger) capacitance ratios and smaller initial element value spreads.

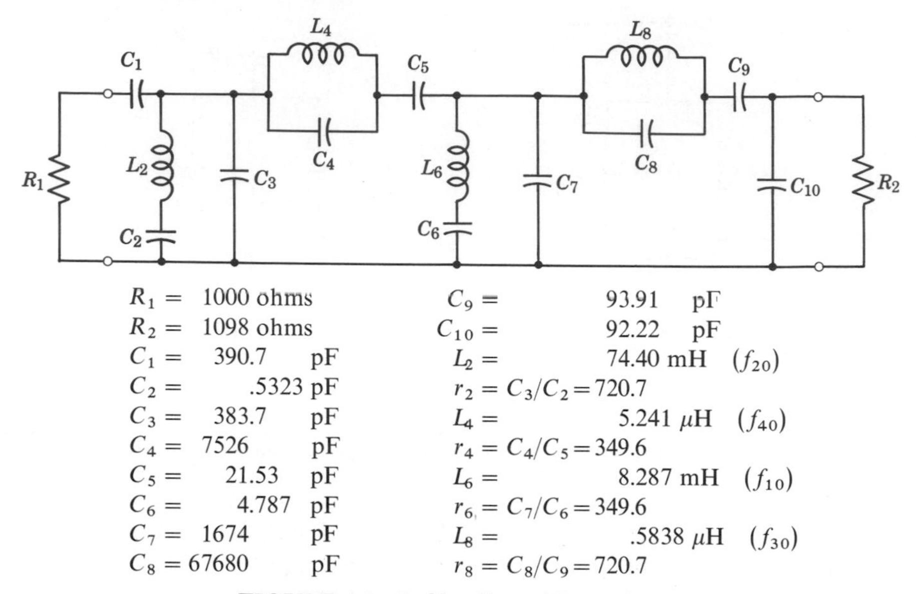

FIGURE 4-6 Ladder filter of Example 4-1.

Before we do that, however, we must modify the structure to accommodate the narrow range of available equivalent crystal element values. The sequence of operations and the final results are far from unique; one possible solution is shown in Figure 4-7.

First C_5 is split into C_5' and C_5'' where $C_5'' = C_4/200$. Next the L_4 C_4 and C_5'' branch is converted into a crystal equivalent with $r = 200$. Now L_4' turns out to be 209.6 mH which is considerably higher than L_2. Therefore parts of C_3 and C_5' are used to reduce the impedance level of all components to the right of them, until $L_4'' = L_2$. But now L_6' becomes 2.941 mH which is too low. Hence C_5^{IV} and part of C_7' are used to transform up the impedance level of the L_6' C_6' branch, and everything to the right of it until L_6'' becomes equal to L_2.

At this stage C_9' is split more or less arbitrarily into three parts. One part is used to convert with L_8' C_8' into the required parallel Foster form. The resulting L_8'' value will turn out to be much too large. Now part of C_7'' and C_9'' are used to reduce the impedance level of this branch until L_8''' also becomes equal to L_2. Finally C_9''' and C_{10}' are used to restore the output impedance level to 1000 ohms.

The last two capacitors could be eliminated if C_9' is split into only two parts, but in such a unique manner that after the parallel Foster form conversion the impedance transformation that makes L_8''' equal to L_2 will simultaneously achieve the right output impedance level.

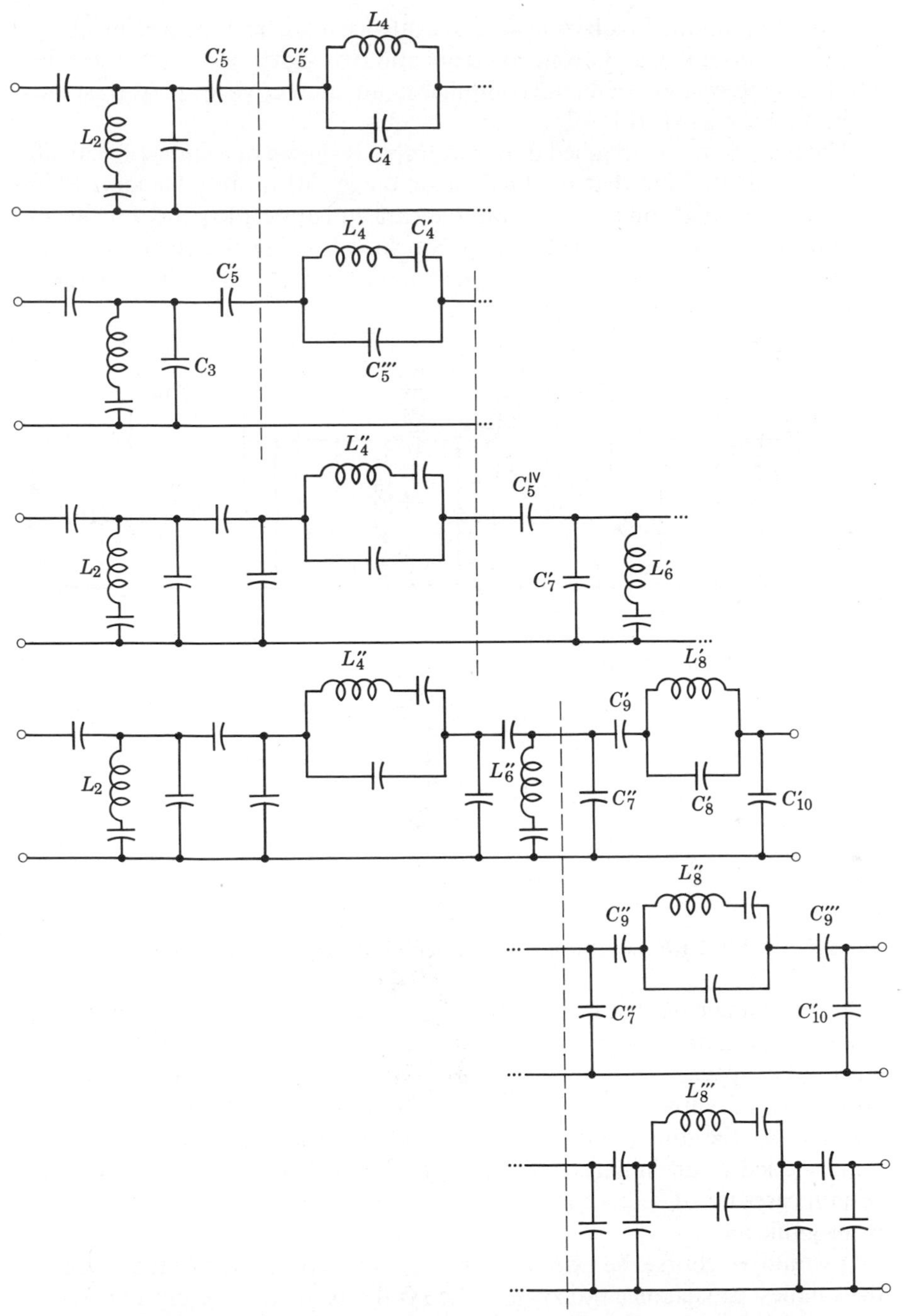

FIGURE 4-7 Modification of the ladder of Figure 4-6 to accept crystals.

The computation involved in such a sequence of operations is elementary, but time consuming and prone to errors and oversights. However, it can be implemented easily on a digital computer in an interactive form. At least one such program exists [6].

The final network obtained in our example is shown in Figure 4-8 with all element values. Note that the final capacitance ratios are all smaller than those of the initial configuration, but they are still quite adequate. Finally an impedance scaling with a factor of 10 or so will bring the element values within the range realizable by crystals and capacitors only. The computed performance is shown in Figure 4-9.

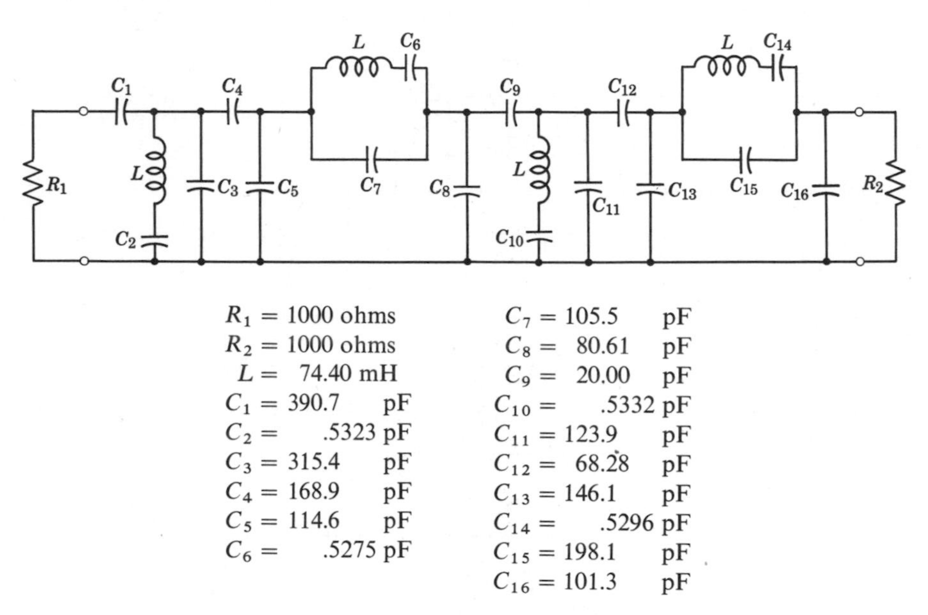

FIGURE 4-8 Final configuration of filter of Example 4-1.

A few comments are in order at this stage. As mentioned above, the sequence of transmission zeros affects the element values, but the worst (smallest) capacitance ratio does not vary much. The selected sequence above had an initial $r = 349.6$ as the smallest ratio, and as such it was the best among the four possible sequences in this case. However, the worst sequence led to an initial ratio of $r = 306.2$, which is not much worse. In certain cases this difference may be important, although in the present one it is insignificant.

It would, of course, be extremely useful if one could predict the value of the smallest capacitance ratio *before* the synthesis. At the present time there is no rigorous way of doing this; however, let us for the moment note the

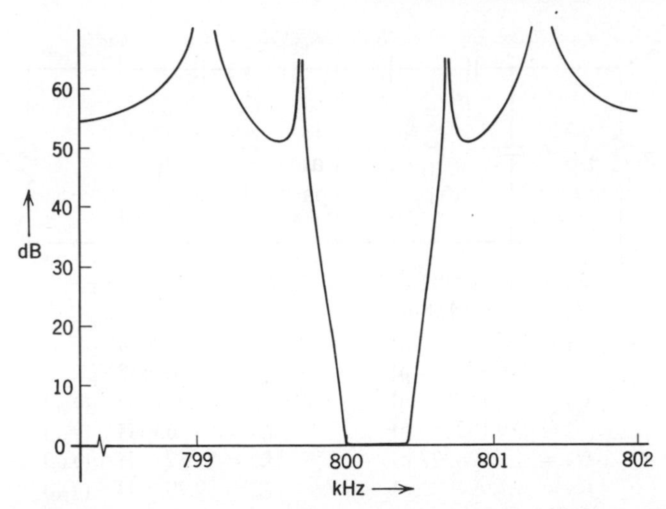

FIGURE 4-9 Computed loss characteristic of the filter of Figure 4-8.

following. If f_0 denotes the midband frequency and f_{max} is the transmission zero farthest away from the midband, then for our example we obtain:

$$\frac{1}{2}\frac{f_0}{f_0 - f_{max}} = 351.0. \tag{4.3}$$

This quantity is very close to the smallest capacitance ratio of the optimum sequence ($r = 349.6$) given above. Note, furthermore, that we did not have to go through several complete syntheses to calculate it.

In the example considered above, the same number of transmission zeros existed in the two stopbands. Let us now see another example, the other extreme, where all the zeros are in one stopband.

Example 4-2. Assume that the passband is identical to our previous example and that we need a 50 dB minimum discrimination in the lower stopband from zero up to 799.93 kHz. There is, however, no requirement to be met in the upper stopband.

Solution: These specifications can be met by four transmission zeros at the frequencies:

$$f_{10} = 794.795 \text{ kHz}$$
$$f_{20} = 799.498 \text{ kHz}$$
$$f_{30} = 799.849 \text{ kHz}$$
$$f_{40} = 799.921 \text{ kHz}.$$

A filter realizing these characteristics was synthesized and is shown in Figure 4-10. The sequence of realization of the zeros was $f_{20}, f_{30}, f_{40}, f_{10}$.

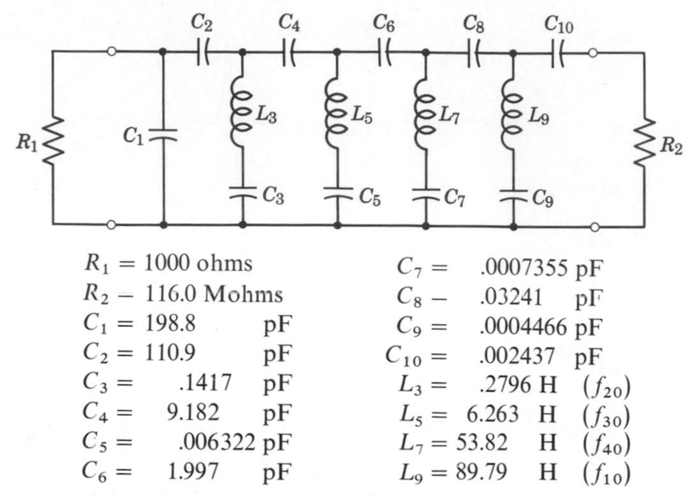

$R_1 = 1000$ ohms	$C_7 = .0007355$ pF
$R_2 = 116.0$ Mohms	$C_8 = .03241$ pF
$C_1 = 198.8$ pF	$C_9 = .0004466$ pF
$C_2 = 110.9$ pF	$C_{10} = .002437$ pF
$C_3 = .1417$ pF	$L_3 = .2796$ H (f_{20})
$C_4 = 9.182$ pF	$L_5 = 6.263$ H (f_{30})
$C_5 = .006322$ pF	$L_7 = 53.82$ H (f_{40})
$C_6 = 1.997$ pF	$L_9 = 89.79$ H (f_{10})

FIGURE 4-10 Ladder filter of Example 4-2.

The element values should not scare us at this stage; they simply indicate that several stages of impedance transformations will have to be performed. But this is needed anyway to distribute the capacitance C_1 across the shunt resonant branches, if we wish to realize the filter using crystals.

The calculation is very simple and if we attempt to make all inductors equal, the structure obtained is that of Figure 4-11.

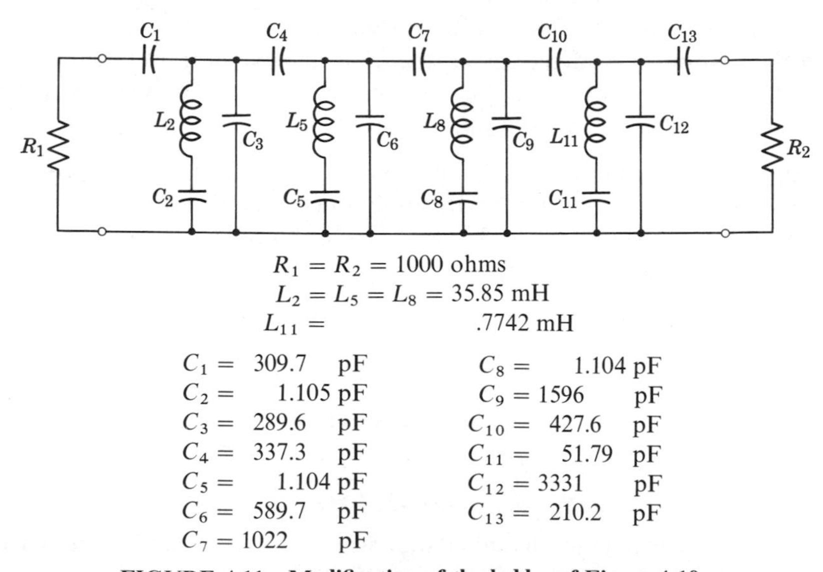

$R_1 = R_2 = 1000$ ohms
$L_2 = L_5 = L_8 = 35.85$ mH
$L_{11} = .7742$ mH

$C_1 = 309.7$ pF	$C_8 = 1.104$ pF
$C_2 = 1.105$ pF	$C_9 = 1596$ pF
$C_3 = 289.6$ pF	$C_{10} = 427.6$ pF
$C_4 = 337.3$ pF	$C_{11} = 51.79$ pF
$C_5 = 1.104$ pF	$C_{12} = 3331$ pF
$C_6 = 589.7$ pF	$C_{13} = 210.2$ pF
$C_7 = 1022$ pF	

FIGURE 4-11 Modification of the ladder of Figure 4-10.

We observe that the last inductor could not be made equal to the others if we wish to have $R_2 = R_1$. But even if we allow $R_2 > R_1$, the last inductor still could not be made equal to the others for $C_9 > 0$. The second, and more important, observation is that the capacitance ratio of the last shunt branch is only 64.3, which is inadequate. It turns out that this ratio cannot be increased past about 71, and then only at the expense of reducing another capacitance ratio to zero. Naturally, one would at this stage investigate whether another sequence of zeros could do any better. There are twelve distinct possible zero sequences, and an exhaustive study would be clearly expensive and time consuming. To simplify matters, one would like to be able to say something about maximum available capacitance ratios before actually performing any impedance transformations, just as we have done in the first example. This can, indeed be done on the basis of Figure 4-12. Investigating a particular resonant shunt branch L-C with no direct shunt capacitance across it, assume that there is a capacitor C_L to ground somewhere to its left. If we were to shift this capacitor all the way to the shunt

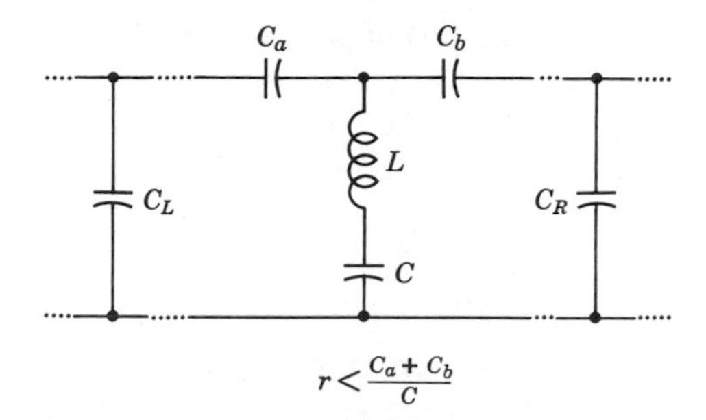

FIGURE 4-12 Maximum available capacitance ratio of series resonant arm in shunt.

branch by one or more impedance transformations, the resulting capacitor across the branch L-C could not be greater than C_a. Hencc a useful upper limit on the achievable capacitance ratio can be written as:

$$r < \frac{C_a}{C}. \tag{4.4}$$

If there is another capacitor to ground on the right of L-C as well, the maximum total shunt capacitance that can be made to shunt this branch must be less than $C_a + C_b$, hence again:

$$r < \frac{C_a + C_b}{C}. \tag{4.5}$$

Similar relationships can be worked out for the dual structure as shown on Figure 4-13.

On this basis one can investigate all twelve possible zero sequences, and the result is that the lowest capacitance ratio must be less than r_{max} where r_{max} ranges from 71.6 all the way to 75.2. In other words, the zero sequence selected has again very little influence on the realizability of the structure.

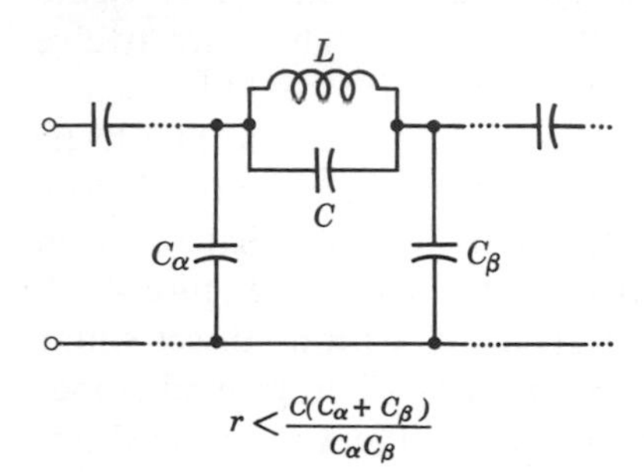

FIGURE 4-13 Maximum available capacitance ratio of parallel resonant arm in series.

This result was obtained at the cost of performing twelve different syntheses. In this process, however, we notice that the offending resonant branch is invariably the one resonating at the transmission zero farthest away from the passband. Again calculating its relative inverse distance, we obtain:

$$\frac{1}{2}\frac{f_0}{f_0 - f_{max}} = 74.0. \tag{4.6}$$

This quantity therefore seems to be an excellent empirical indicator of the maximum of the lowest capacitance ratio for a particular filter. It can be calculated fairly early in the design, and therefore it can prevent a needless and fruitless search for a realizable structure.

The above quantity is apparently independent of the bandwidth, but the bandwidth is implicitly still involved, since clearly $f_0 - f_{max}$ must be greater than half the absolute bandwidth.

The two examples presented here represent two extreme cases of the generally available configurations. The general case can, however, clearly be treated along the lines explained and needs nothing more than a little experience and thought.

In summary, we have demonstrated through examples, how a symmetrical parametric filter in ladder form can be modified until it is realizable by crystals and capacitors alone (but still in a ladder form). In the process we have found a simple empirical rule to help us decide beforehand whether the process has any hope of success. The method is clearly restricted to extremely narrow bandwidths (.1% and less).

4.3 NARROW BAND FILTERS OF MIXED LATTICE-LADDER FORM

The reader familiar with image-parameter crystal filters would immediately recommend the use of lattice sections to solve the problem we encountered in the previous section. This indeed is advisable, since the lattice realization normally results in higher capacitance ratios than the equivalent ladder. The reason for this is that the lattice branches are always obtained by a combination of two or more ladder branches, and this process invariably leads to a relative increase in shunt capacitance values. Now we have the option of realizing the whole filter (if it is symmetrical) by a single lattice, or we can realize parts of it by simpler lattices and connect these (possibly together with ladder parts) in cascade.

For practical reasons the first option, that of a single complicated lattice, is not feasible, because it leads to an extremely sensitive network. For instance, a 60 dB stopband suppression would require a .2% or better balance between the two different branches of the lattice, clearly a difficult proposition.*

Unless the filter is a small one, the correct method is, therefore, to replace selected parts of the ladder by equivalent lattice sections. In order that the lattice branches also be crystal-capacitor realizable, this is done through the equivalence of Figure 4-14. (In the equivalence, we have neglected a possible interchange of the two lattice impedances which is of no practical consequence).

To illustrate the use of this equivalence, let us take the last three branches of the ladder of Figure 4-11 and draw it as shown in Figure 4-15, where

$$C'_{13} = \frac{C_{10}C_{13}}{C_{10}' - C_{13}}. \tag{4.7}$$

Replacing now the symmetrical T section inside the dotted rectangle by its equivalent lattice gives the configuration and element values shown in the same figure. Clearly the capacitance ratio has improved considerably, and the overall structure is now realizable by crystals and capacitors only.

The next step is to inquire if perhaps more complicated ladder structures can be replaced by equivalent lattices. The answer to this is indicated by the equivalence [7] of Figure 4-16. Several methods have been presented for the calculation of the equivalent lattice parameters, but they are all quite involved and the reader should consult the literature [8] for the detailed

* Lattice structures provide attenuation by the balanced-bridge principle (as opposed to ladders, where the loss is due to very large series or very small shunt impedances); hence high loss requires the adjustment and maintenance of precise balance.

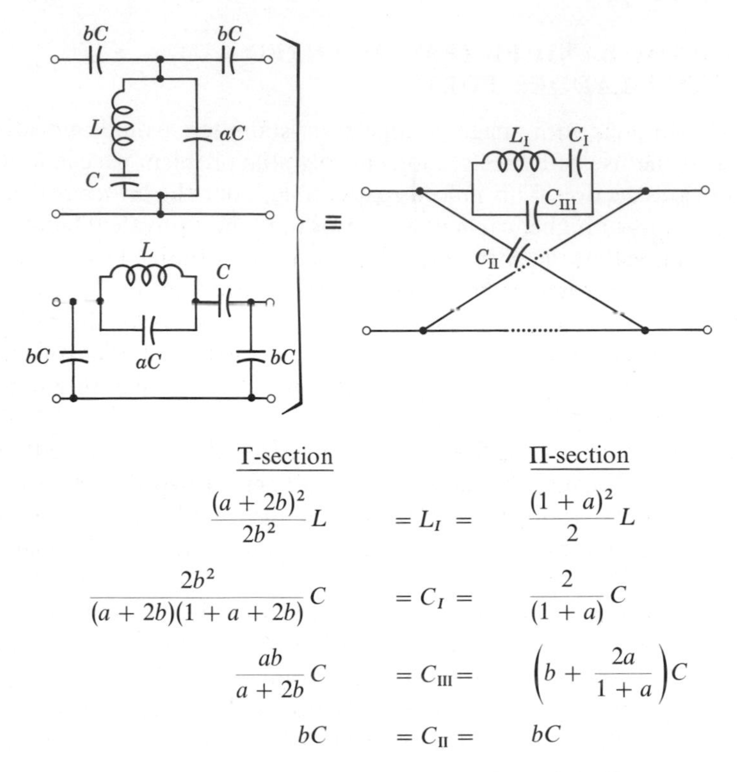

T-section		Π-section
$\frac{(a+2b)^2}{2b^2} L$	$= L_I =$	$\frac{(1+a)^2}{2} L$
$\frac{2b^2}{(a+2b)(1+a+2b)} C$	$= C_I =$	$\frac{2}{(1+a)} C$
$\frac{ab}{a+2b} C$	$= C_{III} =$	$\left(b + \frac{2a}{1+a}\right) C$
bC	$= C_{II} =$	bC

FIGURE 4-14 Lattice equivalent of single-peak ladder sections.

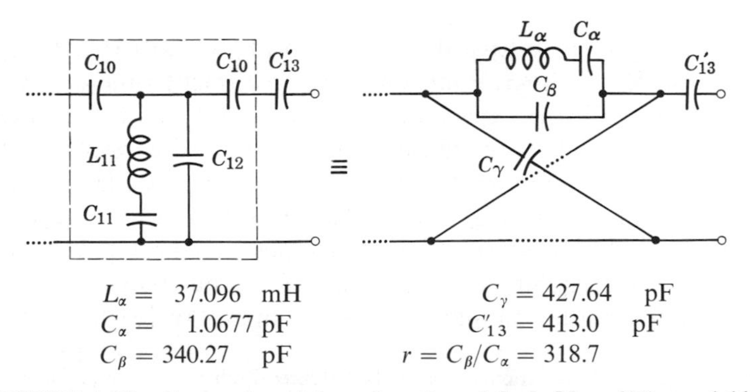

$L_\alpha = 37.096$ mH $C_\gamma = 427.64$ pF

$C_\alpha = 1.0677$ pF $C'_{13} = 413.0$ pF

$C_\beta = 340.27$ pF $r = C_\beta / C_\alpha = 318.7$

FIGURE 4-15 Conversion of the end section of the ladder of Figure 4-11 into a lattice.

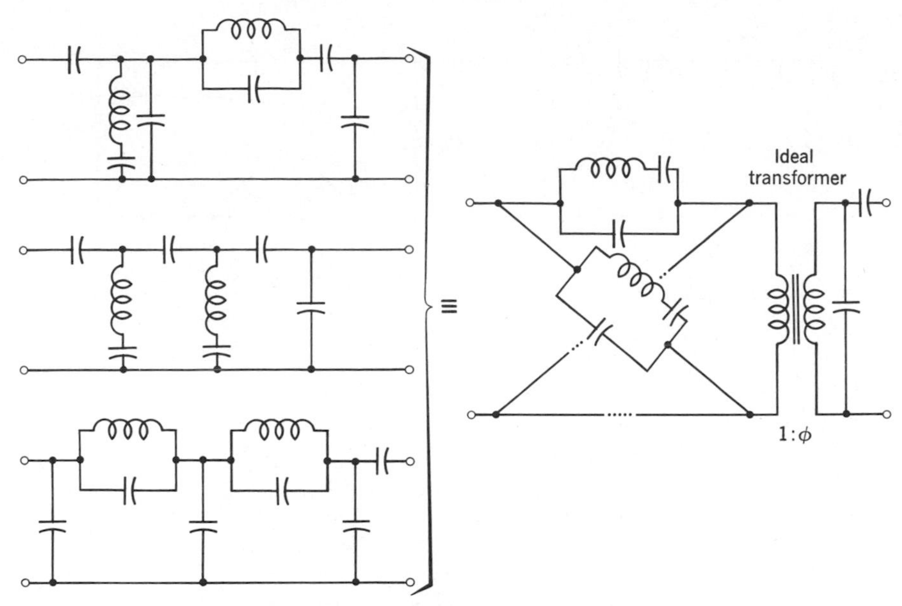

FIGURE 4-16 Lattice equivalent of double-peak ladder sections.

expressions.* Other methods [9, 10] are of the step-by-step type, greatly facilitating understanding, and they may be preferable for hand calculations. It has also been shown [7] that, in general, we cannot go any further. That is to say, a more complicated ladder cannot, in general, be replaced by a symmetrical crystal-capacitor lattice (and possibly an ideal transformer).

As an example, let us consider the filter of Figure 4-6 and try to realize it as two lattices in cascade with possibly some capacitors in between.

Figure 4-17 shows the structure obtained after all ideal transformers have been eliminated and all possible capacitances have been absorbed into the lattices. A quick look at the element values shows that the worst capacitance ratio is increased considerably. The crystal inductances are also improved (increased), although they cannot be made all equal now and neither can the terminations, without a transformer. One can naturally ask whether another peak sequence may, perhaps, result in even better element value distribution, and while this is a possibility, it necessitates the essentially complete design of two other filters. After performing these designs, we do indeed find one that is somewhat better as far as its element value distribution is concerned. The configuration and the element values are shown in Figure 4-18. At

* Bingham's expressions were programmed by this author, and the FORTRAN IV program listing is available on request.

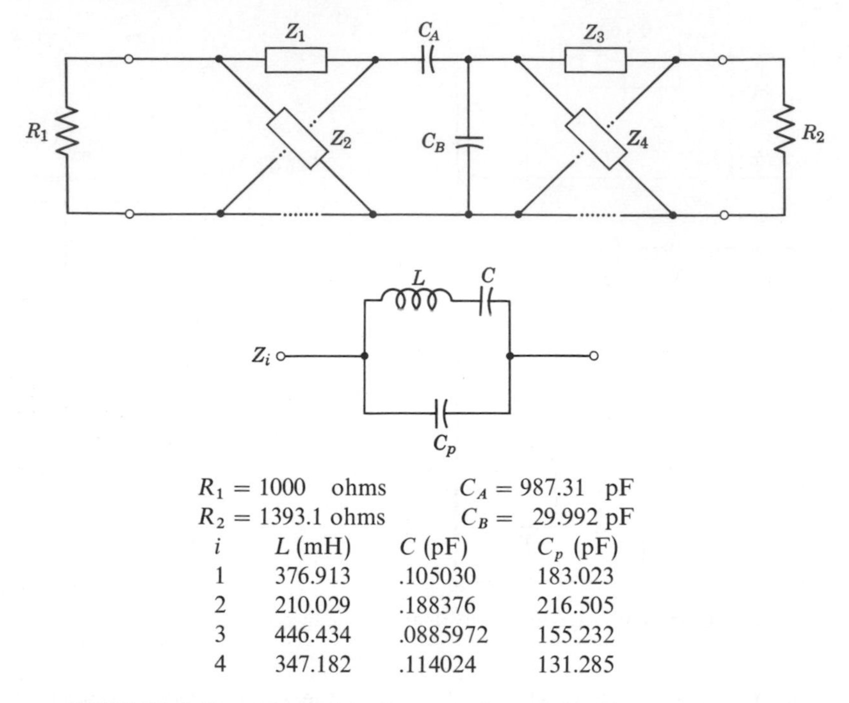

	$R_1 = 1000$ ohms		$C_A = 987.31$ pF
	$R_2 = 1393.1$ ohms		$C_B = 29.992$ pF
i	L (mH)	C (pF)	C_p (pF)
1	376.913	.105030	183.023
2	210.029	.188376	216.505
3	446.434	.0885972	155.232
4	347.182	.114024	131.285

FIGURE 4-17 Cascaded-lattice equivalent of the filter of Figure 4-6.

present, we have no *a priori* way of deciding which one of the peak sequences will lead to the "best" design.

Let us consider now the question of practical realization of structures of mixed ladder and lattice form. The original ladders are naturally common-ground structures, but the structures of Figures 4-17 and 4-18 are "almost" balanced. (They can be made balanced by splitting the series capacitors and putting half of them into the lower lead.) The lattice sections themselves are usually realized by the well-known "half-lattice" equivalent of Figure 4-19 since that contains half as many crystals. The three-winding ideal transformer can be easily realized by a tuned real transformer. As many of these transformers are needed as the number of lattice sections present in the network. However, under certain circumstances, some (but not all) can be eliminated by the use of the equivalence of Figure 4-20. The ideal transformer on the right can then be eliminated by scaling. This equivalence can only be used when there is a series capacitance on one side of the lattice and a shunt capacitance on the other that is larger than or equal to the series one. Unfortunately, neither of the structures of Figures 4-17 and 4-18 meets this requirement. A restricted-capacitive-divider equivalence was developed by

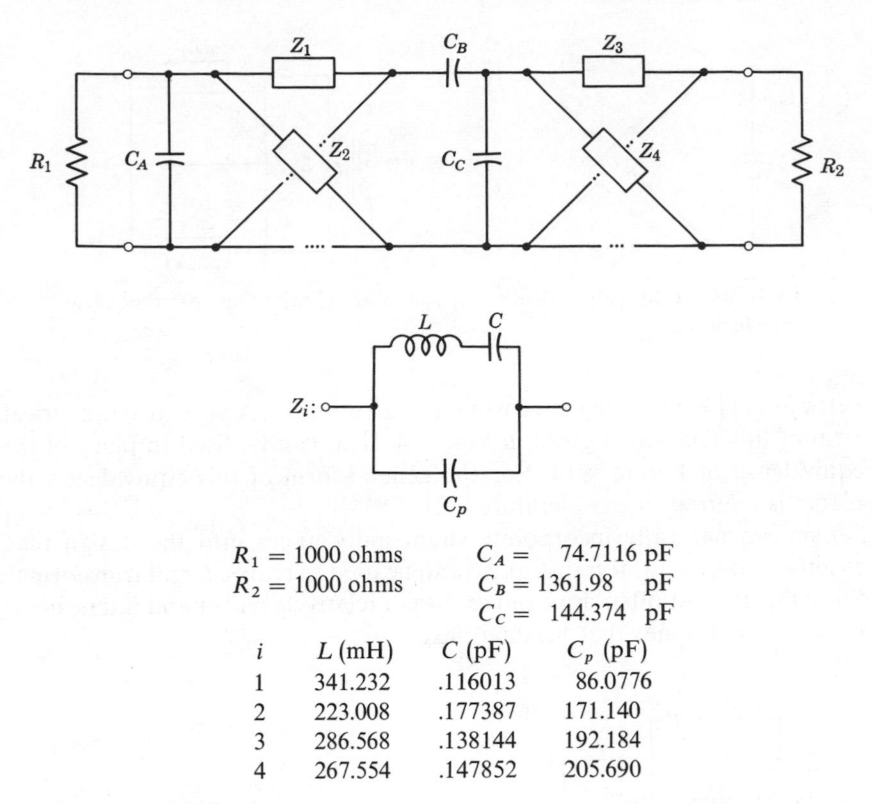

$R_1 = 1000$ ohms $C_A = 74.7116$ pF
$R_2 = 1000$ ohms $C_B = 1361.98$ pF
$C_C = 144.374$ pF

i	L (mH)	C (pF)	C_p (pF)
1	341.232	.116013	86.0776
2	223.008	.177387	171.140
3	286.568	.138144	192.184
4	267.554	.147852	205.690

FIGURE 4-18 Another cascaded-lattice equivalent of the filter of Figure 4-6.

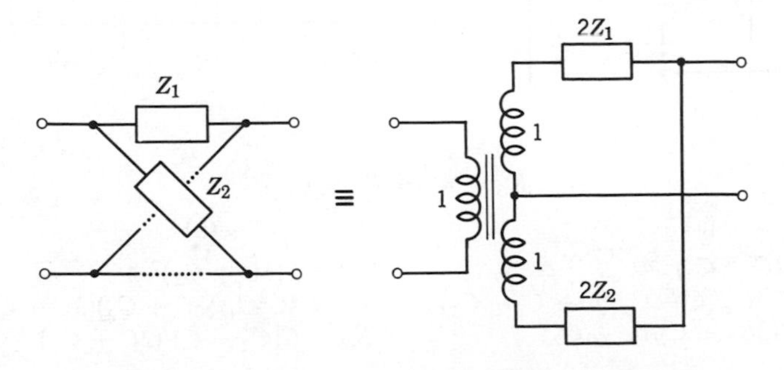

FIGURE 4-19 "Half-lattice" equivalence.

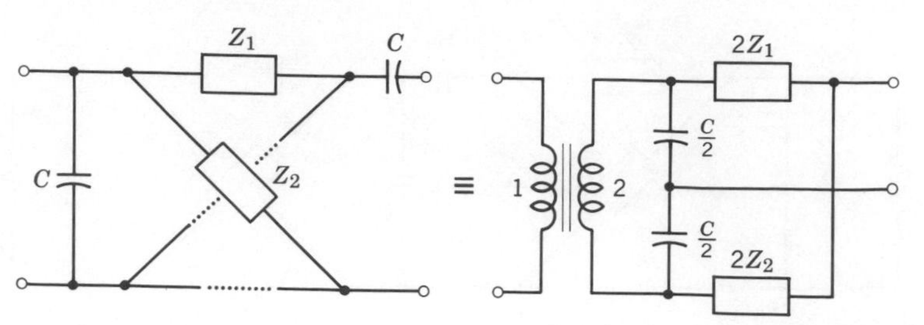

FIGURE 4-20 Equivalence for possible elimination of the ideal transformer.

Fettweis [11] for the single transmission zero lattice. A special symmetrical form of this relation is given in Figure 4-21; it can be used in place of the equivalence of Figure 4-14. For the general form of this equivalence, the reader is referred to the literature [12].

One can naturally incorporate shunt inductances into the design that, together with the ideal transformer, can be used to realize a real transformer. Since this is most often done in the case of relatively wide band filters, it will be considered under that heading next.

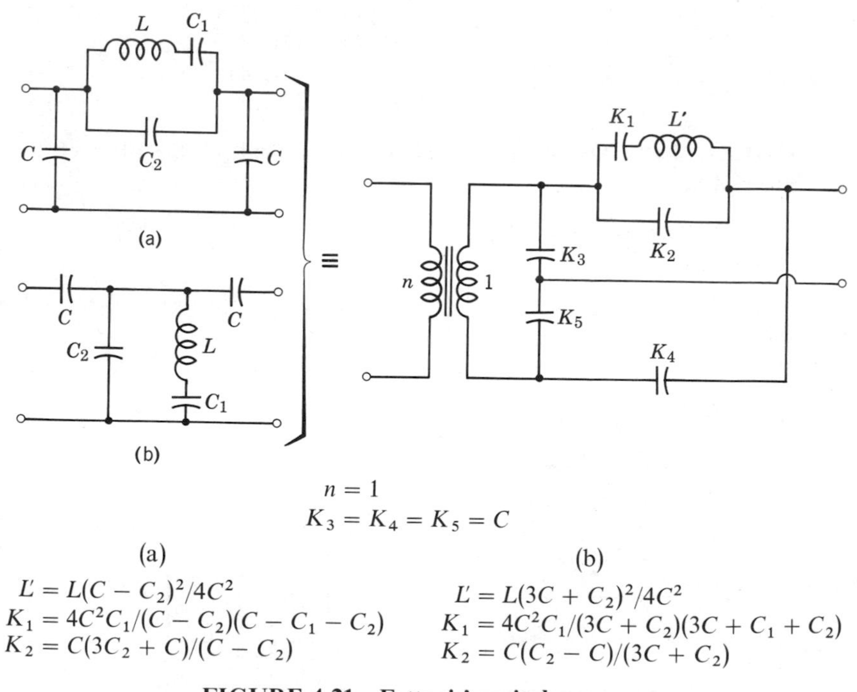

FIGURE 4-21 Fettweis' equivalence.

4.4 WIDE BAND FILTERS

For bandwidths ranging from a fraction of a percent to above 5%, pure ladder filters will not result in structures realizable by crystals mainly because the capacitance ratios will turn out to be too small. Converting parts of the ladder to lattice form helps in increasing these ratios, but a further, rather spectacular, increase can be obtained if one includes (shunt or series) inductors in between lattice sections. It is rather difficult to give a simple physical explanation for this phenomenon, hence we will simply observe it through an example.

First, we observe that while theoretically both series or shunt inductors are acceptable for this purpose, shunt inductors are much preferable for several practical reasons. For one, shunt inductance values will be much more reasonable than those of series inductors in view of the high impedance level needed to accommodate the crystals. For another, shunt inductors can be used as transformers both for the purpose of changing impedance levels and for realizing the "half-lattice" equivalent form of Figure 4-19. Finally, the parasitic parallel capacitances of these inductors can usually be absorbed into existing capacitors if they are in shunt branches. Hence we will assume the use of shunt inductors exclusively.

If, as usual, we first wish to synthesize a ladder network and then convert parts of it into lattice form, we must next consider the appropriate ladder structure. As noted earlier, not more than two finite attenuation peaks can be realized in a single lattice, hence the structure would have to be of the following form. We will start out with a shunt inductor, and follow it by sections fully removing two (usually finite) transmission zeros. This is followed by another shunt inductor, two more full removals of loss peaks and so on, until all peaks are realized. The network is then completed by a last shunt inductor. If we have an odd number of finite loss peaks, then one of the lattices will naturally realize a single peak only. This last one should preferably be as close to the passband as possible.

Consequently, the realization of k finite loss peaks requires n_L lattices, where

$$n_L = \begin{cases} \dfrac{k}{2} & \text{if } k \text{ is even} \\[2ex] \dfrac{k+1}{2} & \text{if } k \text{ is odd} \end{cases} \tag{4-8}$$

embedded between $n_L + 1$ shunt inductors.* Hence the multiplicity of the

* We assume here that all finite loss peaks are to be realized by crystal-capacitor lattices. This is of course not necessary, and the reader will be able to generalize the structure to include LC sections as well, if desired.

transmission zero at zero frequency will be:

$$n_0 = 2n_L - 1 \tag{4.9}$$

while that at infinity will be always

$$n_\infty = 1. \tag{4.10}$$

At this stage, we are again ready for an illustrative example.

Example 4-3. Design a filter which is to have a passband from 200 to 210 kHz (5% bandwidth), with a .25 dB passband loss ripple.

Solution: Preliminary investigations indicate the need for four finite loss peaks, hence we will need two lattices, three shunt inductors, and $n_0 = 5$, $n_\infty = 1$. Next comes the determination of the location of the loss peaks followed by the ladder synthesis. Assuming that we desire the loss peaks to be at:

$$f_{10} = 189.961 \text{ kHz}$$
$$f_{20} = 193.639 \text{ kHz}$$
$$f_{30} = 216.898 \text{ kHz}$$
$$f_{40} = 221.098 \text{ kHz}$$

the appropriate ladder synthesized is that shown in Figure 4-22. Converting the blocks between shunt inductors into lattices, the resulting structure with element values is shown in Figure 4-23. Notice that L_B can be used to

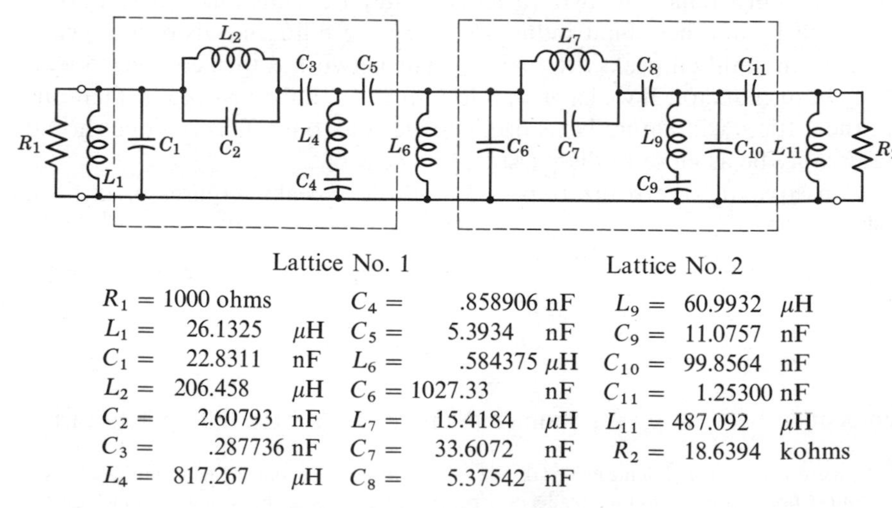

$R_1 = 1000$ ohms	$C_4 = .858906$ nF	$L_9 = 60.9932$ μH
$L_1 = 26.1325$ μH	$C_5 = 5.3934$ nF	$C_9 = 11.0757$ nF
$C_1 = 22.8311$ nF	$L_6 = .584375$ μH	$C_{10} = 99.8564$ nF
$L_2 = 206.458$ μH	$C_6 = 1027.33$ nF	$C_{11} = 1.25300$ nF
$C_2 = 2.60793$ nF	$L_7 = 15.4184$ μH	$L_{11} = 487.092$ μH
$C_3 = .287736$ nF	$C_7 = 33.6072$ nF	$R_2 = 18.6394$ kohms
$L_4 = 817.267$ μH	$C_8 = 5.37542$ nF	

FIGURE 4-22 Ladder filter of Example 4-3.

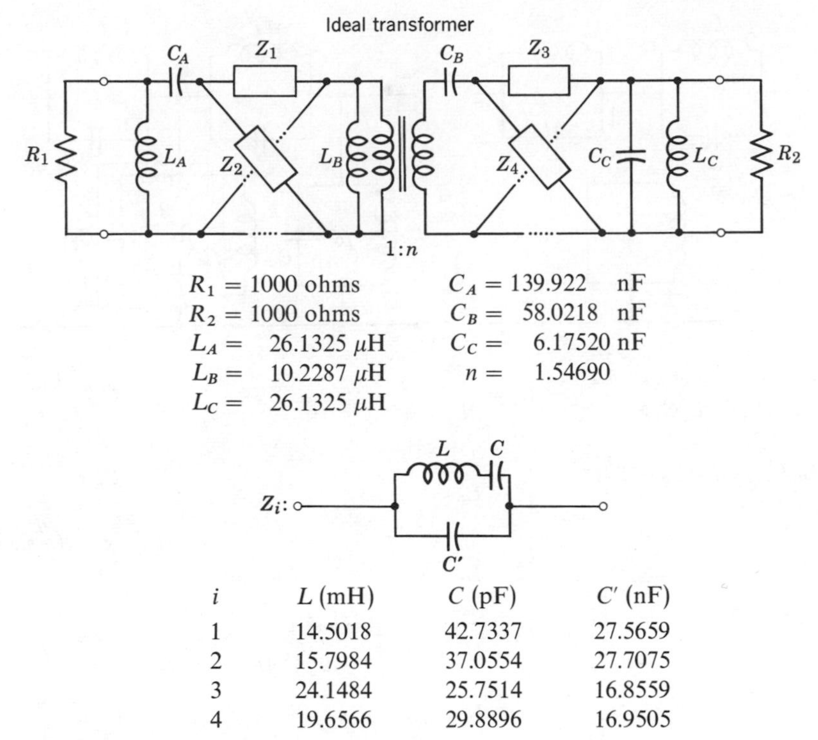

$R_1 = 1000$ ohms $C_A = 139.922$ nF
$R_2 = 1000$ ohms $C_B = 58.0218$ nF
$L_A = 26.1325\ \mu$H $C_C = 6.17520$ nF
$L_B = 10.2287\ \mu$H $n = 1.54690$
$L_C = 26.1325\ \mu$H

i	L (mH)	C (pF)	C' (nF)
1	14.5018	42.7337	27.5659
2	15.7984	37.0554	27.7075
3	24.1484	25.7514	16.8559
4	19.6566	29.8896	16.9505

FIGURE 4-23 Equivalent cascaded lattice realization of filter of Example 4-3.

convert the first lattice into half-lattice form as well as to absorb the remaining ideal transformer. L_C can be used similarly to convert the second lattice, and also to change, together with L_A, the impedance level of the lattices with respect to the terminations. Also, the smallest capacitance ratio is about 570, an order of magnitude greater than could be expected from a filter of the same bandwidth but containing no inductors. Finally, the crystal inductance values are all fairly close, a desirable feature, and this can further be improved upon by the judicious use of transformer ratios.

4.5 VERY WIDE BAND FILTERS

Filters of more than about 10% bandwidth will not be realizable in any of the previous forms. With increasing bandwidth for a while one might be able to realize parts of an otherwise LC filter by crystal-capacitor lattices, but after a while this also becomes impossible. Finally, in very wide band filters, one can only realize a few selected attenuation peaks by embedding crystals in the branches of an LC ladder filter.

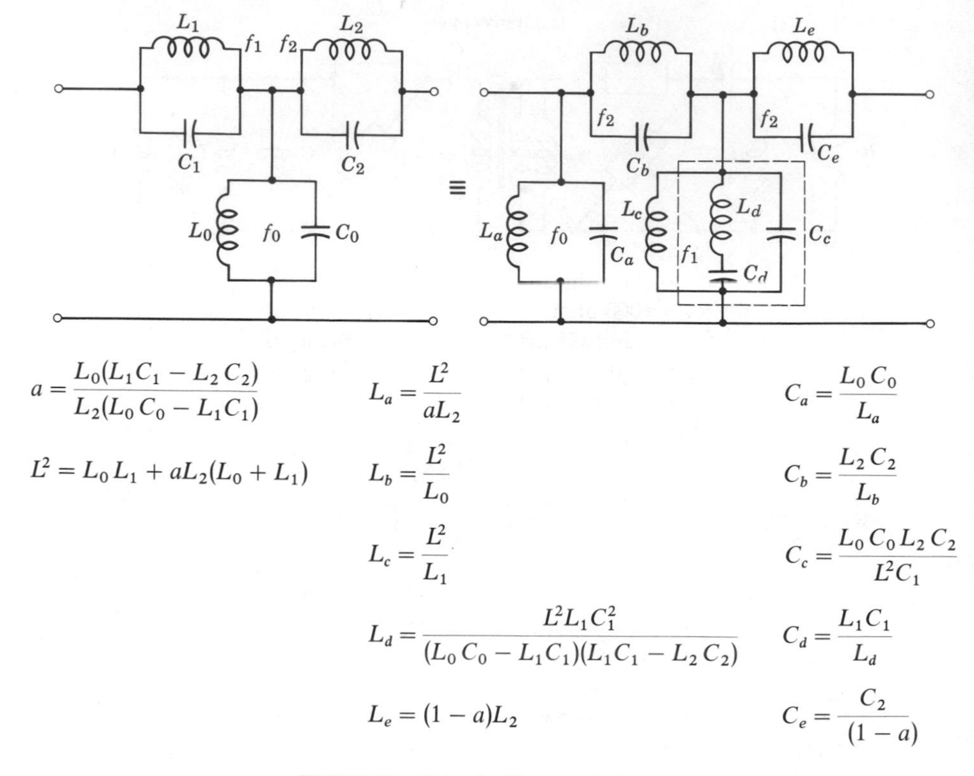

FIGURE 4-24 Ladder equivalence.

The idea is based on the equivalence [13, 14] of Figure 4-24. The equivalent network is realizable if

$$L_0C_0 > L_1C_1 > L_2C_2 \tag{4.11}$$

or

$$L_0C_0 < L_1C_1 < L_2C_2 \tag{4.12}$$

and

$$a \leq 1. \tag{4.13}$$

This means that f_1 and f_2 must be two attenuation peak frequencies on the same side of the passband and f_1 the one closer to it. It is f_1 which is actually realized by the series resonance in the four-element shunt branch on the right, that is, by the crystal. Clearly, to obtain a large capacitance ratio for the crystal, the two peak frequencies should be as close to each other and to the passband as possible. The second condition above is usually readily satisfied since we are considering a wide passband with f_1 and f_2 close to each other, hence normally

$$|L_1C_1 - L_2C_2| \ll |L_0C_0 - L_1C_1|. \tag{4.14}$$

Note that the use of the equivalence will increase the number of components, and that only the closest loss peaks may be realized by crystals. Fortunately this is precisely what is desirable, since, as is so clearly explained by Poschenrieder [13] on physical grounds, the peaks nearest to the passband contribute most severely to the undesirable rounding off near the cut-off of the loss characteristics.

Example 4-4. Consider now the following example. The passband is from 100 kHz to 150 kHz with .25 dB ripple. The loss peaks are selected to be at 80, 151, and 152 kHz, with $n_0 = n_\infty = 2$.

Solution: In order to realize the peak at 151 kHz by a crystal, the appropriate configuration is that of Figure 4-25. Applying the equivalence to the T of parallel resonant circuits, we obtain the network shown on Figure 4-26. The resulting capacitance ratio is about 350, more than adequate. If L_4 is not high enough, L_3 may be built out into a transformer with the crystal connected to the secondary high side.

Figure 4-27 shows the computed performance of the two filters around the upper cut-off frequency, with inductors of $Q = 300$, except the crystal, where we assumed a $Q = 10{,}000$. The effect of a single component (the crystal) is quite marked (1.3 dB) at the 150 kHz cut-off, and furthermore, the loss at 151 kHz is 58 dB for the filter with the crystal, exactly twice that of the filter

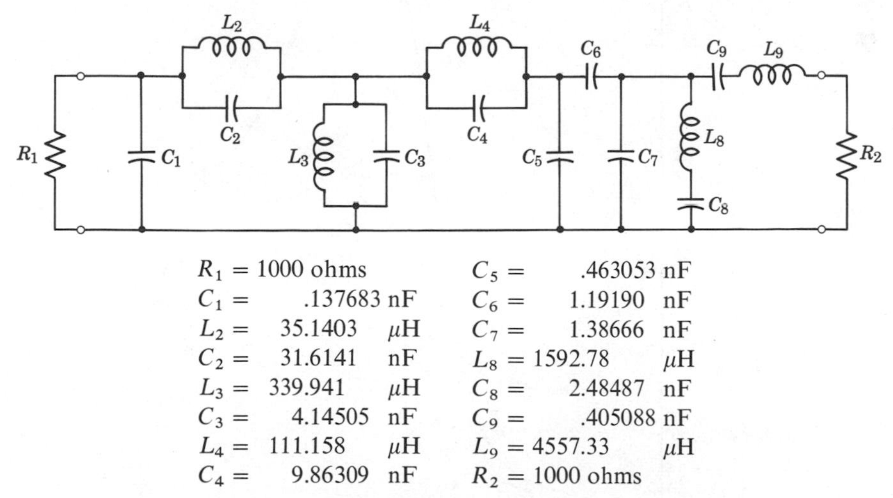

R_1 =	1000 ohms	C_5 =	.463053 nF
C_1 =	.137683 nF	C_6 =	1.19190 nF
L_2 =	35.1403 μH	C_7 =	1.38666 nF
C_2 =	31.6141 nF	L_8 =	1592.78 μH
L_3 =	339.941 μH	C_8 =	2.48487 nF
C_3 =	4.14505 nF	C_9 =	.405088 nF
L_4 =	111.158 μH	L_9 =	4557.33 μH
C_4 =	9.86309 nF	R_2 =	1000 ohms

FIGURE 4-25 Ladder realization of filter of Example 4-4.

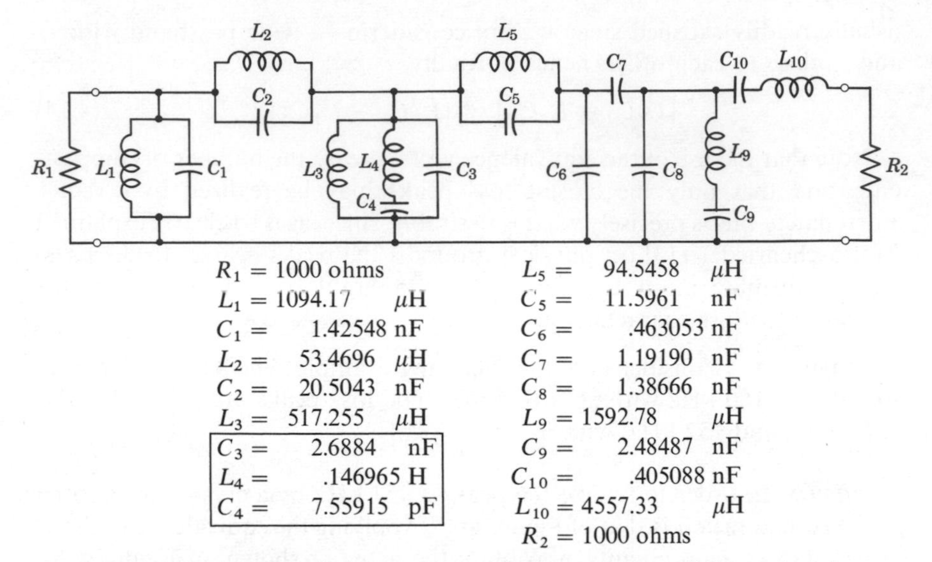

FIGURE 4-26 Equivalent ladder realization of filter of Example 4-4 with crystal in shunt arm.

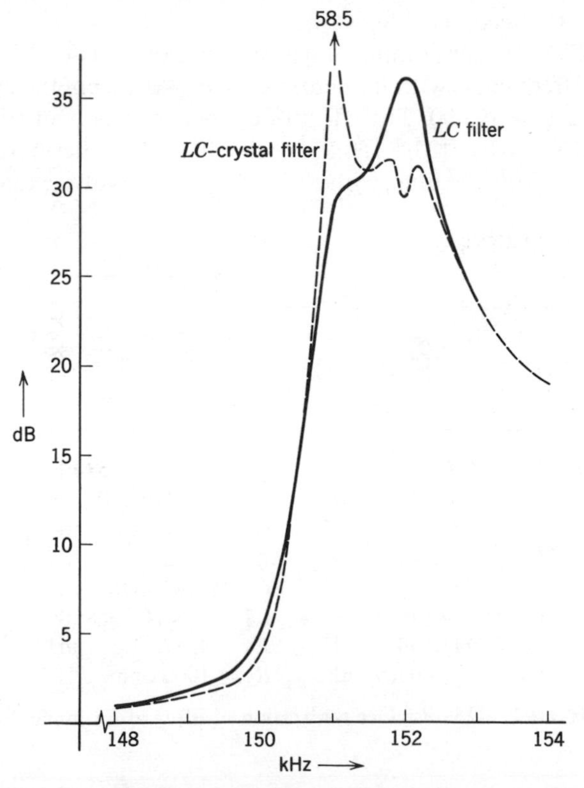

FIGURE 4-27 Computed lossy characteristics of filter realizations of Figures 4-25 and 4-26.

without it. (The slight dip in attenuation around 152 kHz is due to the large difference in Q's and the fact that three consecutive branches all antiresonate at this frequency.)

This is only one of the possible equivalences one may use to obtain structures realizable by crystals. The reader should consult the literature [14, 15] for further useful configurations.

We must point out here that for the proper operation of a filter of this kind, the crystals must not have any significant spurious responses inside the entire passband. Spurious responses could cause sharp unexpected variations in the passband loss of the filter.

4.6 APPROXIMATE METHODS FOR NARROW BAND FILTERS

All the methods considered so far have been exact, but they assume the availability of a general purpose ladder filter synthesis program [16] to provide us with our starting structures. If such a program is not available, we would like to utilize the many tabulated results that have been published. Most of these are tabulations of normalized element values of lowpass filters [17, 18]. For narrow bands, an approximate method has been developed [19, 20] to give us an appropriate configuration.

Without going into the general theory, we can illustrate a simplified version as follows [21–23]. Our starting point will be a polynomial lowpass filter of given (Butterworth, Chebyshev, Bessel, etc.) type, which the conventional lowpass to bandpass transformation

$$P = \frac{s^2 + 1}{\delta s} \tag{4.15}$$

converts into the form illustrated in Figure 4-28. The filter is normalized to 1 ohm termination, and unity center (radian) frequency, while δ is the (fractional) bandwidth. Now we will make use of the impedance inverter illustrated in Figure 4-29 and used extensively in narrow band high-frequency

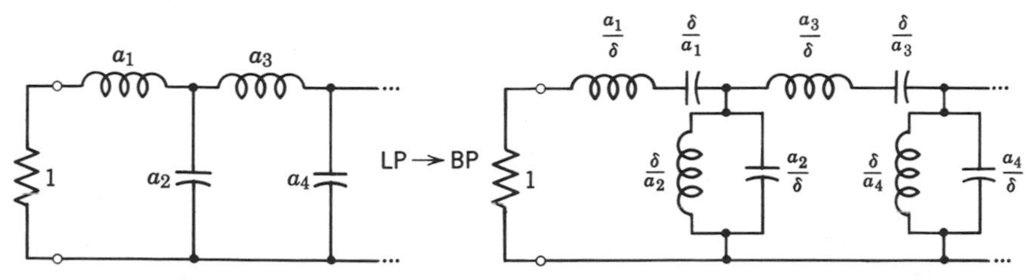

FIGURE 4-28 Lowpass-to-bandpass transformation of polynomial lowpass filter.

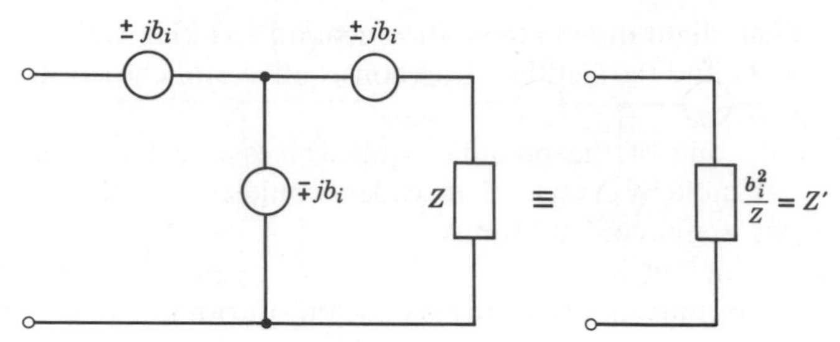

FIGURE 4-29 Ideal reactive impedance inverter.

filter design [24–28]. The circle means a reactive impedance of indicated value that is independent of frequency.

Inserting such inverters in the appropriate places in our bandpass filter, we obtain now the configuration shown in Figure 4-30. Up till now every-

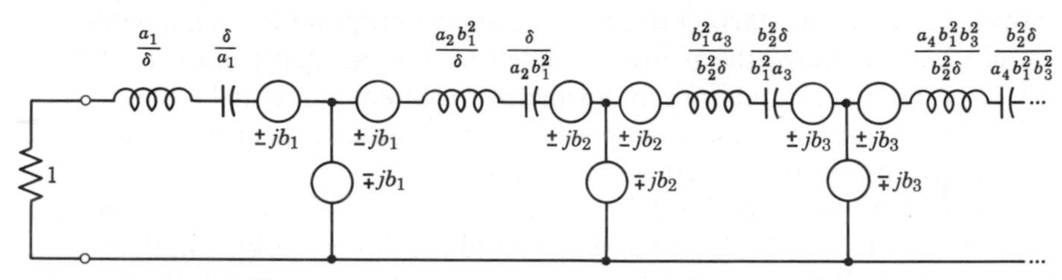

FIGURE 4-30 Equivalent form of the bandpass of Figure 4-28.

thing is exact and the b_i coefficients are arbitrary. If the final realization is to be a cascade of single crystal lattices, the b_i can remain arbitrary. If, however, we wish to use lattices with a crystal in each of their arms, we must select the b_i coefficients such that the series inductors in Figure 4-30 are pairwise equal. The simplest choice is, of course, to make all inductors equal, which leads to:

$$
\begin{aligned}
b_1 &= \sqrt{\frac{a_1}{a_2}} \\
b_2 &= \sqrt{\frac{a_3}{a_2}} \\
b_3 &= \sqrt{\frac{a_3}{a_4}} \\
&\vdots
\end{aligned}
\tag{4.16}
$$

The impedance inverters cannot, of course, be realized exactly, but for narrow band application one can use approximate realizations, one of which

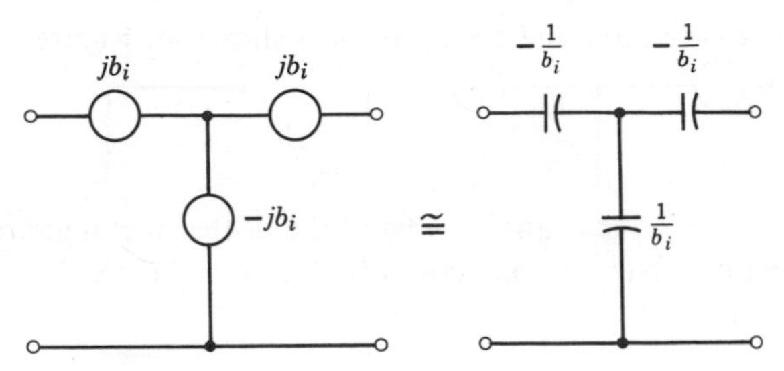

FIGURE 4-31 Approximate (narrow band) realization of reactive impedance inverter.

is shown in Figure 4-31. Using this in our filter, with the above selection of b_i values, we get the approximate realization shown in Figure 4-32. Notice that we introduced a pair of series capacitors $\pm 1/b_2$ at the beginning of the network, to make the indicated series branches (Z_s) identical. Additional capacitance pairs should be introduced at the appropriate places (depending on the b_i values) to make the series branches pairwise identical with positive leftover series capacitors. Note also that the three series capacitors in Z_s can always be combined into a single positive capacitor because δ is a very small number.

The last step before the ladder-lattice conversion is to change the termination together with the first series capacitor $1/b_2$ into the parallel RC combination (Figure 4-32). This equivalence is again approximate, but it holds for narrow bands.

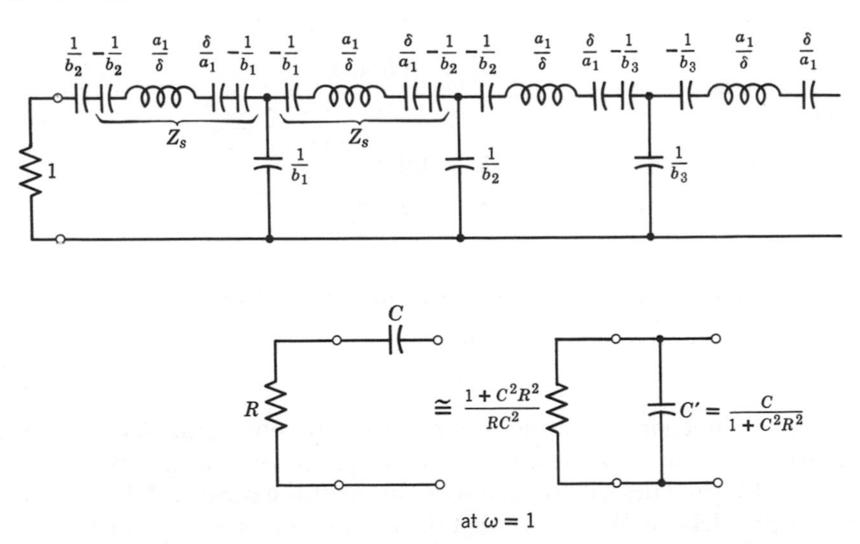

FIGURE 4-32 Approximate realization of the bandpass of Figure 4-30.

At this stage, we have the configuration shown in Figure 4-33, where

$$\frac{1}{C_1} = \frac{a_1}{\delta} - (b_1 + b_2), \tag{4.17}$$

and it is clear that the five branches inside the dotted rectangle represent a symmetrical section that can be converted into a lattice by the use of Bartlett's theorem.

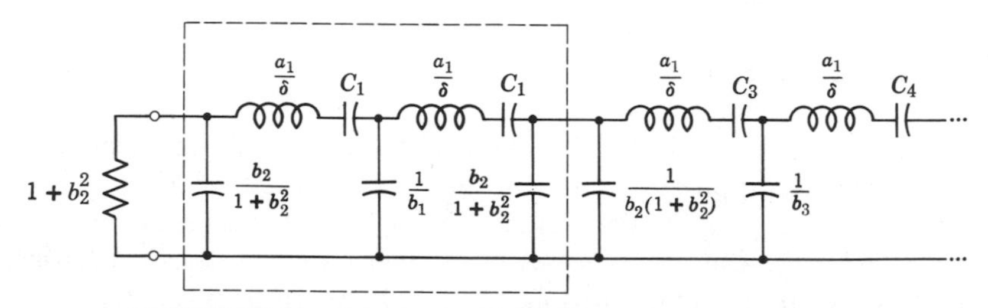

FIGURE 4-33 Preparation for ladder-to-lattice conversion.

Example 4-5. Design a bandpass filter from a fourth-order Bessel polynomial filter.

Solution: We have [29]:

$$a_1 = 1.0598$$
$$a_2 = .5116$$
$$a_3 = .3181$$
$$a_4 = .1104.$$

This gives:

$$b_1 = 1.4393$$
$$b_2 = .7885$$
$$b_3 = 1.6975$$

and the output termination now has a (normalized) value of:

$$R_2 = \frac{b_1^2 b_3^2}{b_2^2} = 9.6011. \tag{4.18}$$

Selecting a fractional bandwidth of $\delta = 10^{-4}$, we obtain the element values shown in Figure 4-34. Here we introduced another pair of capacitors (of value $\pm 1/b_2$ in series with the last series branch) and converted the series RC into a parallel one. We can see that the last four branches, together with an appropriate part of C_4, form another symmetrical section that can be

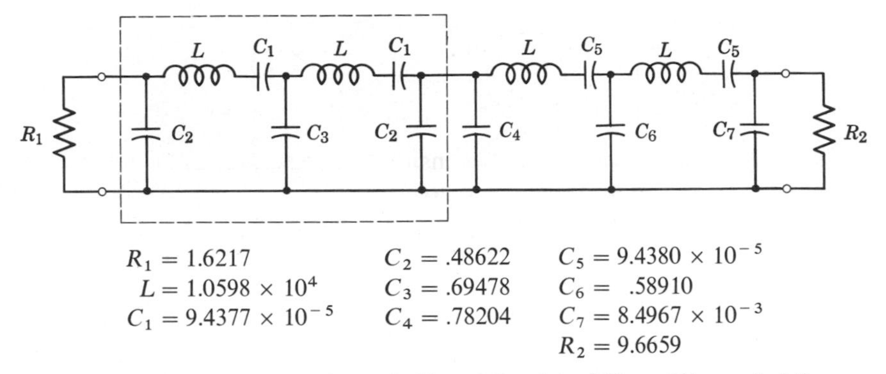

FIGURE 4-34 Approximate ladder realization of filter of Example 4-5.

converted into a second lattice. One problem remains, however, since the branches of this second lattice will have too small capacitance ratios due to the small value of C_7. This can be improved at the cost of making the crystal inductances different in the two lattices and by selecting another value for b_2. We may instead wish to make the two terminations equal. This can be obtained by selecting:

$$\bar{b}_2 = b_1 b_3 = 2.4432, \tag{4.19}$$

and the resulting structure is that shown in Figure 4-35. Here we chose to divide the center capacitor equally between the two lattices. This is not necessary. One may instead divide it to make the capacitance ratios in the two lattices closer. The final lattice network is given in Figure 4-36.

This example has illustrated a number of considerations one can use to select some of the free parameters. For further extensions of the method the reader is asked to consult the literature [20, 26–28, 30].

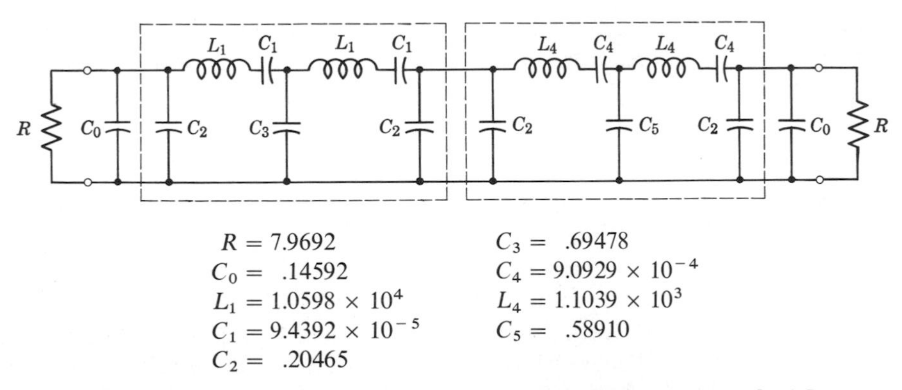

FIGURE 4-35 Another equivalent realization of filter of Example 4-5.

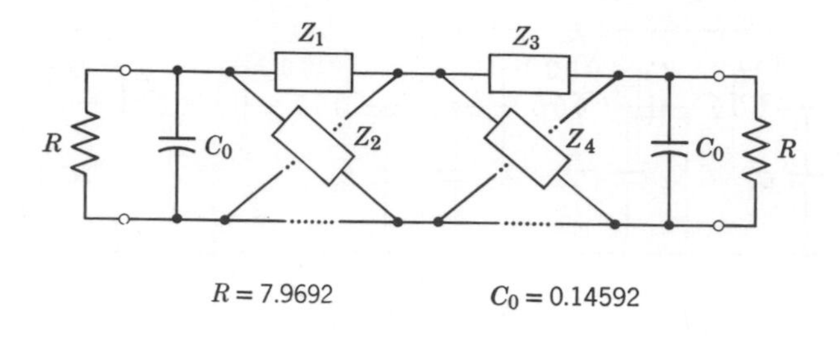

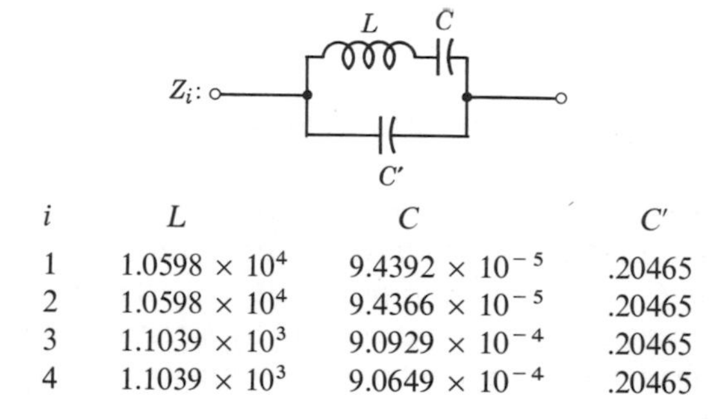

i	L	C	C'
1	1.0598×10^4	9.4392×10^{-5}	.20465
2	1.0598×10^4	9.4366×10^{-5}	.20465
3	1.1039×10^3	9.0929×10^{-4}	.20465
4	1.1039×10^3	9.0649×10^{-4}	.20465

FIGURE 4-36 Cascaded-lattice equivalent of filter of Figure 4-35.

4.7 OTHER FILTER TYPES

Under this heading, we will briefly discuss available design techniques for lowpass, highpass, and band elimination filters.

These are all effectively very wide band filter types and hence not easily suited to crystal realization. In the low and highpass cases, crystals are sometimes used to realize loss peaks extremely close to the passband edge in an otherwise LC ladder filter. Band elimination filters require crystals somewhat more often, especially if the band to be suppressed is extremely narrow. Crystal spurious resonances are, however, likely to cause serious difficulties in all of these filter types.

Lowpass and Highpass Filters

Restricting our discussion to ladder structures obtained through insertion loss design techniques, one could use the equivalence of Figure 4-37 for creating crystal-like branches in lowpass filters [31]. Note that the capacitance ratio depends only on the peak frequencies.

Because of the very limited usefulness of low and high pass crystal filters, we will not consider them any further. The interested reader is referred to the literature [31–34] for further useful equivalences and design methods.

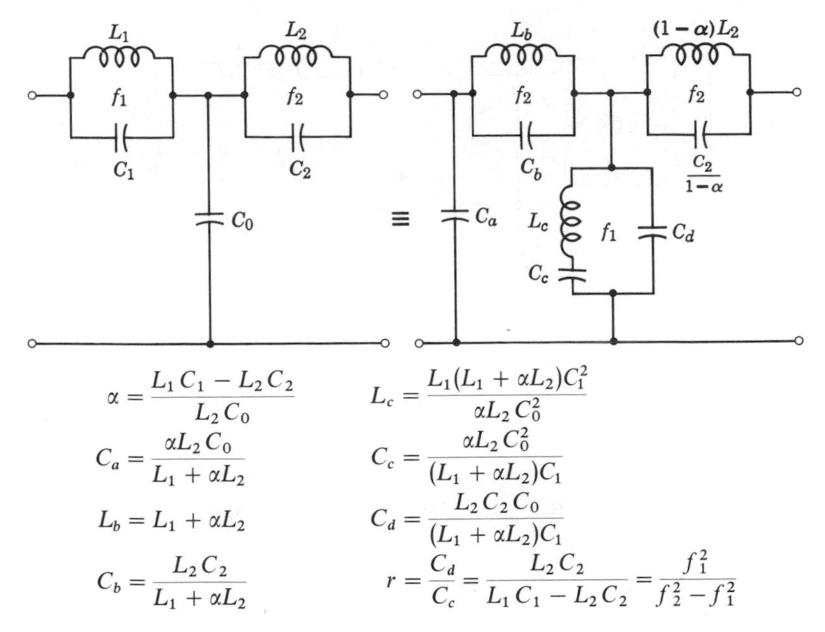

FIGURE 4-37 Ladder equivalence applicable to lowpass filters.

Band Elimination Filters

This filter type is realized more frequently with the help of crystals even though spurious resonances are just as troublesome as in the low and highpass cases. Specifically, if the band to be suppressed is extremely narrow, one needs the excellent stability and very high Q of quartz resonators for a successful realization.

The configuration often used consists of bridged-T allpass sections bridged by an additional series resonant circuit (Figure 4-38), connected in cascade. Each allpass section is a constant resistance two-port characterized by three parameters: R_0 is its impedance, f_c is the frequency at which the phase shift becomes 180°, and b is a parameter indicating the steepness of the phase versus frequency curve around f_c. Now if the $L_s - C_s$ resonant circuit is very "stiff," that is, if

$$\frac{L_s}{C_s} \gg R_0^2, \tag{4.20}$$

its effect can be neglected everywhere except in the immediate vicinity of its resonance, which occurs in the stopband of the filter. Thus the overall network will have passbands at all other frequencies. The $L_s C_s C_1$ triplet is then realized by a piezoelectric crystal.

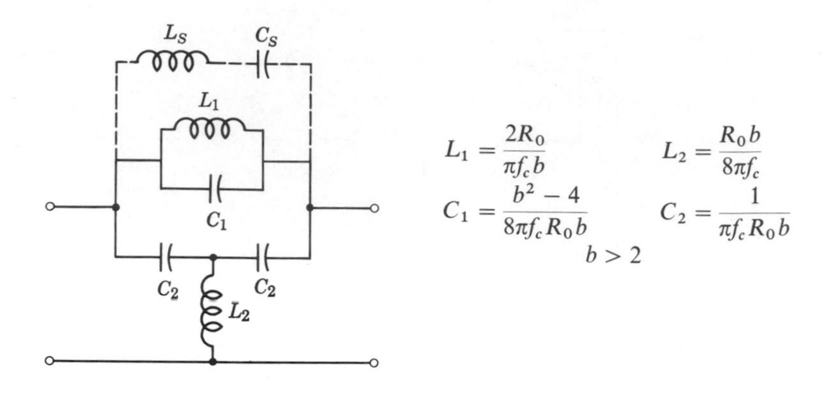

FIGURE 4-38 Band-elimination section.

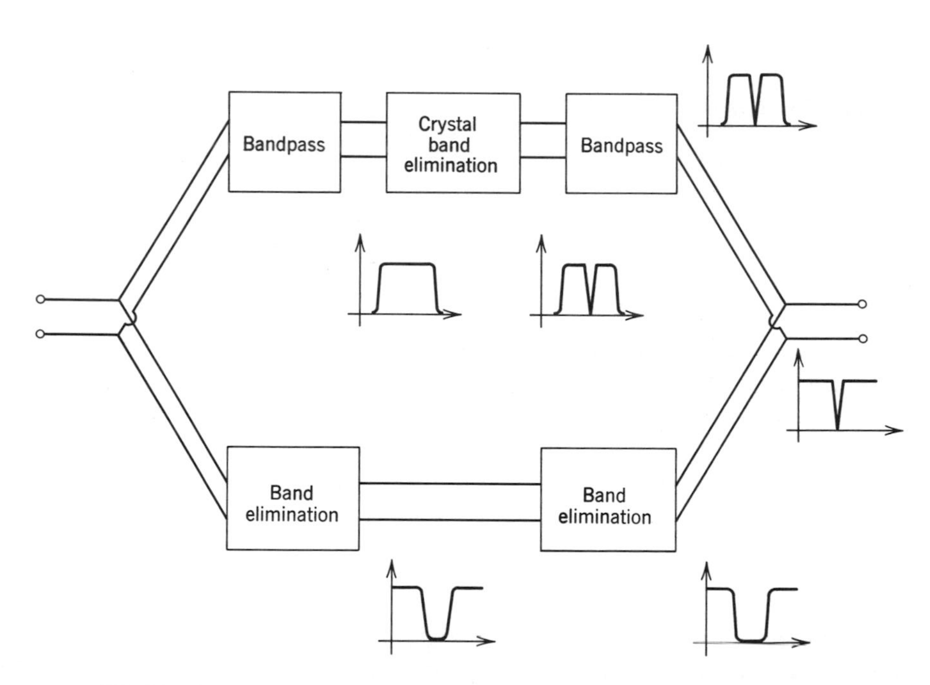

FIGURE 4-39 Scheme to prevent the interference of unwanted modes of crystal resonators.

The insertion loss design of such a structure is complicated by the fact that we have too many adjustable parameters, some of which can consequently be selected arbitrarily. Recently [35] a design technique has been described for this type of filter that is based on the pseudo-reactance concept [19, 20], and it makes good use of the available free parameters to achieve arbitrary prescribed capacitance ratios. For the details of the method, the reader is referred to references [19, 20, 35].

If the frequency bands to be passed by the band elimination filter contain some spurious crystal modes, then steps must be taken to suppress these spurious resonances at the output of the filter. The structure proposed [36, 37] for this is shown in Figure 4-39. The bandpass band-elimination filter pairs at the input and output separate the total frequency band into two bands such that the narrow band to be eliminated is within the passband of the bandpass filter. This bandpass filter is furthermore just wide enough so that the crystals to be used have no spurious responses within its passband. The method is clearly feasible, but expensive.

4.8 SUMMARY

Piezoelectric crystals and ceramic resonators are most often used for the realization of narrow bandpass filters. There is a variety of design techniques available for this filter type, and we have discussed some insertion loss methods in detail. Crystals have also been used in other filter types, and we touched briefly upon the design of wide bandpass filters, lowpass and band elimination filters.

For further design techniques and filter types, we include some additional selected references [12, 32–33, 38–47].

ACKNOWLEDGEMENT

The partial support of the National Science Foundation under Grant GK-14414 is gratefully acknowledged.

PROBLEMS

4.1 Starting from a 5th-order Chebyshev prototype lowpass ladder filter, transform it into a bandpass, and convert that into a crystal-capacitor lattice between two shunt inductors. How many full attenuation peaks are realized by the lattice, and explain why?

4.2 Design a 6th-order narrow band Butterworth bandpass filter as a cascade of lattices using impedance inverters.

4.3 Verify the equivalence given in Figure 4-24.

4.4 Show that for the first of the three equivalences of Figure 4.16, the branches of the equivalent lattice will have larger capacitance ratios than those of the ladder branches.

4.5 Show that a ladder network realizing three or more full attenuation peaks cannot, in general, be converted into a crystal-capacitor lattice.

4.6 Verify the equivalences of Figure 4-21.

REFERENCES

[1] W. P. Mason, "Electrical wave filters employing crystals as elements," *Bell System Tech. J.*, **13**, 405–452 (July, 1934).

[2] W. P. Mason, *Electromechanical Transducers and Wave Filters*, 2nd ed., Van Nostrand, New York, 1948.

[3] D. Indjoudjian and P. Andrieux, *Les Filtres a Cristaux Piezoelectriques*, Gauthier-Villars, Paris, 1953.

[4] D. I. Kosowsky, "High-frequency crystal filter design techniques and applications," *Proc. IRE*, **46**, 419–429 (Feb., 1958).

[5] W. Herzog, *Siebschaltungen mit Schwingkristallen*, 2nd ed., F. Vieweg, Braunschweig, Germany, 1962.

[6] C. L. Semmelman, Private communication.

[7] G. Szentirmai, "On the realization of crystal band pass filters," *IEEE Trans. Circuit Theory*, **CT-11**, 299–301 (June, 1964).

[8] J. A. C. Bingham, "A calculation method for Szentirmai's modified symmetrical lattice circuit," *IEEE Trans. Circuit Theory*, **CT-12**, 284–285 (Sept., 1965).

[9] J. Lang and C. E. Schmidt, "Crystal filter transformations," *IEEE Trans. Circuit Theory*, **CT-12**, 454–457 (Sept., 1965).

[10] E. Christian and G. C. Temes, "On the Szentirmai transformation," *IEEE Trans. Circuit Theory*, **CT-13**, 450–452 (Dec., 1966).

[11] A. Fettweis, "Jaumann structures with capacitive tapping for band-pass crystal filters," *Revue H. F.*, **5**, 116–123 (1962).

[12] A. Fettweis, "Image parameter and effective loss design of symmetrical and antimetrical crystal band-pass filters," *Revue H. F.*, **5**, 378–394 (1963).

[13] W. Poschenrieder, "Steile Quarzfilter grosser Bandbreite in Abzweigschaltung," *NTZ*, **9**, 561–565 (Dec., 1965).

[14] J. E. Colin, "Transformations de quadripoles permettant l'introduction de cristaux piezoelectriques dans les filtres passe-bande en echelle," *Cables et Transm.*, **21**, 124–131 (April, 1967).

[15] W. Haas, "Die Verwendung von Quarzen in Netzwerken, die nach der Betriebsparametertheorie berechnet werden," *Frequenz*, **16**, 161–167 (May, 1962).

[16] G. Szentirmai, "A filter synthesis program," *System Analysis by Digital Computer*, F. F. Kuo and J. F. Kaiser, Ed., John Wiley, New York, 1966, pp. 130–174.
[17] A. I. Zverev, *Handbook of Filter Synthesis*, John Wiley, New York, 1967.
[18] R. Saal, *Der Entwurf von Filtern mit Hilfe des Kataloges normierter Tiefpasse*, Telefunken Gmbh., Backnang, Germany, 1961.
[19] R. F. Baum, "Design of unsymmetrical band-pass filters," *IRE Trans. Circuit Theory*, **CT-4**, 33–40 (June, 1957).
[20] J. K. Skwirzynski, "The concept of pseudo-reactance in the design of narrow-band filters," *Network and Switching Theory*, G. Biorci, Ed., Academic Press, New York, 1968, pp. 341–360.
[21] T. R. O'Meara, "The cascade synthesis of crystal filters with transmission zeros at infinity," *IEEE Trans. Circuit Theory*, **CT-10**, 533–4 (Dec., 1963).
[22] R. C. Smythe, "On the realization of crystal-capacitor tandem lattice bandpass filters," *IEEE Trans. Circuit Theory*, **CT-11**, 170 (March, 1964).
[23] R. C. Smythe, "The synthesis of crystal-capacitor tandem lattice all-pole bandpass filters on the insertion loss basis," *Proc. 18th Ann. Frequency Control Symposium*, 1–22 (1964).
[24] F. S. Atiya, "Theorie der maximal geebneten und quasi-Tschebyscheffschen Filter," *AEÜ*, **7**, 441–450 (Sept., 1953).
[25] S. B. Cohn, "Direct-coupled resonator filters," *Proc. IRE*, **45**, 189–196 (Feb., 1957).
[26] J. Lang, "Narrow-band crystal filter transformations," *Electro-Technol.*, **75**, 34–37 (Aug., 1965).
[27] W. C. Kole, "Designing wide-band crystal filters with Cauer parameters," *EEE*, **15**, 88–91 (June, 1967).
[28] L. C. Hwang, "Synthesis of a four-pole, wide band crystal filter design," *Frequency Technology*, **7**, 15–19 (April, 1969).
[29] L. Weinberg, *Network Analysis and Synthesis*, McGraw-Hill, New York, 1962, p. 619.
[30] A. G. J. Holt and R. L. Gray, "Bandpass crystal filters by transformation of a lowpass ladder," *IEEE Trans. Circuit Theory*, **CT-15**, 492–4 (Dec., 1968).
[31] J. E. Colin, "Mutations des circuits provoquant les pointes d'affaiblissement infini dans les structures de filtres en echelle," *Cables et Transm.*, **12**, 10–22 (Jan., 1958).
[32] J. E. Colin, "De l'introduction de cristaux piezo-electriques dans des filtres passe-bas et passe-haut en echelle (avec extension aux filtres passebande)," *Cables et Transm.*, **16**, 85–93 (April, 1962).
[33] J. E. Colin and P. Allemandou, "Filtres passe-bas et passe-haut symetriques d'adaptation maximale. Application aux filtres passe-bas et passe-haut en echelle a cristaux piezoelectriques," *Cables et Transm.*, **15**, 99–114 (April, 1961).
[34] V. M. Kantor and A. A. Lanne, "The synthesis of low- and high-pass filters from their effective parameters," *Telecommunication*, **21**, 6–14 (April, 1967).
[35] G. Szentirmai, "The synthesis of narrow-band crystal band-elimination filters," *IEEE Trans. Circuit Theory*, **CT-15**, 409–414 (Dec., 1968).

[36] D. W. Robson, U. S. Pat. 3,009.120 (Nov. 14, 1961).
[37] R. O. K. Turvey, U. S. Pat. 3,179.906 (April 20, 1965).
[38] P. van Bastelaer, "The design of band-pass filters with piezoelectric resonators," *Revue H. F.*, **7**, 193–206 (1968).
[39] M. Dishal, "Modern network theory of single-sideband crystal ladder filters," *Proc. IEEE*, **53**, 1205–1216 (Sept., 1965).
[40] W. P. Mason, "Use of piezoelectric crystals and mechanical resonators in filters and oscillators," *Physical Acoustics*, W. P. Mason, Ed., **1**, Pt. A, Academic Press, New York, 1964, pp. 335–415.
[41] A. D. Waren, W. J. Gerber and R. Curran, "Trapped energy modes, network synthesis and the design of quartz filters," *Proc. 19th Ann. Frequency Control Symposium*, 534–564 (1965).
[42] L. S. Lasdon and A. D. Waren, "Optimal design of filters with bounded, lossy elements," *IEEE Trans. Circuit Theory*, **CT-13**, 175–187 (June, 1966).
[43] J. E. Colin, "Formulaire de calcul des filtres passe-bande etroits, a cristaux piezoelectriques identiques, a comportement meplat de l'affaiblissement," *Cables et Transm.*, **22**, 132–135 (April, 1968).
[44] I. Sauerland and W. Blum, "Ceramic IF filters for consumer products," *Spectrum*, **5**, 112–126 (Nov., 1968).
[45] E. Christian and E. Eisenmann, "Considerations for the design of crystal filters," *Proc. 3rd Allerton Conference on Circuit and System Theory*, 806–816 (1965).
[46] G. Szentirmai, "New developments in the theory of filter synthesis," *IEE Conference Publication No. 23*, 1–28 (1966).
[47] D. S. Humpherys, "Active crystal filters," *Electro-Technol.*, **78**, 43–47 (July, 1966).

5

Mechanical Bandpass Filters

R. A. Johnson
Collins Radio
Newport Beach, California

5.1 PRELIMINARY CONSIDERATIONS

The primary difference between a mechanical bandpass filter and the filters described in the preceding chapters is that, in the case of mechanical filters, both the resonant elements and the coupling elements are mechanical. This is opposed to the case of conventional crystal filters, for instance, where the resonators are mechanical vibrating elements, but all coupling between resonators is accomplished electrically through the use of capacitors, inductors, or wire conductors. What mechanical and crystal filters do have in common is the reason for their existence, namely, that they are able to provide narrow bandwidth filtering without excessive loss, temperature drift, or aging. Constant modulus iron-nickel alloy resonators have Q (quality factor) values of 10,000 to 20,000, and quartz resonators have Q values as high as 150,000 or more. This means that it is possible to build an alloy-resonator mechanical filter that has a bandwidth as narrow as 500 Hz, at a center frequency of 455 kHz, without experiencing excessive rounding of the passband response due to losses. At this same frequency, the center frequency shift with temperature may be as low as 50 Hz over a 100°C temperature range.

Basically the mechanical filter is a device that converts an electrical signal into mechanical vibrations by means of an input transducer as shown in Figure 5-1. Within the mechanical network, which is composed of resonators and coupling elements (most often short wires), the filtering takes place. The filtered signal is then converted back into electrical energy by means of the output transducer.

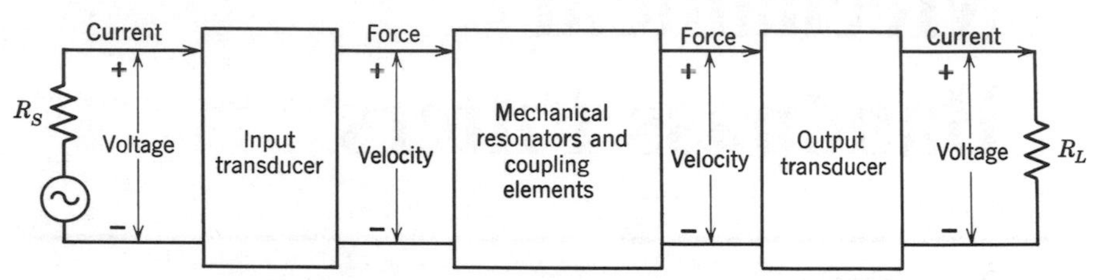

FIGURE 5-1 Electromechanical bandpass filter.

Interest in mechanical filters as frequency-selective devices was stimulated by the work of Campbell and Wagner on electrical bandpass filters. Probably the most notable early work was done by Maxfield and Harrison in the area of recording and reproducing music and speech [1]. By applying electrical-network theory, through the use of electromechanical analogs, to the design of a phonograph, they were able to improve the frequency response in a very dramatic way as shown in Figure 5-2. The network was

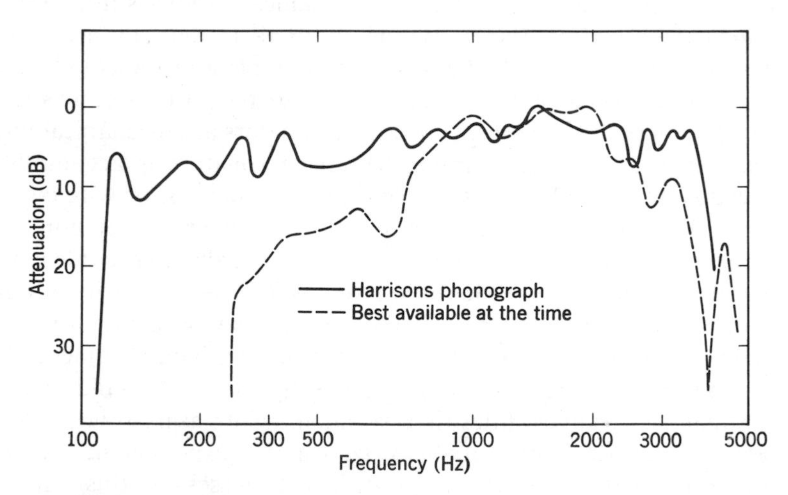

FIGURE 5-2 Improvement in phonograph design through the use of electromechanical analogs (Maxfield and Harrison).

treated as a system of lumped capacitors and inductors where the needle compliance (inverse of stiffness) was represented by capacitance and the needle arm mass by inductance, and so on.

Although patents describing various spring-mass type mechanical filters were issued, and electrical filter theory was applied to the design of various acoustic devices, there was actually very little work done in the area of designing practical mechanical filters for frequency selection, for instance in radios or telephone communications equipment, until the 1940's. In 1942 Warren Mason, who a few years previously had invented the wide-band crystal filter, published the book, *Electromechanical Transducers and Wave Filters*, in which he described a flexural-resonator, spring-coupled mechanical filter [2]. Although during this same period he had also filed patents on resonant rod (mechanical transmission line) filters, it was not until 1947 that a really practical device was invented and manufactured.

Previous to 1947, it was impossible (except at low frequencies) to use electrical filters in high performance single-sideband equipment, although they were still being used in conventional AM receivers. A great deal of progress had been made on the design of low/intermediate frequency (100 kHz) crystal filters, but at the more often used 455 kHz frequency these filters were extremely large and costly. Motivated by this need, Adler developed the wire-coupled plate-resonator type mechanical filter [3] shown in Figure 5-3(a). The first and last plates act as electromechanical transducers as well as extensional mode resonators. The iron-nickel alloy resonator used is not only very stable, but it exhibits magnetostrictive properties as

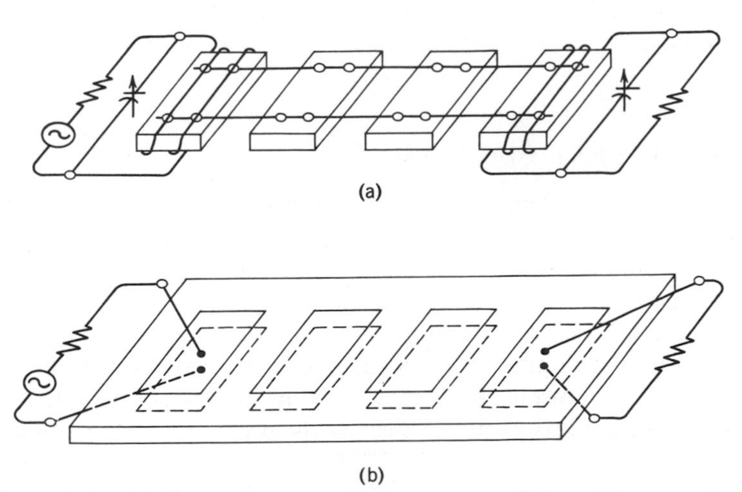

FIGURE 5-3 (a) Adler's plate-wire mechanical filter and (b) monolithic mechanical filter.

well. This means that current through the coil surrounding the plate produces a magnetic field which causes the plate to change dimensions, and, therefore, to vibrate with a relatively large amplitude near its resonant frequency. The mechanical energy is then transferred from plate to plate by means of the wires connecting the plates. When the wires are less than one eighth acoustic wavelength long, they act as springs. Strains in the last resonator produce a magnetic field which induces a voltage in the output coil.

Within a very short time following the development of the plate-wire mechanical filter, both disk-wire and rod-neck type mechanical filters were developed. The disk-wire filters were similar in operation to the plate-wire filters, whereas the rod-neck filters were designed as cascaded transmission lines or more specifically, as half-wavelength rods coupled by quarter-wavelength rods or "necks" [3].

With the exception of using wires to couple nonadjacent resonators in order to produce attenuation poles, there were no major breakthroughs in mechanical filter technology until 1965 when Sykes and Beaver published the results of their work on monolithic filters [4]. The monolithic or quartz-mechanical filter is simply an array of electrode pairs deposited on a quartz substrate as shown in Figure 5-3(b). This type of filter is very similar to the wire-coupled plate type mechanical filter both in appearance and operation. The input signal through the piezoelectric effect sets up shear-mode standing waves between the first electrode pair, and subsequently, between all of the electrode pairs. This energy is coupled between resonators through the nonelectroded regions in the form of an exponentially decaying signal. The electrode pairs act as the resonators, the nonelectroded regions as the coupling elements. The last electrode pair is used as the output transducer.

5.2 EQUIVALENT CIRCUITS

The subject of equivalent circuits is of major importance to the mechanical filter designer. Most often the design engineer has been trained in electrical network theory and has very little background in dynamics or acoustics. In addition, most literature dealing with filters, and most computer programs used for the analysis and design of filters, are written in electrical terms. Therefore, it is desirable that the mechanical filter designer be able to convert his electromechanical system to an electrical equivalent circuit model with a complexity only as great as the problem to be solved dictates.

Mechanical Schematic Diagrams

In order to be able to find an electrical equivalent circuit, it is necessary either to be able to write directly differential equations that describe the mechanical network, which can then be transformed to electrically analogous equations, or to construct a mechanical schematic diagram and transform mass to capacitance, velocity to voltage, etc., with or without the help of differential equations. Let us take as examples the simple single-degree-of-freedom systems shown in Figure 5-4.

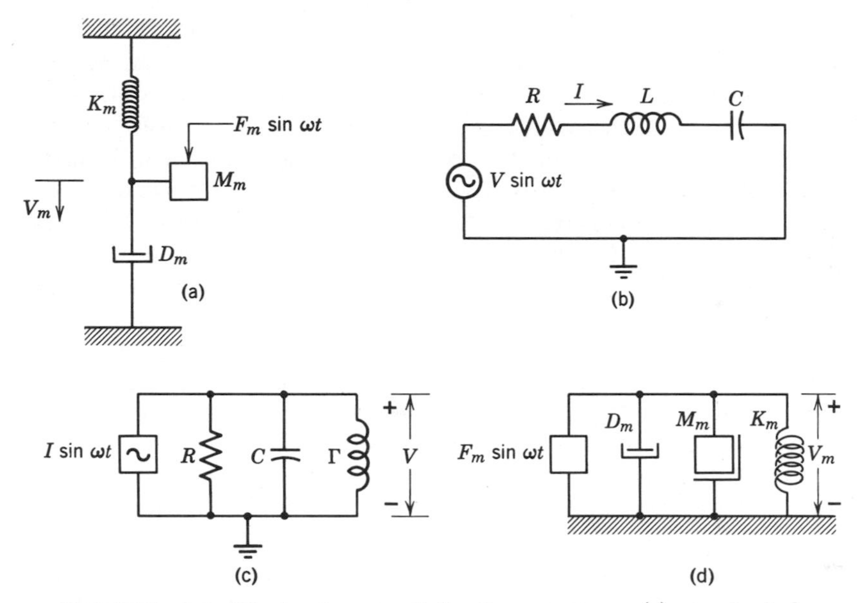

FIGURE 5-4 Single degree of freedom systems: (a) mechanical pictorial diagram, (b) series connected electrical schematic diagram, (c) parallel schematic diagram, and (d) mechanical schematic diagram.

Figure 5-4(a) is a pictorial diagram of a linear mechanical system composed of a spring (K_m), mass (M_m), and dashpot (D_m). Figures 5-4(b) and 5-4(c) are schematic diagrams of series and parallel resonator electrical networks. Each of the systems has two storage elements; K_m and M_m in the mechanical case, L and C in the electrical networks. The energy dissipating elements are D_m and R, the network variables are force (F_m), velocity (V_m), current (I), and voltage (V).

The differential equations describing the three systems are,

$$M_m \frac{dV_m}{dt} + D_m V_m + K_m \int_{t_0}^{t} V_m \, dt = F_m \sin \omega t, \tag{5.1}$$

$$L\frac{dI}{dt} + RI + \frac{1}{C}\int_{t_0}^{t} I\,dt = V \sin \omega t, \quad \text{and} \tag{5.2}$$

$$C\frac{dV}{dt} + \frac{1}{R}V + \Gamma\int_{t_0}^{t} V\,dt = I \sin \omega t. \tag{5.3}$$

Note that the above integro-differential equations have the same form, i.e., a simple substitution of constant and variable names would make all three equations identical. Because the equations have the same form, in the broad sense of the term, these systems are "analogous."

Next, rather than writing differential equations, let us construct a mechanical schematic diagram that can by inspection be converted to an electrically analogous circuit [5]. The first step in the construction of the diagram is to select a set of variables such as force and velocity and to draw a pictorial representation of the system such as that shown in Figure 5-4(a). Next, the two terminals of each element are identified. Those that move together are connected. All terminals that are "clamped" to a reference such as ground must also be tied together. In the case of a mass, one of its terminals is always connected to a ground or inertial reference. The resultant mechanical schematic diagram for the simple spring-mass system is shown in Figure 5-4(d). Note that in the pictorial diagram the elements are connected in series, whereas, in this particular case, in the schematic diagram they are connected in parallel. This, as one can see, is due to the mass having one of its terminals connected to ground. Note that the mechanical schematic diagram of Figure 5-4(d) can be converted to the analogous electrical network of Figure 5-4(c) by a simple substitution of element and variable names as in (5.1) and (5.3).

From the above example, we see that the mechanical schematic diagram should not be confused with a pictorial diagram of the system. The pictorial diagram is used to give us a better feeling or picture of the physical system. The schematic diagram includes both the variables and element constants, and, therefore, can be used as a direct aid in writing equations to describe the network or translating the mechanical network into its electrical analog.

Let us carry these concepts one step further. Mechanical vibrating systems operating at frequencies greater than 100 Hz are usually composed of distributed elements such as plate, disk or bar resonators, and coupling wires, rather than simple springs and masses. Even so, we can most often approximate these distributed elements by lumped springs and masses. Let us consider the case of the flexural-resonator coupling-wire filter, shown in Figure 5-5, as an example.

The first step in analyzing this system is to find a spring-mass representation of the flexural element near its resonant frequency. This can be done by (1) calculating its resonant frequency, and (2) determining its "equivalent

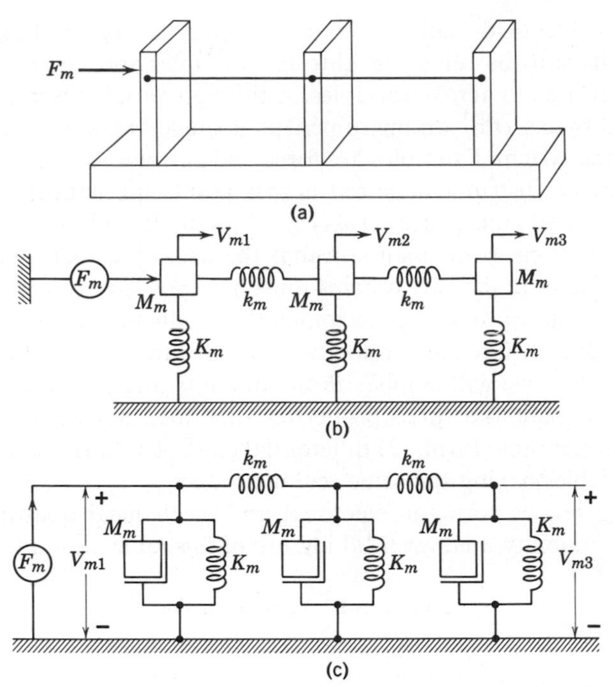

FIGURE 5-5 Three resonator, flexural bar-wire mechanical filter: (a) physical structure, (b) pictorial diagram, and (c) schematic diagram.

mass" at the point where the wire is attached (at the resonant frequency). The term equivalent mass refers to the apparent mass of the resonator or the mass of an equivalent spring-mass system. In the case of the flexural resonator, the equivalent mass in the direction of the force F_m increases as the point of the wire attachment moves toward the base of the structure. Figure 5-5(b) shows the pictorial representation of the structure. All that remains to be done in this simple case is to "ground" the masses and we have the mechanical schematic diagram shown in Figure 5-5(c).

Electromechanical Analogies

Although there is an abundance of literature on the subject of analogies, much of it is confusing because of the limited scope covered by most authors. The reason for this difficulty is that there is no unique analogy between electrical and mechanical systems. That is to say, there is no unique analogy if we imply that two systems described by differential equations of the same form are analogous. Many feel that this definition of the word analogy is not complete, because they would require that the topologies of the two networks be identical. This is the same thing as saying that "through-variables" have to be related to through-variables and "across-variables" have to be related to across-variables. A through-variable is a quantity that is measured by inserting an instrument in a single break at a point (or in a branch) of the system. Examples are force and current. An across-variable is measured by an instrument placed across points (or nodes) in a system; hence velocity and voltage are across-variables [6]. In addition to the topological considerations, some maintain that the scalar products of across- and through-variables in the two systems must have the same dimensions for the systems to be analogous. As an example, the product of current and voltage and the product of force and velocity both have the dimensions of power.

In this chapter we will emphasize the mobility analogy where the analogous systems share the three properties described above: (1) differential equations of the same form, (2) differential equations of the same topology, and (3) variables having the same scalar product.

Table 5-1 relates common electrical and mechanical quantities on the basis of the mobility analogy. Making use of this table allows us to convert

TABLE 5-1 Mobility analogy

Parameter \ System		Electrical	Mechanical
Variable	Across	Voltage (V)	Velocity (V_m)
	Through	Current (I)	Force (F_m)
Network Parameters		Conductance (G) Inductance (L) Inductance^{-1} (Γ) Capacitance (C)	Resistance (D_m) Compliance (C_m) Stiffness (K_m) Mass (M_m)
Immittances		Impedance (Z) Admittance (Y) Short Circuit ($Z = 0$) Open Circuit ($Y = 0$)	Mobility (Z_m) Immobility (Y_m) Clamped Point ($Z_m = 0$) Free Point ($Y_m = 0$)
Topology		Loop Node Series Connection Parallel Connection	Loop Node Series Connection Parallel Connection

directly equations (or schematic diagrams) describing the mechanical system into a set of equations describing the analogous electrical system (or vice versa) by simply replacing V_m by V or F_m by I, etc. Note that we have designated immobility (F_m/V_m) by Y_m which becomes the symbol for the analogous electrical quantity when the subscript is deleted.

Transducer Equivalent Circuits

We have now found a means of representing a mechanical system as an analogous electrical network [7, 8]. The question that is often asked at this point is, "What value of capacitance in farads corresponds to X grams of mass?" etc. This question has meaning only in an electromechanical system, i.e., a system where we are trying to relate equivalent (analogous) electrical parameters to actual electrical parameters. In fact, in an electrical-input, electrical-output system, the question that should be asked is "What are the values of the equivalent or analogous electrical parameters in relationship to the external electrical values?" The answer to both questions involves the electromechanical transducer through which we "see" mechanical parameters as electrical parameters with electrical units. As the degree of coupling between the electrical and mechanical networks is varied, the values of the equivalent electrical network parameters as seen from the electrical terminals are changed. Therefore, we can state that there is no absolute conversion of mechanical to electrical values other than that which is relative to the degree of transducer electromechanical coupling.

What this means, in the case of an actual device, can be illustrated by the piezoelectric transducer shown in Figure 5-6(a). A piezoelectric material is one that changes its dimensions in the presence of an electric field, which means that when the ends of the transducer of Figure 5-6(a) are plated with a conductive material, and an alternating voltage is applied across these terminals, the bar will vibrate. If we were to make measurements at the electrical terminals, we would find that the driving-point reactance (We will assume, for now, that we have a lossless system) has a zero and a pole in the region of the mechanical resonant frequency of the bar. This is shown in Figure 5-6(b). In addition, if the resonator is prevented from vibrating, or if measurements are made at frequencies that are far removed from any resonance, the network will look like a capacitor C_0. Based on the measurement of the pole and zero frequencies, as well as C_0, the network of Figure 5-6(c) can be constructed.

Earlier in this chapter we saw that a flexual-mode resonator could be represented by a lumped spring-mass system. In the same way, the bar of Figure 5-6(a) can be represented by a spring, mass, and dashpot (mechanical dissipation element), each having one terminal grounded and the other terminal connected to a common point. From Figure 5-6(a) we see that force

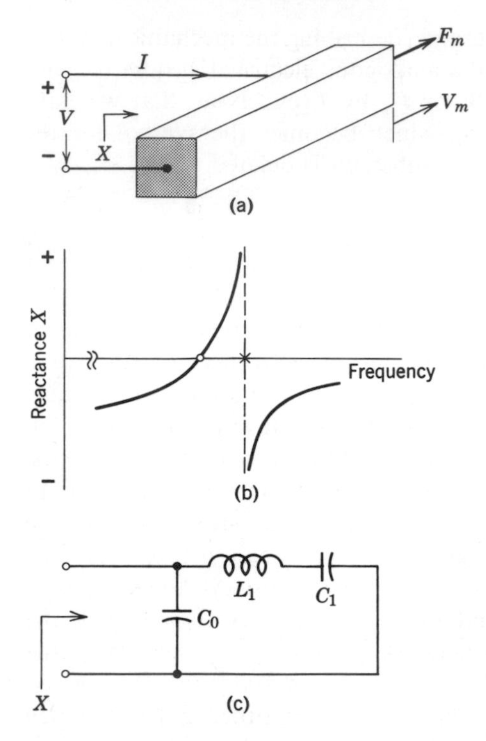

FIGURE 5-6 Piezoelectric transducer: (a) pictorial diagram, (b) frequency response near resonance, and (c) electrical driving point reactance network.

F_m and velocity V_m measurements can be made at one end of the transducer (The other end can be free or fixed) in order to determine the lumped mechanical-element values.

Having made both electrical and mechanical measurements on the transducer, the equivalent circuit of Figure 5-7(a) can be constructed.* The superscript E of the compliance C_m^E denotes a constant electric field, which corresponds to the case where the compliance is measured under a condition where the electrical terminals are shorted (an electric field equal to zero). Between the electrical and mechanical sides of the network is an electromechanical gyrator which is used to connect electrical and mechanical elements.

The gyrator is needed because a voltage (across-variable) on the electrical side generates an electric field that, in turn, generates, under clamped conditions, a force (through-variable) on the mechanical side and vice versa. The

*It is important to note that force F_m is a through-variable and velocity V_m in an across-variable. This is not taken into account in most of the older literature, but it is becoming more widely accepted. The difficulty in having two systems of analogies can be understood by trying to use voltage as a through-variable in solving some problems and as an across-variable in solving others.

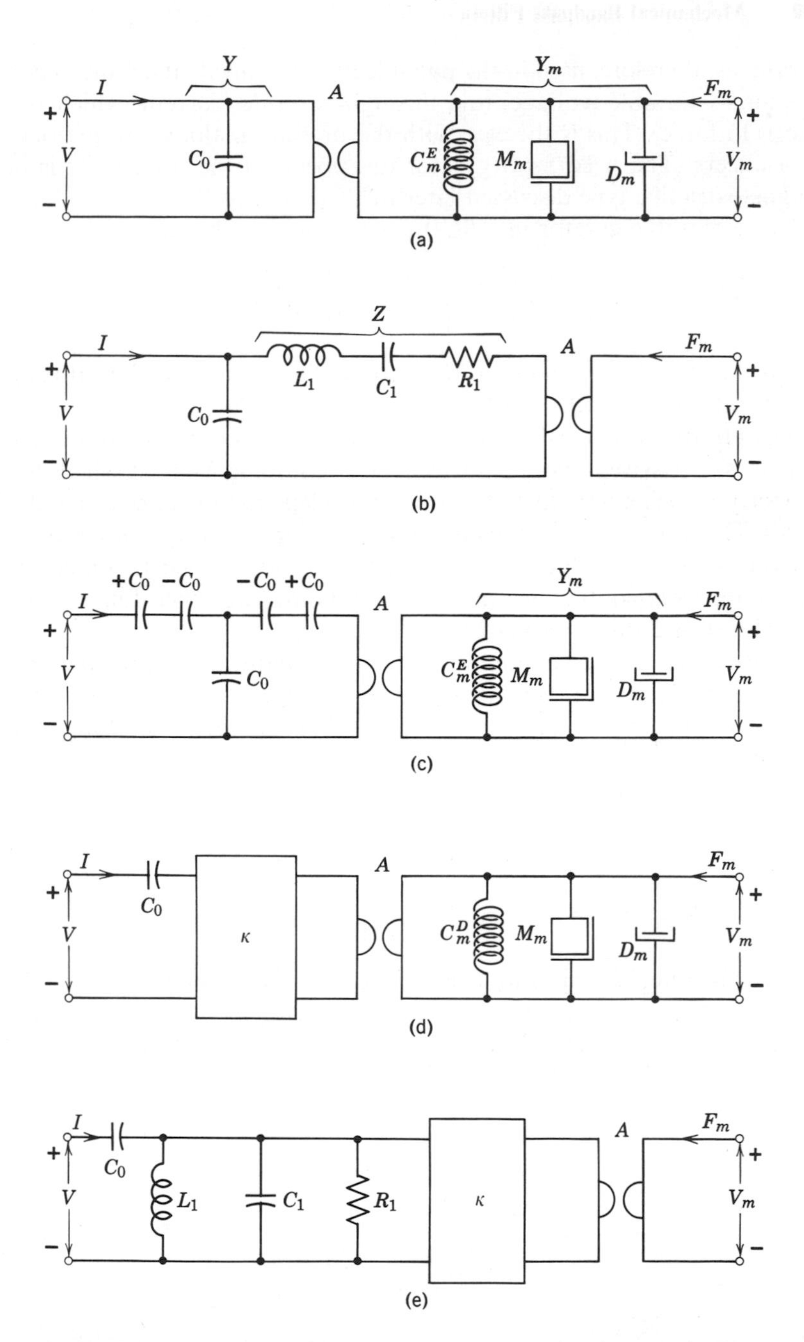

FIGURE 5-7 Electric-field transducer equivalent circuits based on the voltage-velocity mobility analogy.

gyrator is, therefore, used in the pure electrical equivalent circuit to convert the across-variable (voltage) to a through-variable (current, which is analogous to force). This is the case with the present analogy and piezoelectric transducers. The need for a gyrator disappears if the transducer is of the magnetostrictive type discussed later in this section.

Describing the gyrator in *ABCD* matrix form we have,

$$\begin{bmatrix} V \\ I \end{bmatrix} = \begin{bmatrix} 0 & A \\ A^{-1} & 0 \end{bmatrix} \begin{bmatrix} V_m \\ F_m \end{bmatrix}, \tag{5.4}$$

where the reciprocity condition $AD - BC = 1$ is replaced by the antireciprocity condition $AD - BC = -1$.

The electromechanical gyrator is much like its electrical counterpart in that it is a passive, lossless, antireciprocal coupling element which has the property of being able to transform shunt elements to series elements, and series elements to shunt elements. This inversion property is necessary if the network of Figure 5-7(a) is capable of being transformed to Figure 5-6(c), which represented the electrical driving point reactance function of the transducer and single resonator system.

An expression relating the electrical and mechanical variables of the entire network of Figure 5-7(a) can be found by multiplying the *ABCD* matrices of the electrical, electromechanical, and mechanical sections. Doing this we obtain,

$$\begin{aligned} \begin{bmatrix} V \\ I \end{bmatrix} &= \begin{bmatrix} 1 & 0 \\ Y & 1 \end{bmatrix} \begin{bmatrix} 0 & A \\ A^{-1} & 0 \end{bmatrix} \begin{bmatrix} 1 & 0 \\ Y_m & 1 \end{bmatrix} \begin{bmatrix} V_m \\ F_m \end{bmatrix} \\ &= \begin{bmatrix} Y_m A & A \\ (A^{-1} + YY_m A) & YA \end{bmatrix} \begin{bmatrix} V_m \\ F_m \end{bmatrix}. \end{aligned} \tag{5.5}$$

Transforming (5.5) into its y-parameter form we obtain,

$$\begin{bmatrix} I \\ F_m \end{bmatrix} = \begin{bmatrix} Y & A^{-1} \\ -A^{-1} & Y_m \end{bmatrix} \begin{bmatrix} V \\ V_m \end{bmatrix} \tag{5.6}$$

where

$$Y = \left.\frac{\partial I}{\partial V}\right|_{V_m = 0}$$

$$A^{-1} = \left.\frac{\partial I}{\partial V_m}\right|_{V=0} = -\left.\frac{\partial F_m}{\partial V}\right|_{V_m=0}$$

and

$$Y_m = \left.\frac{\partial F_m}{\partial V_m}\right|_{V=0}.$$

If we move the gyrator to the right hand side of the mechanical resonator Y_m, we obtain the network of Figure 5-7(b). Note that the shunt elements in the mechanical circuit become series elements in the electrical equivalent circuit. In *ABCD* matrix notation we can write,

$$\begin{bmatrix} 0 & A \\ A^{-1} & 0 \end{bmatrix} \begin{bmatrix} 1 & 0 \\ Y_m & 1 \end{bmatrix} = \begin{bmatrix} 1 & Z \\ 0 & 1 \end{bmatrix} \begin{bmatrix} 0 & A \\ A^{-1} & 0 \end{bmatrix}. \tag{5.7}$$

After multiplying and equating terms, we obtain the result that

$$Z = A^2 Y_m.$$

If the transducer is allowed to vibrate freely, the mechanical terminals of the circuit of Figure 5-7(b) are free (open). This open circuit is reflected across the gyrator as a short circuit, and the network of Figure 5-7(b) becomes that of Figure 5-6(c).

The first two equivalent circuits of Figure 5-7 can be converted to networks (Figures 5-7(c) and (d)) that show the transducer capacitance C_0 as a series capacitor and the mechanical network in terms of parallel shunt elements. The significance of the last circuit of Figure 5-7 is that the resultant electrical elements have the same topology as those in the corresponding mechanical network. The circuit is therefore both an equivalent circuit and the electrically analogous network (in the strict sense).

To obtain the network of Figure 5-7(c) we add $+C_0$ and $-C_0$ in series to both the right and left hand side of C_0 in Figure 5-7(a). The T section composed of $\pm C_0$'s represents an inverter, that is, the two-port κ shown in Figure 5-7(d) having the characteristics,

$$\begin{bmatrix} V_1 \\ I_1 \end{bmatrix} = \begin{bmatrix} 0 & j\kappa \\ j/\kappa & 0 \end{bmatrix} \begin{bmatrix} V_2 \\ I_2 \end{bmatrix}, \tag{5.8}$$

where $\kappa = 1/\omega C_0$. This inverter has the property that it resembles an ideal inverter only over a narrow frequency range where $\kappa \sim$ constant.

Although the gyrator of (5.4) and the inverter of (5.8) both convert series elements to shunt, etc., the inverter in this case has two electrical ports, whereas the electromechanical gyrator has one electrical and one mechanical. Note also that by moving the right side C_0 to the right of the gyrator (which from (5.7) results in the capacitor being transformed into a compliance) and combining compliances, the resultant compliance is C_m^D. This transformation in *ABCD* parameter notation makes use of (5.7) and can be written,

$$\begin{bmatrix} 1 & 1/\omega C_0 \\ 0 & 1 \end{bmatrix} \begin{bmatrix} 0 & A \\ A^{-1} & 0 \end{bmatrix} \begin{bmatrix} 1 & 0 \\ 1/\omega C_m^E & 1 \end{bmatrix} = \begin{bmatrix} 0 & A \\ A^{-1} & 0 \end{bmatrix} \begin{bmatrix} 1 & 0 \\ 1/\omega C_m & 1 \end{bmatrix} \begin{bmatrix} 1 & 0 \\ 1/\omega C_m^E & 1 \end{bmatrix}$$

$$= \begin{bmatrix} 0 & A \\ A^{-1} & 0 \end{bmatrix} \begin{bmatrix} 1 & 0 \\ 1/\omega C_m^D & 1 \end{bmatrix}, \tag{5.9}$$

where $C_m = A^2 C_0$ and $C_m^D = (1/C_m + 1/C_m^E)^{-1}$. The superscript D signifies that the mechanical parameter in question was measured under electrical open-circuit conditions. Therefore, the mechanical resonant frequency of the transducer is a function of C_m^D under open-circuit conditions, and C_m^E when the electrical terminals are shorted. Since C_m^D is less than C_m^E, the natural mechanical resonant frequency of the open-circuited transducer is greater than when the terminals are shorted (Remember that the $C_m M_m$ product is analogous to the LC product)*.

If both the gyrator and inverter are moved to the right of the mechanical network, we obtain an electrical equivalent circuit, Figure 5-7(e), that has the same topology as the original mechanical network of Figure 5-7(a). We will make considerable use of this equivalent circuit in sections to follow.

One of the most commonly used transducers makes use of the magnetostrictive or Joule effect, which relates to a change in the dimensions of a material as the result of an impressed magnetic field. The magnetostrictive transducer shown in Figure 5-8(a) is composed of a coil, a permanent magnet (to provide bias flux in order to prevent frequency doubling), and a ferrite rod. In the region of the acoustical half-wavelength resonance of the bar, the electrical driving-point reactance first goes through a pole and then a zero, at a slightly higher frequency. The transducer coil inductance L_0 and the pole-zero-producing mechanical bar represented by L_1 and C_1 are shown in Figure 5-8(b). This simplified driving-point impedance network can be expanded into a two-port in much the same way as was done in the electric-field transducer case.

When using mobility-type equivalent circuits where force is a through-variable and is analogous to current, the equivalent circuit of the magnetic field transducer is somewhat simpler than that of the electric field transducer. The reason for this is that the two-port which describes the electrical-mechanical conversion has the form of an ideal transformer rather than a gyrator. In $ABCD$ matrix terms,

$$\begin{bmatrix} V \\ I \end{bmatrix} = \begin{bmatrix} \eta & 0 \\ 0 & \eta^{-1} \end{bmatrix} \begin{bmatrix} V_m \\ F_m \end{bmatrix}. \tag{5.10}$$

The reciprocal nature $(AD - BC = 1)$ of this transformation is consistent with the fact that when an electromechanical device is built with an electric field transducer on one end and a magnetic field transducer on the other, the composite system is antireciprocal. This method is used to design isolators where most of the energy transmission is in only one direction. When the

*This fact is used in the tuning of monolithic mechanical filters where each electrode-pair resonance is isolated, by open circuiting all other electrode pairs, and then measured and its frequency adjusted to a prescribed value. The open-circuited electrode pairs are resonant above the filter passband and have only a second-order effect on the pair being measured.

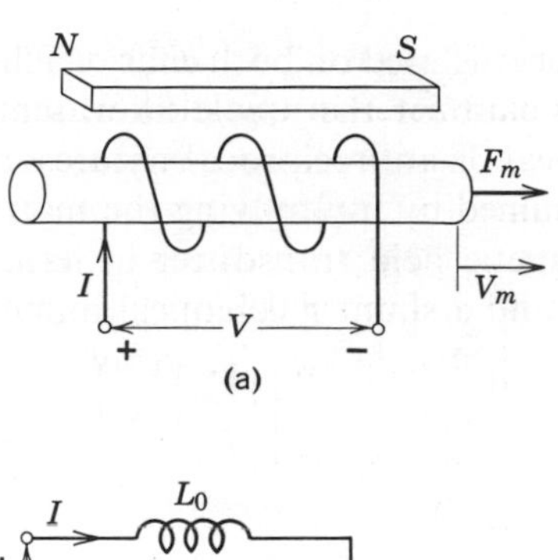

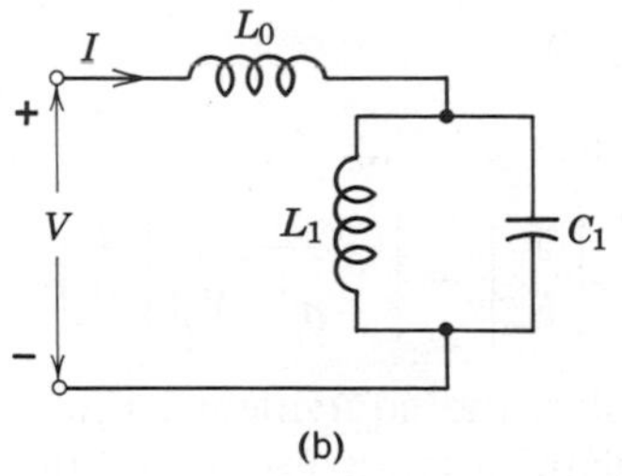

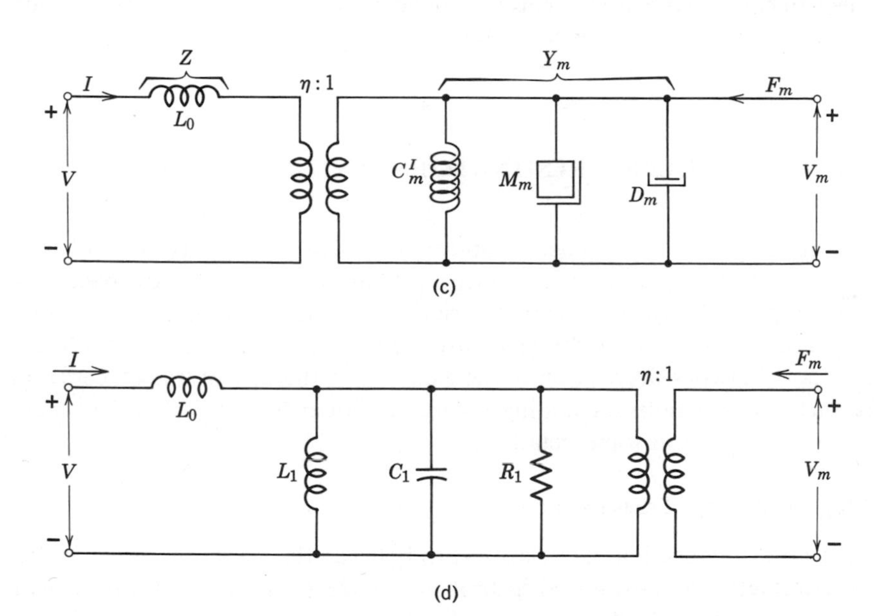

FIGURE 5-8 Magnetostrictive transducer: (a) pictorial diagram, (b) simplified electrical reactance network, and (c) and (d), equivalent circuits based on the voltage-velocity mobility analogy.

same type of transducer is used on both ends of a filter, it will be a reciprocal device which means that for the whole filter, input to output, reciprocity applies. The reciprocal or antireciprocal nature of a combination of transducers can be determined by multiplying the matrixes of (5.4) and (5.10).

Describing a magnetic field transducer in terms of η, a series electrical impedance $Z = sL_0$ and a shunt mechanical immobility Y_m we can write

$$\begin{bmatrix} V \\ I \end{bmatrix} = \begin{bmatrix} 1 & Z \\ 0 & 1 \end{bmatrix} \begin{bmatrix} \eta & 0 \\ 0 & \eta^{-1} \end{bmatrix} \begin{bmatrix} 1 & 0 \\ Y_m & 1 \end{bmatrix} \begin{bmatrix} V_m \\ F_m \end{bmatrix} = \begin{bmatrix} (\eta + Y_m Z \eta^{-1}) & Z\eta^{-1} \\ Y_m \eta^{-1} & \eta^{-1} \end{bmatrix} \begin{bmatrix} V_m \\ F_m \end{bmatrix} \tag{5.11}$$

or in "h" parameter form,

$$\begin{bmatrix} V \\ F_m \end{bmatrix} = \begin{bmatrix} Z & \eta \\ -\eta & Y_m \end{bmatrix} \begin{bmatrix} I \\ V_m \end{bmatrix}. \tag{5.12}$$

Equation (5.12) describes the equivalent circuit shown in Figure 5-8(c). Moving the electromechanical transformer to the right hand side of the mechanical network transforms the mechanical elements to electrical ones, but it does not change the topology.

5.3 TRANSDUCERS, RESONATORS, AND COUPLING ELEMENTS

Having considered the general subject of equivalent circuits, we will now look more closely at specific electromechanical and mechanical elements, namely piezoelectric and magnetostrictive transducers, mechanical resonators, and coupling wires. In each case we will be concerned with finding a two-port network description of the element. In the case of transducers and resonators, equations for finding resonant frequencies and equivalent mass will be discussed in some detail.

Magnetostrictive Transducers

In the previous section, the relationship between the mechanical and electrical parameters was expressed in terms of an electromechanical transformer (η) or a gyrator (A). A very convenient parameter which relates the coil inductance L_0 and the electrical equivalent inductance L_1 of the transducer rod is the effective electromechanical coupling coefficient k_{em} (which we shall define). If relationships between the equivalent mass and compliance of the transducer rod and the other mechanical elements in the filter can be found, we shall have no difficulty in finding an electrical equivalent circuit representing the entire electromechanical filter.

Defining the electromechanical coupling coefficient as the ratio of the energy stored in the mechanical circuit to the total input energy we obtain [2, 7],

$$\frac{k_{em}^2}{1-k_{em}^2} = \frac{L_1}{L_0} = \frac{f_s^2}{f_p^2} - 1, \tag{5.13}$$

or in a typical low-coupling coefficient ($k_{em}^2 \ll 1$), highly stable magnetostrictive transducer,

$$k_{em}^2 \approx \frac{L_1}{L_0} \approx \frac{2(f_s - f_p)}{f_p}. \tag{5.14}$$

The subscripts s and p in (5.13) and (5.14) correspond to the series and parallel resonant frequencies (zeros and poles of the driving-point impedance) of the network of Figure 5-8(b). A typical Fe-Ni-Co rod-type magnetostrictive transducer will have a coupling coefficient between 0.10 and 0.15.

The next step in characterizing a mechanical filter in terms of its electrical equivalent circuit is to take a detailed look at the transducer rod. This will include a description of its mechanical two-port parameters, its resonant frequency, as well as its equivalent mass and compliance (which can be converted to electrical parameters through the use of the coupling coefficient).

A long thin rod vibrating in a longitudinal mode is analogous to an electromagnetic transmission line. The rod can be considered to have a certain value of mass per unit length shunted to ground and a corresponding value of compliance per unit length acting in series. A rod of length l, and cross-sectional area A is shown in Figure 5.9.

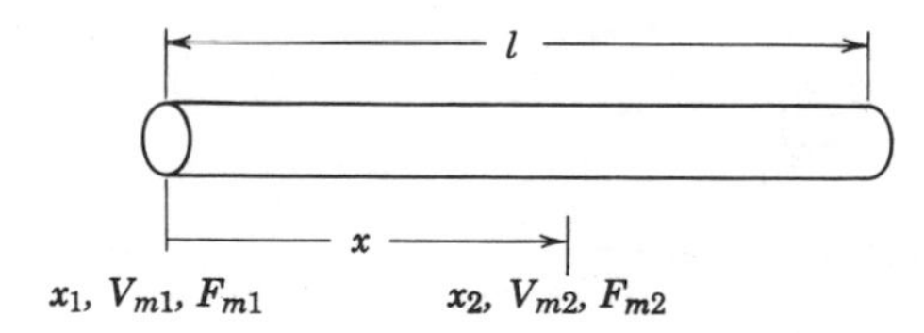

FIGURE 5-9 Magnetostrictive transducer rod (transmission line).

As in the case of the electromagnetic line [2], the transmission matrix can be written in terms of a propagation constant β, the distance x from one end, and the characteristic impedance (actually mobility in the mechanical case) Z_{m0}. Thus for the lossless case,

$$\begin{bmatrix} V_{m1} \\ F_{m1} \end{bmatrix} = \begin{bmatrix} \cos \beta_m x & jZ_{m0} \sin \beta_m x \\ \dfrac{j}{Z_{m0}} \sin \beta_m x & \cos \beta_m x \end{bmatrix} \begin{bmatrix} V_{m2} \\ F_{m2} \end{bmatrix}$$

$$\beta_m = \frac{\omega}{v_p}, \; v_p = \sqrt{\frac{E_m}{\rho_m}}, \; Z_{m0} = \frac{1}{A\sqrt{\rho_m E_m}}, \tag{5.15}$$

where ω is the angular frequency, E_m is Young's modulus, v_p is the velocity of propagation of extensional waves in the rod, and ρ_m is the density of the material. The equations for β_m, v_p, and Z_{m0} are used in the derivation of a number of expressions in this chapter.

If one end of the rod is free (open circuited), i.e., $F_{m2} = 0$ when $x = l$, we can write the driving-point mobility expression at the opposite end as

$$Z_m = \frac{V_m}{F_m} = \frac{\cos \beta_m l}{\dfrac{j}{Z_{m0}} \sin \beta_m l} = -jZ_{m0} \cot \beta_m l.$$

The natural resonances of the rod occur at open-circuit driving-point mobility poles (open-circuit impedance poles in the electrical case) which correspond to conditions where $\beta_m l = n\pi$. Thus

$$\beta_m l = \frac{\omega l}{\sqrt{\dfrac{E_m}{\rho_m}}} = n\pi;\, n = 1, 2, 3, \ldots$$

or solving for ω,

$$\omega_n = \frac{n\pi}{l}\sqrt{\frac{E_m}{\rho_m}};\, n = 1, 2, 3, \ldots. \tag{5.16}$$

The next problem is that of finding the equivalent mass of the bar at one end with the other end free. The equivalent mass M_{eq} is defined as the lumped-mass equivalent of the distributed parameter rod at a given point and in the region of a specified natural frequency. Therefore, its value can be determined by dividing the total kinetic energy of the system by one-half of the velocity squared at the point. Therefore,

$$M_{eq} = \frac{\frac{1}{2}\int_0^l (V_{m1} \cos \beta_m x)^2 \rho_m A\, dx}{V_{m1}^2}.$$

Making use of the fact that $\beta_m l = n\pi$ we obtain,

$$M_{eq} = \frac{\rho_m l A}{2} = \frac{M_{\text{static}}}{2} \tag{5.17}$$

$$C_{eq} = \frac{1}{\omega_n^2 M_{eq}} = \frac{2l}{\pi^2 A E_m}.$$

Equations (5.17) can be used to describe the equivalent mass and compliance at the end of a half-wavelength resonator. The resonator may be either a transducer, which is attached to a tuned disk or rod, or one of many other resonators in a cascade of resonant elements and coupling elements (See Figure 5-19).

In the case of more complex vibrating systems such as thick disks, the kinetic energy term will include velocity components in all directions. In addition, the equivalent mass has meaning only in terms of the direction of the velocity at the point. Therefore, if the velocity in a particular direction is zero, the equivalent mass with reference to that direction is infinite.

Piezoelectric Ceramic Transducers

A widely used piezoelectric ceramic transducer is the composite or Langevin type shown in Figure 5-10. The Langevin transducer is composed of a thin piezoelectric ceramic sandwiched between two temperature-stable metal rods. The primary purpose of the metal is to improve the temperature stability of the composite resonator. Although the electromechanical coupling is reduced by the metal bars, the overall coupling coefficient still remains fairly high because of the very high coupling factor of the ceramic itself.

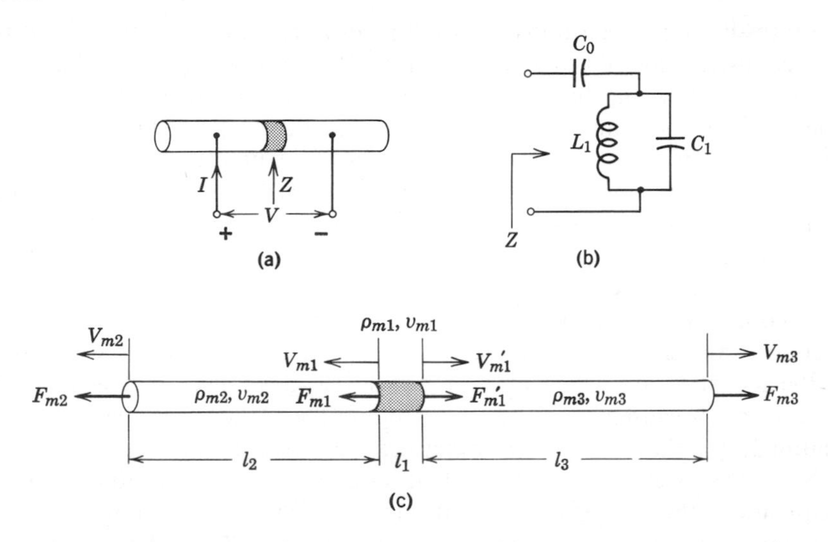

FIGURE 5-10 Langevin transducer: (a) pictorial diagram, (b) simplified electrical equivalent circuit, and (c) diagram showing acoustic properties.

In order to be able to represent the Langevin transducer by a simple equivalent circuit, such as that shown in Figure 5-10(b) (where C_0 is the static capacity of the transducer and L_1, C_1 represents the mechanical resonator), we must be able to represent our transducer by a network that takes into account conditions at both ends of the bar as well as at the ceramic, metal boundaries. In Figure 5-10(c) we have again treated

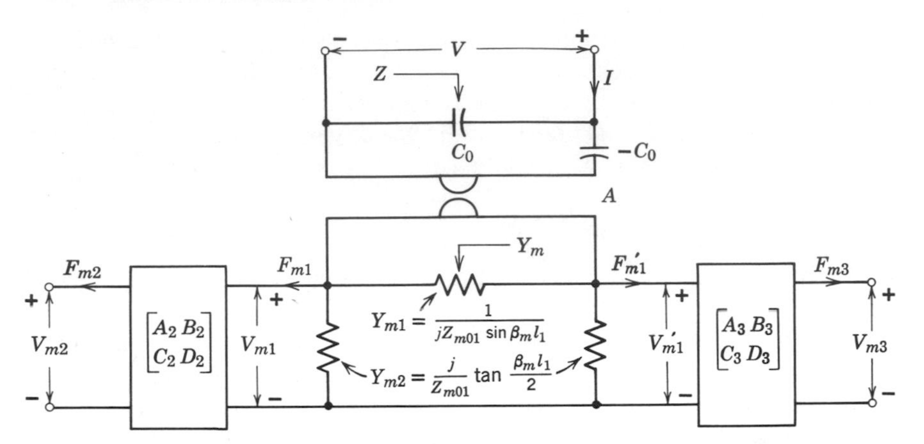

FIGURE 5-11 Langevin transducer equivalent circuit based on the mobility analogy.

the transducer rod as a transmission line, or as a cascade of lines of density ρ_{mi}, acoustic velocity v_{pi}, and characteristic mobility $Z_{m0i} = (\rho_{mi} v_{pi} A_i)^{-1}$. An equivalent circuit that takes the four mechanical ports (boundaries) and one electrical port into account is shown in Figure 5-11. This equivalent circuit is different from that shown in most of the literature on transducers [2, 8] in that we have consistently represented force as a through-variable and velocity as an across-variable. This made the use of the gyrator A necessary as in Figure 5-7(a) of the previous section. The $-C_0$ is the result of representing the invertor of Figure 5-7(d) in its lumped-element form and then combining the input series C_0 and $-C_0$. This representation is often used for transducers that have the applied electric field in the same direction as the acoustic wave propagation. These devices are called "stiffened" mode transducers because of the effect of bringing the $+C_0$ across the gyrator in Figure 5-7(c), thus raising the mechanical resonant frequency.

Now that we have an equivalent circuit we can find the resonant frequency of the composite bar or its electromechanical coupling, which, in analogy to the magnetostrictive case, is related to C_0 and C_1 of Figure 5-10(b). One of the simplest yet most important cases to consider is where the ceramic disk is sandwiched between two identical metal rods. Thus $l_2 = l_3$, $\rho_{m2} = \rho_{m3}$, etc. This case is shown in the schematic diagram of Figure 5-12. The open-circuit mechanical resonant frequency of the composite bar is where $Y_m = 0$ or from the electrical terminals were $Z = \infty$.

We can begin our analysis by writing an expression for $A_2 B_2 C_2 D_2$, namely,

$$\begin{bmatrix} V_{m1} \\ F_{m1} \end{bmatrix} = \begin{bmatrix} \cos \alpha_m l_2 & jZ_{m02} \sin \alpha_m l_2 \\ \dfrac{j \sin \alpha_m l_2}{Z_{m02}} & \cos \alpha_m l_2 \end{bmatrix} \begin{bmatrix} V_{m2} \\ F_{m2} \end{bmatrix}. \tag{5.18}$$

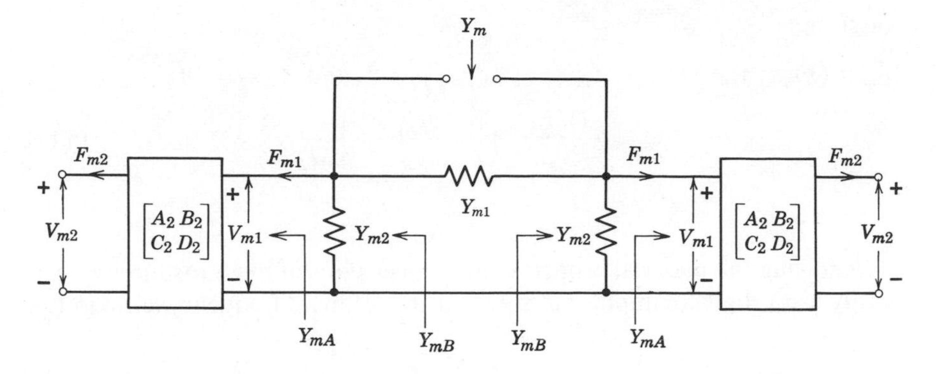

FIGURE 5-12 Langevin transducer equivalent circuit for the case where $l_2 = l_3$ (See Figure 5-10(c)).

The input immobility when $F_{m2} = 0$ is, from (5.18),

$$Y_{mA} = \frac{1}{Z_{mA}} = \frac{C_2}{A_2} = \frac{j \sin \alpha_m l_2}{Z_{m02} \cos \alpha_m l_2} = j \frac{\tan \alpha_m l_2}{Z_{m02}}.$$

Y_{mB} is the parallel combination of Y_{m2} and Y_{mA} or

$$Y_{mB} = j \frac{\tan \frac{\beta_m l_1}{2}}{Z_{m01}} + j \frac{\tan \alpha_m l_2}{Z_{m02}}.$$

Since the network is symmetrical we can write

$$Y_m = Y_{m1} + \frac{Y_{mB}}{2}$$

or

$$Y_m = -j \frac{1}{Z_{m01} \sin \beta_m l_1} + j \frac{\tan \frac{\beta_m l_1}{2}}{2Z_{m01}} + j \frac{\tan \alpha_m l_2}{2Z_{m02}},$$

which after making use of some trigonometric identities, becomes

$$Y_m = -j \frac{1}{2Z_{m01} \tan \frac{\beta_m l_1}{2}} + j \frac{\tan \alpha_m l_2}{2Z_{m02}}.$$

At resonance ($Y_m = 0$),

$$\tan \frac{\beta_m l_1}{2} = \frac{Z_{m02}}{Z_{m01} \tan \alpha_m l_2},$$

or since

$\beta_m = \omega/v_{p1}$, $\alpha_m = \omega/v_{p2}$, $Z_{m01} = (A_1\,\rho_{m1}\,v_{p1})^{-1}$, and $Z_{m02} = (A_2\,\rho_{m2}\,v_{p2})^{-1}$.

$$\tan\frac{\omega_0 l_1}{2v_{p1}} = \left(\frac{\rho_{m1}\,v_{p1}}{\rho_{m2}\,v_{p2}}\right)\frac{1}{\tan\dfrac{\omega_0 l_2}{v_{p2}}}. \tag{5.19}$$

Knowing the material properties and dimensions of the transducer we can easily use a digital computer to solve for the resonant frequency ω_0 in (5.19).

Resonant Elements

Many of the concepts discussed in the previous sections on transducers can be applied to the study of resonators. In fact, the transducer rod usually acts as part of the end resonant element or in some cases actually is the first or last resonator in the filter structure. An example of the latter case is the plate-wire filter discussed in the first section of this chapter, where the end plate is both the transducer and end resonator. Another case is that of the disk-wire filter driven with either a magnetostrictive ferrite rod or a ceramic transducer. Both ferrite and ceramic transducers can be either directly attached to the disk-resonator or attached through a small diameter coupling wire. Both situations are shown in Figure 5-13. The transducer vibrates in a half-wavelength longitudinal mode, the disk resonator in a flexural or bending mode; the shaded portions vibrate in phase.

The transducer-disk system of Figure 5-13(a) can be represented near resonance by the electrical equivalent circuit of Figure 5-13(b). Note that the two resonators could be combined into a single-parallel-tuned circuit. In the wire-coupled case of Figure 5-13(c) and (d), the coupling wire may be of such a size as to allow the transducer rod to act as the end resonant element in the filter and the end disk to act as the second resonator. In either case, it is necessary to know the resonant frequency and equivalent mass (which is proportional to capacity) of the disk resonator with respect to the transducer rod.

Rather than attempt to solve the very complex acoustical problem of calculating equivalent mass and the resonant frequency of the disk, both of which can be easily measured in the lab, let us look instead at the problem of modelling the resonator to take into account some of its unwanted modes of vibration. More specifically, we will consider the problem of modelling a lossless disk resonator in terms of a lumped-element equivalent circuit, having already measured the frequency and equivalent mass of all modes near the filter passband.

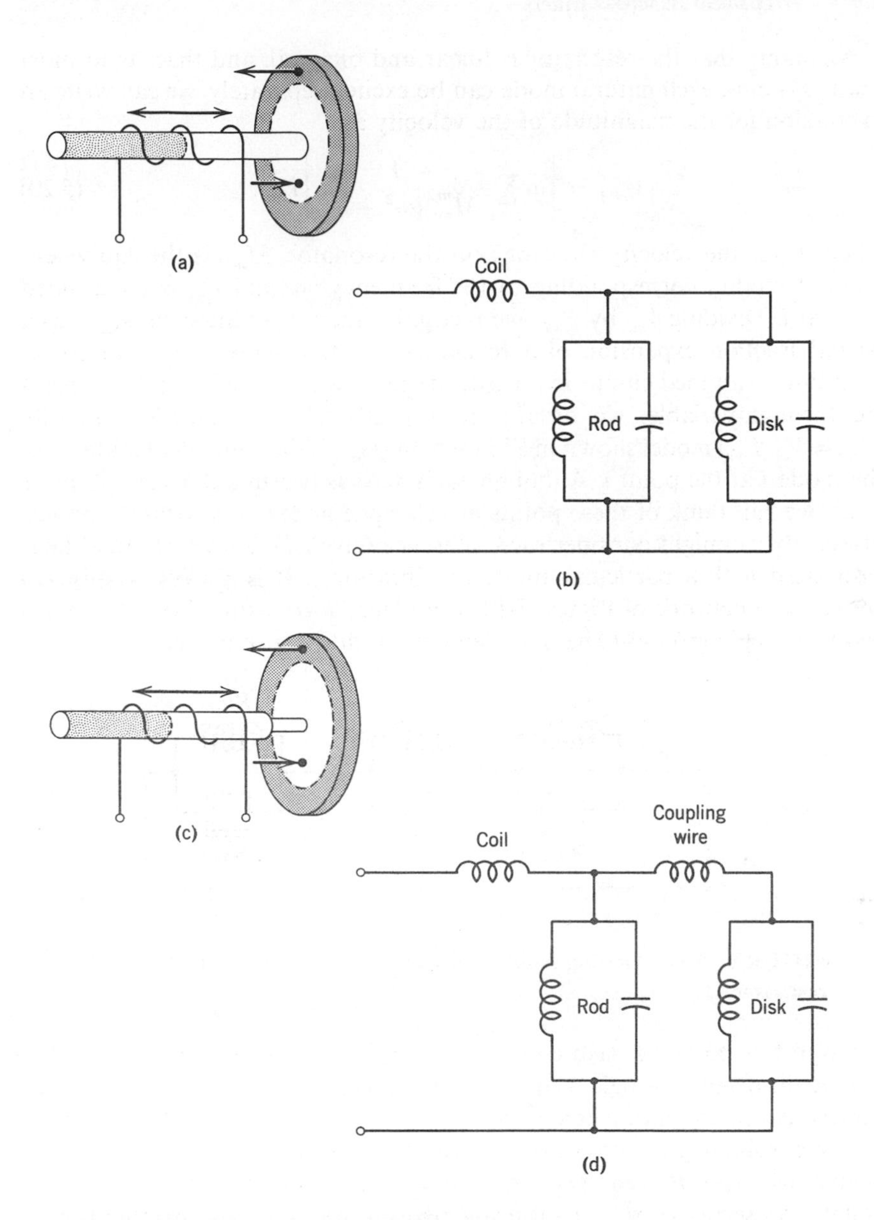

FIGURE 5-13 Magnetostrictive transducer and end resonator assemblies: (a) and (b), direct connection; (c) and (d), wire coupling.

Assuming that the resonator is linear and bilateral, and that, as in most linear systems, each natural mode can be excited separately, we can write an expression for the magnitude of the velocity as,

$$|V_{mi}| = \left[\omega \sum_{j=1}^{\infty} \frac{1}{M_{mij}(\omega^2 - \omega_j^2)}\right] F_{mi}, \tag{5.20}$$

where V_{mi} is the velocity at point i on the resonator, M_{mij} is the equivalent mass of mode j corresponding to the frequency ω_j, and F_{mi} is the applied force at i. Dividing V_{mi} by F_{mi}, we recognize the remaining expression as a partial fraction expansion of a reactance function with each term corresponding to a tuned circuit in a Foster-type network. Assuming that force is the through-variable, we obtain from (5.20) the driving-point mobility ($Z_{mi} = V_{mi}/F_{mi}$) model shown in Figure 5-14. K_{mij} is the equivalent stiffness of the mode j at the point i. Although each mass is tied to a different point or node, we can think of these points as reference nodes or separate grounds. Physically we might consider each reference (ground) as a set of nodal lines associated with a particular mode of vibration j. It is always possible to convert the network of Figure 5-14 to a Cauer form with all of the masses tied to a single ground (The first element would be a shunt mass).

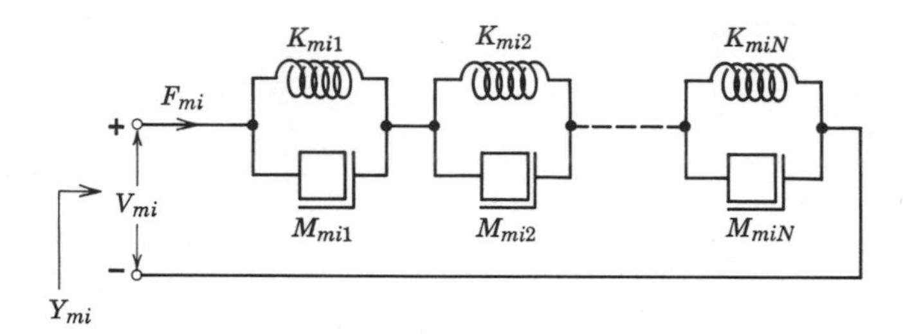

FIGURE 5-14 Driving-point equivalent circuit of a mechanical resonator.

Up to this point we have considered the disk as a one-port, i.e., we have only considered a single point on the disk. A more general lumped-parameter representation can be derived from Figure 5-15. In Figure 5-15(a) we are considering not only two flexural modes of vibration, but two points of measurement as well. The first mode is a 2-diameter mode where the nodal lines separate 180° out-of-phase regions of the disk (Remember that in a network composed only of L's and C's, all steady-state currents and voltages assume two conditions with respect to one another; namely they are in-phase or 180° out-of-phase).

The points corresponding to the velocities V_{m11} and V_{m21} are therefore moving in opposite directions. In the case of the second mode at frequency

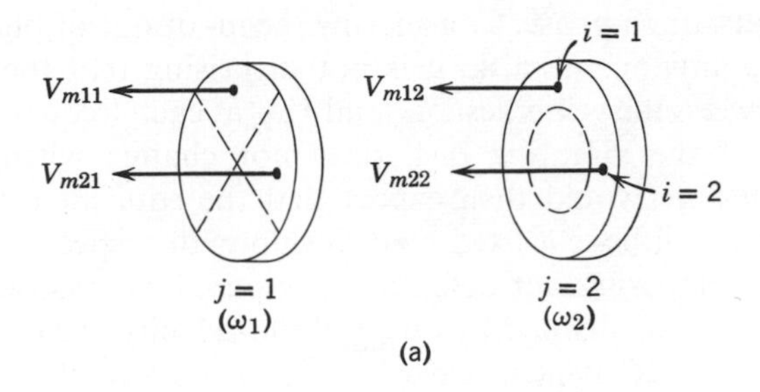

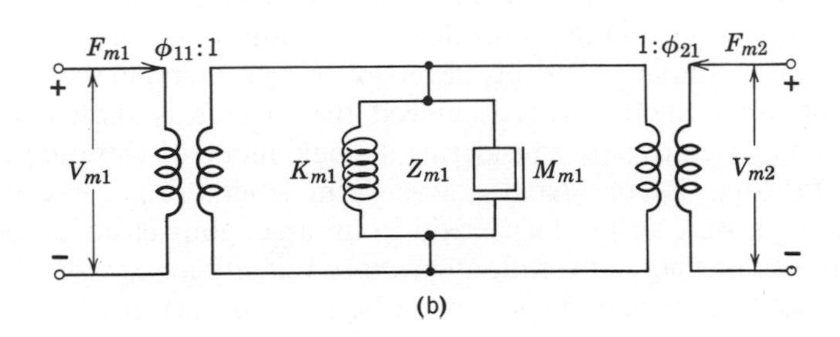

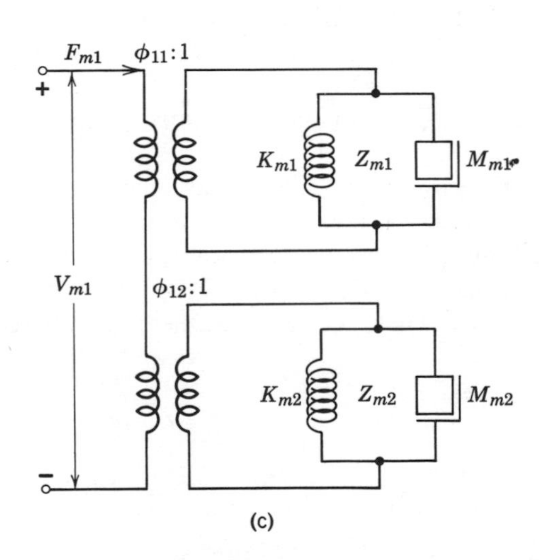

FIGURE 5-15 Two-port double mode resonator: (a) two nodal diameter and one nodal circle modes of vibration, (b) the disk as a two-port near frequency ω, and (c) the driving-point immittance schematic diagram at point (port) $i = 1$.

ω_2, the two points are in-phase. Considering the in- or out-of-phase relationships between points on the disk, it is not surprising that there is a fixed relationship between the velocities V_{m1j} and V_{m2j} at each frequency. Since the kinetic energy of the vibrating disk does not change when we change measuring points, we would then expect that the equivalent mass at any point on the disk will have a fixed relationship with respect to that at any other point. Remembering our definition of equivalent mass as the kinetic energy of the resonator divided by one-half the velocity squared at a point, we can conclude that equivalent masses in our network models are inversely proportional to the square of their respective velocities. We would expect that these constraints on the equivalent masses and velocities would lead to the use of ideal coupling elements (levers or transformers) in the mechanical equivalent-circuit model. This is indeed the case as is shown in Figure 5-15(b), where we are only considering a single mode of vibration ω_1. The velocity ratio ϕ_{ij} is considered a lever arm length ratio in mechanical networks or a transformer-turns ratio in an analogous electrical network. Therefore defining ϕ_{ij} as the ratio of primary velocity to secondary velocity, we can write a set of equations to describe the two-port of Figure 5-15(b);

$$\begin{bmatrix} V_{m1} \\ V_{m2} \end{bmatrix} = \begin{bmatrix} \phi_{11}^2 Z_{m1} & \phi_{11}\phi_{21} Z_{m1} \\ \phi_{11}\phi_{21} Z_{m1} & \phi_{21}^2 Z_{m1} \end{bmatrix} \begin{bmatrix} F_{m1} \\ F_{m2} \end{bmatrix} \tag{5.21}$$

where Z_{m1} is the mobility of the parallel tuned circuit.

Since it has been necessary to make use of ideal transformers in the two-port problem, possibly we can also use them in driving-point immittance models. Considering a single point (or port) $i = 1$ and two modes $j = 1$, 2 of the disk of Figure 5-15(a), it seems logical that instead of representing the network in the manner shown in Figure 5-14, the circuit of Figure 5-15(c) can be used in its place. The velocity V_{m1} is, therefore, related to the force F_{m1} by

$$V_{m1} = (\phi_{11}^2 Z_{m1} + \phi_{12}^2 Z_{m2}) F_{m1}. \tag{5.22}$$

Making use of the results of (5.21) and (5.22), we can express the velocities of an M-port, N-mode resonator as a function of the forces by the set of equations

$$\begin{bmatrix} V_{m1} \\ V_{m2} \\ \vdots \\ V_{mi} \\ \vdots \\ V_{mM} \end{bmatrix} = \begin{bmatrix} z_{m11} & z_{m12} & \cdots & z_{m1i} & \cdots & z_{m1M} \\ z_{m21} & z_{m22} & \cdots & z_{m2i} & \cdots & z_{m2M} \\ \vdots & \vdots & & \vdots & & \vdots \\ z_{mi1} & z_{mi2} & \cdots & z_{mii} & \cdots & z_{miM} \\ \vdots & \vdots & & \vdots & & \vdots \\ z_{mM1} & z_{mM2} & & z_{mMi} & & z_{mMM} \end{bmatrix} \begin{bmatrix} F_{m1} \\ F_{m2} \\ \vdots \\ F_{mi} \\ \vdots \\ F_{mM} \end{bmatrix}$$

where

$$z_{mkl} = \sum_{j=1}^{N} \phi_{kj}\phi_{lj} Z_{mj}.$$

A generalized resonator model having M ports and N natural modes is shown in Figure 5-16.

The M ports in Figure 5-16 may be considered points of attachment of the transducer as well as coupling wires on the disk. Having found lumped-element equivalent circuits for various transducers, we will next consider the coupling wires.

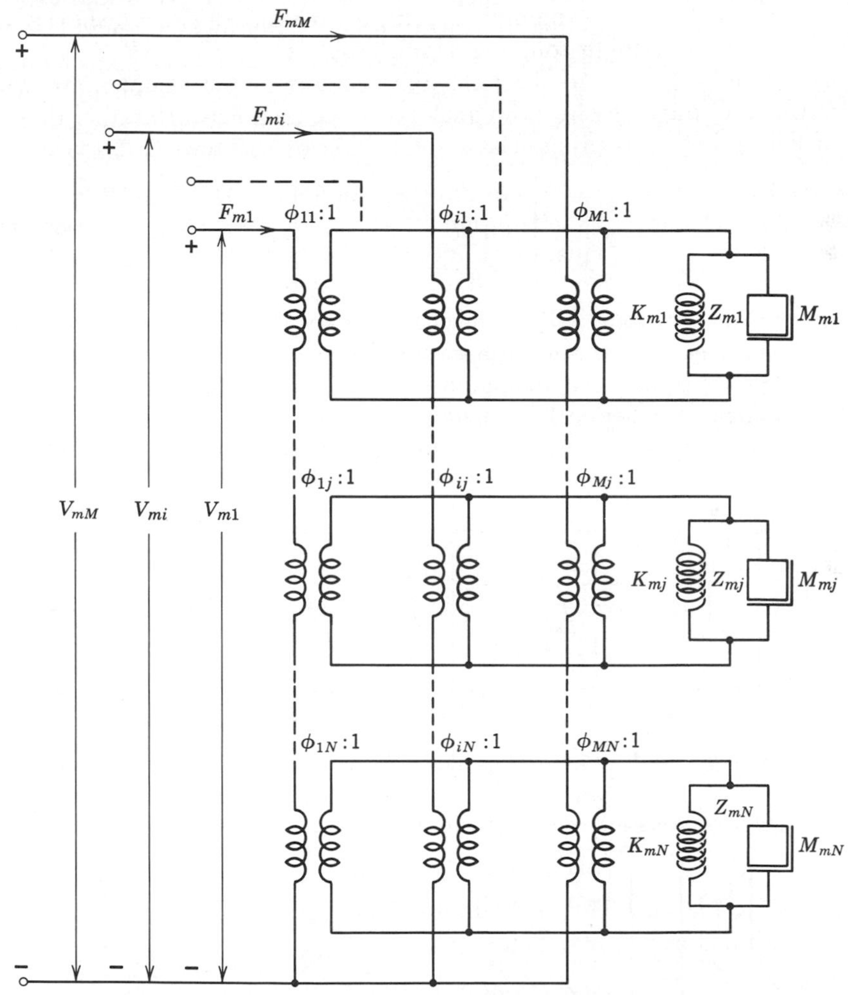

FIGURE 5-16 Generalized schematic diagram of an M-port, N-mode resonator.

Coupling Wires

Most mechanical filters make use of small diameter wires to couple energy from resonator to resonator. If the wires are driven either longitudinally or in torsion, the equivalent circuit representation is that of a transmission line, as in the case of the ferrite- and ceramic-transducer rods. If the coupling mode is flexural, then both a bending moment and an angular velocity must be added to the translational force and velocity. If both flexure and longitudinal modes are present, as in the case of coupling thick disks, the equivalent circuits become extremely complex. In this section we will consider only the longitudinal-mode case, and more specifically the case of either a short wire or a quarter-wavelength coupling wire.

As an introduction to the following sections of this chapter, we will describe the coupling wire as an analogous electrical network rather than a mechanical line or a spring-mass system. First we can rewrite (5.15) as

$$\begin{bmatrix} V_1 \\ I_1 \end{bmatrix} = \begin{bmatrix} \cos \beta l & jZ_0 \sin \beta l \\ j\dfrac{\sin \beta l}{Z_0} & \cos \beta l \end{bmatrix} \begin{bmatrix} V_2 \\ I_2 \end{bmatrix} \tag{5.23}$$

where l is the wire length, Z_0 is the characteristic impedance of the line, and β is the propagation constant (phase shift per unit length). Keeping in mind the relationships between analogous electrical and mechanical systems (i.e., the topologies are identical, capacitance is analogous to mass, inductance to compliance, etc.), we can make use of the pi network shown in Figure 5-17(a).

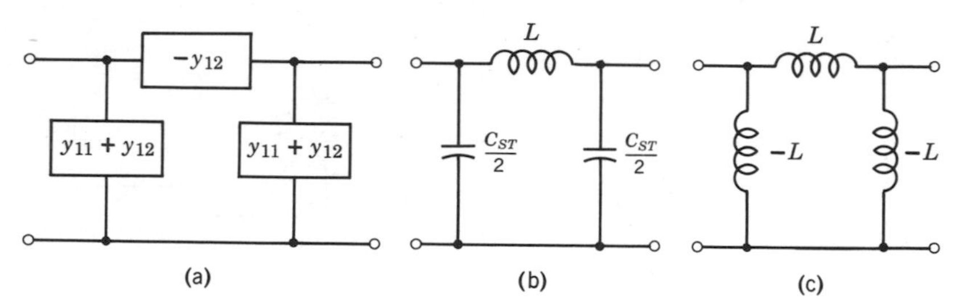

FIGURE 5-17 Pi-type coupling-wire descriptions: (a) generalized two-port, (b) short transmission line equivalent circuit, and (c) quarter-wavelength equivalent circuit.

The y's are the short-circuited admittance parameters of the two-port and can be formed from (5.23). Since

$$y_{11} = \frac{D}{B} = \frac{\cot \beta l}{jZ_0}; \; y_{12} = -\frac{1}{B} = \frac{j}{Z_0 \sin \beta l},$$

we can use a half-angle formula to obtain the shunt-arm admittances,

$$y_{11} + y_{12} = \frac{j \tan(\beta l/2)}{Z_0}.$$

When the coupling wire is less than one-eighth wavelength ($\tan(\beta l/2) \approx \beta l/2$), the shunt arm can be replaced by a capacitance $C_{st}/2$ which is analogous to one-half of the static mass of the wire. The series arm can be replaced, as shown in Figure 5-17(b), by an inductance L which is analogous to the compliance $C_m = l/AE_m$.

A very useful case to consider is one when the coupling wire is a quarter wavelength. It is not difficult to show that the equivalent circuit of Figure 5-17(c) approximates a quarter-wavelength line by equating matrix elements in (5.23) and the *ABCD* matrix of Figure 5-17(c). Doing this we find that

$$L = \frac{Z_0}{\omega}, \tag{5.24}$$

or in mechanical terms, using the definitions of Z_0 etc. in (5.15),

$$C_m = \frac{2l}{\pi A E_m}. \tag{5.25}$$

Comparing (5.8), (5.23), and (5.24), we see that for the quarter-wavelength line $\kappa = Z_0 = \omega L$.

5.4 NARROW-BAND MONOTONIC LOSS RESPONSE DESIGNS

The procedure for designing a mechanical filter is basically that of designing an electrical filter with various mechanical constraints kept in mind. Most of the design is done in electrical terms because of (1) the vast body of literature dealing with the design of LC filters as compared to that dealing with the synthesis of coupled mechanical systems, and (2) the fact that most network analysis and synthesis computer programs are written in terms of electrical parameters. The need for keeping the mechanical constraints in mind is necessary because of cost considerations and physical realizability. As an example, in a disk-wire filter it is most economical to keep all of the disk resonators the same size. Therefore, the design method used may be a direct synthesis of the equal-resonator filter or involve a series of transformations that will accomplish the same result.

A more fundamental restriction than that of relative element values is the fact that the topology of mechanical filters is quite limited, either by choice (economics again) or because of physical realizability. Therefore, if we have an excellent lattice-synthesis computer program, it may be of no direct use because there may be no way to realize a lattice mechanically, or at least not without using electrical transformers. However, it may be possible to transform the lattice network to a ladder network or a bridged-T network that is realizable with mechanical resonators and coupling wires. Therefore, the subject of network transformations is extremely important to the mechanical filter designer.

The designer has available to him both insertion-loss and image-parameter synthesis techniques. A knowledge of image-parameter design is very helpful to the designer, because it provides him with a simple way of checking mechanical realizability of an electrical design having perhaps an unusual topology. Once realizability has been established, an insertion-loss design computer program can be written to find element values that will result in a response closer to optimum.

Lowpass to Bandpass Transformations

One insertion-loss technique used to design monotonic-ladder filters is based on the series of transformations shown in Figure 5-18. The design

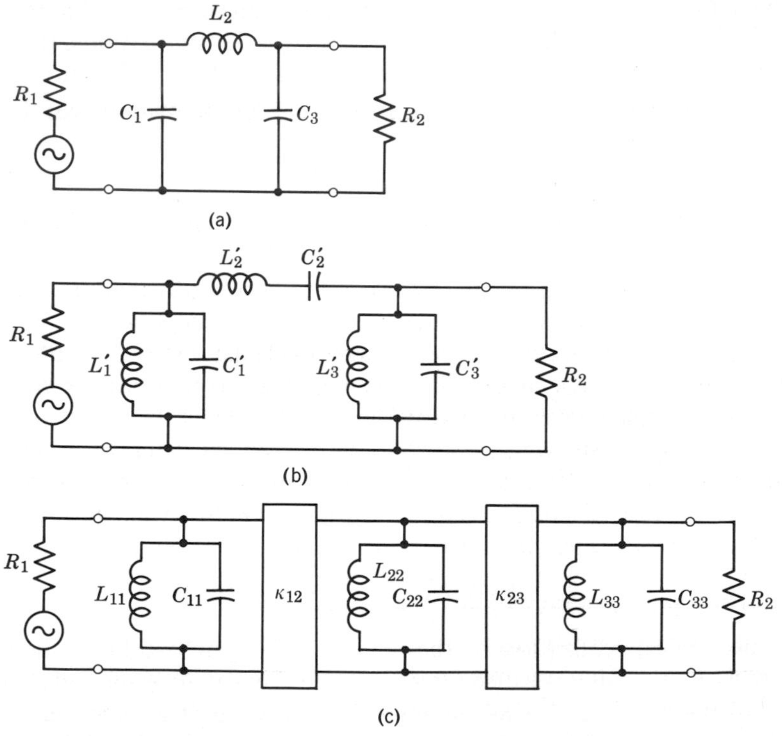

FIGURE 5-18 Lowpass to inverter-coupled bandpass filter transformations.

problem is this: based on a lowpass, equal-ripple prototype, design a bandpass-ladder mechanical filter having resonators with equal equivalent masses. The required input data are the number of resonators n, the pass-band ripple, the bandwidth BW, and the center frequency f_0. From the value of ripple and the number of resonators we can calculate the lowpass element values (The necessary equations are found in reference [9]). If we make use of tables in the literature we often find the lowpass network described in terms of normalized lowpass coupling coefficients, k. The element values and the coupling coefficients of a lowpass filter having a cut-off frequency ω_c are related by

$$k_{i,\,i+1} = \frac{1}{\omega_c \sqrt{C_i L_{i+1}}}\,; i = 1, n-1,$$

where C_1 can be found from a normalized Q value q_1 and the terminating resistances R_1 and R_2 as follows [10]:

$$q_1 = R_1 C_1 \omega_c; \qquad q_2 = R_2 C_n \omega_c.$$

The network element values of Figure 5-18(a) can be converted to those of Figure 5-18(b) by standard lowpass-to-bandpass transformations. Inverters can then be placed between the resonators and the network of Figure 5-18(c) obtained. We can avoid some of this work by making use of the equations,

$$\kappa_{i,\,i+1} = \frac{R_1}{(k_{i,\,i+1})q_1} \tag{5.26}$$

and

$$C_{11} = \frac{q_1}{R_1\,BW}; L_{11} = \frac{1}{\omega_0^2 C_{11}}\,; \omega_0 = 2\pi f_0 \tag{5.27}$$

where κ is the reactance of the inverter matrix element $j\kappa$ of (5.8), BW is the bandwidth in rad/sec, and C_{11} and L_{11} are the capacitance and inductance of the equal-valued shunt resonators.

Through the use of (5.26) and (5.27), we are able to calculate the electrical element values of a bandpass filter composed of equal shunt resonators coupled by inverters. The question we must now ask is how do we realize the resonators and inverters with mechanical elements?

Realization as a Cascade of Half-Wavelength Rods

One mechanical realization of the network of Figure 5-18(c) is shown in Figure 5-19. This filter is called a rod-neck mechanical filter, the rods being a half-wavelength long, the coupling necks being a quarter-wavelength long. We have previously discussed the fact that over a narrow band of frequencies a half-wavelength resonator (or transducer rod) can be replaced by a

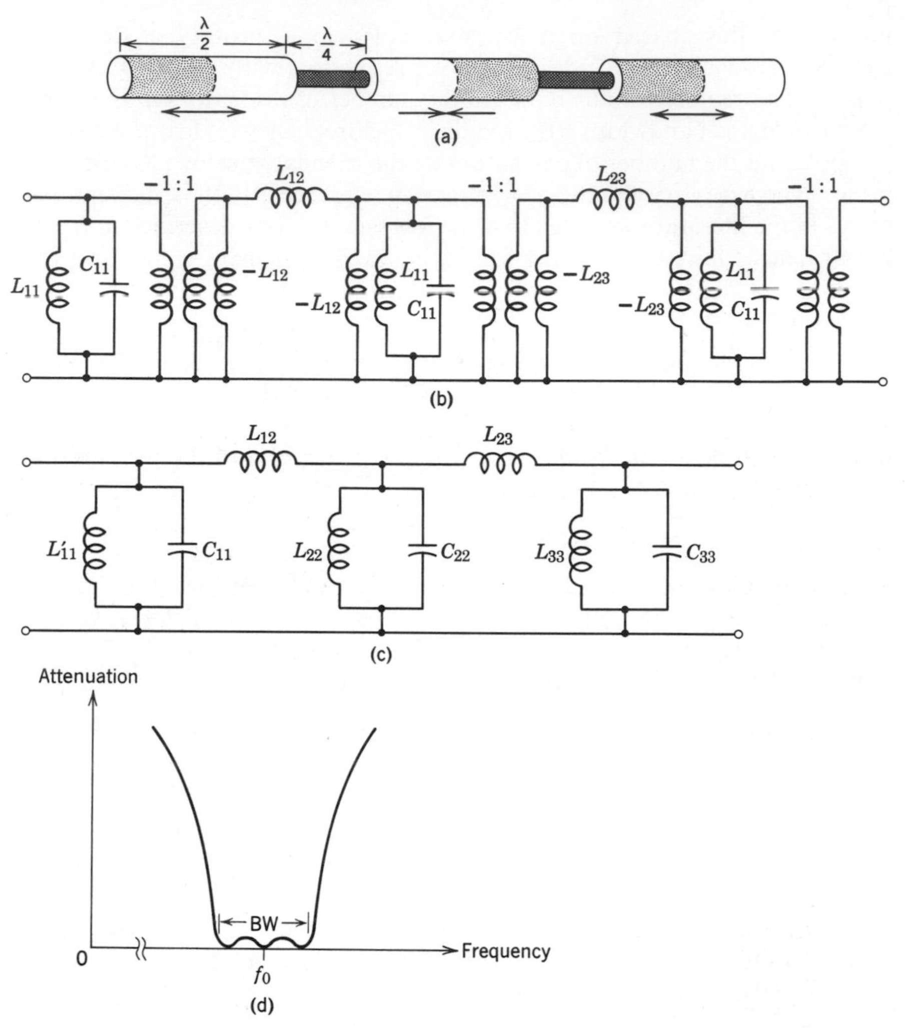

FIGURE 5-19 Rod-neck mechanical filter: (a) pictorial diagram, (b) and (c), lumped-element schematic diagrams, and (d) frequency response.

simple parallel spring and mass described by (5.17). To be more exact, the *ABCD* description of a half-wavelength resonator is

$$\begin{bmatrix} V_{m1} \\ F_{m1} \end{bmatrix} = \begin{bmatrix} 1 & 0 \\ 1/Z_{m1} & 1 \end{bmatrix} \begin{bmatrix} -1 & 0 \\ 0 & -1 \end{bmatrix} \begin{bmatrix} V_{m2} \\ F_{m2} \end{bmatrix}, \tag{5.28}$$

where the mobility expressed in terms of the equivalent compliance and mass is

$$Z_{m1} = j\omega C_{m1} + \frac{1}{j\omega M_{m1}} = j\frac{C_{m1}}{\omega}[\omega^2 - \omega_0^2] \tag{5.29}$$

where from (5.17)

$$M_{m1} = \frac{\rho_{m1} l_1 A_1}{2}; \qquad C_{m1} = \frac{2l_1}{\pi^2 A_1 E_{m1}}.$$

In other words, a half-wavelength resonator can be represented by a shunt resonator and a phase inverter as shown in Figure 5-19(b).

The impedance inverter κ_{12} of Figure 5-18(c) can be realized with a quarter-wavelength rod or line having the matrix description,

$$\begin{bmatrix} V_{m1} \\ F_{m1} \end{bmatrix} = \begin{bmatrix} 0 & j\kappa_{12} \\ j/\kappa_{12} & 0 \end{bmatrix} \begin{bmatrix} V_{m2} \\ F_{m2} \end{bmatrix}$$

where from (5.25)

$$\kappa_{12} = \omega_0 C_{m12} = \frac{2\omega_0 l_{12}}{\pi A_{12} E_{m12}}. \tag{5.30}$$

Assuming that both the resonator and coupling element are made from the same material, we can use (5.29) and (5.30) to find the ratio of their cross-sectional areas. Thus

$$\frac{A_{12}}{A_1} = \frac{\left(\dfrac{2\omega_0 l_{12}}{\kappa_{12}\pi E_{m12}}\right)}{\left(\dfrac{2l_1}{C_{m1}\pi^2 E_{m1}}\right)} = \frac{\left(\dfrac{2\omega_0 l_{12}}{\kappa_{12}\pi E_{m12}}\right)}{\left(\dfrac{2l_1 \omega_0^2 M_{m1}}{\pi^2 E_{m1}}\right)} = \frac{\pi}{2\omega_0 \kappa_{12} M_{m1}}$$

with $l_1 = 2l_{12}$.

Since M_{m1} is analogous to C_{11} of (5.27), we can write, using (5.26)

$$\kappa_{12} M_{m1} = \frac{1}{k_{12} BW}$$

or

$$\frac{A_{12}}{A_1} = \frac{\pi}{2}\left(\frac{BW}{\omega_0}\right) k_{12},$$

or in general,

$$\frac{A_{i,\,i+1}}{A_1} = \frac{\pi}{2}\left(\frac{BW}{\omega_0}\right) k_{i,\,i+1}. \tag{5.31}$$

From (5.31) we see that the cross-sectional area of the coupling wire is proportional to the fractional bandwidth and the coupling coefficient $k_{i,\,i+1}$. In the case of equal-ripple designs, the coupling is greatest at the ends and decreases toward the center of the filter, becoming an almost constant value near the center for designs having a large number of resonators.

In the electrical equivalent circuit of Figure 5-19(b), we have replaced the quarter-wavelength inverters with pi networks composed of $\pm L_{i,\,i+1}$ values. Eliminating the phase inverter transformers by moving them all to the right and then combining the shunt negative coupling elements with the resonator inductors, we obtain the equivalent circuit of Figure 5-19(c) which has the frequency response shown in Figure 5-19(d).

In a design of this type (where the resonators are coupled with quarter-wavelength lines), the resonators are tuned to the center frequency of the band by varying their length. The lengths of the coupling elements are not critical, but the diameters must be held very close or the coupling will be incorrect. For a narrow-bandwidth design the coupling necks become quite small in diameter, as we can see from (5.31), and the filter becomes very fragile. In addition, at low frequencies the resonators become quite long (almost 1 in. at 100 kHz in the case of iron-nickel alloys), and the length of the filter shown in Figure 5-19(a) becomes impractical. As a means of both increasing the strength of the coupling wires (necks) and reducing the length of the filter, we can design the mechanical filter as shown in Figure 5-20(a).

The first resonator A of the folded design of Figure 5-20(a) also acts as a transducer of the Langevin type and vibrates in a half-wavelength longitudinal mode. The coupling wires D and E between the resonators are very short, in fact, less than one-eighth wavelength long. This corresponds to the case shown in Figure 5-17(b) where the coupling inductance L is analogous to the compliance $C_m = l/AE_m$, and the shunt capacitors are analogous to one-half the static mass of the wire. The three-resonator, two-coupling wire filter is shown in the form of an electrical equivalent circuit in Figure 5-20(b). The coupling inductance L_{12} is now directly proportional to the length of the wire and inversely proportional to the cross-sectional area. The required coupling-wire resonator cross-sectional area ratio can be found by making use of the equation

$$\frac{C_{m11}}{C_{m12}} = \frac{L_{11}}{L_{12}} = k_{12}\left(\frac{BW}{\omega_0}\right),$$

which can be easily derived from (5.26) and (5.27) where $\kappa_{12} = \omega_0 L_{12}$. Since,

$$C_{m12} = \frac{l_{12}}{A_{12} E_{m12}} \qquad \text{and} \qquad C_{m11} = \frac{2l_1}{\pi^2 A_1 E_{m1}},$$

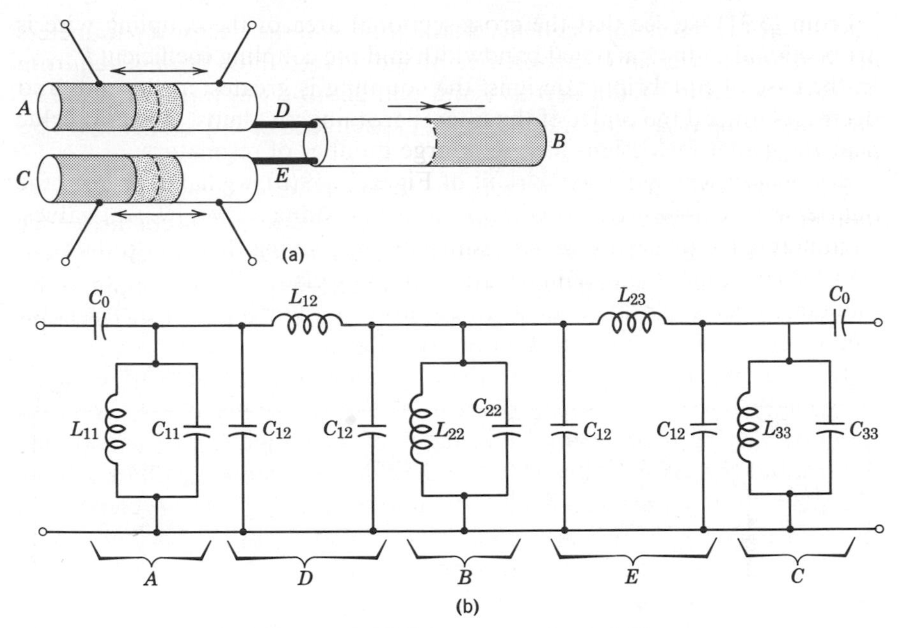

FIGURE 5-20 Folded-line mechanical filter.

we can write (for the case where $E_{m1} = E_{m12}$),

$$\frac{A_{12}}{A_1} = \frac{\pi^2}{2}\left(\frac{BW}{\omega_0}\right)\left(\frac{l_{12}}{l_1}\right)k_{12}. \tag{5.32}$$

Comparing (5.32) to (5.31), we see that the ratio of the cross-sectional areas is now proportional to the wire length, or, if we solve for the bandwidth BW, that the bandwidth is inversely proportional to the coupling-wire length.

Besides the fact that no phase inverters are needed in the equivalent circuit of Figure 5-20(b), we can also conclude after a comparison with Figure 5-19(b) that the resonators in the short-wire and the quarter-wavelength cases must be tuned to different frequencies. From (5.27) we see that in the quarter-wavelength case all resonators are tuned to the center frequency. Since the short-wire case is also derived from the same lowpass prototype, the resonator frequencies can be found by comparing the effect of C_{12} of Figure 5-20(b) and $-L_{12}$ of Figure 5-19(b). In other words, after combining the C_{ij}'s and $-L_{ij}$'s with their adjacent resonator capacitors or inductors, the resultant resonant frequencies should be identical. When the transducer is electrically tuned, the static capacitance or coil inductance will have no effect on the frequency of the end mechanical resonator, and what we have said about the resonator frequencies still applies. If the electrical parameters are not tuned externally, capacitance or inductance must be removed from the end resonators, and a narrow-band parallel-to-series transformation of the terminating resistance and the shunt capacitance or inductance must be performed.

The basic concepts that have been outlined for designing filters with monotonic loss response can be applied to a wide variety of mechanical filter types [11]. Figure 5-21 shows a number of designs, some of which we have already discussed. The nodal patterns and directions of motion apply to the case where the coupling elements are very short and the system is vibrating

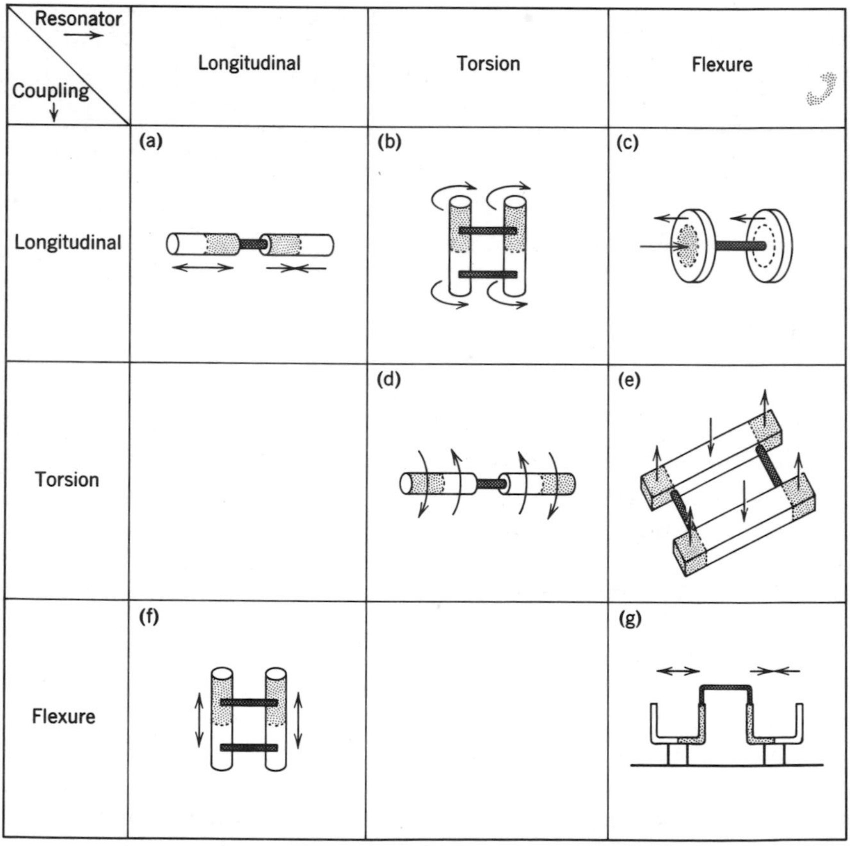

Resonator → / Coupling ↓	Longitudinal	Torsion	Flexure
Longitudinal	(a)	(b)	(c)
Torsion		(d)	(e)
Flexure	(f)		(g)

FIGURE 5-21 Various mechanical filter configurations using longitudinal, torsion, and flexural resonators and coupling elements. Displacements in the shaded areas are in-phase at the lower natural mode.

at its lower natural mode. All of the configurations shown are practical devices that have been manufactured in the United States, Europe, or Japan for a number of years. Some modes, such as the contour modes (radial, concentric shear, and face shear) which have also been used in practice, have been omitted. In addition, we could have included the plate-wire and folded-rod-wire designs in Figure 5-21(a), as well as a number of other longitudinal and flexural mode devices.

The two flexural-resonator designs shown in Figures 5-21(e) and (g) are most often used at low frequencies; from a few hundred Hz to 50 kHz [12]. The other filters shown in Figure 5-21 are used in the frequency range of 50 kHz to 500 kHz. The fractional bandwidths of all designs (ratio of bandwidth to center frequency) vary from 0.1% to 10%, the lower limit being a function of resonator Q and temperature stability, while the upper limit is fixed by unwanted modes and the electromechanical-coupling coefficient of the transducer. Greater selectivity is achieved in each case by adding more resonators. Of course, adding more resonators increases the cost of the device. In addition, as more resonators are added, the delay varies more in the passband, and in most cases the passband ripple also increases (due to practical problems in manufacturing). In order to keep the number of resonators to a minimum, methods have been developed to realize attenuation poles at finite frequencies.

5.5 FINITE ATTENUATION POLES

A number of methods have been used to realize finite attenuation poles. These include the use of resonant coupling elements (producing poles in the series arms of a ladder), multiple resonator modes (resulting in zeros of the shunt arms of a ladder), and bridging wires to couple nonadjacent resonators [13]. The latter method has been by far the most effective, and it will be discussed in detail in this section.

Figure 5-22(a) shows a 3-disk-resonator section where one of the coupling wires is used to bridge the center disk. Below the lower edge of the filter passband frequency, all three resonators vibrate in-phase. At frequencies above the upper passband edge, the center resonator vibrates out-of-phase with respect to the end resonators as the result of each resonator experiencing a 180° phase shift across the passband. Because of the attenuation contributed by the center resonator and its out-of-phase condition, there is a frequency in the upper stopband where the sum of the forces through the 3 coupling wires and the bridging wire are equal to zero, and no motion is developed in the end resonator, i.e., an attenuation pole is produced as shown in Figure 5-22(b).

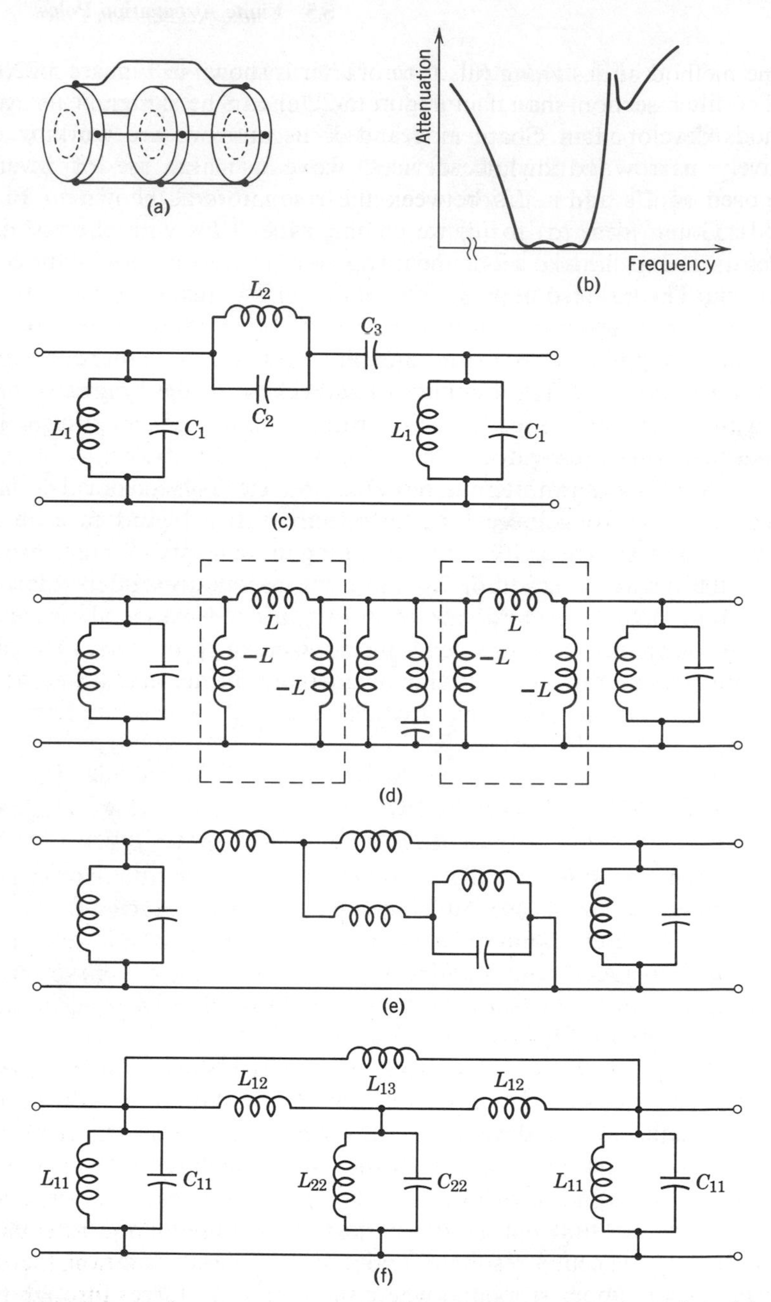

FIGURE 5-22 Single resonator bridging without phase inversion: (a) disk-wire filter section, (b) frequency response, and (c) to (f), network transformations.

One method of designing this type of filter is shown in Figures 5-22(c) to (f). The filter section shown in Figure 5-22(c) can be designed using the methods developed in Chapters 2 and 3. Because we are working with relatively narrow bandwidth devices, we can make use of inverters composed of L's and $-L$'s between the resonators as shown in Figure 5-22(d). Going from (d) to (e) we combine the $-L$'s with the resonator inductors and then make a 3-element-reactance transformation of the center shunt arm. The last step in going from (e) to (f) is to make a simple T to pi inductance transformation. In this case all element values are positive; the tuned circuits represent the disk resonators, while the inductors correspond to the compliance of the coupling wires. Because of the symmetry of the nodal-circle flexural mode, all three wires coupling adjacent disks can be represented by a single inductor L_{12}.

In order to realize an attenuation pole on the low-frequency side of the passband, a slightly different procedure can be used. In this case we start with a network such as that shown in Figure 5-23(a). A three-element reactance transformation of the series arm results in Figure 5-23(b). The next step involves inverting the center resonator, which leaves a phase inverter and two impedance inverters between the two end resonators. (The inversion of an arm involves starting with an ideal transformer in cascade with the arm. The ideal transformer is then converted to a phase inverter (half-wavelength line) and two impedance inverters (quarter-wavelength lines). One of the impedance inverters is then moved to the other side of the arm, thus inverting the arm.) Normally, in a simple ladder network, we can ignore the phase inverter, but if the transformation takes place under a bridging arm this is no longer true. Therefore, in the case we are considering, the phase inverter is brought out from under the bridging inductor as shown in Figure 5-23(c). The bridging inductor can now be considered to have a phase inverter associated with it, or we can think in terms of the bridging wire being attached to a point on the last resonator that is out of phase with respect to the point to which the adjacent-resonator coupling wires are attached. The first case can be realized as shown in Figure 5-23(d), where the bridging wire is greater than one-half wavelength long but less than a full wavelength. The phase inverter represents the additional half-wavelength. If we associate the phase inverter with the end resonator we can realize the filter in the manner shown in Figure 5-23(e).

The torsional-rod-resonator, longitudinal-wire-coupling mechanical filter is used in the 200 kHz to 250 kHz frequency range. A broad range of bandwidths can be covered by varying the position of the coupling wires with respect to the nodal lines. As in any transmission line resonator, the velocity (or voltage) is proportional to $\sin(2\pi x/\lambda)$ where x is the distance from a nodal line. From a previous discussion, the impedance or equivalent

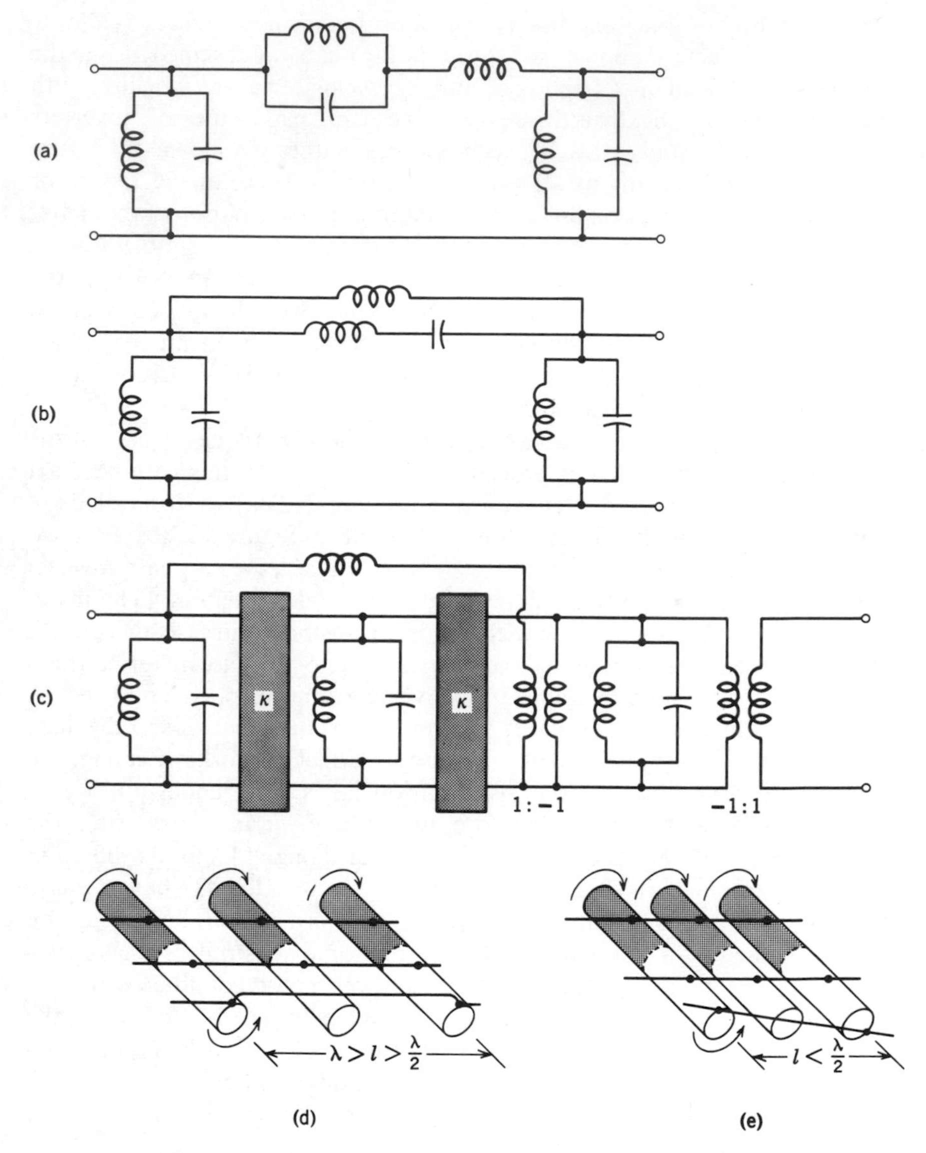

FIGURE 5-23 Single resonator-bridging transformations for the case where the attenuation pole is in the lower stopband. Alternate rod-wire realizations are shown.

mass of the resonator is therefore proportional to $1/\sin^2(2\pi x/\lambda)$. For a particular bandwidth, the coupling-wire diameter can therefore be increased as the wire is moved toward the nodal line, or decreased as the wire is moved toward the end of the resonator. This is the kind of flexibility the designer often looks for when choosing a mechanical configuration.

Still another set of transformations is needed when a coupling wire is used to bridge across two resonators in order to realize attenuation poles on both sides of the filter passband. A very useful transformation that can be used for both lowpass and bandpass filters is shown in Figure 5-24(a). The element values in the two networks are related by the equations,

$$\eta^* = 1 + \frac{1}{\eta}$$

$$y_1^* = -\eta y_2$$

$$z_2^* = \frac{1}{(1+\eta)(\eta y_2 + (1+\eta)/z_1)}$$

$$y_3^* = (1+\eta)y_2. \tag{5.33}$$

Starting with either the network shown on the top of Figure 5-24(b) or its lowpass equivalent, it is not difficult to obtain the two-disk bridging section shown on the bottom. A necessary condition is that the bandpass filter have a symmetrical response, i.e., that it have a lowpass equivalent. With the exception of the transformation of Figure 5-24(a), all the steps in the over-all transformation have been discussed in regards to the single-disk bridging cases. Note that in this case the use of inverters under the bridging wire have again resulted in the need for phase inversion associated with either the bridging wire or the end resonator. Figure 5-24(c) shows the frequency response of the filter. The attenuation poles are at real, finite frequencies, and the delay response within the passband is roughly parabolic. If the phase inverter is not included in the realization, the resultant response (after adjusting frequencies and coupling) is that shown in Figure 5-24(d). This is the result of the attenuation poles moving off the $j\omega$-axis into the right half of the s-plane, making delay compensation possible.

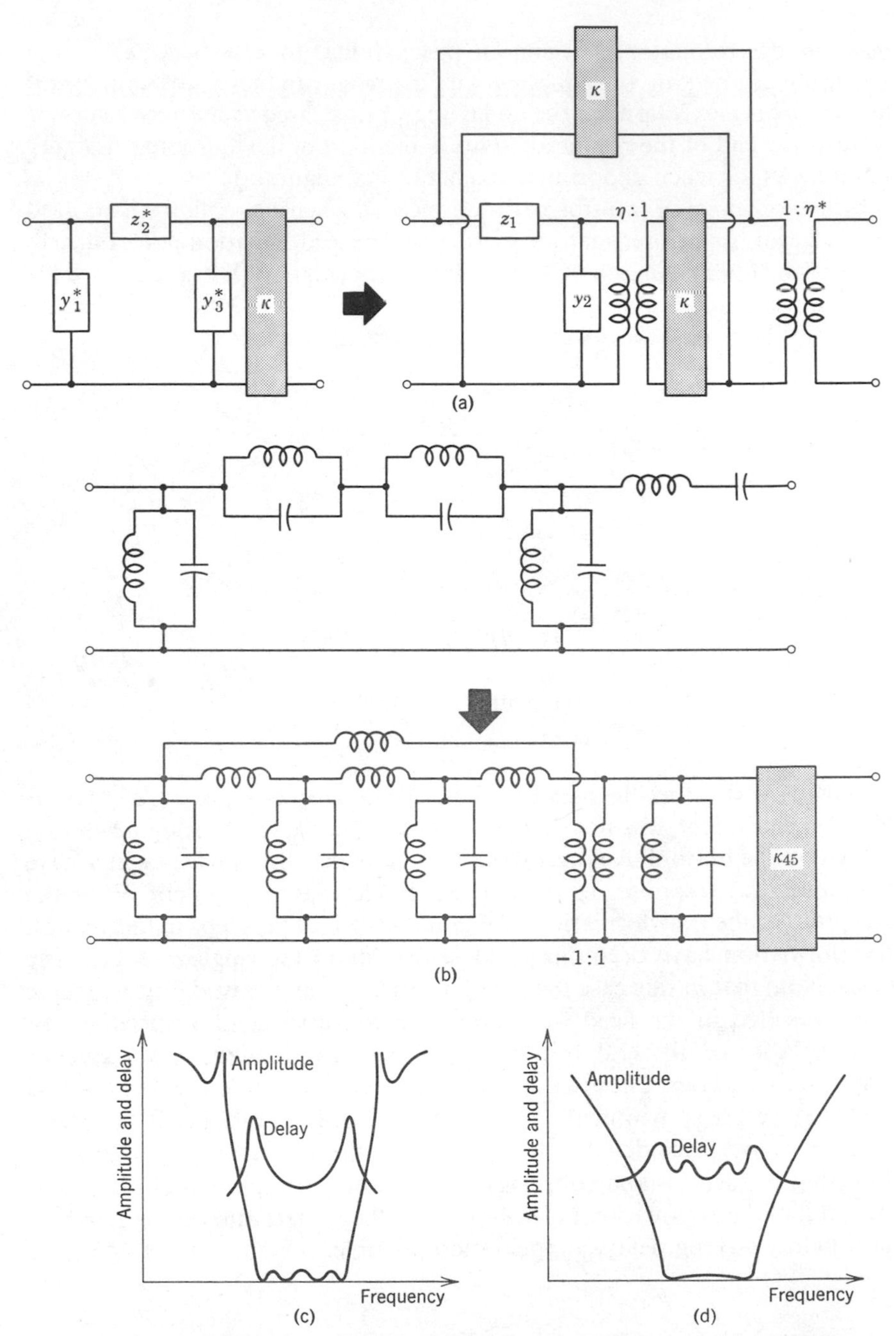

FIGURE 5-24 Two resonator-bridging transformations: (a) general transformation, (b) bandpass ladder to mechanical filter topology, and (c) and (d), bridging with and without phase inversion.

5.6 MONOLITHIC MECHANICAL FILTERS

Earlier in this chapter it was mentioned that a bandpass filter described as a monolithic or quartz-mechanical filter could be constructed from a quartz substrate and an array of electrode pairs. Figure 5-25(a) shows an example of a two-electrode-pair monolithic filter and Figure 5-25(b) describes the vibration of this filter at one of its natural resonances. The particular mode of vibration shown in (b) is the thickness twist (TT), where the direction of wave

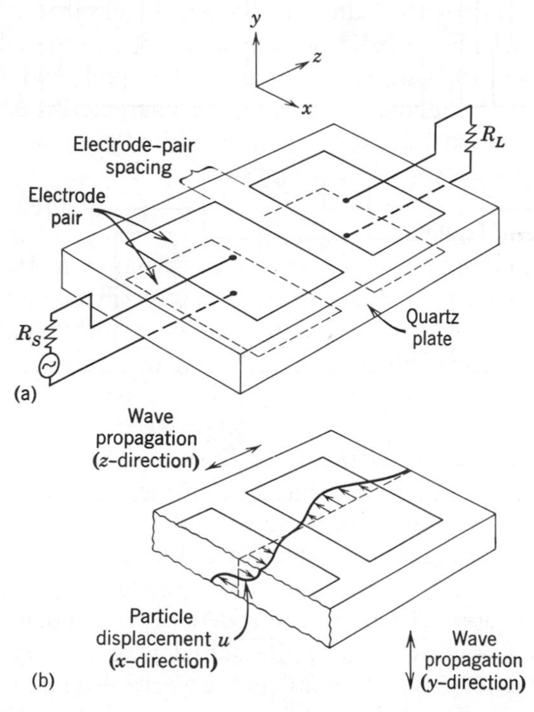

FIGURE 5-25 Two resonator monolithic filter: (a) pictorial diagram and (b) cutaway view showing particle displacements and wave propagation directions for the thickness-twist (TT) mode of vibration.

propagation down the length z of the plate is perpendicular to the direction of particle displacement x. The acoustic (or mechanical if you like) wave is generated by the voltage across the input electrode pair as the result of the fact that the plate is piezoelectric. It should also be noted that not only is there a wave propagated in the length direction but in the thickness direction (y) as well.

The waves propagated in the length direction are of two types, freely propagating waves (standing waves) and exponentially decaying waves.

Within the nonattenuating electrode region we find standing wave patterns, and hence the electrode pairs act like resonators. In the region between the electrode pairs, the acoustic wave decays, thus decoupling the two resonators. Therefore, the system of two electrode pairs, coupled (or perhaps we should say decoupled) by the unplated region between the electrodes, has two natural resonances. The frequency difference between these two modes is a function of the amount of electrode plating as well as the spacing or distance between the two pairs of electrodes. When electrically terminated by the proper source and load impedances, the electroded plate becomes a bandpass filter having a bandwidth proportional to the difference between the natural frequencies, i.e., a function of plating and spacing between electrode pairs. Equations describing the characteristics of a monolithic filter are very closely related to equations found in optics, microwaves, and quantum theory.

Single Resonator System

A single-electrode-pair acoustic resonator is shown in Figure 5-26(a) [14]. The electrodes are often mixtures or layers of silver, chromium, and gold deposited on a quartz plate. The differences in both thickness and density across the electroded and nonelectroded regions make an exact analysis of this system difficult. Adding to the difficulty is the fact that quartz is a nonisotropic material, i.e., the acoustical properties of the plate are different in each direction. Rather than attempting to analyze the actual system, we can derive a good basic understanding of the single resonator system by instead looking at the plate of Figure 5-26(b).

The resonator of Figure 5-26(b) has two different mass density (ρ_e and ρ_s) regions. The subscript e represents the electroded region, the subscript s the surrounding region. The term electroded region, though somewhat of a misnomer, is helpful in keeping the actual physical system in sight. The thickness of the plate is d, the width of the electroded region is $2b$. The mode of vibration we will consider is the thickness-twist shown in Figure 5-25(b). The solution to the single-electrode-pair case, in terms of finding natural frequencies, is as follows:

1. The wave equation is written for each region of the plate.
2. Based on certain boundary conditions, preliminary solutions are assumed for each region.
3. The solutions are substituted into the wave equations and the results evaluated.
4. A general solution for each region of the plate is assumed.
5. A set of simultaneous equations is written for the entire system. From these the natural frequencies are found.
6. The frequency solutions are interpreted physically.

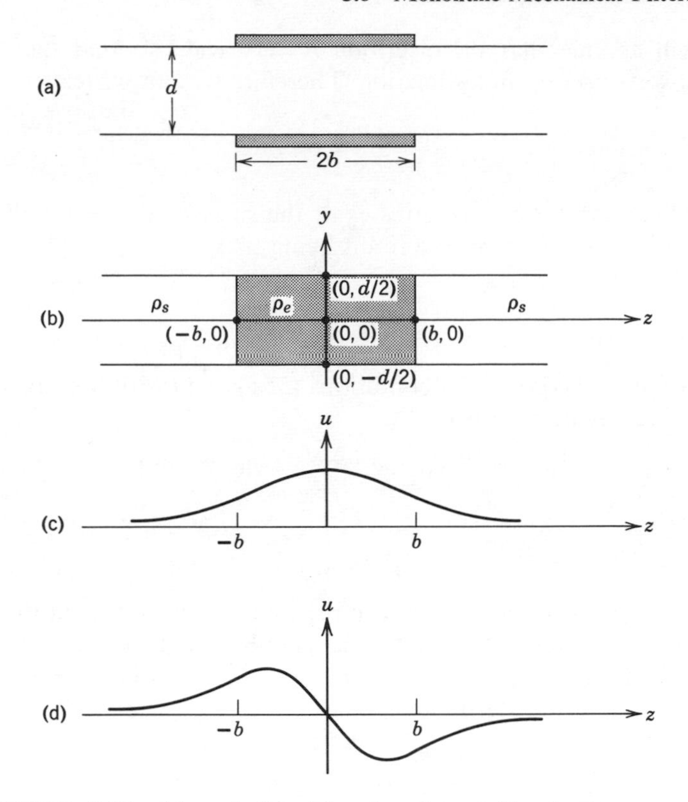

FIGURE 5-26 (a) and (b), Plated and variable density acoustic (monolithic) resonators, and (c) and (d), particle displacement at the two lowest natural modes.

Because we are considering the thickness-twist mode, the wave equation for each region will be that of a shear wave, namely,

$$\nabla^2 u_e = \left(\frac{1}{v_e^2}\right)\frac{\partial^2 u_e}{\partial t^2}; \qquad |z| \le b, \tag{5.34}$$

$$\nabla^2 u_s = \left(\frac{1}{v_s^2}\right)\frac{\partial^2 u_s}{\partial t^2}; \qquad |z| > b,$$

where u is the displacement in the x-direction. The velocity v is the square root of the shear modulus divided by the density ρ. ∇^2 is the operator,

$$\nabla^2 = \frac{\partial^2}{\partial x^2} + \frac{\partial^2}{\partial y^2} + \frac{\partial^2}{\partial z^2}.$$

We will assume that the electrode is very wide so that there are no standing waves set up in x-direction. Therefore we can write,

$$\frac{\partial u}{\partial x} = 0. \tag{5.35}$$

Since there are no normal stresses at the surfaces $y = \pm d/2$, the strains must also be equal to zero; the result being that

$$\left.\frac{\partial u}{\partial y}\right|_{y=\pm d/2} = 0. \tag{5.36}$$

As a result of the boundary conditions (5.35) and (5.36), we can assume a solution (see Figure 5-25(b))

$$u = \sin \beta_m y[Ae^{j(\alpha_m z - \omega t)} + Be^{j(\alpha_m z - \omega t)}] \tag{5.37}$$

where

$$\beta_m = \pi/d. \tag{5.38}$$

In (5.37) and (5.38), α_m and β_m are propagation constants in the z and y directions respectively. Note that β_m is fixed by the thickness of the plate.

The next step in the solution is to substitute (5.37) into the wave equations (5.34) resulting in the algebraic equations

$$\alpha_m^2 = \frac{\omega^2}{v_e^2} - \beta_m^2 = \alpha_e^2 - \beta_m^2;\ |z| \le b$$

$$\alpha_m^2 = \frac{\omega^2}{v_s^2} - \beta_m^2 = \alpha_s^2 - \beta_m^2;\ |z| > b \tag{5.39}$$

where $\alpha_e = \omega/v_e$ and $\alpha_s = \omega/v_s$.

If we assume that $\rho_e > \rho_s$, then $v_e < v_s$, and there is a range of frequencies where α_m is positive in the electroded region and imaginary (α_m^2 is negative) in the surrounding region. The limits of this range are ω_s and ω_e where

$$\omega_s = \frac{\pi v_s}{d}$$

$$\omega_e = \frac{\pi v_e}{d}. \tag{5.40}$$

Therefore, below the frequency ω_s the propagation constant α_m in the surrounding region is imaginary and becomes an attenuation constant,

$$\alpha_m = \pm j\gamma_m;\ -b > z > b;\ \omega < \omega_s.$$

Next, a general solution for each region of the plate will be considered.

Within the frequency range ω_e to ω_s, the wave propagation is attenuation free in the electroded region but attenuated in the surrounding region. This leads to the possibility that there will be reflections at the boundary ($z = \pm b$), and standing waves will be set up in the electroded region. Since we are considering an infinite plate, all waves that leave the electroded region propagate outward from this region, never to return. Therefore, our general solutions can be written,

$$
\begin{aligned}
u_e &= [Ae^{-j\alpha_m z} + Be^{+j\alpha_m z}]\sin \beta_m y \\
u_s &= [Ce^{-\gamma_m z}]\sin \beta_m y;\ z > b \\
u_s &= [De^{\gamma_m z}]\sin \beta_m y;\ z < -b.
\end{aligned}
\tag{5.41}
$$

We can find the natural frequencies of the plate by writing a set of simultaneous equations for the entire system. At the boundary between the electrode and surrounding regions ($z = \pm b$), two continuity equations must be satisfied, namely that the amplitude and slope (with respect to z) of the displacement be continuous. Therefore from (5.41) at $z = \pm b$ we can write,

$$
\begin{bmatrix}
e^{-j\alpha_m b} & e^{j\alpha_m b} & -e^{-\gamma_m b} & 0 \\
-j\alpha_m e^{-j\alpha_m b} & j\alpha_m e^{j\alpha_m b} & \gamma_m e^{-\gamma_m b} & 0 \\
e^{j\alpha_m b} & e^{-j\alpha_m b} & 0 & -e^{-\gamma_m b} \\
-j\alpha_m e^{j\alpha_m b} & j\alpha_m e^{-j\alpha_m b} & 0 & -\gamma_m e^{-\gamma_m b}
\end{bmatrix}
\begin{bmatrix} A \\ B \\ C \\ D \end{bmatrix} = 0. \tag{5.42}
$$

A solution to the above set of homogeneous equations exists only if the determinant formed from the matrix elements, called the secular determinant, is equal to zero. Using Laplace's development we obtain for the secular determinant

$$2j\alpha_m \gamma_m[2(\cos 2\alpha_m b)] + (\alpha_m^2 - \gamma_m^2)[-2j(\sin 2\alpha_m b)] = 0$$

or

$$\tan 2\alpha_m b = \frac{2\alpha_m \gamma_m}{\alpha_m^2 - \gamma_m^2}, \tag{5.43}$$

where from (5.38), (5.39), and (5.40)

$$
\begin{aligned}
\alpha_m &= \frac{\pi}{d}\left[\frac{\omega^2 - \omega_e^2}{\omega_e^2}\right]^{1/2} \\
\gamma_m &= \frac{\pi}{d}\left[\frac{\omega_s^2 - \omega^2}{\omega_s^2}\right]^{1/2}.
\end{aligned}
$$

Frequencies that satisfy (5.43) can be found by making use of iterative techniques involving both the right and left sides of the equation for various values of ω. This can be done with relative ease on a digital computer.

Equation (5.43) can be written as

$$\tan \alpha_m b = \frac{\gamma_m}{\alpha_m} \tag{5.44}$$

or, alternatively, as

$$\tan \alpha_m b = -\frac{\alpha_m}{\gamma_m}. \tag{5.45}$$

The solutions of (5.44) and (5.45) correspond to the curves shown in Figures 5-26(d) and (e), which result from solving for the coefficients A, B, C, and D in (5.42) at the two natural resonant frequencies of this system.

Two-Resonator Systems

The complexity of the general solution in the single-resonator case could have been reduced if we had first assumed a symmetric solution (Figure 5-26(d)) and had equated amplitudes and slopes at only one boundary, and then assumed an antisymmetric solution and again had equated amplitudes and slopes at only one boundary. The result would have been (5.44) and (5.45).

In the case of the two-resonator system of Figure 5-25, we can assume that the two lowest natural modes will have the amplitude distributions shown in Figure 5-27. Because of symmetry, five equations are needed to describe the system. Amplitude and slope equations are written at $\pm b$ and an amplitude equation at $b + s/2$. One of the equations can be eliminated because in the symmetrical case, the amplitudes (i.e., the coefficients C and D shown in

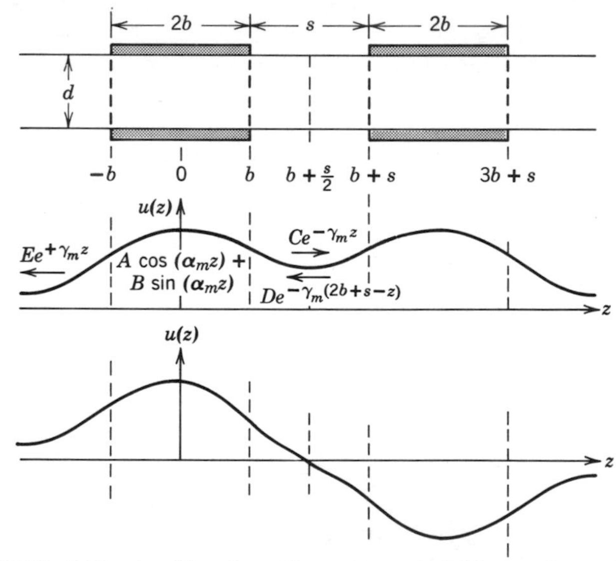

FIGURE 5-27 Double electrode pair monolithic mechanical filter showing the physical structure and the particle displacement at the two lowest natural modes of the system.

Figure 5-27) of the two waves in between the electrodes are equal in magnitude and sign, and in the antisymmetrical case they are equal in magnitude but opposite in sign (thus zero amplitude in the center). Solving the secular determinant of the system we obtain the symmetrical and antisymmetrical solutions,

$$2\alpha_m b = \tan^{-1}\left[\frac{-2\alpha_m \gamma_m}{(\gamma_m^2 - \alpha_m^2) - (\gamma_m^2 + \alpha_m^2)e^{-\gamma_m s}}\right] \tag{5.46}$$

$$2\alpha_m \gamma_m = \tan^{-1}\left[\frac{-2\alpha_m \gamma_m}{(\gamma_m^2 - \alpha_m^2) + (\gamma_m^2 + \alpha_m^2)e^{-\gamma_m s}}\right] \tag{5.47}$$

where α_m and γ_m are as defined in the single electrode-pair system. Although no closed-form solution is available for these equations, they can easily be solved on a digital computer. Using the normalized frequency variable $\Omega = (\omega - \omega_e)/(\omega_s - \omega_e)$, we obtain the curves shown in Figure 5-28(a). Note

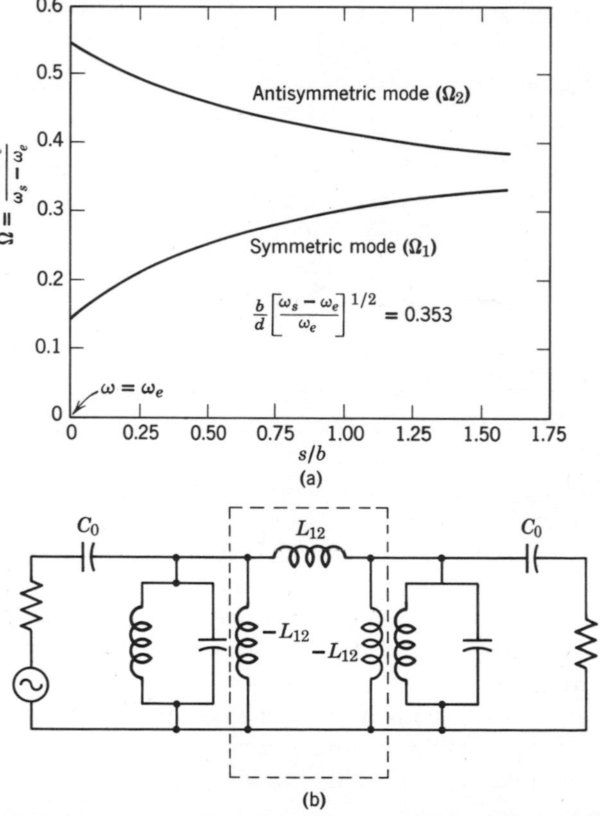

FIGURE 5-28 Double electrode pair monolithic filter: (a) lowest natural frequencies versus ratio of electrode spacing and half electrode width, and (b) electrical equivalent circuit.

that as the spacing becomes very large the two natural frequencies Ω_1 and Ω_2 converge to a value approximately equal to $(\Omega_1 + \Omega_2)/2$. If we were to show a curve of natural resonant frequencies versus coupling inductance for the network of Figure 5-28(b), we would obtain essentially the same curve. The capacitor C_0 corresponds to the static capacity of an electrode pair, the parallel tuned circuits represent the electrode-pair regions, and the inverter composed of positive and negative inductors represents the region between the electrode pairs.

By generalizing the results shown in this section, it is possible to design n-resonator, quartz-mechanical filters in much the same way as we design conventional mechanical filters. The equations relating to the networks of Figure 5-18 apply directly to the monolithic case. The relationship between the coupling in the electrical equivalent circuit and the acoustic system can be established either analytically through the use of equations such as (5.46) and (5.47) or experimentally by measuring the bandwidth of coupled pairs in the laboratory.

5.7 SUMMARY

A considerable portion of this chapter was devoted to the subject of equivalent circuits. This emphasis was for the purpose of creating a bridge between electroacoustics and electrical-circuit theory. In other words, an attempt was made to translate electroacoustic theory and measurements into an electrical engineering language. This was done at the expense of neglecting detailed acoustic problems (except for the monolithic filter case), which are important but probably not as important as being able to develop models that can be used for both analysis and synthesis of complete systems.

In terms of network models, the concepts that are probably of greatest importance involve equivalent mass and electromechanical coupling. The electromechanical coupling coefficient, for instance, relates the electrical network parameters to the mechanical in a very simple way ((5.14) is an example). Coupling between mechanical resonators involves a relationship between the compliance of the coupling wires and the apparent or equivalent mass of the mechanical resonant elements at the point of coupling. A thorough understanding of these relationships is basic to the understanding of mechanical filters.

The derivations of the monolithic filter equations are of broad importance because of their applicability to almost any complex vibration problem and to problems in other fields as well.

PROBLEMS

5.1 For the filter shown in Figure 5-29, sketch and label the mechanical schematic diagram and mobility analog, calculate the natural resonant frequencies, and represent the network as a 4-element driving-point immobility F_{m1}/V_{m1} (Foster form).

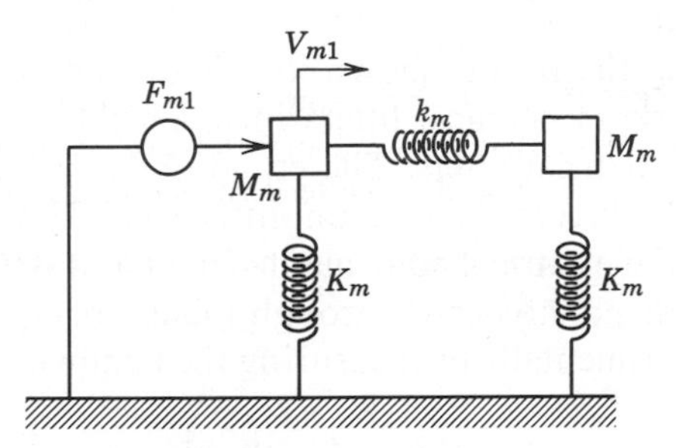

5.2 Sketch 3 alternate mobility representations of the magnetostrictive transducer equivalent circuit shown in Figure 5-8. Gyrators and inverters can be added.

5.3 For the case of an electric field transducer such as that shown in Figure 5-7, express the mechanical natural resonant frequency under short-circuit conditions ($V = 0$) in terms of C_m^D, M_m, C_0 and each of the following: (a) the gyrator impedance A, (b) the electromechanical transformer η, and (c) the electromechanical coupling coefficient k_{em}.

5.4 Find the equivalent mass of the half-wavelength bar shown in Figure 5-30 by (a) equating the slopes of the driving point immobilities at resonance and (b) by the method suggested in the text.

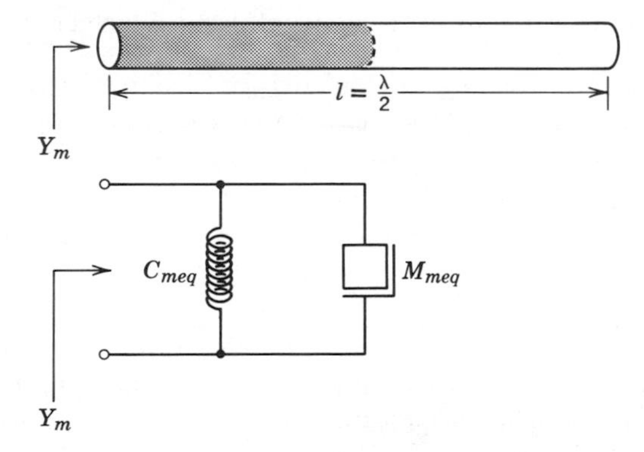

5.5 Sketch the mechanical equivalent circuit of the disk-resonator, coupling-wire mechanical filter shown in Figure 5-31. Consider only two natural resonances

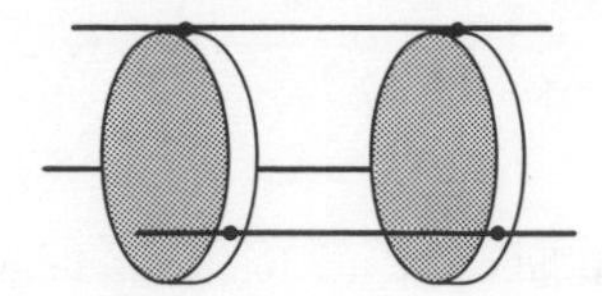

per disk. Use Figure 5-16 as a reference. Assume that the coupling wire is short enough that static mass effects can be ignored, i.e., that the coupling wires act as simple springs.

5.6 Making use of (5.15) show that the shunt arm of a lumped-element pi equivalent circuit of a short wire (less than one-eighth wavelength) is equal to one-half the static mass of the wire, and that the series arm is equal to the compliance l/AE_m.

5.7 Sketch and label lumped-element mechanical and electrical equivalent circuits representing the filter of Figure 5-32.

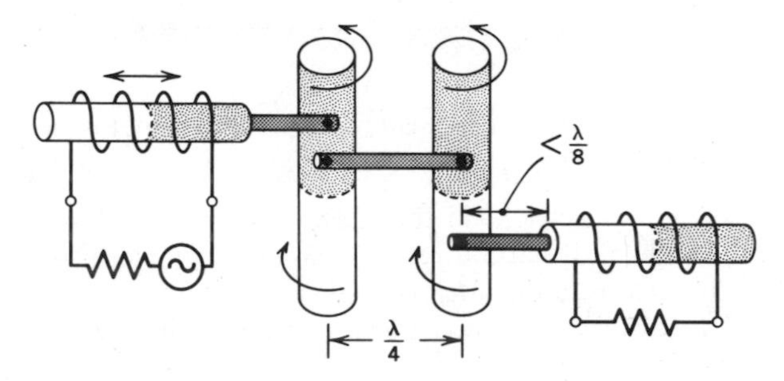

5.8 Calculate the element values of the electrical equivalent circuit of Figure 5-33 which represents a disk-wire mechanical filter. The center frequency is 455 kHz, the bandwidth at 0.1 dB is 3 kHz. Lowpass parameters corresponding to 0.1 dB ripple are $q_1 = q_4 = 1.34$, $k_{12} = k_{34} = 0.690$, and $k_{23} = 0.542$. Assume that a magnetostrictive transducer is used and that it is tuned at the center frequency

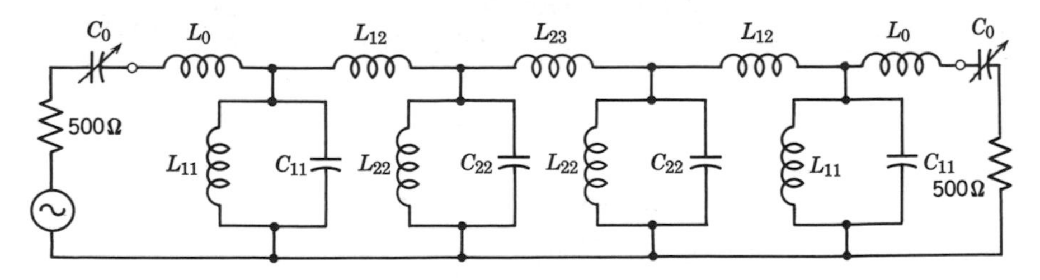

with a series capacitor. The electromechanical coupling coefficient $k_{em} = 0.10$ is high enough so that the effect of the transducer on the frequency response can be ignored. All of the disk resonators have the same equivalent mass. The equivalent mass of the transducer is one-half that of the disks. Terminate the filter in 500 ohms.

5.9 Write a series of equations (that can be easily coded in a computer language) for the transformation of the network of Figure 5-22(c) to the network of Figure 5-22(f). Application of Norton transformations to the coupling elements is often helpful in realizing equal-size (capacitor value) resonators.

5.10 Derive (5.33).

5.11 Derive (5.44) and (5.45) separately. By assuming symmetry (or antisymmetry), the complexity of the secular determinant is reduced to a 2×2.

5.12 Derive (5.46) and (5.47). Derive each equation separately so as to be able to assume symmetry or antisymmetry in order to reduce the complexity of the problem. Use the single electrode-pair case as a model, although you may want to represent the trapped waves with sines or cosines rather than exponentials.

5.13 Sketch the electrical equivalent circuit of a 4-resonator monolithic filter. Use the mobility analogy. Draw a pictorial diagram of the filter and discuss the relationship of the electrode spacings to the coupling inductance of the equivalent circuit.

REFERENCES

[1] J. P. Maxfield and H. C. Harrison, "Methods of high quality recording and reproducing of music and speech based on telephone research," *Bell System Tech. J.*, **5**, 493–523 (July, 1926).

[2] W. P. Mason, *Electromechanical Transducers and Wave Filters*, (Sec. Ed., 1948), D. Van Nostrand, New York, 1942.

[3] J. C. Hathaway and D. F. Babcock, "Survey of mechanical filters and their applications," *Proc. I.R.E.*, **45**, No. 1, 5–16 (Jan., 1957).

[4] R. A. Sykes and W. D. Beaver, "High frequency monolithic filters with possible application to single frequency and single side band use," *Proc. Ann. Freq. Control Sympos.*, Atlantic City, 288–308 (April, 1966).

[5] H. M. Trent, "On the construction of schematic diagrams for mechanical systems," *J. Acoust. Soc. Am.*, **30**, 795–800 (August, 1958).

[6] H. M. Trent, "Isomorphisms between oriented linear graphs and lumped physical systems," *J. Acous. Soc. Am.*, **27**, 500–527 (May, 1955).

[7] R. S. Woollett, "Effective coupling factor of single-degree-of-freedom transducers," *J. Acous. Soc. Am.*, **40**, No. 5, 1112–1123 (May, 1966).

[8] H. W. Katz, Ed., *Solid State Magnetic and Dielectric Devices*, John Wiley, New York, 1959.

[9] R. Saal and E. Ulbrich, "On the design of filters by synthesis," *IRE Trans. Circuit Theory*, **CT-5**, No. 4, 284–327 (Dec., 1958).

[10] International Telephone and Telegraph Corporation, *Reference Data for Radio Engineers*, Howard W. Sams, Indianapolis, 1969.

[11] R. A. Johnson, M. Börner and M. Konno, "Mechanical filters—a review of progress," *IEEE Trans. Sonics Ultrasonics*, **SU-18**, No. 3, 155–170 (July, 1971).

[12] M. Konno, C. Kusakabe, and Y. Tomikawa, "Electromechanical filter composed of transversely vibrating resonators for low frequencies," *J. Acous. Soc. Am.*, **41**, No. 4, 953–961 (April, 1967).

[13] R. A. Johnson and R. J. Teske, "A mechanical filter having general stopband characteristics," *IEEE Trans. Sonics Ultrasonics*, **SU-13**, No. 2, 41–48 (July, 1966).

[14] W. Shockley, D. R. Curran, and D. J. Koneval, "Trapped-energy modes in quartz filter crystals," *J. Acous. Soc. Am.*, **41**, No. 4, 981–993 (April, 1967).

6

Computer-Aided Circuit Optimization

J. W. Bandler
McMaster University
Hamilton, Ontario

This chapter deals with formulations and methods which can be implemented in the ever increasing number of situations when the classical synthesis approach, whether analytic or numerical, is inappropriate. When the so-called closed-form solution is, for some reason, out of the question, the modern approach is to use efficient, iterative, automatic optimization methods to achieve a design that meets or exceeds certain requirements. Not infrequently, exact methods may be used to great advantage in providing the initial feasible design for optimization.

In order to make the mathematics tractable, usable synthesis methods are usually restricted to ideal *commensurate* networks. As soon as we have to take into account active devices, a narrow range of element values, parasitic effects, high frequency operation, nonlinearities, frequency-dependent elements, *noncommensurate* elements (e.g., mixed lumped and distributed elements, uniformly distributed transmission lines with unequal or variable lengths, etc.), elements characterized by measurement data, response constraints, and so on, classical methods of design provide, at best, only approximate answers. In some cases these answers adequately approximate the solution to the actual design problem, but in many cases they do not.

The author is not advocating numerical methods for their own sake. Generally speaking, for the same job, iterative methods require more computation time than more specialized methods which do not require iteration (if they are available). Computing time is not, however, the only criterion an engineer has to consider. For example, in deciding whether or not to devote his own time to deriving an *analytic* algorithm, as distinct from a *numerical* algorithm, he also has to ask himself how often the algorithm would be used, how well it would represent real situations, how widely applicable it would be, and last but not least, how accurate the numerical results would be. After all, as engineers, we are ultimately working toward producing meaningful numbers as the solutions to realistic design problems.

Methods for *automating* the optimal design process will be emphasized. *Ad hoc* cut-and-try techniques using a general purpose analysis program are discouraged, particularly for filter design problems with anything other than the simplest of design specifications and a handful of variable parameters. The pitfalls are the same as with automated methods, the strategy for dealing with them is inevitably less sophisticated, and in the long run it will almost certainly cost more. It is desirable that the decision making process should, as far as possible, be left to the computer.

Poor or unacceptable results in computer-aided circuit optimization (or with any design process) are felt to be most likely due to bad preparation of the problem, a lack of understanding of the hazards that can be encountered, and the wrong choice of algorithm. This chapter, therefore, attempts to show how problems within the scope of filter design may be formulated effectively as optimization problems, to explain the differences between these formulations, to indicate appropriate optimization methods, and to indicate how the results might be interpreted. Details of optimization algorithms, proofs of convergence, etc., are beyond the scope of this work. Adequate references to the original papers and relevant text books will permit the reader to investigate these for himself.

Following a section on basic concepts which are essential for an understanding of optimization theory, a formulation and description of typical objectives and objective functions is presented. Constraints and some methods of dealing with them are discussed in fair detail in the next section, including the conditions for a constrained minimum. Minimax approximation, including conditions for a minimax optimum, is then dealt with. This is followed by sections on one-dimensional search methods, direct search methods, and methods using gradient information. Least pth approximation comes naturally after a discussion of gradient methods. A fairly long section is devoted to the adjoint network method of gradient evaluation.

6.1 BASIC CONCEPTS

The problem of optimization may be stated as follows. Minimize the scalar *objective function** U where

$$U \triangleq U(\boldsymbol{\phi}) \tag{6.1}$$

subject to the *inequality constraints*

$$\mathbf{g}(\boldsymbol{\phi}) \geq \mathbf{0} \tag{6.2}$$

and *equality constraints*

$$\mathbf{h}(\boldsymbol{\phi}) = \mathbf{0}. \tag{6.3}$$

In (6.1) to (6.3), $\boldsymbol{\phi}$ is a vector of k independent variables or parameters,† thus

$$\boldsymbol{\phi} \triangleq \begin{bmatrix} \phi_1 \\ \phi_2 \\ \vdots \\ \phi_k \end{bmatrix} \tag{6.4}$$

defining a k-dimensional space. In general, we might have m inequality constraints and s equality constraints so that

$$\mathbf{g}(\boldsymbol{\phi}) \triangleq \begin{bmatrix} g_1(\boldsymbol{\phi}) \\ g_2(\boldsymbol{\phi}) \\ \vdots \\ g_m(\boldsymbol{\phi}) \end{bmatrix} \tag{6.5}$$

and

$$\mathbf{h}(\boldsymbol{\phi}) \triangleq \begin{bmatrix} h_1(\boldsymbol{\phi}) \\ h_2(\boldsymbol{\phi}) \\ \vdots \\ h_s(\boldsymbol{\phi}) \end{bmatrix} \tag{6.6}$$

The *feasible region* R is defined by all vectors $\boldsymbol{\phi}$ satisfying (6.2) and (6.3). This may be written

$$R \triangleq \{\boldsymbol{\phi} \,|\, \mathbf{g}(\boldsymbol{\phi}) \geq \mathbf{0}, \mathbf{h}(\boldsymbol{\phi}) = \mathbf{0}\}. \tag{6.7}$$

* Also called *cost function*, *performance index*, or *error criterion*.

† Typically element values, residues, critical frequencies, etc.

R is said to be *closed* if, as in (6.2), equalities are allowed. If no equalities are allowed it is said to be *open.* A *proper* minimum of U located by a vector $\check{\boldsymbol{\phi}}$ on the *response hypersurface* generated by $U(\boldsymbol{\phi})$ is such that

$$\check{U} \triangleq U(\check{\boldsymbol{\phi}}) < U(\boldsymbol{\phi}) \tag{6.8}$$

for any feasible $\boldsymbol{\phi}$ close but not equal to $\check{\boldsymbol{\phi}}$.* Since we cannot generally guarantee to find a *global minimum,* we usually have to resign ourselves to a consideration of *local minima.* Our objective then is to find a feasible $\check{\boldsymbol{\phi}}$, if it indeed exists, such that

$$U(\check{\boldsymbol{\phi}}) = \min_{\boldsymbol{\phi} \in R} U(\boldsymbol{\phi}).$$

Figure 6-1 is an illustration of the problem in two dimensions, and it contains a number of features usually encountered in optimization problems. Note that only inequality constraints, i.e., constraints of the form of (6.2), are indicated.

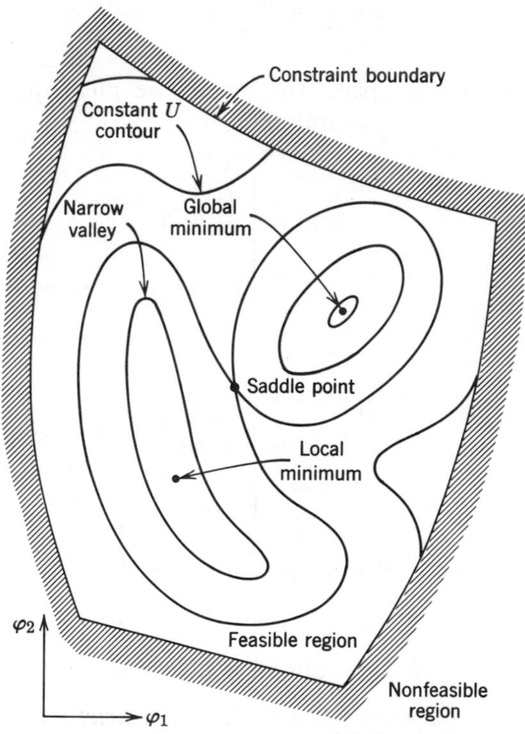

FIGURE 6-1 Some features encountered in optimization problems.

* $U(\check{\boldsymbol{\phi}}) \leq U(\boldsymbol{\phi})$ can also define a minimum, but $\check{\boldsymbol{\phi}}$ may then be nonunique.

Examples of *unimodal, multimodal, strictly concave,* and *strictly convex* functions of one variable are shown in Figure 6-2. A unimodal function for our purposes is one having a unique optimum in the feasible region. It may or may not be continuous with continuous derivatives. A *strictly convex*

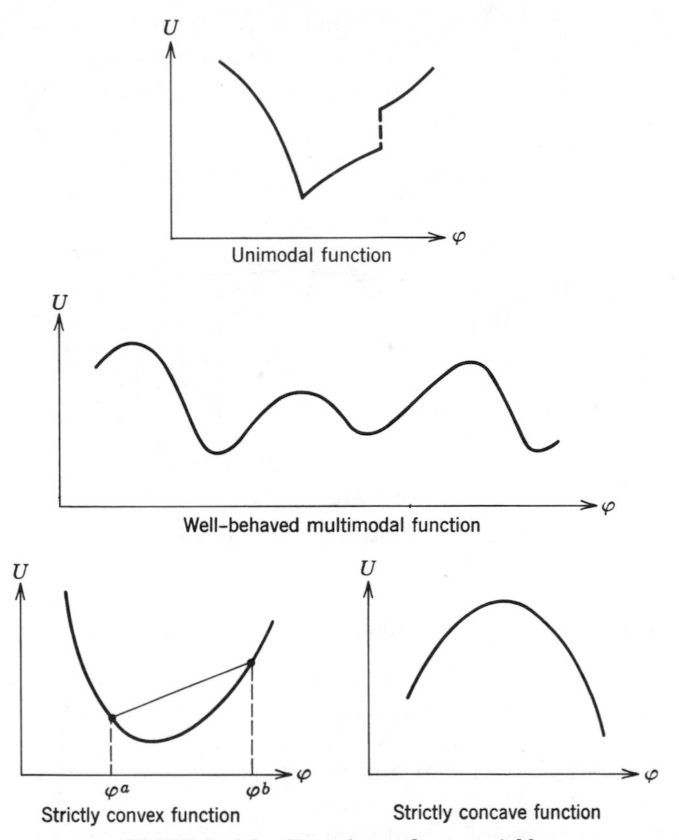

FIGURE 6-2 Functions of one variable.

function is one which can only be overestimated by a linear interpolation between two points on its surface. Thus, for $\boldsymbol{\phi}^a \neq \boldsymbol{\phi}^b$,

$$U(\boldsymbol{\phi}^a + \lambda(\boldsymbol{\phi}^b - \boldsymbol{\phi}^a)) < U(\boldsymbol{\phi}^a) + \lambda(U(\boldsymbol{\phi}^b) - U(\boldsymbol{\phi}^a))$$
$$0 < \lambda < 1 \tag{6.9}$$

for a strictly convex function. See Figure 6-3. A *strictly concave* function is one whose negative is strictly convex. Note that if we omit strictly, then we imply that equality of the function and a linear interpolation can occur, i.e., (6.9) would have to admit equalities.

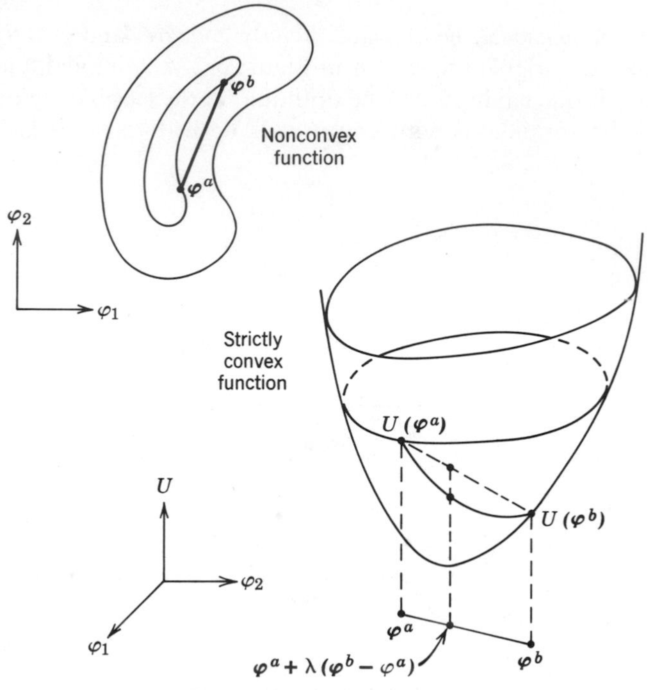

FIGURE 6-3 Illustration of convexity.

A region R is convex if for all $\boldsymbol{\phi}^a, \boldsymbol{\phi}^b \in R$ all points

$$\boldsymbol{\phi} = \boldsymbol{\phi}^a + \lambda(\boldsymbol{\phi}^b - \boldsymbol{\phi}^a)$$
$$0 \le \lambda \le 1 \tag{6.10}$$

lie in R. Illustrations of convex and nonconvex regions are given in Figure 6-4.

The first three terms of a multidimensional Taylor series expansion of $U(\boldsymbol{\phi})$ are given by

$$U(\boldsymbol{\phi} + \Delta\boldsymbol{\phi}) = U(\boldsymbol{\phi}) + \nabla U^T \Delta\boldsymbol{\phi} + \tfrac{1}{2}\Delta\boldsymbol{\phi}^T \mathbf{H}\, \Delta\boldsymbol{\phi} + \cdots \tag{6.11}$$

where the vector

$$\Delta\boldsymbol{\phi} \triangleq \begin{bmatrix} \Delta\phi_1 \\ \Delta\phi_2 \\ \vdots \\ \Delta\phi_k \end{bmatrix} \tag{6.12}$$

contains k parameter increments, $\Delta\boldsymbol{\phi}^T$ is the transposed (row) vector,

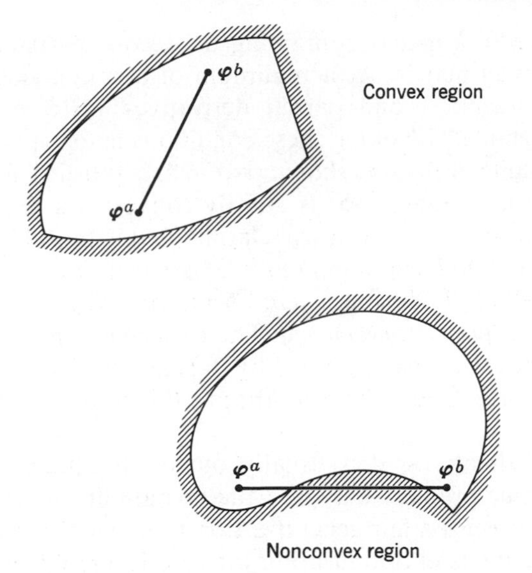

FIGURE 6-4 Convex and nonconvex regions.

$$\nabla U \triangleq \begin{bmatrix} \dfrac{\partial U}{\partial \phi_1} \\ \dfrac{\partial U}{\partial \phi_2} \\ \vdots \\ \dfrac{\partial U}{\partial \phi_k} \end{bmatrix} \tag{6.13}$$

is a vector containing the first partial derivatives of the objective function called the *gradient vector*, and

$$\mathbf{H} \triangleq \begin{bmatrix} \dfrac{\partial^2 U}{\partial \phi_1^2} & \dfrac{\partial^2 U}{\partial \phi_1\, \partial \phi_2} & \cdots & \dfrac{\partial^2 U}{\partial \phi_1\, \partial \phi_k} \\ \dfrac{\partial^2 U}{\partial \phi_2\, \partial \phi_1} & \dfrac{\partial^2 U}{\partial \phi_2^2} & \cdots & \dfrac{\partial^2 U}{\partial \phi_2\, \partial \phi_k} \\ \vdots & \vdots & & \vdots \\ \dfrac{\partial^2 U}{\partial \phi_k\, \partial \phi_1} & \dfrac{\partial^2 U}{\partial \phi_k\, \partial \phi_2} & \cdots & \dfrac{\partial^2 U}{\partial \phi_k^2} \end{bmatrix} \tag{6.14}$$

is a symmetric $k \times k$ matrix containing the second partial derivatives and called the *Hessian* matrix. At a minimum of a continuous function with continuous first and second partial derivatives $\nabla U(\check{\boldsymbol{\phi}}) = \mathbf{0}$ and $\mathbf{H}(\check{\boldsymbol{\phi}})$ is positive semidefinite.* Invoking these conditions in (6.11) but with $\mathbf{H}$ taken as positive definite, it may be shown that (6.8) is satisfied, implying that we have a proper minimum. $U(\boldsymbol{\phi})$ is strictly convex in a region where $\mathbf{H}$ is positive definite as may be seen by relating (6.9) with (6.11).

The problem formulated in (6.1) to (6.3) is called a *mathematical programming* problem. If all the functions are linear, we have *linear programming*; if not, we have *nonlinear programming*. The term *convex programming* is often used to describe the problem defined by (6.1) and (6.2) when $U(\boldsymbol{\phi})$ is convex and $\mathbf{g}(\boldsymbol{\phi})$ is concave. Under these conditions R is convex, and $\check{U}$ is the global minimum.

In practical situations, it is usually out of the question to determine whether a specific problem falls into the domain of convex programming. Nevertheless, it seems a fair generalization to make, that the most reliable and efficient methods of optimization for practical problems are invariably those which invoke some of the nice properties of convex programming in their proofs of convergence. The better methods usually have built-in safeguards for dealing with the hazards of more general nonlinear programming problems while substantially retaining their desirable convergence features. Note that essentially unconstrained problems are regarded as special cases in the above discussion.

6.2 SOME OBJECTIVES AND OBJECTIVE FUNCTIONS

Optimization by Solving Nonlinear Equations

Classically, to find $\check{U}$ we must in general solve k nonlinear equations in k unknowns, namely

$$\nabla U = \mathbf{0}.$$

Denoting this set of equations $\mathbf{f}(\boldsymbol{\phi}) = \mathbf{0}$ where

$$\mathbf{f}(\boldsymbol{\phi}) \triangleq \begin{bmatrix} f_1(\boldsymbol{\phi}) \\ f_2(\boldsymbol{\phi}) \\ \vdots \\ f_k(\boldsymbol{\phi}) \end{bmatrix}, \tag{6.15}$$

we could define a new objective function

$$U(\boldsymbol{\phi}) = \mathbf{f}^T \mathbf{f} \tag{6.16}$$

*It should be noted that $\mathbf{H}$ might not be positive definite, even in some cases when $U(\boldsymbol{\phi})$ is strictly convex.

to be minimized. A minimum of value zero would imply that the solution to $\mathbf{f}(\boldsymbol{\phi}) = \mathbf{0}$ had been found. Now, using a Taylor series expansion

$$\mathbf{f}(\boldsymbol{\phi} + \Delta\boldsymbol{\phi}) = \mathbf{f}(\boldsymbol{\phi}) + \mathbf{J}\,\Delta\boldsymbol{\phi} + \cdots \tag{6.17}$$

where

$$\mathbf{J} \triangleq \begin{bmatrix} \dfrac{\partial f_1}{\partial \phi_1} & \dfrac{\partial f_1}{\partial \phi_2} & \cdots & \dfrac{\partial f_1}{\partial \phi_k} \\ \dfrac{\partial f_2}{\partial \phi_1} & \dfrac{\partial f_2}{\partial \phi_2} & \cdots & \dfrac{\partial f_2}{\partial \phi_k} \\ \vdots & \vdots & & \vdots \\ \dfrac{\partial f_k}{\partial \phi_1} & \dfrac{\partial f_k}{\partial \phi_2} & \cdots & \dfrac{\partial f_k}{\partial \phi_k} \end{bmatrix} \tag{6.18}$$

is a $k \times k$ *Jacobian* matrix. The well-known *Newton-Raphson* method of solution is based on the hope that, if we evaluate $\mathbf{f}$ and $\mathbf{J}$ at $\boldsymbol{\phi}$, then the incremental change

$$\Delta\boldsymbol{\phi} = -\mathbf{J}^{-1}\mathbf{f}(\boldsymbol{\phi}) \tag{6.19}$$

brings one closer to the solution. (In Section 6.8 these ideas are extended).

Quadratic Objective Function

Consider the quadratic objective function

$$U(\boldsymbol{\phi}) = \tfrac{1}{2}\boldsymbol{\phi}^T\mathbf{A}\boldsymbol{\phi} + \mathbf{b}^T\boldsymbol{\phi} + c \tag{6.20}$$

where

$\mathbf{A}$ is a $k \times k$ constant symmetric matrix,
$\mathbf{b}$ is a constant vector with k components,
c is a constant.

In this case, it is readily shown that

$$\nabla U = \mathbf{A}\boldsymbol{\phi} + \mathbf{b}$$
$$\mathbf{H} = \mathbf{A}.$$

A stationary point of $U(\boldsymbol{\phi})$ can be found by solving the linear equations

$$\mathbf{A}\boldsymbol{\phi} + \mathbf{b} = \mathbf{0}.$$

If $\mathbf{A}$ is nonsingular, the point is unique and can be found in a finite number of operations. The term *quadratic convergence* (Fletcher [1] prefers the term "property Q") is used to describe the convergence properties of optimization methods, which guarantee to find the minimum of a quadratic function in a finite number of steps. Such methods can be expected to be very efficient in minimizing functions adequately representable by positive-definite quadratic forms in the vicinity of a minimum. Some of them are discussed in later sections.

Error Criteria

Most electrical network design problems can be formulated as approximation problems. Let us, therefore, introduce a weighted error or deviation between a specified function and an approximating function as

$$e(\phi, \psi) \triangleq w(\psi)[F(\phi, \psi) - S(\psi)] \tag{6.21}$$

where

$S(\psi)$ is the real or complex specified function,
$F(\phi, \psi)$ is the real or complex approximating function,
$w(\psi)$ is a weighting function,
ψ is an independent variable,
ϕ represents the adjustable parameters.

Thus $F(\phi, \psi)$ may be a network response, $S(\psi)$ may be the desired response, and ψ may be frequency or time. See Figure 6-5.

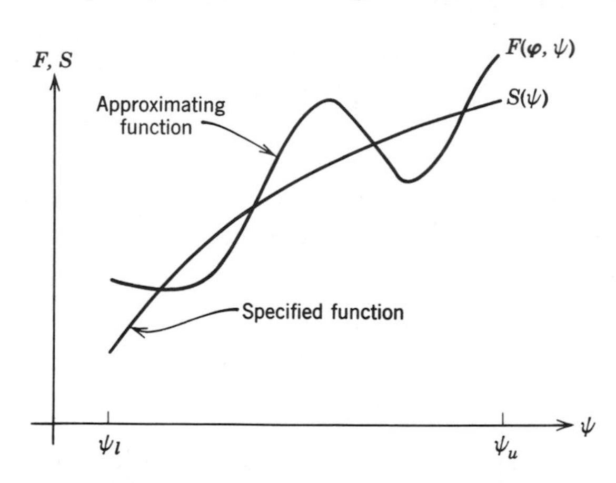

FIGURE 6-5 An approximation problem.

We may define a *norm*

$$\|e\|_p \triangleq \left\{\int_{\psi_l}^{\psi_u} |e(\phi, \psi)|^p \, d\psi\right\}^{1/p}, \qquad 1 \le p \le \infty \tag{6.22}$$

for the continuous case; and a *norm*

$$\|\mathbf{e}\|_p \triangleq \left\{\sum_i |e_i(\boldsymbol{\phi})|^p\right\}^{1/p}, \qquad \begin{array}{c} i \in I \\ 1 \le p \le \infty \end{array} \tag{6.23}$$

for the discrete case, where

$$\mathbf{e}(\boldsymbol{\phi}) \triangleq \begin{bmatrix} e_1(\boldsymbol{\phi}) \\ e_2(\boldsymbol{\phi}) \\ \vdots \\ e_n(\boldsymbol{\phi}) \end{bmatrix}, \tag{6.24}$$

$$e_i(\boldsymbol{\phi}) \triangleq e(\boldsymbol{\phi}, \psi_i) = w(\psi_i)[F(\boldsymbol{\phi}, \psi_i) - S(\psi_i)] \tag{6.25}$$

and

$$I \triangleq \{1, 2, \ldots, n\}. \tag{6.26}$$

Thus, I is an index set relating to discrete values of ψ on an interval $[\psi_l, \psi_u]$, which is closed and finite.

Now for well-behaved functions

$$\max_{[\psi_l, \psi_u]} |e(\boldsymbol{\phi}, \psi)| = \lim_{p \to \infty} \left\{\frac{1}{\psi_u - \psi_l} \int_{\psi_l}^{\psi_u} |e(\boldsymbol{\phi}, \psi)|^p \, d\psi\right\}^{1/p} \tag{6.27}$$

when $|e(\boldsymbol{\phi}, \psi)|$ is defined on $[\psi_l, \psi_u]$. If $|e(\boldsymbol{\phi}, \psi)|$ is continuous on a finite interval $[\psi_l, \psi_u]$, then (6.27) is certainly valid. Similarly,

$$\max_i |e_i(\boldsymbol{\phi})| = \lim_{p \to \infty} \left\{\sum_i |e_i(\boldsymbol{\phi})|^p\right\}^{1/p}, \qquad i \in I. \tag{6.28}$$

Suppose we formulate an objective function as

$$U = \int_{\psi_l}^{\psi_u} |e(\boldsymbol{\phi}, \psi)|^p \, d\psi \tag{6.29}$$

for the continuous case and

$$U = \sum_{i \in I} |e_i(\boldsymbol{\phi})|^p \tag{6.30}$$

for the discrete case. The minimization of the U of (6.29) or (6.30) is called *least* p*th approximation.* A minimum for the continuous case is called a best approximation with respect to $\|e\|_p$, defined in (6.22). A minimum for the discrete case is called a best approximation with respect to $\|\mathbf{e}\|_p$, defined in (6.23). Now $\|e\|_\infty$ and $\|\mathbf{e}\|_\infty$ are called *Chebyshev* or *uniform* norms. Because of the consequences of (6.27) and (6.28), minimization with respect to $\|e\|_\infty$ or $\|\mathbf{e}\|_\infty$ is widely referred to as *minimax approximation.* Least pth approximation tends to minimax approximation as $p \to \infty$.

A word of caution concerning the weighting function $w(\psi)$ and the index p is in order. Clearly their purpose is to emphasize or deemphasize the difference between $F(\boldsymbol{\phi}, \psi)$ and $S(\psi)$. Thus, an optimum with respect to one weighting function or value of p may not be an optimum with respect to another. Large errors will be emphasized by large values of p. If one knew in advance where these large errors would be, $w(\psi)$ might also be used to emphasize them. The use of $w(\psi)$ to do this is a poor approach, however, and should be discouraged.

6.3 CONSIDERATION OF CONSTRAINTS

It is rare to find any network design problem which is unconstrained. When physical considerations indicate that the optimum will lie in the interior of the feasible region, the designer is lucky and should take advantage of it. Often this will not be possible, and steps have to be taken to ensure that a realizable and practical design will be achieved. One of the great advantages of computer-aided circuit optimization is that, if the design problem has been properly formulated, a feasible design can always be achieved assuming the initial design is feasible.

Constraints in network design can take a variety of forms. They can include upper and lower bounds on parameters; they can include nonnegativity requirements on network elements. The topology, overall size, the suppression of unwanted modes of operation, considerations for parasitic effects whether reactive or lossy, and the stability of active devices can all result in constraints on parameters. Response constraints such as constraints on the phase while the amplitude is optimized can also occur.

Most network designers seem to treat constraints as an afterthought, and then complain that the optimization process gave them negative resistors, etc. Their faith in automated optimization methods is shattered as a result. The author would like to stress that a thorough consideration should be given to the constraints *before* the selection of an optimization strategy.

In this section we will look at some methods of converting constrained problems into essentially unconstrained ones. For other methods of nonlinear programming, the reader should refer elsewhere [2, 3, 4].

Transformations for Parameter Constraints

Various upper and lower bounds on the variable parameters are probably the most common kinds of constraints [5]. In Table 6-1 we show some simple parameter constraints falling into this class with appropriate transformations. It is useful to distinguish between constraints defining open and closed feasible regions. If the optimum is expected to lie away from the boundary or if it is desired to discourage the solution from getting too close, the former type might be chosen.

TABLE 6-1 Simple parameter constraints and transformations

Constraint	Transformation
$\phi_i \geq 0$	$\phi_i = \phi_i'^2$
$\phi_i > 0$	$\phi_i = \exp \phi_i'$
$\phi_i \geq \phi_{li}$	$\phi_i = \phi_{li} + \phi_i'^2$
$\phi_i > \phi_{li}$	$\phi_i = \phi_{li} + \exp \phi_i'$
$-1 \leq \phi_i \leq 1$	$\phi_i = \sin \phi_i'$
$0 \leq \phi_i \leq 1$	$\phi_i = \sin^2 \phi_i'$
$0 < \phi_i < 1$	$\phi_i = \dfrac{\exp \phi_i'}{1 + \exp \phi_i'}$
$\phi_{li} \leq \phi_i \leq \phi_{ui}$	$\phi_i = \phi_{li} + (\phi_{ui} - \phi_{li})\sin^2 \phi_i'$
	$\phi_i = \frac{1}{2}(\phi_{li} + \phi_{ui}) + \frac{1}{2}(\phi_{ui} - \phi_{li})\sin \phi_i'$
$\phi_{li} < \phi_i < \phi_{ui}$	$\phi_i = \phi_{li} + (\phi_{ui} - \phi_{li}) \dfrac{\exp \phi_i'}{1 + \exp \phi_i'}$
	$\phi_i = \phi_{li} + \dfrac{1}{\pi}(\phi_{ui} - \phi_{li})\cot^{-1} \phi_i'$ for $0 < \cot^{-1} \phi_i' < \pi$

More General Considerations

Parameter constraints of the form

$$\phi_{li} \leq \phi_i \leq \phi_{ui} \tag{6.31}$$

can if necessary be written as

$$\begin{aligned} \phi_i - \phi_{li} &\geq 0 \\ \phi_{ui} - \phi_i &\geq 0 \end{aligned} \tag{6.32}$$

in order to fit them into the scheme of (6.2). Frequency- or time-dependent constraints may be put into the form

$$c_j(\boldsymbol{\phi}, \psi) \geq 0 \tag{6.33}$$

where j denotes some jth function, or at discrete points on the ψ-axis into the form

$$c_j(\boldsymbol{\phi}, \psi_i) \geq 0 \tag{6.34}$$

where i denotes an ith sample point. The form of (6.34) is preferrable to that of (6.33), since it allows us to consider a finite rather than an infinite number of constraints.

We might eliminate a number of constraints on physical or logical grounds if, for instance,

1. $U(\boldsymbol{\phi}) \rightarrow \infty$ as $g_i(\boldsymbol{\phi}) \rightarrow 0$. The attenuation of a filter becomes infinite, for example, if a zero valued element short circuits the structure;
2. Some $h_i(\boldsymbol{\phi}) = 0$ can be explicitly written as $\phi_j = f(\phi_1, \phi_2, \ldots, \phi_{j-1}, \phi_{j+1}, \ldots, \phi_k)$. In this case we can optimize with $k - 1$ parameters;
3. $g_i(\boldsymbol{\phi})$ is known *a priori* to be positive.

Our design problem may be so complicated that we cannot easily find an initial design to serve as a feasible starting point in the optimization process. We could try to find one by unconstrained optimization by minimizing

$$-\sum_{i=1}^{m} w_i g_i(\boldsymbol{\phi}) + \sum_{j=1}^{s} h_j^2(\boldsymbol{\phi}) \qquad w_i \begin{cases} = 0 & g_i(\boldsymbol{\phi}) \geq 0 \\ > 0 & g_i(\boldsymbol{\phi}) < 0. \end{cases} \tag{6.35}$$

If the minimum is zero we have a feasible point. Failure to converge to zero does not necessarily mean that a feasible point does not exist.

Having obtained a feasible starting point we might decide to simply reject nonfeasible points if they are obtained during optimization. Equivalently we might set $U(\boldsymbol{\phi})$ to a most unattractive value if any violation occurs. Alternatively, we could add the term

$$\sum_{i=1}^{m} w_i g_i^2(\boldsymbol{\phi}) + \sum_{j=1}^{s} h_j^2(\boldsymbol{\phi}) \qquad w_i \begin{cases} = 0 & g_i(\boldsymbol{\phi}) \geq 0 \\ > 0 & g_i(\boldsymbol{\phi}) < 0 \end{cases} \tag{6.36}$$

to the objective function. The objective function is not penalized as long as the constraints are satisfied. This procedure does not, unfortunately, always insure a strictly feasible solution.

The simple approaches just described have other disadvantages also. Discontinuities in the new function or its derivatives may be introduced. Steep walls or valleys may be formed at the boundary of the feasible region which can drastically slow down the optimization process. A method which simply rejects nonfeasible points can easily terminate at a false minimum [6].

Sequential Unconstrained Minimization Techniques

One of the best known and most highly developed of the sequential unconstrained minimization techniques (SUMT) will be briefly outlined here [7]. Consider first the problem of minimization subject to inequality constraints defined in (6.1) and (6.2). Let

$$P(\boldsymbol{\phi}, r) \triangleq U(\boldsymbol{\phi}) + rG(\mathbf{g}) \tag{6.37}$$

where $G(\mathbf{g})$ is continuous for $\mathbf{g} > \mathbf{0}$ and $G(\mathbf{g}) \to \infty$ for any $g_i(\boldsymbol{\phi}) \to 0$, and where $r > 0$. Two possible candidates for $G(\mathbf{g})$ immediately suggest themselves, namely

$$G(\mathbf{g}) = \sum_{i=1}^{m} \frac{1}{g_i(\boldsymbol{\phi})}, \tag{6.38}$$

and

$$G(\mathbf{g}) = -\sum_{i=1}^{m} \log g_i(\boldsymbol{\phi}). \tag{6.39}$$

Let us denote the interior of the region R of feasible points by R°, where

$$R^\circ \triangleq \{\boldsymbol{\phi} \,|\, \mathbf{g}(\boldsymbol{\phi}) > \mathbf{0}\} \tag{6.40}$$

and

$$R \triangleq \{\boldsymbol{\phi} \,|\, \mathbf{g}(\boldsymbol{\phi}) \geq \mathbf{0}\}. \tag{6.41}$$

The procedure is to select a $\boldsymbol{\phi}$ and a value of r, initially $\boldsymbol{\phi}^0 \in R^\circ$ and $r_1 > 0$, respectively, and minimize the function P of (6.37). The form of this equation is such that one would expect the minimum, namely $\check{\boldsymbol{\phi}}(r_1)$, to lie in R°. Repeat the procedure for different values of r such that

$$r_1 > r_2 > \cdots r_j > 0 \qquad \text{and} \qquad \lim_{j \to \infty} r_j = 0, \tag{6.42}$$

each minimization being started at the previous minimum. The minimization of $P(\boldsymbol{\phi}, r_2)$ would be started at $\check{\boldsymbol{\phi}}(r_1)$, and so on.

The effect of the penalty is reduced every time the parameter r is reduced, so it is reasonable to expect that, under suitable conditions

$$\lim_{j \to \infty} \check{\boldsymbol{\phi}}(r_j) = \check{\boldsymbol{\phi}}$$

since by (6.42)

$$\lim_{j \to \infty} r_j = 0$$

so that

$$\lim_{j \to \infty} U[\check{\boldsymbol{\phi}}(r_j)] = \check{U}$$

the constrained minimum. A minimum of P should always be available in R°, so any nonfeasible point that may be encountered can be rejected. This safeguard should not be overlooked, for obvious reasons.

This procedure is termed an interior point unconstrained minimization technique, and it requires an initial $\boldsymbol{\phi}^0 \in R^{\circ}$. If one is not available, the following approach may be adopted. Let

$$S \triangleq \{s \mid g_s(\boldsymbol{\phi}) \leq 0, \quad s \in \{1, 2, \ldots, m\}\}$$

$$T \triangleq \{t \mid g_t(\boldsymbol{\phi}) > 0, \quad t \in \{1, 2, \ldots, m\}\}.$$

Now define a

$$P(\boldsymbol{\phi}, r) = -\sum_{s \in S} g_s(\boldsymbol{\phi}) + r \sum_{t \in T} G_t(g_t(\boldsymbol{\phi})) \tag{6.43}$$

to be minimized for a sequence of r values satisfying (6.42). The implications of (6.43) are that any satisfied constraints are prevented from becoming violated while an attempt to satisfy the rest is being made. As soon as any constraint is satisfied the corresponding index is transferred from S to T, and the procedure repeated. When S becomes empty we have obtained a $\boldsymbol{\phi}^0 \in R^{\circ}$ and the solution process of the problem can commence.

To prove convergence one must invoke the requirements for convex programming (See Section 6.1). In practice, however, the conditions may be difficult to verify even if they hold. Nevertheless, the method should work successfully on a wide variety of practical problems for which convergence is not readily proved. Bad initial choices of r and $\boldsymbol{\phi}$ will slow down convergence. Too large a value of r_1 may render the first few minima of P to be relatively independent of U, whereas too small a value may render the penalty ineffective except near the boundary where elongated valleys with steep sides are produced. Because of this and the fact that a sequence of unconstrained problems has to be solved, efficient gradient methods are generally required.

A reduction factor of 10 for the values of r is probably as good as any once the process has started. The arbitrariness of this can be somewhat alleviated by using the SUMT method without the r parameters [7, 8].

To include equality constraints the term

$$\frac{1}{r^{1/2}} \sum_{j=1}^{s} h_j^2(\boldsymbol{\phi}) \tag{6.44}$$

can be added to the right hand side of (6.37). Clearly, as $r \to 0$, $\mathbf{h}(\boldsymbol{\phi})$ must approach $\mathbf{0}$ or a minimum will not be reached.

The reader is referred to a number of selected references which discuss or extend SUMT [7, 8, 9, 10]. A lucid discussion is found in Chapter 5 of Kowalik and Osborne [8].

Conditions for a Constrained Minimum

Necessary conditions which a stationary point $\boldsymbol{\phi}^\circ$ must satisfy in the problem of minimizing $U(\boldsymbol{\phi})$ subject to $\mathbf{g}(\boldsymbol{\phi}) \geq \mathbf{0}$ can be formulated. Assume $U(\boldsymbol{\phi})$ and $\mathbf{g}(\boldsymbol{\phi})$ to be differentiable in the neighborhood of a feasible stationary point $\boldsymbol{\phi}^\circ$, then

$$\nabla U(\boldsymbol{\phi}^\circ) = \sum_{i=1}^{m} u_i \nabla g_i(\boldsymbol{\phi}^\circ) \tag{6.45}$$

and

$$\mathbf{u}^T \mathbf{g}(\boldsymbol{\phi}^\circ) = 0 \tag{6.46}$$

where

$$\mathbf{u} \triangleq \begin{bmatrix} u_1 \\ u_2 \\ \vdots \\ u_m \end{bmatrix} \geq \mathbf{0}.$$

These necessary conditions can be interpreted as follows: $\nabla U(\boldsymbol{\phi}^\circ)$ is a nonnegative linear combination of the gradients $\nabla g_i(\boldsymbol{\phi}^\circ)$ of those constraints which are active* at $\boldsymbol{\phi}^\circ$.

Under the conditions of convex programming, i.e., if $U(\boldsymbol{\phi})$ is convex, $\mathbf{g}(\boldsymbol{\phi})$ is concave, and R° is nonempty, the conditions become sufficient for $\boldsymbol{\phi}^\circ$ to be $\check{\boldsymbol{\phi}}$, the constrained minimum. The relations (6.45) and (6.46) are called the Kuhn-Tucker relations [11]. An interpretation is sketched in Figure 6-6. Note that if we have been using a reliable optimization method, and if the relations are satisfied, we can be reasonably sure that a local minimum has been attained even if the convexity requirements are not met.

For a detailed treatment of the Kuhn-Tucker relations, including their derivation and a discussion of the constraint qualification which must also hold, the reader is referred to an appropriate book such as Zangwill [4].

* A constraint $g_i(\boldsymbol{\phi}) \geq 0$ is active at $\boldsymbol{\phi}^\circ$ if $g_i(\boldsymbol{\phi}^\circ) = 0$.

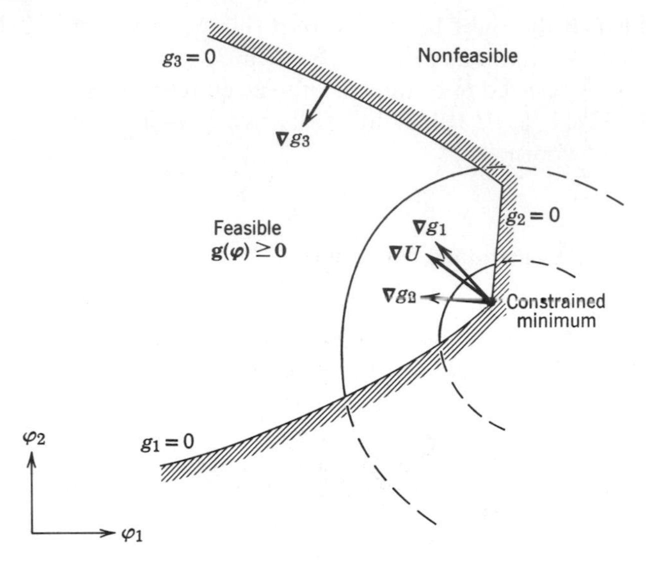

FIGURE 6-6 Sufficient conditions for a constrained minimum, $u_1 > 0$, $u_2 > 0$, $u_3 = 0$.

6.4 MINIMAX APPROXIMATION

Classically, minimax approximation (See Section 6.2 for definitions) has implied the selection of the coefficients of a suitable polynomial or rational function so that it fits some desired specification (usually continuous on a closed interval) in an optimal equal-ripple manner. The Remez method and its generalizations are notable examples of iterative processes for obtaining best approximations using polynomials and rational functions.

The number of practical problems in filter design which can be solved by the classical approach is certainly diminishing in comparison with those that need solving, notwithstanding progress in transformations in the frequency variable, Richard's transformation for transmission-line networks, and so on. This section will, therefore, emphasize less specialized methods applicable to a wider range of practical design problems. Recent references are available which discuss in detail methods well-suited to polynomials and rational functions in the context of filter design [12, 13, 14, 15].

Formulation in Terms of Inequality Constraints

Figure 6-7 illustrates a typical filter design problem. We would like to find the (constrained) parameters of a suitable network so that certain passband and stopband specifications are met or exceeded. Assuming the approximat-

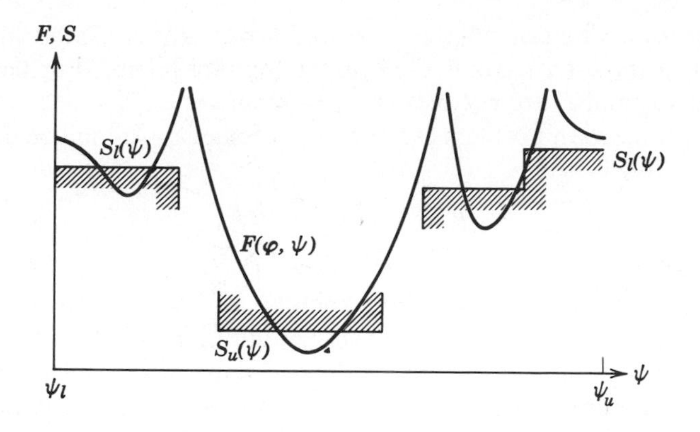

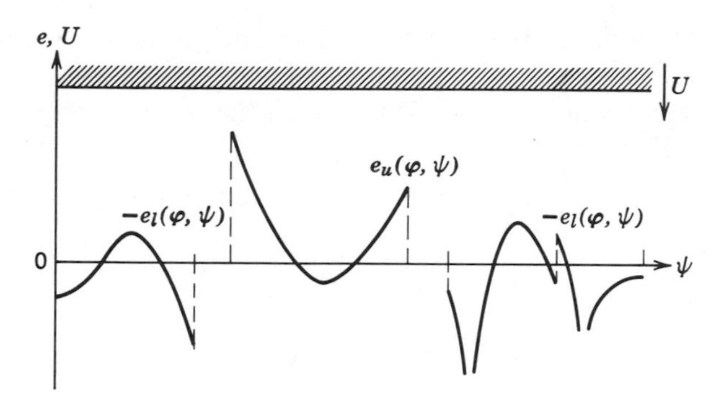

FIGURE 6-7 Typical filter design problem (the specifications are violated).

ing function and the specifications are real, let the error functions e_u and e_l be given by

$$\begin{aligned} e_u(\boldsymbol{\phi}, \psi) &\triangleq w_u(\psi)[F(\boldsymbol{\phi}, \psi) - S_u(\psi)] \\ e_l(\boldsymbol{\phi}, \psi) &\triangleq w_l(\psi)[F(\boldsymbol{\phi}, \psi) - S_l(\psi)] \end{aligned} \tag{6.47}$$

so that

$$\begin{aligned} e_{ui}(\boldsymbol{\phi}) &\triangleq e_u(\boldsymbol{\phi}, \psi_i), \qquad i \in I_u \\ e_{li}(\boldsymbol{\phi}) &\triangleq e_l(\boldsymbol{\phi}, \psi_i), \qquad i \in I_l. \end{aligned} \tag{6.48}$$

This is simply a generalization of (6.21), (6.25), and (6.26), where the symbols have the same meaning. In the present case of (6.47) and (6.48), the subscript u refers to the upper or *passband* specification, the l to the lower or *stopband* specification.

Since approximation in the time and other domains can also be formulated in these terms, ψ is used rather than frequency. Furthermore, the index sets I_u and I_l are not necessarily disjoint.

The optimization problem can now be specified as: minimize the quantity U subject to

$$U \geq e_{ui}(\boldsymbol{\phi}), \qquad i \in I_u$$
$$U \geq -e_{li}(\boldsymbol{\phi}), \qquad i \in I_l \tag{6.49}$$

and also to all other constraints, such as on $\boldsymbol{\phi}$. Observe that U is an *additional* independent variable. As shown in Figure 6-7 it may be visualized as a level or ceiling which is forced down on the deviations e_u and $-e_l$.

At a minimum at least one constraint in (6.49) must be an equality. Otherwise U can be lowered without violation. Further if

1. $\breve{U} < 0$, the minimum amount by which the network response exceeds the specifications is maximized;
2. $\breve{U} > 0$, the maximum amount by which the network response violates the specifications is minimized.

For loss or phase equalization, or time-domain approximation, for example, we might have only one specification, namely, $S(\psi)$. To treat these special cases we simply drop the subscripts u and l in (6.47) to (6.49) and the objective is equivalent to minimizing

$$U = \max_{i \in I} |e_i(\boldsymbol{\phi})|. \tag{6.50}$$

The weighting functions in (6.47) serve the following purpose. If one is much larger than the other, it emphasizes the deviation associated with it at the expense of the rest of the response if the specifications are violated. When the specifications are satisfied (we can now set the weighting function effectively to infinity if required), effort is switched to the rest of the response.

Methods for Minimax Approximation

An approach successfully implemented in optimal filter design [16] and reviewed by Waren, Lasdon, and Suchman [17] is to use sequential unconstrained minimization. We could, for example, define

$$P(\boldsymbol{\phi}, U, r) = U + r\left\{\sum_{i \in I_u} \frac{w_{ui}}{U - e_{ui}(\boldsymbol{\phi})} + \sum_{i \in I_l} \frac{w_{li}}{U + e_{li}(\boldsymbol{\phi})} + \text{other terms}\right\} \tag{6.51}$$

where the other terms might include parameter constraints. Note that in (6.51) the elements of $\mathbf{g}(\boldsymbol{\phi})$ include

$$\frac{1}{w_{ui}}[U - e_{ui}(\boldsymbol{\phi})] \geq 0, \qquad i \in I_u$$

$$\frac{1}{w_{li}}[U + e_{li}(\boldsymbol{\phi})] \geq 0, \qquad i \in I_l. \tag{6.52}$$

Further, it should be remembered that U is an independent variable. The appropriate formulations described in Section 6.3 are thus applicable to minimax approximation.

Ishizaki and Watanabe [18] have described a method in many respects similar to the more recent one by Osborne and Watson [19], which applies linear programming iteratively to achieve a best approximation in the minimax sense. Let us concern ourselves with the objective suggested by (6.50). This should not, however, be taken to imply that the method is less general than the one already outlined.

Linearizing $e_i(\boldsymbol{\phi})$, which is taken as real, at some point $\boldsymbol{\phi}^j$ the problem becomes one of minimizing U subject to

$$\begin{aligned} &\frac{1}{w_i}[U - e_i(\boldsymbol{\phi}^j) - \nabla e_i^T(\boldsymbol{\phi}^j)\,\Delta\boldsymbol{\phi}^j] \geq 0 \\ &\qquad\qquad\qquad\qquad\qquad\qquad\qquad i = 1, 2, \ldots, n > k \\ &\frac{1}{w_i}[U + e_i(\boldsymbol{\phi}^j) + \nabla e_i^T(\boldsymbol{\phi}^j)\,\Delta\boldsymbol{\phi}^j] \geq 0 \end{aligned} \tag{6.53}$$

and other (linearized) constraints. Noting that the variables for linear programming should all be nonnegative, and imposing a rather practical constraint that the elements of $\boldsymbol{\phi}$ should not change sign we have the linear program in $\mathbf{x} \triangleq [x_1\, x_2 \cdots x_{k+1}]^T$ such as to

$$\text{minimize } U = x_{k+1}$$

subject to

$$\pm\left\{e_i(\boldsymbol{\phi}^j) + \nabla e_i^T(\boldsymbol{\phi}^j)\begin{bmatrix}\phi_1^j x_1 - \phi_1^j \\ \phi_2^j x_2 - \phi_2^j \\ \vdots \\ \phi_k^j x_k - \phi_k^j\end{bmatrix}\right\} \leq x_{k+1},\ i = 1, 2, \ldots, n > k \tag{6.54}$$

$$\mathbf{x} \geq \mathbf{0}$$

where

$$x_i \triangleq \frac{\Delta\phi_i^j}{\phi_i^j} + 1, \qquad i = 1, 2, \ldots, k.$$

The solution produces a direction given by $\Delta\boldsymbol{\phi}^j$. Next we find α^j such that $\max_i |e_i(\boldsymbol{\phi}^j + \alpha^j \Delta\boldsymbol{\phi}^j)|$ is a minimum, set $\boldsymbol{\phi}^{j+1} = \boldsymbol{\phi}^j + \alpha^j \Delta\boldsymbol{\phi}^j$ and repeat the process. For conditions for convergence the reader is referred to the original papers [18, 19]. Other linearized constraints can also be considered [14, 15]. Clearly such an approach is directly applicable to linear functions such as polynomials, for which $k + 1$ equal extrema results at the optimum.

Bandler, Srinivasan, and Charalambous [20] have described a descent type of algorithm for minimax approximation which also employs linear programming. Basically, the algorithm attempts to find a locally optimal downhill direction for the problem of minimizing U, where

$$U = \max_{i \in I} f_i(\boldsymbol{\phi}), \tag{6.55}$$

where the $f_i(\boldsymbol{\phi})$ are real nonlinear differentiable functions generally. Linearizing $f_i(\boldsymbol{\phi})$ and letting

$$J \triangleq \{i \,|\, f_i(\boldsymbol{\phi}) = \max_i f_i(\boldsymbol{\phi}), \qquad i \in I\} \tag{6.56}$$

we can obtain, at some feasible point $\boldsymbol{\phi}^j$, the first-order changes

$$\Delta f_i(\boldsymbol{\phi}^j) = \nabla f_i^T(\boldsymbol{\phi}^j)\, \Delta\boldsymbol{\phi}^j, \qquad i \in J. \tag{6.57}$$

In order for $\Delta\boldsymbol{\phi}^j$ to define a descent direction for $\max_{i \in I} f_i(\boldsymbol{\phi})$ we must have

$$\nabla f_i^T(\boldsymbol{\phi}^j)\, \Delta\boldsymbol{\phi}^j < 0, \qquad i \in J.$$

Consider

$$\Delta\boldsymbol{\phi}^j = -\sum_{i \in J} \alpha_i^j \, \nabla f_i(\boldsymbol{\phi}^j) \tag{6.58}$$

$$\sum_{i \in J} \alpha_i^j = 1 \tag{6.59}$$

$$\alpha_i^j \geq 0, \qquad i \in J, \tag{6.60}$$

which suggests the linear program:

$$\text{maximize } \alpha_{r+1}^j \geq 0 \tag{6.61}$$

subject to

$$-\nabla f_i^T(\boldsymbol{\phi}^j) \sum_{i \in J} \alpha_i^j \, \nabla f_i(\boldsymbol{\phi}^j) \leq -\alpha_{r+1}^j, \qquad i \in J \tag{6.62}$$

plus (6.59) and (6.60), where it is assumed that J has r elements.

Observe that J should be nonempty, and that if J has only one element, we obtain the steepest descent direction for the corresponding maximum of the $f_i(\boldsymbol{\phi})$. The solution to the linear program provides $\Delta\boldsymbol{\phi}^j$. We then find γ^j corresponding to the minimum value of $\max_{i \in I} f_i(\boldsymbol{\phi}^j + \gamma^j \Delta\boldsymbol{\phi}^j)$. $\boldsymbol{\phi}^{j+1}$ is set to $\boldsymbol{\phi}^j + \gamma^j \Delta\boldsymbol{\phi}^j$ and the procedure is repeated.

In practice, we will not have a set of $f_i(\phi)$ identically equal to the maximum value. An appropriate tolerance must, therefore, be introduced into (6.56) and a more suitable selection procedure for the elements of J formulated. For further details the original paper should be consulted [20]. It can be proved that the algorithm will, if correctly implemented, converge to the minimax solution.

Example 6-1. Figure 6-8 shows an example of minimax approximation [21]. The objective was to find

$$\check{U} = \min_{\varphi} \left\{ \max_{[f_l, f_u]} |\rho(\phi, f)| \right\}$$

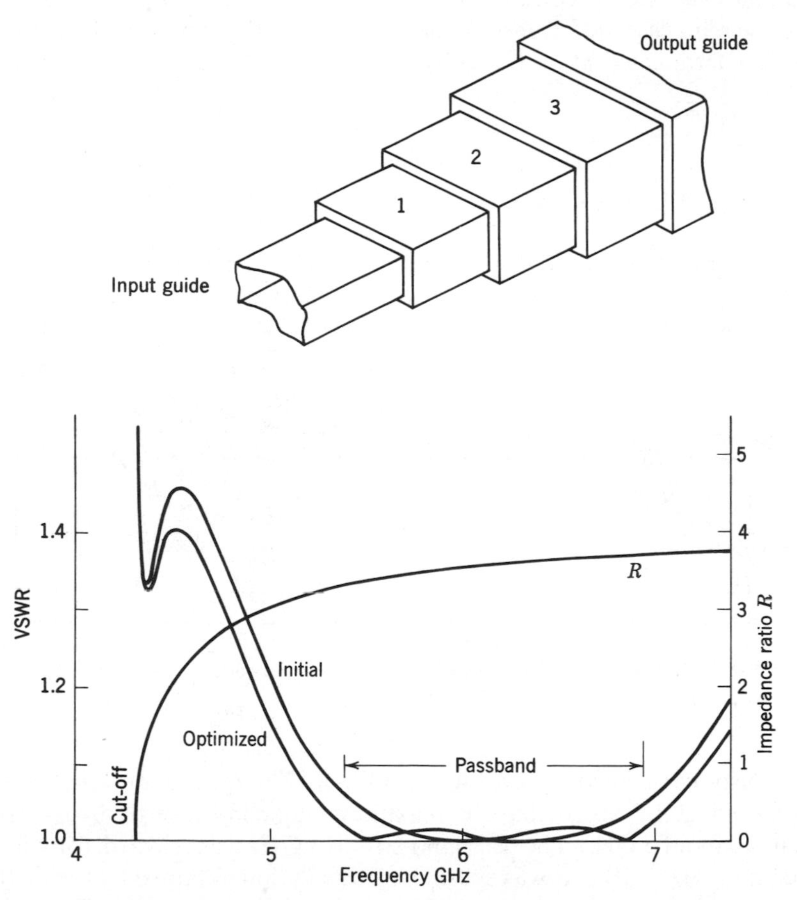

FIGURE 6-8 Example of constrained minimax approximation.

for the 3-section inhomogeneous rectangular waveguide impedance transformer, where ρ is the reflection coefficient, and f is frequency in GHz. The parameters $\boldsymbol{\phi}$ to be varied were the actual geometrical dimensions of the sections. The lower and upper band edges were $f_l = 5.4$ GHz and $f_u = 6.95$ GHz. It should be noted that (1) both input and output waveguides had different cut-off frequencies so that an exact synthesis was not possible, (2) severe constraints were placed on the parameters for a variety of physical reasons, (3) discontinuity susceptances could be taken directly into account, and (4) the razor search method [22] (See Section 6.6) was employed. The reader is referred to Bandler [21] for further details of this type of problem and for some other numerical results.

Example 6-2. Let us consider in a little more detail, the optimization of a seven-section cascaded transmission-line filter of the type shown in Figure 6-9. It is terminated at each end by

$$R_g(\omega) = R_L(\omega) = \frac{377}{\sqrt{1 - (f_c/f)^2}}$$

where f is frequency in GHz and $f_c = 2.077$ GHz. The frequency variation of the terminations is thus like that of rectangular waveguides operating in the H_{10} mode with cut-off frequency 2.077 GHz. This interesting problem was

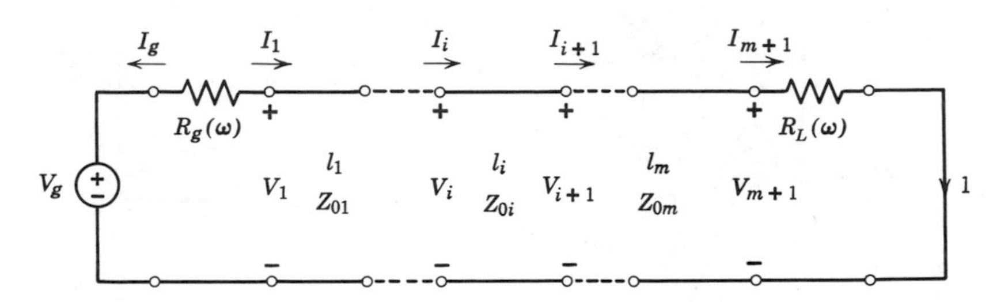

FIGURE 6-9 Cascaded transmission-line filter between frequency-variable resistors.

previously considered by Carlin and Gupta [23]. All section lengths were kept fixed at 1.5 cm so that the maximum stopband insertion loss would occur at about 5 GHz. The passband 2.16 to 3 GHz was selected, for which a maximum of 0.4 dB loss was specified. The solution obtained by the method of Carlin and Gupta was used as the initial design as shown by Figure 6-10.

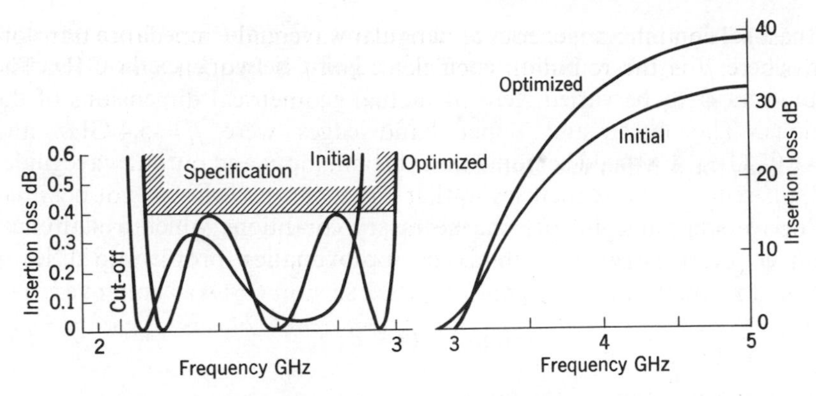

FIGURE 6-10 Comparison between the initial and optimized responses of the filter of Figure 6-9.

As optimized by Bandler and Lee-Chan [24], the problem was to minimize $\max_i f_i(\boldsymbol{\phi})$ where

$$f_i(\boldsymbol{\phi}) = \begin{cases} \frac{1}{2}[\,|\rho_i(\boldsymbol{\phi})|^2 - r^2] & \text{in the passband} \\ \frac{1}{2}[1 - |\rho_i(\boldsymbol{\phi})|^2] & \text{in the stopband} \end{cases}$$

$$\boldsymbol{\phi} = \begin{bmatrix} Z_{01} \\ Z_{02} \\ \vdots \\ Z_{07} \end{bmatrix}$$

r is the reflection coefficient magnitude corresponding to an insertion loss of 0.4 dB, and $\rho_i(\boldsymbol{\phi})$ is the reflection coefficient of the filter at the ith frequency point. In particular, 22 uniformly spaced frequencies were selected from the passband and a single frequency, namely, 5 GHz for the stopband. The appropriately optimized response is shown in Figure 6-10. These results have also been reproduced by the method of Bandler, Srinivasan, and Charalambous [20], using

$$\boldsymbol{\phi} = \begin{bmatrix} Z_{01} \\ Z_{02} \\ Z_{03} \\ Z_{04} \end{bmatrix}$$

and letting $Z_{05} = Z_{03}$, $Z_{06} = Z_{02}$, $Z_{07} = Z_{01}$.

The method used to analyze the filter at each frequency is suggested in Figure 6-9. A load current of 1 amp was assumed and a simple *ABCD* matrix analysis was carried out to find all the other voltage and current variables shown (V_g will, of course, be generally complex and frequency dependent in this case). The appropriate partial derivatives were obtained from *one* such analysis per frequency point, using the adjoint network method (Section 6.9).

Conditions for a Minimax Optimum

To derive some insight into the necessary conditions which a stationary point $\boldsymbol{\phi}^\circ$ must satisfy in a minimax approximation problem [25], let us reduce it to the form

$$\text{minimize } U = \phi_{k+1} \tag{6.63}$$

subject to constraints of the form

$$\phi_{k+1} \geq f_i(\boldsymbol{\phi}), \qquad i = 1, 2, \ldots, m. \tag{6.64}$$

Rewriting the constraints as

$$g_i(\boldsymbol{\phi}) \triangleq \phi_{k+1} - f_i(\boldsymbol{\phi}) \geq 0, \qquad i = 1, 2, \ldots, m, \tag{6.65}$$

allows us to apply the Kuhn-Tucker relations (Section 6.3). Assuming U and the $f_i(\boldsymbol{\phi})$ to be differentiable in the neighborhood of $\boldsymbol{\phi}^\circ$, we have at $\boldsymbol{\phi} = \boldsymbol{\phi}^\circ$

$$\begin{gathered} \begin{bmatrix} \nabla U \\ \dfrac{\partial U}{\partial \phi_{k+1}} \end{bmatrix} = \sum_{i=1}^{m} u_i \begin{bmatrix} \nabla \\ \dfrac{\partial}{\partial \phi_{k+1}} \end{bmatrix} (\phi_{k+1} - f_i(\boldsymbol{\phi})) \\ \mathbf{u}^T \mathbf{g} = 0, \end{gathered} \tag{6.66}$$

where $\mathbf{u}$ is defined by (6.46). But

$$\begin{gathered} \nabla U = \nabla \phi_{k+1} = \mathbf{0} \\ \frac{\partial U}{\partial \phi_{k+1}} = 1 \\ \frac{\partial f_i(\boldsymbol{\phi})}{\partial \phi_{k+1}} = 0 \end{gathered} \tag{6.67}$$

everywhere. Furthermore, at least one constraint must be an equality. For convenience, assume the first m_0 constraints are equalities. Then

$$\begin{bmatrix} \mathbf{0} \\ 1 \end{bmatrix} = \sum_{i=1}^{m_0} u_i \begin{bmatrix} -\nabla f_i(\boldsymbol{\phi}^\circ) \\ 1 \end{bmatrix} \tag{6.68}$$

since

$$u_i = 0, \qquad i = m_0 + 1, m_0 + 2, \ldots, m.$$

Alternatively, the necessary conditions may be written as

$$\sum_{i=1}^{m_0} u_i \nabla f_i(\boldsymbol{\phi}^\circ) = \mathbf{0}$$

$$\sum_{i=1}^{m_0} u_i = 1$$

$$u_i \geq 0, \qquad i = 1, 2, \ldots, m_0. \tag{6.69}$$

An interpretation of these relations is sketched in Figure 6-11. Under the conditions of convex programming, the $f_i(\boldsymbol{\phi})$ would have to be convex, and the conditions become sufficient for $\boldsymbol{\phi}^\circ$ to be $\check{\boldsymbol{\phi}}$, the minimax optimum. Often m_0 will be equal to $k + 1$, but this is not a general requirement. The reader should observe the correspondence between (6.58) to (6.60) for $\Delta\boldsymbol{\phi}^j = \mathbf{0}$ with (6.69). More insight into these relations, in particular as they relate to filter problems, should be gained by referring to Bandler [25].

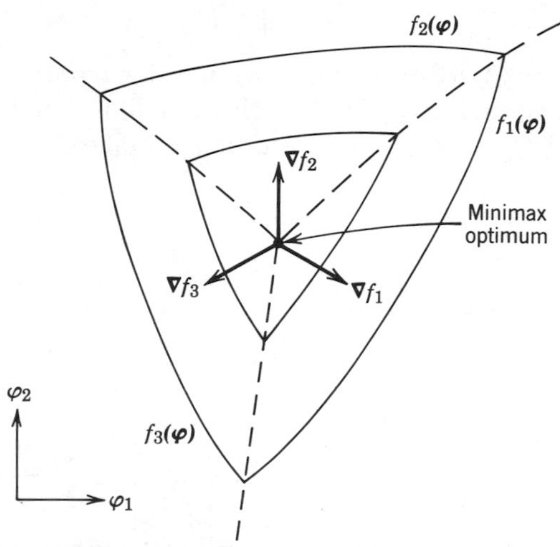

FIGURE 6-11 Sufficient conditions for a minimax optimum, $u_1 > 0$, $u_2 > 0$, $u_3 > 0$.

6.5 ONE-DIMENSIONAL SEARCH METHODS OF MINIMIZATION

Three main possible reasons spring to mind for investigating the optima of functions of one variable. The obvious one is that this might be the problem we are given. The second is that the multidimensional method we are using may call for a one-dimensional search for a minimum in some feasible downhill direction.* The third is that we may be dealing with an approximation problem for which the extrema of the error function are required during an optimization process.

* That is, in a feasible direction for which U is decreasing.

Powerful methods are available for functions known to be unimodal on an interval. We can broadly distinguish two classes, first the *elimination* methods which chop away subintervals not containing the optimum in an efficient manner with no assumptions except unimodality; second the *approximation* or *interpolation* methods which assume the function is smooth and well-represented by a low-order polynomial near the optimum.

Without loss of generality and to simplify discussions we will assume we have a function U of a single variable ϕ.

Elimination Methods

At the start of the jth iteration of a search for a minimum of a unimodal function suppose we have an *interval of uncertainty* I^j where, referring to Figure 6-12,

$$I^j \triangleq u - l \tag{6.70}$$

with $\phi_u^j = u$, $\phi_l^j = l$. Further, we have two interior points $\phi_a^j = a$ and $\phi_b^j = b$ at which we have evaluated the objective function. Let $U(a)$ and $U(b)$ be denoted U_a and U_b, respectively. Note that we take

$$l < a < b < u. \tag{6.71}$$

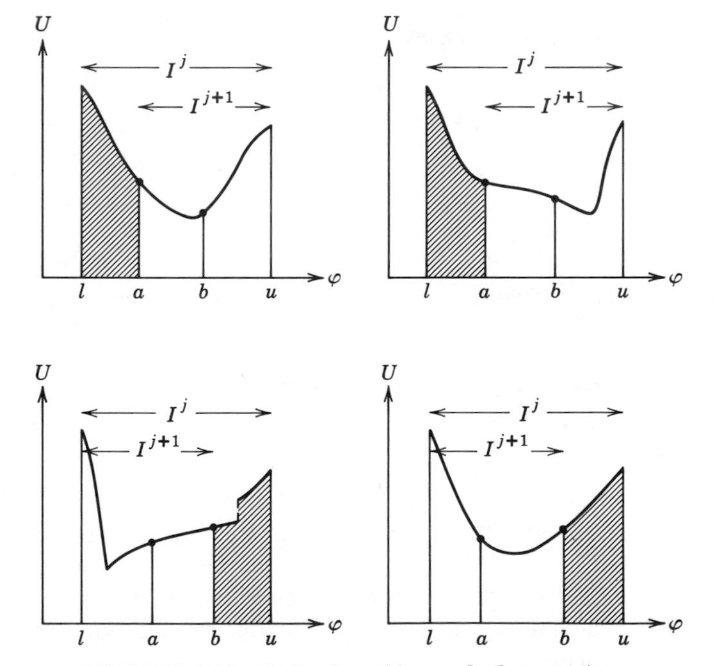

FIGURE 6-12 Reduction of interval of uncertainty.

Two conclusions can be drawn:

1. If $U_a > U_b$, the minimum lies in $[a, u]$ and $I^{j+1} = u - a$.
2. If $U_a < U_b$, the minimum lies in $[l, b]$ and $I^{j+1} = b - l$.

The difference between two well-known and efficient methods, the Fibonacci search and the Golden Section search, is in how these interior points are located. Let us discuss the slightly less efficient but simpler Golden Section search method. The reader is referred elsewhere for more detailed accounts of the various methods [6, 8, 26].

Whatever the outcome of comparing U_a and U_b, we want

$$I^{j+1} = u - a = b - l, \tag{6.72}$$

which is achieved by symmetrical placement of a and b on $[l, u]$. We want to minimize I^{j+1} and use one of the points in our new interval again which leads to

$$I^{j+2} = u - b = a - l. \tag{6.73}$$

Combining (6.70) to (6.73)

$$I^j = I^{j+1} + I^{j+2}. \tag{6.74}$$

To reduce the interval of uncertainty by a constant factor τ at each iteration:

$$\frac{I^j}{I^{j+1}} = \frac{I^{j+1}}{I^{j+2}} = \tau. \tag{6.75}$$

Equations (6.74) and (6.75) lead to

$$\tau^2 = \tau + 1, \tag{6.76}$$

the solution of relevance being $\tau = 1/2(1 + \sqrt{5}) \cong 1.618034$. The division of a line according to (6.74) and (6.75) is called the Golden Section of a line.

At the jth iteration of this scheme

$$\begin{aligned} \phi_a^j &= \frac{1}{\tau^2} I^j + \phi_l^j \\ \phi_b^j &= \frac{1}{\tau} I^j + \phi_l^j \end{aligned} \qquad j = 1, 2, 3, \ldots. \tag{6.77}$$

Note that each iteration except the first involves only one function evaluation due to symmetry. Depending on the outcome of the jth iteration, the appropriate quantities are set for the $(j + 1)$th iteration and the procedure repeated. After n function evaluations

$$\frac{I^1}{I^n} = \tau^{n-1}. \tag{6.78}$$

For a desired accuracy of σ, n should be chosen such that

$$\tau^{n-2} < \frac{\phi_u^1 - \phi_l^1}{\sigma} \leq \tau^{n-1}. \tag{6.79}$$

It is readily shown that Golden Section provides an interval of uncertainty only about 17% greater than Fibonacci search for large n. The latter method also has the disadvantage that the number of function evaluations needs to be fixed in advance.

It is also possible to construct a scheme described by Temes [15] whereby the initial interval of uncertainty does not have to be fixed in advance. This scheme has been used with the method of Bandler, Srinivasan and Charalambous [20] (Section 6.4).

Interpolation Methods

There are several interpolation methods, including quadratic and cubic, which are available [8, 26, 27, 28]. A rather straightforward method suggested by Davies, Swann, and Campey [8, 26] will be described here. The method does not require a unimodal interval containing the minimum to be known in advance, but the unimodality restriction should hold.

Evaluate $U^i \triangleq U(\phi^0 + \alpha^i s)$ for

$$\alpha^0 = 0$$

$$\alpha^i = \sum_{j=1}^{i} 2^{j-1}\,\delta, \qquad i = 1, 2, \ldots \tag{6.80}$$

where s determines the negative gradient direction, i.e.,

$$s \triangleq \left. \frac{-\dfrac{\partial U}{\partial \phi}}{\left|\dfrac{\partial U}{\partial \phi}\right|} \right|_{\phi=\phi^0} \tag{6.81}$$

and $\delta > 0$, e.g., 1% of ϕ^0, is a convenient increment. Thus α^i is a positive step in the direction of decreasing U. When, for some i,

$$U(\alpha^i) > U(\alpha^{i-1}), \tag{6.82}$$

evaluate U^{i+1} at

$$\alpha^{i+1} = \alpha^{i-1} + (\alpha^{i-1} - \alpha^{i-2}). \tag{6.83}$$

It should be clear that we now have four uniformly spaced points on the α axis, namely, α^{i-2}, α^{i-1}, α^{i+1}, and α^i in order of increasing α. Note that $i \geq 2$.

$$\begin{aligned} &\text{If} \quad U(\alpha^{i+1}) < U(\alpha^{i-1}), \quad \text{let} \quad a = \alpha^{i-1},\ b = \alpha^{i+1},\ c = \alpha^i. \\ &\text{If} \quad U(\alpha^{i+1}) > U(\alpha^{i-1}), \quad \text{let} \quad a = \alpha^{i-2},\ b = \alpha^{i-1},\ c = \alpha^{i+1}. \end{aligned} \tag{6.84}$$

It is easily shown that the minimum of a quadratic fitted at a, b, and c is at

$$\alpha_{\min} = b + \frac{(b - a)(U_a - U_c)}{2(U_a - 2U_b + U_c)}. \tag{6.85}$$

Evaluation of U at $\alpha_{\min}$ gives the estimate of the minimum and completes one stage of the method. A new stage with reduced δ can be started at b or $\alpha_{\min}$, whichever corresponds to a smaller U.

6.6 DIRECT SEARCH METHODS OF MINIMIZATION

Direct search methods as interpreted by this author are methods which do not depend explicitly on evaluation or estimation of the gradient vector of the objective function. Such methods have enjoyed fairly wide use in network optimization [6, 21, 22, 29]. To what extent they will remain competitive, however, in the light of currently available methods of evaluating derivatives (See Section 6.9), remains to be seen.

One of the simplest methods is the one-at-a-time method. As Figure 6-13 shows, this process basically consists of letting one parameter vary until no improvement is obtained, and then another one, and so on. Progress is fairly slow on valleys not oriented in the direction of any coordinate axis.

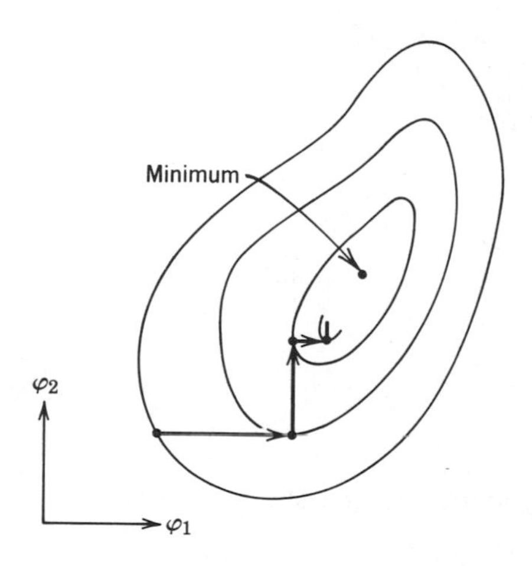

FIGURE 6-13 One-at-a-time search.

Obviously we need to consider more efficient methods. Two widely used methods will be reviewed, namely the pattern search method of Hooke and Jeeves [30] and the simplex method of Nelder and Mead [31]. Other well-known methods are Rosenbrock's method [32], the Powell-Zangwill method [28, 33], and the method of Davies, Swann, and Campey [26]. These methods are discussed in some of the general references [1, 6, 8, 26, 34].

Pattern Search

An advantage the *pattern search* method has over the one-at-a-time method is that it attempts to detect the presence of a valley and align a direction of search along it. The tactics employed by pattern search will be explained by means of the example shown in Figure 6-14.

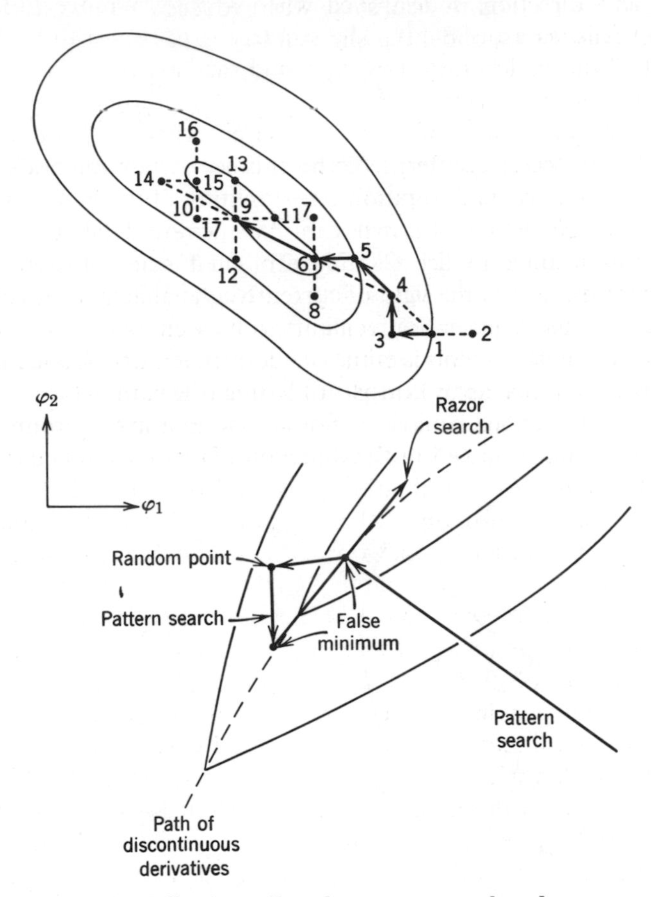

FIGURE 6-14 Following valleys by pattern search and razor search.

The first *base point* $\mathbf{b}^1$ is taken as the starting point $\boldsymbol{\phi}^1$. A series of *exploratory moves* from $\boldsymbol{\phi}^1$ is initiated to find the second base point. In the example, ϕ_1 is incremented leading us to $\boldsymbol{\phi}^2$. Now $U^2 > U^1$ so $\boldsymbol{\phi}^2$ is rejected, and ϕ_1 is incremented in the opposite direction to $\boldsymbol{\phi}^3$. Exploration with ϕ_1 is over. $U^3 < U^1$ so $\boldsymbol{\phi}^3$ is retained and exploration with ϕ_2 begins. $U^4 < U^3$ so $\boldsymbol{\phi}^4$ is retained in place of $\boldsymbol{\phi}^3$. The first set of exploratory moves is complete, and so $\boldsymbol{\phi}^4$ becomes the second base point $\mathbf{b}^2$. In the expectation that our success would be repeated we make a *pattern move* to $\boldsymbol{\phi}^5 = 2\mathbf{b}^2 - \mathbf{b}^1$, which is in the direction $\mathbf{b}^2 - \mathbf{b}^1$. By another set of exploratory moves we try to find the most promising point in the vicinity of $\boldsymbol{\phi}^5$. Here, this point is $\boldsymbol{\phi}^6$ which becomes the third base point $\mathbf{b}^3$, since $U^6 < U^4$. The search continues with a pattern move in the direction $\mathbf{b}^3 - \mathbf{b}^2$ to $\boldsymbol{\phi}^9$.

The pattern direction is destroyed when a pattern move followed by exploration fails, as around $\boldsymbol{\phi}^{14}$. The strategy is to return to the previous base point. If the exploratory moves around the base point fail, as around $\boldsymbol{\phi}^9$, the parameter increments are reduced and the procedure is restarted at that point. The search may be terminated either when the parameter increments fall below prescribed levels or the number of function evaluations or running time have reached upper limits.

The *razor search* method of Bandler and Macdonald [22] is a development of pattern search suited to direct optimization in the minimax sense without using derivatives. The name was suggested by the fact that "razor sharp" valleys are, in general, generated by an attempt to minimize functions of the form of (6.50). Paths of discontinuous derivatives are found along the bottom of such valleys, as indicated in Figures 6-11 and 6-14.

An investigation of the behavior of pattern search in the optimization of cascaded noncommensurate transmission lines acting as impedance transformers between resistive terminations was carried out [29]. It was observed that pattern search failed only when a sharp valley whose contours lay entirely within a quadrant of the coordinate axes was encountered. In that case no improvement was possible by searching parallel to these axes.

The razor search method makes a random move from a point where pattern search fails (assuming a false minimum) and uses pattern search to return to the path of discontinuous derivatives. (See Figure 6-14.) When pattern search fails again, an attempt is made to establish a pattern in the apparent downhill direction and resume with pattern search. The results shown in Figure 6-8 were produced by the razor search method [21].

An observation worth making here is that manual network optimization in the minimax sense, using an interactive system and employing, say, the one-at-a-time method, can easily terminate at a false minimum. A false minimum in the present context is a point representing a possibly equal-ripple response but which is not a local optimum in the minimax sense.

The Simplex Method

In simplex methods of nonlinear optimization, the objective function is evaluated at the $k + 1$ vertices of a *simplex* in k-dimensional space. In two dimensions, for example, we would have a triangle, for three dimensions a tetrahedron. An attempt is then made to replace the point with the greatest objective function value by another point.

A method having very desirable valley-following properties is the one due to Nelder and Mead [31]. The basic move is to *reflect* the point having the greatest function value in the centroid of the simplex formed by the remaining points. If the reflected point results in a function value lower than the current lowest, an *expansion* is attempted. Otherwise the point is retained if it results in a function value lower than the second highest. *Contraction* is attempted if reflection fails. Finally, *shrinking* of the simplex about the vertex corresponding to the lowest function value occurs following an unsuccessful attempt at contraction. Some of these moves are illustrated in Figure 6-15.

An example of the simplex strategy is shown in Figure 6-16. Observe that $\boldsymbol{\phi}^4$, $\boldsymbol{\phi}^6$, $\boldsymbol{\phi}^8$, $\boldsymbol{\phi}^9$, and $\boldsymbol{\phi}^{10}$ have resulted from reflection; $\boldsymbol{\phi}^5$ from expansion; and $\boldsymbol{\phi}^7$ and $\boldsymbol{\phi}^{11}$ from contraction. The reader should follow the strategy through carefully to ensure his understanding of it. Its desirable valley-following properties result from its ability to align elongated simplexes in the

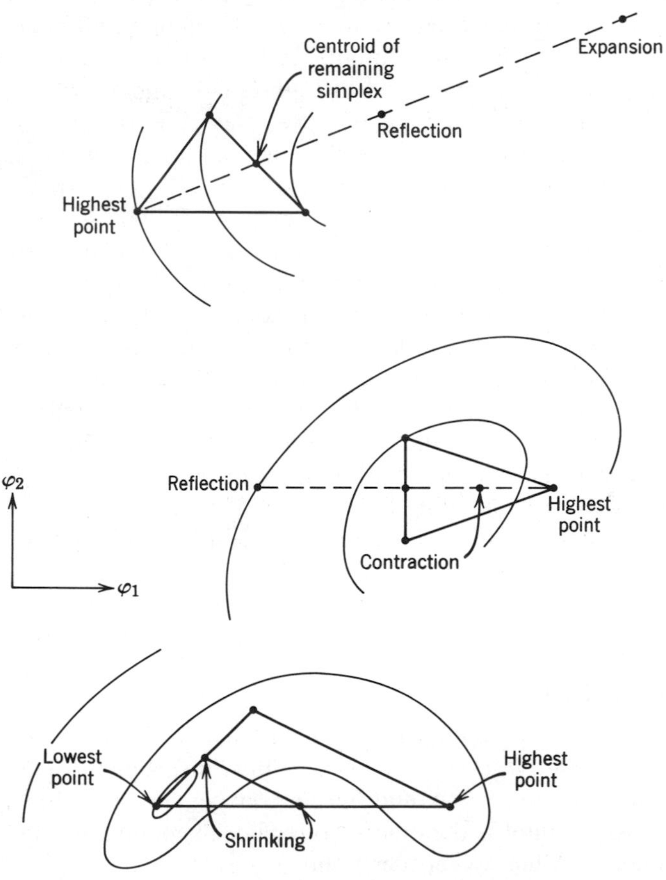

FIGURE 6-15 Examples of moves made by the simplex method.

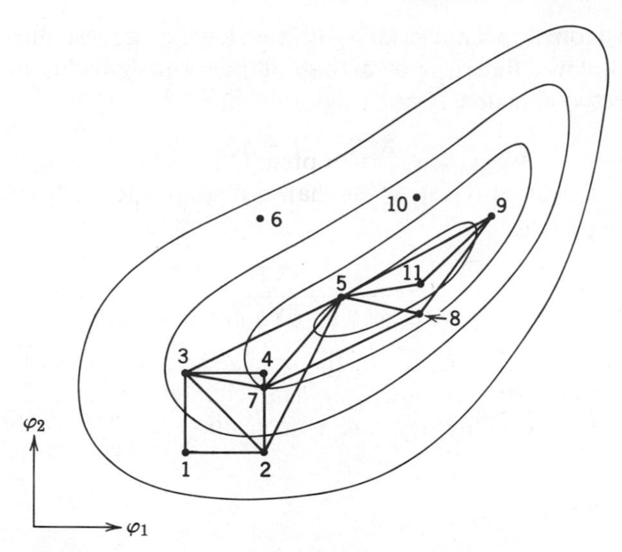

FIGURE 6-16 Optimization by the simplex method.

direction of the valleys. In particular, repeated success, for example, if a long straight valley is being followed, tends to increase the size of the moves, whereas repeated failure, for example, if a bend in the valley is encountered, tends to cause a decrease in the size of the moves.

It has been claimed to the author on a number of occasions that, unlike some other direct search methods, the simplex method can be successfully employed for minimax approximation. In the author's experience the simplex method is no less infallible than pattern search, for example. The principal fallacy in the argument is the assumption that, if the method requires no derivative information, it can necessarily handle problems with discontinuous derivatives.

6.7 GRADIENT METHODS OF MINIMIZATION

We turn our attention now to a class of minimization methods which require derivatives. By and large the most efficient algorithms currently available rely on evaluation of the gradient vector [1, 5, 6, 8, 14, 15, 26, 27, 35].

Steepest Descent

At the jth iteration of most gradient methods, we proceed to

$$\boldsymbol{\phi}^{j+1} = \boldsymbol{\phi}^j + \alpha^j \mathbf{s}^j \tag{6.86}$$

where $\mathbf{s}^j$ is (hopefully) a downhill direction of search and $\alpha^j > 0$ is a scale factor chosen to minimize $U(\boldsymbol{\phi}^j + \alpha^j \mathbf{s}^j)$. One-dimensional minimization methods suitable for this purpose were discussed in Section 6.5.

The most obvious choice for $\mathbf{s}^j$ is the *steepest descent* direction at $\boldsymbol{\phi}^j$, defined as follows. Referring back to (6.11), we note that a first-order change in the objective function is given by

$$\Delta U = \nabla U^T \, \Delta\boldsymbol{\phi}. \tag{6.87}$$

If $\Delta\boldsymbol{\phi} = \alpha\mathbf{s}$, where $\alpha > 0$ is fixed and $\|\mathbf{s}\| = 1$*, then it is easy to show that the $\mathbf{s}$ minimizing ΔU is

$$\mathbf{s} = -\frac{\nabla U}{\|\nabla U\|} \tag{6.88}$$

The $\mathbf{s}$ in (6.88) is the negative of the *normalized gradient* vector. Although $-\nabla U/\|\nabla U\|$ provides the greatest local change, success of the steepest descent method is highly dependent on scaling. As Figure 6-17 shows, the

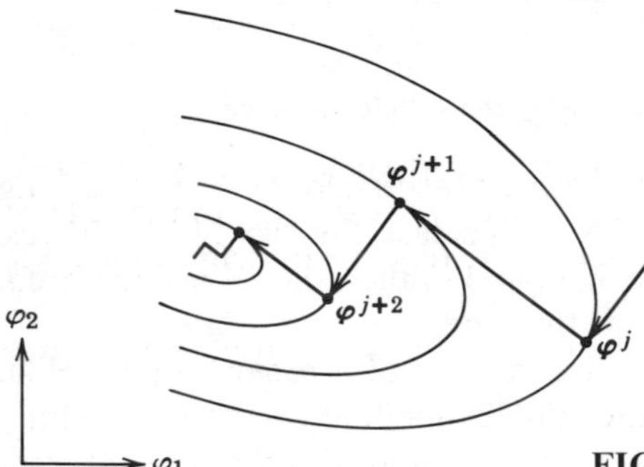

FIGURE 6-17 A steepest-descent strategy.

first few iterations may give good reduction in U, but subsequently the method usually deteriorates rapidly into oscillations, and progress becomes very slow.

The Newton Method

This method was already mentioned in Section 6.2 in the context of solution of nonlinear equations. Differentiating the Taylor series (6.11)

$$\nabla U(\boldsymbol{\phi} + \Delta\boldsymbol{\phi}) = \nabla U(\boldsymbol{\phi}) + \mathbf{H}\,\Delta\boldsymbol{\phi} + \cdots. \tag{6.89}$$

For $\boldsymbol{\phi} + \Delta\boldsymbol{\phi}$ to be the minimizing point $\check{\boldsymbol{\phi}}$, $\nabla U(\boldsymbol{\phi} + \Delta\boldsymbol{\phi})$ should be $\mathbf{0}$ so that, neglecting higher-order terms,

$$\Delta\boldsymbol{\phi} = -\mathbf{H}^{-1}\,\nabla U. \tag{6.90}$$

* The expression $\|\cdot\|$ is the *Euclidean* norm. It has the form of (6.23) with $p = 2$.

This incremental change takes us to the minimum in only one iteration if we are dealing with a quadratic function (Section 6.2). It is instructive to compare (6.90) with (6.19).

When U is not quadratic, we could try the iterative scheme

$$\boldsymbol{\phi}^{j+1} = \boldsymbol{\phi}^j - \mathbf{H}^{-1}\,\nabla U^j \tag{6.91}$$

where $\mathbf{H}^{-1}$ is the inverse of the Hessian matrix at the jth iteration. This scheme has, however, several disadvantages. $\mathbf{H}$ must be positive definite otherwise divergence could occur. In particular, $-\mathbf{H}^{-1}\,\nabla U^j$ might not point downhill. To counteract these possibilities, the modification

$$\boldsymbol{\phi}^{j+1} = \boldsymbol{\phi}^j - \alpha^j\mathbf{H}^{-1}\,\nabla U^j \tag{6.92}$$

can be employed where α^j is chosen to minimize U^{j+1}. This might also be ineffective: α^j may have to be negative; $\mathbf{H}$ may be locally singular. Finally, the computation of $\mathbf{H}$ and its inverse are time consuming.

Conjugate Directions

Certain gradient methods which exploit the properties of *conjugate directions* associated with quadratic functions and do not explicitly evaluate $\mathbf{H}$ or its inverse are highly effective. Before discussing them let us define conjugate directions.

The directions $\mathbf{u}_i$ and $\mathbf{u}_j$ are said to be conjugate with respect to a positive definite matrix $\mathbf{A}$ if

$$\mathbf{u}_i^T\mathbf{A}\mathbf{u}_j = 0, \qquad i \neq j. \tag{6.93}$$

In Figure 6-18, a two-dimensional interpretation of conjugate directions is given. Methods which generate such directions will minimize a quadratic

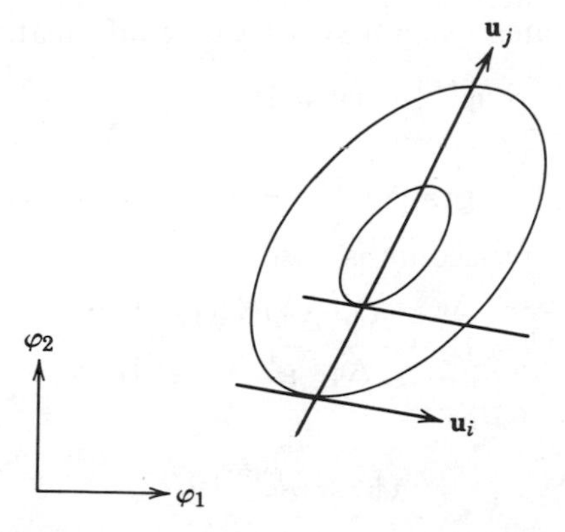

FIGURE 6-18 An illustration of two conjugate directions.

function in a finite number of iterations. It is evident that one linear minimization along each direction in turn locates the minimum.

Three well-known methods which use conjugate directions are the *conjugate gradient* method described by Fletcher and Reeves [35], the Fletcher-Powell-Davidon method [27], and the Powell-Zangwill method [28, 33] which does not require derivatives (See also references [1, 8]).

The Conjugate Gradient Method

The direction of search $\mathbf{s}^j$ is given by [35]

$$\mathbf{s}^j = -\nabla U^j + \beta^j \mathbf{s}^{j-1} \tag{6.94}$$

where

$$\beta^j = \frac{(\nabla U^j)^T \nabla U^j}{(\nabla U^{j-1})^T \nabla U^{j-1}}, \tag{6.95}$$

and, initially, $\beta^0 = 0$. Thus the first iteration is in the direction of steepest descent. Apart from round-off errors, the procedure will terminate at the minimum of a quadratic in at most k iterations. In general, however, it is recommended that $k + 1$ iterations be completed before restarting the procedure.

The Fletcher-Powell-Davidon Method

Redefining $\mathbf{H}$ as any positive definite matrix, we have [27]

$$\mathbf{s}^j = -\mathbf{H}^j \nabla U^j. \tag{6.96}$$

Note that $\mathbf{H}^j$ is the jth approximation to the *inverse* of the Hessian matrix. Initially, $\mathbf{H}^0$ is the unit matrix, and again we have the steepest descent direction.

$\mathbf{H}$ is continually updated using first derivative information such that

$$\boldsymbol{\phi}^{j+1} - \boldsymbol{\phi}^j = \mathbf{H}^{j+1} \mathbf{g}^j \tag{6.97}$$

where

$$\mathbf{g}^j = \nabla U^{j+1} - \nabla U^j.$$

The following updating procedure is used:

$$\mathbf{H}^{j+1} = \mathbf{H}^j + \frac{\Delta\boldsymbol{\phi}^j \, \Delta\boldsymbol{\phi}^{j^T}}{\Delta\boldsymbol{\phi}^{j^T} \mathbf{g}^j} - \frac{\mathbf{H}^j \mathbf{g}^j \mathbf{g}^{j^T} \mathbf{H}^j}{\mathbf{g}^{j^T} \mathbf{H}^j \mathbf{g}^j} \tag{6.98}$$

where

$$\Delta\boldsymbol{\phi}^j = \alpha^j \mathbf{s}^j,$$

and α^j is found by a one-dimensional search (Section 6.5).

Fletcher and Powell prove by induction that if $\mathbf{H}^j$ is positive definite then $\mathbf{H}^{j+1}$ is also positive definite. $\mathbf{H}^0$, being the unit matrix, is clearly positive definite. On a quadratic function it is further proved that $\mathbf{H}^k$ is the inverse of the Hessian matrix and $\nabla U^k = \mathbf{0}$, apart from round-off errors. Both the proof of convergence and success in practice depend on accurate location of the minimum in the linear searches. If necessary, $\mathbf{H}$ may be reset to the unit matrix.

This method is still generally acknowledged to be the best general purpose gradient optimization method.

6.8 LEAST *p*th APPROXIMATION

The material in this section could equally well have been treated under gradient methods. It is useful, however, to distinguish between these problems since special techniques are available for least *p*th approximation.

For objective functions in the form of (6.29) and (6.30) we can write

$$\nabla U = \int_{\psi_l}^{\psi_u} \operatorname{Re}\{p\,|e(\boldsymbol{\phi}, \psi)|^{p-2} e^*(\boldsymbol{\phi}, \psi)\, \nabla e(\boldsymbol{\phi}, \psi)\}\, d\psi \tag{6.99}$$

for the continuous case and

$$\nabla U = \sum_{i \in I} \operatorname{Re}\{p\,|e_i(\boldsymbol{\phi})|^{p-2} e_i^*(\boldsymbol{\phi})\, \nabla e_i(\boldsymbol{\phi})\} \tag{6.100}$$

for the discrete case. If the appropriate derivatives, namely ∇e, are available, we could proceed to optimize with a suitable gradient method (Section 6.7).

In more complicated situations we can envisage a linear combination of functions in the form (6.29) and (6.30), for example,

$$U = \alpha_1 U_1 + \alpha_2 U_2 + \cdots. \tag{6.101}$$

Simultaneous approximation of more than one response specification might be posed in this way (See Section 6.9). The factors α_1, α_2, etc. would be given values commensurate with the importance of U_1, U_2, etc.

Temes and Zai [15, 36] have extended the well-known least squares method of Gauss [6, 8, 14] to a *least* p*th method*. Since the former method falls out as a special case, the latter method will be briefly described. For definiteness, assume the objective function is of the form (with real $e_i(\boldsymbol{\phi})$)

$$U = \sum_{i=1}^{n} [e_i(\boldsymbol{\phi})]^p \tag{6.102}$$

where $n > k$ and p is any positive even integer. Then

$$\nabla U = \sum_{i=1}^{n} p e_i^{p-1}\, \nabla e_i \tag{6.103}$$

and

$$\mathbf{H} = \nabla(\nabla U)^T = \sum_{i=1}^{n} [pe_i^{p-1} \nabla(\nabla e_i)^T + p(p-1)e_i^{p-2} \nabla e_i(\nabla e_i)^T]. \quad (6.104)$$

Now assume that the first term may be neglected in comparison with the second. This really corresponds to a linearization of $e_i(\boldsymbol{\phi})$. Then

$$\mathbf{H} \approx \sum_{i=1}^{n} p(p-1)e_i^{p-2} \nabla e_i(\nabla e_i)^T.$$

This can be rewritten as

$$\mathbf{H} \approx p(p-1)\mathbf{A}^T\mathbf{B}\mathbf{A} \quad (6.105)$$

where

$$\mathbf{A} \triangleq [\nabla e_1 \quad \nabla e_2 \quad \cdots \quad \nabla e_n]^T$$

and

$$\mathbf{B} \triangleq \begin{bmatrix} e_1^{p-2} & 0 & \cdots & 0 \\ 0 & e_2^{p-2} & \cdots & 0 \\ \vdots & \vdots & & \vdots \\ 0 & 0 & \cdots & e_n^{p-2} \end{bmatrix}.$$

Letting

$$\boldsymbol{\epsilon} \triangleq [e_1^{p-1} \quad e_2^{p-1} \quad \cdots \quad e_n^{p-1}]^T,$$

(6.103) becomes

$$\nabla U = p\mathbf{A}^T\boldsymbol{\epsilon}. \quad (6.106)$$

Using the step given by the Newton method (6.90),

$$\Delta\boldsymbol{\phi} = -(p-1)^{-1}(\mathbf{A}^T\mathbf{B}\mathbf{A})^{-1}\mathbf{A}^T\boldsymbol{\epsilon}. \quad (6.107)$$

Under suitable conditions, it can be shown that $\Delta\boldsymbol{\phi}$ points in the downhill direction. The modified Newton procedure

$$\boldsymbol{\phi}^{j+1} = \boldsymbol{\phi}^j - \alpha^j(p-1)^{-1}(\mathbf{A}^T\mathbf{B}\mathbf{A})^{-1}\mathbf{A}^T\boldsymbol{\epsilon} \quad (6.108)$$

is recommended where α^j is chosen to minimize U^{j+1}.

Damping techniques similar to those used in the Gauss method are applicable [8, 14]. Define, for example,

$$U = \sum_{i=1}^{n} [e_i(\boldsymbol{\phi})]^p + \lambda\, \Delta\boldsymbol{\phi}^T\, \Delta\boldsymbol{\phi}. \quad (6.109)$$

Then

$$\mathbf{H} \approx p(p-1)\mathbf{A}^T\mathbf{B}\mathbf{A} + 2\lambda\mathbf{I}_k \quad (6.110)$$

and

$$\Delta\boldsymbol{\phi} = -p[p(p-1)\mathbf{A}^T\mathbf{B}\mathbf{A} + 2\lambda\mathbf{I}_k]^{-1}\mathbf{A}^T\boldsymbol{\epsilon} \tag{6.111}$$

It may be shown that the convergence and downhill properties are preserved and that for $\lambda > 0$ the step is no larger than the undamped step. As $\lambda \to 0$ the process is undamped, while for $\lambda \to \infty$ the step is in the steepest descent direction. The introduction of α^j to permit a linear search as in (6.108) is also possible.

Example 6-3. An example of least pth approximation [37] compared with minimax approximation is depicted in Figure 6-19. The structure is the

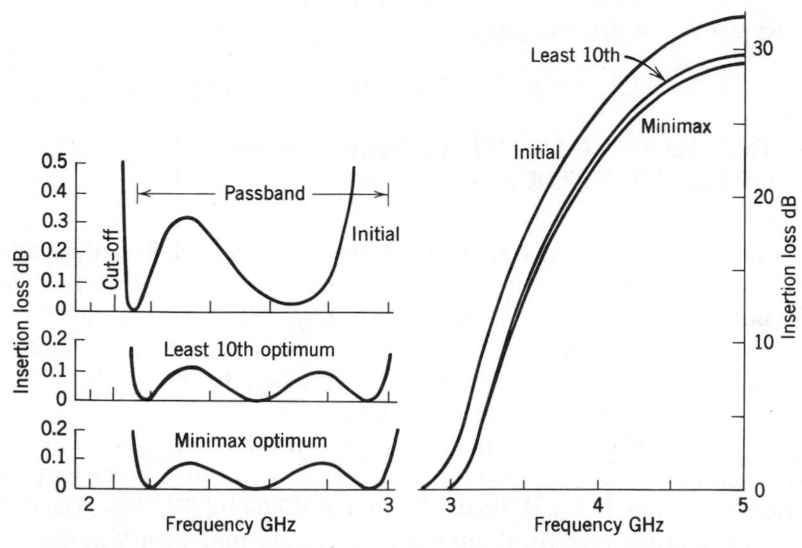

FIGURE 6-19 Example of least 10th approximation compared with minimax approximation in optimizing the passband of the filter of Figure 6-9.

seven-section cascade of transmission lines acting as a filter discussed in Section 6.4. The problem here was to see how small the passband insertion loss could be made under the constraints of the problem (if R_g and R_L were frequency independent, or the lengths were allowed to vary, the answer would be trivial).

A least pth objective function was set up with $p = 10$, using 51 uniformly spaced points in the passband. The objective function was of the form

$$U = \sum_{i=1}^{n} \frac{1}{p} |\rho_i(\boldsymbol{\phi})|^p.$$

The Fletcher-Powell-Davidon method (Section 6.7) was used, the required first derivatives being obtained from *one* network analysis using the adjoint network method (Section 6.9).

Compare the almost equal-ripple passband response obtained with a maximum insertion loss of about 0.1 dB with the equal-ripple response (maximum insertion loss 0.086 dB) produced by minimax approximation. The latter solution was obtained by Bandler and Lee-Chan [24] using a gradient algorithm with quadratic interpolation used to locate the ripple extrema.

The main conclusion to be reached from this example is that acceptable results can be achieved with relatively moderate values of p. Unless special precautions are taken to avoid ill-conditioning, the use of values of p much greater than 10 is discouraged.

6.9 THE ADJOINT NETWORK METHOD OF GRADIENT EVALUATION

The *adjoint network method* can be used to great advantage in evaluating the gradient vector of objective functions related to gain, insertion loss, reflection coefficient, or any other desired response. A very broad class of networks can be treated by this method. As will be seen, no more than two complete network analyses are required to evaluate the gradient vector regardless of the number of variable parameters.

Director and Rohrer have discussed the concept of the adjoint network and indicated its relevance to automated design of networks in the frequency and time domains [38, 39]. In the frequency domain [39], they considered reciprocal and nonreciprocal, lumped, linear, and time-invariant elements. We will restrict ourselves here to the frequency domain, review Director and Rohrer's results, and extend them to least pth and minimax approximation. Some uniformly distributed elements will also be included [37, 40].

Adjoint Networks And Network Sensitivities

Let

$$\mathbf{v} \triangleq \begin{bmatrix} v_1 \\ v_2 \\ \vdots \\ v_b \end{bmatrix} \tag{6.112}$$

contain all the branch voltages in a network and

$$\mathbf{i} \triangleq \begin{bmatrix} i_1 \\ i_2 \\ \vdots \\ i_b \end{bmatrix} \tag{6.113}$$

contain all the corresponding branch currents (using associated reference directions*). $\mathbf{v}$ and $\mathbf{i}$ must satisfy Kirchhoff's voltage and current laws, respectively. Then Tellegen's theorem states [41]

$$\mathbf{v}^T\mathbf{i} = 0. \tag{6.114}$$

As long as the topologies are the same, $\mathbf{v}$ can refer to one network and $\mathbf{i}$ to another (See the example in Figure 6-20). Let us, therefore, imagine we have

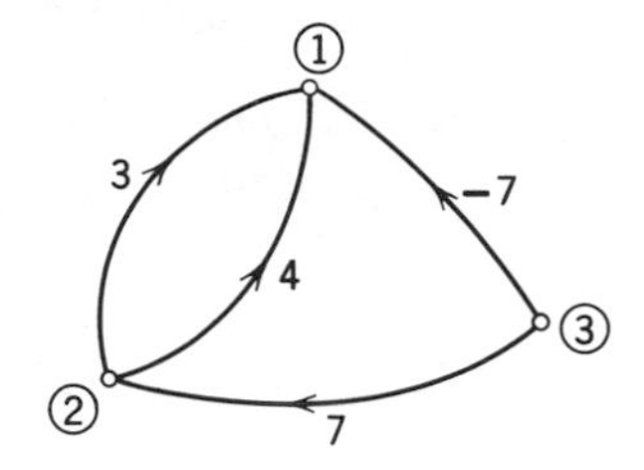

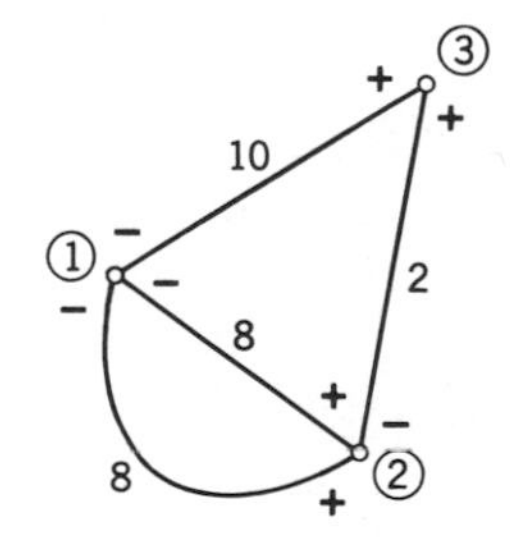

FIGURE 6-20 Illustration of Tellegen's theorem applied to two networks of the same topology. Observe that

$$\mathbf{v}^T\mathbf{i} = 24 + 32 - 70 + 14 = 0.$$

Since the nature of the elements is immaterial, they are replaced by branches.

two networks, the original one which is to be optimized and a topologically equivalent adjoint network. As mentioned earlier we will confine ourselves to a consideration of linear, time-invariant networks in the *frequency domain.* Variables V and I will thus denote phasors associated with the original

* With associated reference directions, the current always enters a branch at the plus sign and leaves at the minus sign.

network, and $\hat{V}$ and $\hat{I}$ the corresponding phasors associated with the adjoint network. By Tellegen's theorem

$$\mathbf{V}_B^T\hat{\mathbf{I}}_B = 0$$
$$\mathbf{I}_B^T\hat{\mathbf{V}}_B = 0 \tag{6.115}$$

where the subscript B implies that the associated vectors contain all corresponding complex branch voltages and currents. Perturbing elements in the original network we have

$$\Delta\mathbf{V}_B^T\hat{\mathbf{I}}_B = 0 \tag{6.116a}$$

$$\Delta\mathbf{I}_B^T\hat{\mathbf{V}}_B = 0 \tag{6.116b}$$

since Kirchhoff's voltage and current laws must also be applicable to $\Delta\mathbf{V}_B$ and $\Delta\mathbf{I}_B$. Subtracting (6.116b) from (6.116a)

$$\Delta\mathbf{V}_B^T\hat{\mathbf{I}}_B - \Delta\mathbf{I}_B^T\hat{\mathbf{V}}_B = 0. \tag{6.117}$$

Figure 6-21 shows N-port original and adjoint elements characterized in terms of open-circuit impedance matrices $\mathbf{Z}$ and $\hat{\mathbf{Z}}$, respectively. Letting $\mathbf{V}$, $\mathbf{I}$,

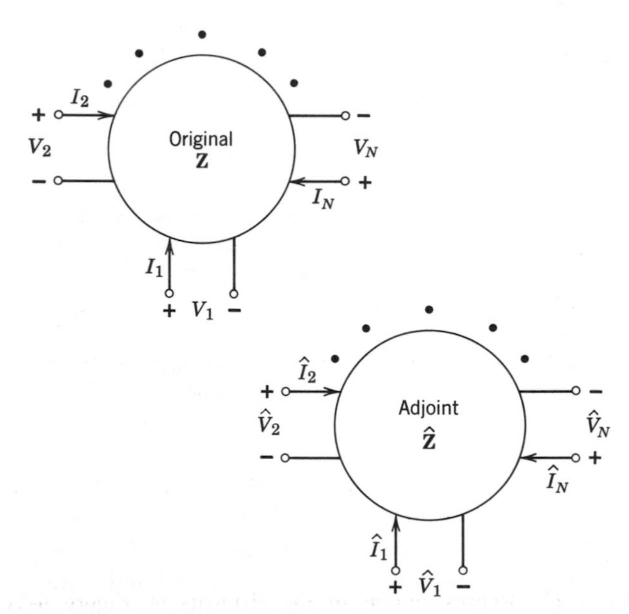

FIGURE 6-21 Original and adjoint elements represented by impedance matrices. In general, many such elements suitably connected form the original and adjoint networks.

$\hat{\mathbf{V}}$, and $\hat{\mathbf{I}}$ denote N-element vectors containing the relevant port variables

$$\mathbf{V} = \mathbf{Z}\mathbf{I} \tag{6.118}$$

$$\hat{\mathbf{V}} = \hat{\mathbf{Z}}\hat{\mathbf{I}}. \tag{6.119}$$

Perturbing the parameters in the original element and neglecting higher-order terms

$$\Delta\mathbf{V} = \Delta\mathbf{Z}\mathbf{I} + \mathbf{Z}\,\Delta\mathbf{I}. \tag{6.120}$$

As indicated by Figure 6-22, the port variables can be thought of as equivalent branch variables, so that, substituting (6.120) into (6.117) we see that

$$(\mathbf{I}^T\,\Delta\mathbf{Z}^T + \Delta\mathbf{I}^T\mathbf{Z}^T)\hat{\mathbf{I}} - \Delta\mathbf{I}^T\hat{\mathbf{V}}$$

reduces to

$$\mathbf{I}^T\,\Delta\mathbf{Z}^T\hat{\mathbf{I}} \tag{6.121}$$

if

$$\hat{\mathbf{Z}} \equiv \mathbf{Z}^T. \tag{6.122}$$

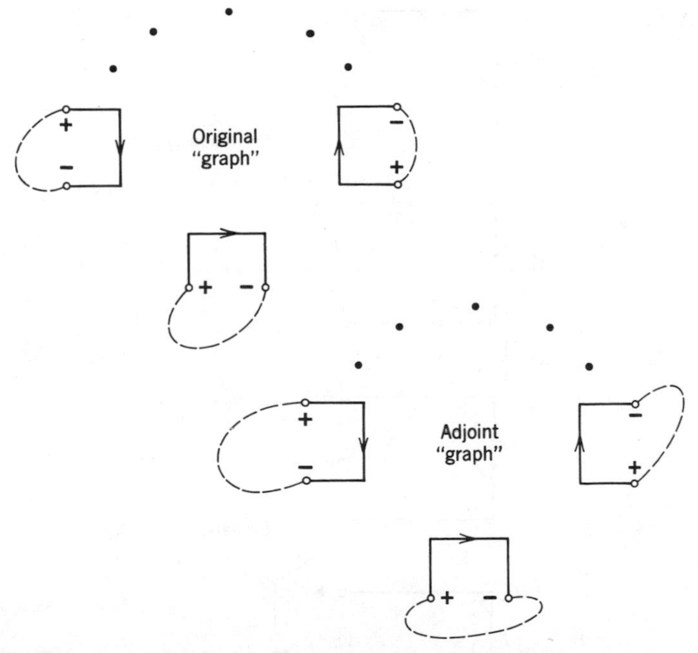

FIGURE 6-22 Representation of the elements of Figure 6-21 for application of Tellegen's theorem. Some *i*th equivalent branch might consist of an impedance z_{ii} in series with voltage generators of value $z_{ij}I_j$, $j = 1, 2, \ldots$. See, for example, Figure 6-25.

This defines the adjoint element. Observe that expression (6.121), the only term in (6.117) relating to the N-port element, does not contain $\Delta\mathbf{I}$ or $\Delta\mathbf{V}$. Further, note that the adjoint of a reciprocal element is identical to the original, since $\mathbf{Z}^T = \mathbf{Z}$.

Next define voltage and current excitation vectors and response vectors as in Figure 6-23. In keeping with the present notation, the hat " ∧ " will distinguish the corresponding quantities for the adjoint network. Terms in (6.117) associated with the excitations and responses are

$$\Delta\mathbf{V}_V^T\hat{\mathbf{I}}_V - \Delta\mathbf{I}_V^T\hat{\mathbf{V}}_V + \Delta\mathbf{V}_I^T\hat{\mathbf{I}}_I - \Delta\mathbf{I}_I^T\hat{\mathbf{V}}_I$$

which reduces to

$$-\Delta\mathbf{I}_V^T\hat{\mathbf{V}}_V + \Delta\mathbf{V}_I^T\hat{\mathbf{I}}_I \tag{6.123}$$

since $\Delta\mathbf{V}_V$ and $\Delta\mathbf{I}_I$ become zero when the excitations are held fixed.

Clearly, any network may be thought of as consisting of the interconnection of a number of multiport elements. Thus, several terms of the form of expression (6.121) can appear in (6.117).

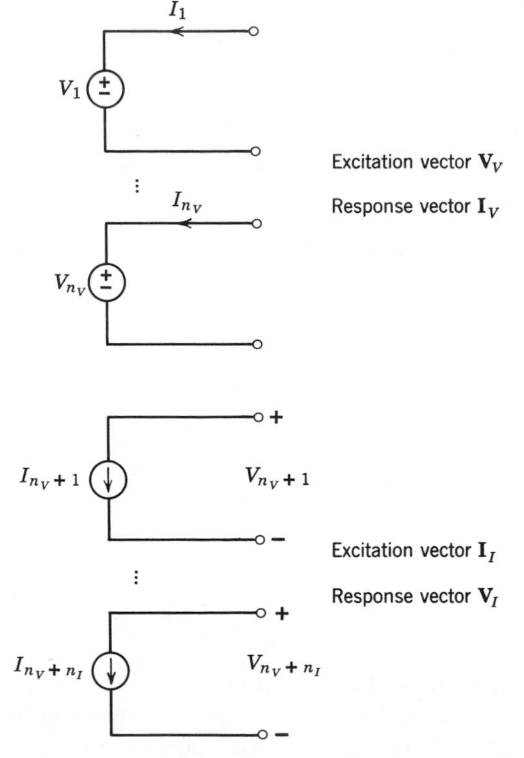

FIGURE 6-23 Port excitation and response vectors.

For an admittance matrix representation we can show that

$$-\mathbf{V}^T \,\Delta\mathbf{Y}^T\hat{\mathbf{V}} \tag{6.124}$$

corresponds to expression (6.121). $\mathbf{Y}^T$ is the admittance matrix of the adjoint. Things are slightly more complicated for the hybrid matrix. If we take

$$\begin{bmatrix}\mathbf{I}_a\\ \mathbf{V}_b\end{bmatrix} = \begin{bmatrix}\mathbf{Y} & \mathbf{A}\\ \mathbf{M} & \mathbf{Z}\end{bmatrix}\begin{bmatrix}\mathbf{V}_a\\ \mathbf{I}_b\end{bmatrix}, \tag{6.125}$$

then the corresponding relation for the adjoint is

$$\begin{bmatrix}\hat{\mathbf{I}}_a\\ \hat{\mathbf{V}}_b\end{bmatrix} = \begin{bmatrix}\mathbf{Y}^T & -\mathbf{M}^T\\ -\mathbf{A}^T & \mathbf{Z}^T\end{bmatrix}\begin{bmatrix}\hat{\mathbf{V}}_a\\ \hat{\mathbf{I}}_b\end{bmatrix}. \tag{6.126}$$

The expression corresponding to (6.121) can be shown to be

$$[\mathbf{V}_a^T \quad \mathbf{I}_b^T]\begin{bmatrix}-\Delta\mathbf{Y}^T & \Delta\mathbf{M}^T\\ -\Delta\mathbf{A}^T & \Delta\mathbf{Z}^T\end{bmatrix}\begin{bmatrix}\hat{\mathbf{V}}_a\\ \hat{\mathbf{I}}_b\end{bmatrix}. \tag{6.127}$$

To summarize the results of the above discussion, we note that (6.117) can be written in the form

$$\Delta\mathbf{I}_V^T\hat{\mathbf{V}}_V - \Delta\mathbf{V}_I^T\hat{\mathbf{I}}_I = \mathbf{G}^T\,\Delta\boldsymbol{\phi} \tag{6.128}$$

where $\mathbf{G}$ is a vector of sensitivity components related to the adjustable parameters of the network, namely $\boldsymbol{\phi}$. Equation (6.128) basically relates changes in port responses due to changes in element values.

Figure 6-24 shows the results of a direct application of the formulas (6.121) and (6.124) to three commonly used elements. Table 6-2 summarizes sensitivity expressions for some commonly used lumped and distributed elements. An element consisting of a single branch is simply viewed as a one-port element.

Consider, for example, a uniformly distributed line (Figure 6-25) having characteristic impedance Z_0 and electrical length θ. Since the element is reciprocal:

$$\hat{\mathbf{Z}} = \mathbf{Z}^T = \mathbf{Z} = Z_0\begin{bmatrix}\coth\theta & \operatorname{csch}\theta\\ \operatorname{csch}\theta & \coth\theta\end{bmatrix}. \tag{6.129}$$

Invoking expression (6.121) we obtain

$$\begin{aligned}\mathbf{I}^T\,\Delta\mathbf{Z}^T\hat{\mathbf{I}} &= \mathbf{I}^T\left(\Delta Z_0\begin{bmatrix}\coth\theta & \operatorname{csch}\theta\\ \operatorname{csch}\theta & \coth\theta\end{bmatrix} - \frac{Z_0\,\Delta\theta}{\sinh\theta}\begin{bmatrix}\operatorname{csch}\theta & \coth\theta\\ \coth\theta & \operatorname{csch}\theta\end{bmatrix}\right)^T\hat{\mathbf{I}}\\ &= \left(\frac{\Delta Z_0}{Z_0}\mathbf{ZI} - \frac{\Delta\theta}{\sinh\theta}\begin{bmatrix}0 & 1\\ 1 & 0\end{bmatrix}\mathbf{ZI}\right)^T\hat{\mathbf{I}} = \frac{\Delta Z_0}{Z_0}\mathbf{V}^T\hat{\mathbf{I}} - \frac{\Delta\theta}{\sinh\theta}\mathbf{V}^T\begin{bmatrix}0 & 1\\ 1 & 0\end{bmatrix}\hat{\mathbf{I}}.\end{aligned} \tag{6.130}$$

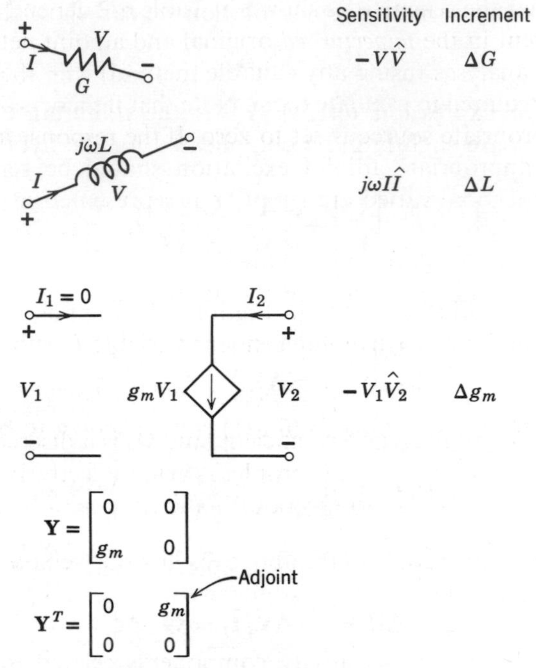

FIGURE 6-24 Sensitivities for three common elements: a resistor of conductance G, an inductor of inductance L and a voltage-controlled current source with transfer conductance g_m.

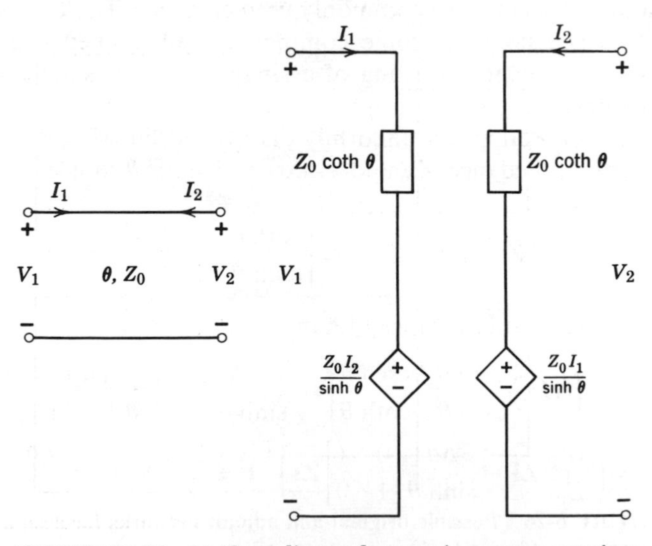

FIGURE 6-25 Uniform line and convenient representation.

Observe that the sensitivities shown in Table 6-2 depend on currents and voltages present in the *unperturbed* original and adjoint networks. At most, two network analyses (using any suitable method) will, therefore, yield the information required to evaluate them. Note that if there is no excitation at a port, the appropriate source is set to zero. If the response at a port is of no interest, the appropriate adjoint excitation should be zero. Elements or parameters not to be varied are simply not represented in **G** or $\boldsymbol{\phi}$.

An Application to Minimax Approximation

Consider the situation depicted in Figure 6-26. Suppose we are given the problem: minimize a positive independent variable U subject to

$$U \geq f(\boldsymbol{\phi}, \omega_i) \triangleq |\rho(\boldsymbol{\phi}, j\omega_i)|^2, \qquad \omega_i \in \Omega_d \tag{6.131}$$

where ρ is the input reflection coefficient, and Ω_d is a discrete set of frequencies in the band of interest. This problem then is effectively to minimize the maximum magnitude of the reflection coefficient over a band [37, 42]. Now

$$\rho = \frac{Z_{in} - R_g}{Z_{in} + R_g} = 1 - \frac{2R_g}{Z_{in} + R_g} = 1 + \frac{2R_g I_g}{V_g} \tag{6.132}$$

so that

$$\begin{aligned}\nabla f(\boldsymbol{\phi}, \omega_i) &= \text{Re}\{2\rho^*(\boldsymbol{\phi}, j\omega_i)\, \nabla\rho(\boldsymbol{\phi}, j\omega_i)\} \\ &= \text{Re}\left\{\frac{4R_g}{V_g}\, \rho^*(\boldsymbol{\phi}, j\omega_i)\, \nabla I_g(\boldsymbol{\phi}, j\omega_i)\right\}.\end{aligned} \tag{6.133}$$

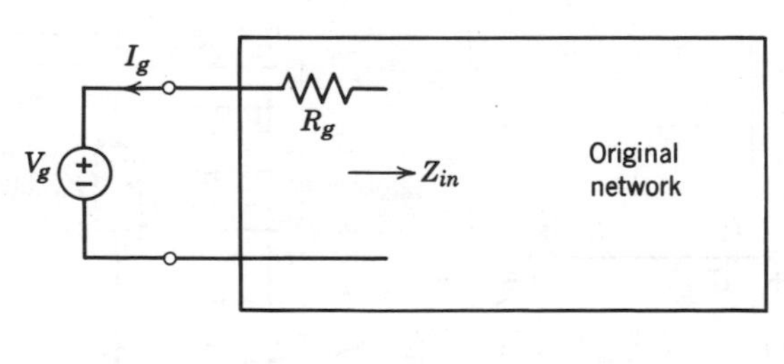

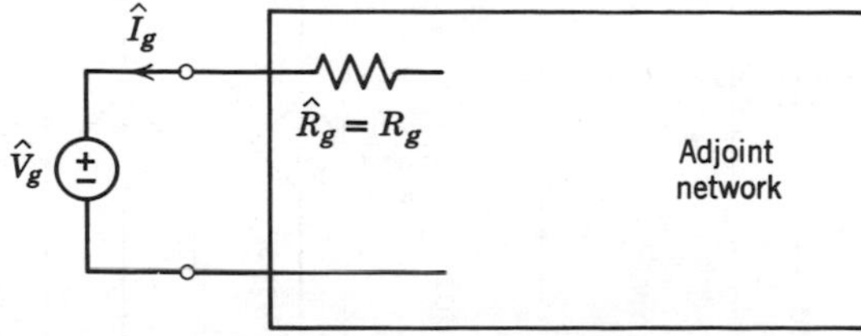

FIGURE 6-26 Possible original and adjoint networks for design on the reflection coefficient basis.

TABLE 6-2 Sensitivity Expressions for Some Lumped and Distributed Elements

Element	Equation		Sensitivity (component of **G**)	Increment (component of $\Delta\boldsymbol{\phi}$)
	Original	Adjoint		
Resistor	$V = RI$ $I = GV$	$\hat{V} = R\hat{I}$ $\hat{I} = G\hat{V}$	$I\hat{I}$ $-V\hat{V}$	ΔR ΔG
Inductor	$V = j\omega LI$ $I = \frac{1}{j\omega}\Gamma V$	$\hat{V} = j\omega L\hat{I}$ $\hat{I} = \frac{1}{j\omega}\Gamma\hat{V}$	$j\omega I\hat{I}$ $-\frac{1}{j\omega}V\hat{V}$	ΔL $\Delta\Gamma$
Capacitor	$V = \frac{1}{j\omega}SI$ $I = j\omega CV$	$\hat{V} = \frac{1}{j\omega}S\hat{I}$ $\hat{I} = j\omega C\hat{V}$	$\frac{1}{j\omega}I\hat{I}$ $-j\omega V\hat{V}$	ΔS ΔC
Transformer	$\begin{bmatrix} V_1 \\ I_2 \end{bmatrix} = \begin{bmatrix} 0 & n \\ -n & 0 \end{bmatrix} \begin{bmatrix} I_1 \\ V_2 \end{bmatrix}$	$\begin{bmatrix} \hat{V}_1 \\ \hat{I}_2 \end{bmatrix} = \begin{bmatrix} 0 & n \\ -n & 0 \end{bmatrix} \begin{bmatrix} \hat{I}_1 \\ \hat{V}_2 \end{bmatrix}$	$V_2\hat{I}_1 + I_1\hat{V}_2$	Δn
Gyrator	$\mathbf{V} = \begin{bmatrix} 0 & \alpha \\ -\alpha & 0 \end{bmatrix}\mathbf{I}$	$\hat{\mathbf{V}} = \begin{bmatrix} 0 & -\alpha \\ \alpha & 0 \end{bmatrix}\hat{\mathbf{I}}$	$I_1\hat{I}_2 - I_2\hat{I}_1$	$\Delta\alpha$
Voltage controlled voltage source	$\begin{bmatrix} I_1 \\ V_2 \end{bmatrix} = \begin{bmatrix} 0 & 0 \\ \mu & 0 \end{bmatrix} \begin{bmatrix} V_1 \\ I_2 \end{bmatrix}$	$\begin{bmatrix} \hat{I}_1 \\ \hat{V}_2 \end{bmatrix} = \begin{bmatrix} 0 & -\mu \\ 0 & 0 \end{bmatrix} \begin{bmatrix} \hat{V}_1 \\ \hat{I}_2 \end{bmatrix}$	$V_1\hat{I}_2$	$\Delta\mu$
Voltage controlled current source	$\mathbf{I} = \begin{bmatrix} 0 & 0 \\ g_m & 0 \end{bmatrix}\mathbf{V}$	$\hat{\mathbf{I}} = \begin{bmatrix} 0 & g_m \\ 0 & 0 \end{bmatrix}\hat{\mathbf{V}}$	$-V_1\hat{V}_2$	Δg_m

Table 6-2 *Continued*

Current controlled voltage source	$\mathbf{V} = \begin{bmatrix} 0 & 0 \\ r_m & 0 \end{bmatrix} \mathbf{I}$	$\hat{\mathbf{V}} = \begin{bmatrix} 0 & r_m \\ 0 & 0 \end{bmatrix} \hat{\mathbf{I}}$	$I_1 \hat{I}_2$	Δr_m
Current controlled current source	$\begin{bmatrix} V_1 \\ I_2 \end{bmatrix} = \begin{bmatrix} 0 & 0 \\ \beta & 0 \end{bmatrix} \begin{bmatrix} I_1 \\ V_2 \end{bmatrix}$	$\begin{bmatrix} \hat{V}_1 \\ \hat{I}_2 \end{bmatrix} = \begin{bmatrix} 0 & -\beta \\ 0 & 0 \end{bmatrix} \begin{bmatrix} \hat{I}_1 \\ \hat{V}_2 \end{bmatrix}$	$-I_1 \hat{V}_2$	$\Delta\beta$
Short circuited uniformly distributed line	$V = Z_0 \tanh \theta I$	$\hat{V} = Z_0 \tanh \theta \hat{I}$	$\tanh \theta I \hat{I}$ $Z_0 \operatorname{sech}^2 \theta I \hat{I}$	ΔZ_0 $\Delta\theta$
	$I = Y_0 \coth \theta V$	$\hat{I} = Y_0 \coth \theta \hat{V}$	$-\coth \theta V \hat{V}$ $Y_0 \operatorname{csch}^2 \theta V \hat{V}$	ΔY_0 $\Delta\theta$
Open circuited uniformly distributed line	$V = Z_0 \coth \theta I$	$\hat{V} = Z_0 \coth \theta \hat{I}$	$\coth \theta I \hat{I}$ $-Z_0 \operatorname{csch}^2 \theta I \hat{I}$	ΔZ_0 $\Delta\theta$
	$I = Y_0 \tanh \theta V$	$\hat{I} = Y_0 \tanh \theta \hat{V}$	$-\tanh \theta V \hat{V}$ $-Y_0 \operatorname{sech}^2 \theta V \hat{V}$	ΔY_0 $\Delta\theta$
Uniformly distributed line	$\mathbf{V} = Z_0 \begin{bmatrix} \coth\theta & \operatorname{csch}\theta \\ \operatorname{csch}\theta & \coth\theta \end{bmatrix} \mathbf{I}$	same as original network equation but with $\hat{\mathbf{V}}$ and $\hat{\mathbf{I}}$ replacing $\mathbf{V}$ and $\mathbf{I}$ respectively	$\frac{1}{Z_0} \mathbf{V}^T \hat{\mathbf{I}}$ $-\frac{1}{\sinh\theta} \mathbf{V}^T \begin{bmatrix} 0 & 1 \\ 1 & 0 \end{bmatrix} \hat{\mathbf{I}}$	ΔZ_0 $\Delta\theta$
	$\mathbf{I} = Y_0 \begin{bmatrix} \coth\theta & -\operatorname{csch}\theta \\ -\operatorname{csch}\theta & \coth\theta \end{bmatrix} \mathbf{V}$		$-\frac{1}{Y_0} \mathbf{I}^T \hat{\mathbf{V}}$ $-\frac{1}{\sinh\theta} \mathbf{I}^T \begin{bmatrix} 0 & 1 \\ 1 & 0 \end{bmatrix} \hat{\mathbf{V}}$	ΔY_0 $\Delta\theta$

Table 6-2 *Continued*

Lossless transmission line	$\mathbf{V} = -jZ_0 \begin{bmatrix} \cot \beta l & \csc \beta l \\ \csc \beta l & \cot \beta l \end{bmatrix} \mathbf{I}$	same as original network equation but with $\hat{\mathbf{V}}$ and $\hat{\mathbf{I}}$ replacing $\mathbf{V}$ and $\mathbf{I}$ respectively	$\frac{1}{Z_0} \mathbf{V}^T \hat{\mathbf{I}}$	ΔZ_0
			$-\frac{\beta}{\sin \beta l} \mathbf{V}^T \begin{bmatrix} 0 & 1 \\ 1 & 0 \end{bmatrix} \hat{\mathbf{I}}$	Δl
	$\mathbf{I} = -jY_0 \begin{bmatrix} \cot \beta l & -\csc \beta l \\ -\csc \beta l & \cot \beta l \end{bmatrix} \mathbf{V}$		$-\frac{1}{Y_0} \mathbf{I}^T \hat{\mathbf{V}}$	ΔY_0
			$-\frac{\beta}{\sin \beta l} \mathbf{I}^T \begin{bmatrix} 0 & 1 \\ 1 & 0 \end{bmatrix} \hat{\mathbf{V}}$	Δl
Uniform RC line	as for uniformly distributed line with $Z_0 = \sqrt{\frac{R}{sC}}$ and $\theta = \sqrt{sRC}$		$\frac{1}{2R} \mathbf{V}^T \begin{bmatrix} 1 & -\frac{\theta}{\sinh \theta} \\ -\frac{\theta}{\sinh \theta} & 1 \end{bmatrix} \hat{\mathbf{I}}$	ΔR
			$-\frac{1}{2C} \mathbf{V}^T \begin{bmatrix} 1 & \frac{\theta}{\sinh \theta} \\ \frac{\theta}{\sinh \theta} & 1 \end{bmatrix} \hat{\mathbf{I}}$	ΔC

From (6.128)

$$\Delta I_g \hat{V}_g = \mathbf{G}^T \, \Delta \boldsymbol{\phi}. \tag{6.134}$$

Hence

$$\Delta I_g = \left(\frac{1}{\hat{V}_g} \mathbf{G}^T \right) \Delta \boldsymbol{\phi} = \nabla I_g^T \, \Delta \boldsymbol{\phi},$$

so that

$$\nabla I_g = \frac{1}{\hat{V}_g} \mathbf{G}, \tag{6.135}$$

and, finally,

$$\nabla f(\boldsymbol{\phi}, \omega_i) = \text{Re}\left\{ \frac{4R_g}{V_g \hat{V}_g} \rho^*(\boldsymbol{\phi}, j\omega_i) \mathbf{G}(\boldsymbol{\phi}, j\omega_i) \right\}. \tag{6.136}$$

Observe that we are at liberty to set $\hat{V}_g = V_g$. If the original network is reciprocal so that the adjoint network is identical to the original, we need perform only one network analysis to obtain $\nabla f(\boldsymbol{\phi}, \omega_i)$.

An Application to Least *p*th Approximation

It can be shown that if there are n_V independent voltage sources and n_I independent current sources

$$\mathbf{G} = \sum_{i=1}^{n_V} \hat{V}_i \nabla I_i - \sum_{i=n_V+1}^{n_V+n_I} \hat{I}_i \nabla V_i. \tag{6.137}$$

Suppose we are given the objective function [37, 39, 42],

$$U = \sum_{i=1}^{n_V+n_I} \int_\Omega |e_i(\boldsymbol{\phi}, j\omega)|^p \, d\omega, \tag{6.138}$$

where Ω defines a frequency range of interest and where $e_i(\boldsymbol{\phi}, j\omega)$ is an ith function of the form of (6.21) such that

$$F_i(\boldsymbol{\phi}, j\omega) \triangleq \begin{cases} I_i(\boldsymbol{\phi}, j\omega), & i = 1, 2, \ldots, n_V \\ V_i(\boldsymbol{\phi}, j\omega), & i = n_V + 1, \ldots, n_V + n_I. \end{cases} \tag{6.139}$$

Equation (6.138) thus represents a summation of functions of the form of (6.101). The specified functions $S_i(j\omega)$ correspond to desired response currents and voltages. In general, $F_i(\boldsymbol{\phi}, j\omega)$, $S_i(j\omega)$, and hence $e_i(\boldsymbol{\phi}, j\omega)$ may be complex. Now, from (6.99)

$$\nabla U = \sum_{i=1}^{n_V+n_I} \int_\Omega \text{Re}\{p \, |e_i(\boldsymbol{\phi}, j\omega)|^{p-2} w_i(\omega) e_i^*(\boldsymbol{\phi}, j\omega) \, \nabla F_i(\boldsymbol{\phi}, j\omega)\} \, d\omega. \tag{6.140}$$

Comparing (6.137), (6.139), and (6.140), we see that if the adjoint network excitations are taken as

$$p\,|e_i(\boldsymbol{\phi}, j\omega)|^{p-2} w_i(\omega) e_i^*(\boldsymbol{\phi}, j\omega) = \begin{cases} \hat{V}_i(j\omega) & i = 1, 2, \ldots, n_V \\ -\hat{I}_i(j\omega) & i = n_V + 1, \ldots, n_V + n_I, \end{cases} \tag{6.141}$$

then

$$\nabla U = \int_{\Omega} \mathrm{Re}\{\mathbf{G}\}\, d\omega. \tag{6.142}$$

The corresponding expression for the discrete case is

$$\nabla U = \sum_{\Omega_d} \mathrm{Re}\{\mathbf{G}\} \tag{6.143}$$

where Ω_d is the discrete set of frequencies.

An Application to Group Delay Computation

In group delay computations we are essentially interested in sensitivities with respect to frequency ω [43]. This parameter is different from others that we have considered in that it is common throughout the network. Specifically, let us distinguish variables associated with some jth element of an n-element network by the subscript j. Then, assuming only ω is varied, (6.128) can be written as

$$\Delta \mathbf{I}_V^T \hat{\mathbf{V}}_V - \Delta \mathbf{V}_I^T \hat{\mathbf{I}}_I = \sum_{j=1}^{n} [\mathbf{V}_{aj}^T \mathbf{I}_{bj}^T] \begin{bmatrix} -\Delta \mathbf{Y}_j^T & \Delta \mathbf{M}_j^T \\ -\Delta \mathbf{A}_j^T & \Delta \mathbf{Z}_j^T \end{bmatrix} \begin{bmatrix} \hat{\mathbf{V}}_{aj} \\ \hat{\mathbf{I}}_{bj} \end{bmatrix}, \tag{6.144}$$

if each element, for complete generality, is characterized by an appropriate hybrid matrix. Using the rule that $\Delta x = (\partial x/\partial \omega)\, \Delta\omega$, where x is any quantity depending on ω, (6.144) can be more appropriately written

$$\sum_{i=1}^{n_V} \hat{V}_i \frac{\partial I_i}{\partial \omega} - \sum_{i=n_V+1}^{n_V+n_I} \hat{I}_i \frac{\partial V_i}{\partial \omega} = \sum_{j=1}^{n} G_{\omega j} \tag{6.145}$$

where

$$G_{\omega j} \triangleq [\mathbf{V}_{aj}^T \mathbf{I}_{bj}^T] \begin{bmatrix} -\dfrac{\partial \mathbf{Y}_j^T}{\partial \omega} & \dfrac{\partial \mathbf{M}_j^T}{\partial \omega} \\ -\dfrac{\partial \mathbf{A}_j^T}{\partial \omega} & \dfrac{\partial \mathbf{Z}_j^T}{\partial \omega} \end{bmatrix} \begin{bmatrix} \hat{\mathbf{V}}_{aj} \\ \hat{\mathbf{I}}_{bj} \end{bmatrix}. \tag{6.146}$$

If, in particular, the kth port is to be investigated, and this happens to be a current-excited port,* then (6.145) reduces to

$$-\hat{\mathbf{I}}_k \frac{\partial V_k}{\partial \omega} = \sum_{j=1}^{n} G_{\omega j} \tag{6.147}$$

if all adjoint excitations except $\hat{I}_k$ are set to zero. Evaluation of the sensitivity expression $G_{\omega j}$ is accomplished by the results of two network analyses. The sensitivity formulas from Table 6-2 may be used if appropriate, since

$$\begin{bmatrix} -\dfrac{\partial \mathbf{Y}_j^T}{\partial \omega} & \dfrac{\partial \mathbf{M}_j^T}{\partial \omega} \\ -\dfrac{\partial \mathbf{A}_j^T}{\partial \omega} & \dfrac{\partial \mathbf{Z}_j^T}{\partial \omega} \end{bmatrix} = \sum_r \begin{bmatrix} -\dfrac{\partial \mathbf{Y}_j^T}{\partial \phi_r} & \dfrac{\partial \mathbf{M}_j^T}{\partial \phi_r} \\ -\dfrac{\partial \mathbf{A}_j^T}{\partial \phi_r} & \dfrac{\partial \mathbf{Z}_j^T}{\partial \phi_r} \end{bmatrix} \frac{\partial \phi_r}{\partial \omega}, \tag{6.148}$$

where the subscript r denotes some rth parameter in the jth element with respect to which a sensitivity expression is already available.

Consider, for example, $\theta = j\omega l/c = j\beta l$ where c is the velocity of propagation. Then the ω-sensitivity of a lossless transmission line is jl/c times the θ-sensitivity shown in Table 6-2. Consider an inductor as a second example. The lefthand side of (6.148) reduces immediately to jL using $Z = j\omega L$.

Finally, to compute the group delay $T_G(\omega)$ we note that

$$T_G(\omega) = -\operatorname{Im}\left\{\frac{1}{V_k}\frac{\partial V_k}{\partial \omega}\right\}, \tag{6.149}$$

where it is assumed that all sources have constant, frequency-independent phase angles. For convenience, letting the excitation $\hat{I}_k = 1/V_k$,

$$T_G(\omega) = \operatorname{Im}\left\{\sum_{j=1}^{n} G_{\omega j}\right\}. \tag{6.150}$$

Equation (6.150) is also valid for calculations of group delay if the kth port is a voltage-excited port.* All one has to remember is to set all adjoint excitations to zero except $\hat{V}_k$ which is set to $-1/I_k$.

Extensions and Other Applications

An important point to remember about the adjoint network method is that the analysis of the adjoint, in general, can take considerably less effort than the analysis of the original network. If $\mathbf{Y}_n$ is, for example, the nodal admittance matrix of the original network, and its inverse $\mathbf{Y}_n^{-1}$ has been computed, then we can use the result $(\mathbf{Y}_n^T)^{-1} = (\mathbf{Y}_n^{-1})^T$. For a further discussion of possible computational efficiency, the reader is referred to Director [44].

* The value of the excitation could, of course, be zero.

Extensions to second-order sensitivities have been formulated [45], including group-delay sensitivities [43]. Of particular interest to filter designers are the recent applications of the adjoint network concept to the computation of dissipation-induced loss distortion in both lumped and distributed networks [46, 47]. Further extensions include the exploitation of the adjoint network concept in first- and second-order sensitivity computation using wave variables rather than voltages and currents [48, 49, 50]. These results should also be of interest to filter designers.

6.10 SUMMARY

A wide range of topics in the field of computer-aided circuit optimization has been discussed. Formulations and methods suitable for automated design, when the classical approach is inappropriate, have been stressed. The formulation of objective functions from design objectives has been discussed, including least pth and minimax. Methods of dealing with parameter and response constraints by means of transformations or penalties have been considered in some detail. Minimax approximation through linear programming and nonlinear programming has been discussed. Efficient one-dimensional methods and multidimensional gradient and direct search methods have been reviewed. Least pth approximation has been considered, with emphasis on gradient methods of solution. Finally, the adjoint network method of evaluating derivatives for design in the frequency domain was reviewed.

Most computer centers should have linear programming routines, and at least one efficient gradient algorithm, available as library programs, and possibly other methods also. It is hoped that this chapter has gone a reasonable way towards helping the network designer formulate his problems effectively so that he can take full advantage of the available computer programs.

ACKNOWLEDGEMENT

The cooperation of R. E. Seviora of the University of Toronto, particularly in the section on adjoint networks, is much appreciated. Dr. E. Della Torre of McMaster University provided considerable constructive criticism.

PROBLEMS

6.1 (a) Prove that ΔU is maximized in the direction of ∇U for a given step size.

(b) Use the multidimensional Taylor series expansion to show that a turning point of a convex differentiable function is a global minimum.

6.2 (a) If $\mathbf{g}(\boldsymbol{\phi})$ is concave, verify that $\mathbf{g}(\boldsymbol{\phi}) \geq \mathbf{0}$ describes a convex feasible region.
(b) Under what conditions could equality constraints be included in convex programming?

6.3 Find suitable transformations for the following constraints so that we can use unconstrained optimization.
(a) $0 \leq \phi_1 \leq \phi_2 \leq \cdots \leq \phi_i \leq \cdots \leq \phi_k$.
(b) $0 < l \leq \phi_2/\phi_1 \leq u$
$\phi_1 > 0$
$\phi_2 > 0$.

6.4 Derive (6.99) and (6.100).

6.5 Derive the sensitivity expression (6.124) from first principles.

6.6 Derive the entries of Table 6-2 relating to:
(a) A voltage controlled voltage source.
(b) An open-circuited uniformly distributed line.
(c) A uniform RC line.

6.7 Verify that the adjoint network may be characterized by the hybrid matrix description in (6.126).

6.8 Obtain the adjoint network in terms of an *ABCD* or chain matrix characterization of a two-port. Find sensitivity expressions in these terms for some of the entries of Table 6-2.

6.9 Consider the problem of minimizing

$$U = \phi_3(\phi_1 + \phi_2)^2$$

subject to

$$g_1 = \phi_1 - \phi_2^2 \geq 0$$
$$g_2 = \phi_2 \geq 0$$
$$h = (\phi_1 + \phi_2)\phi_3 - 1 = 0.$$

Is this a convex programming problem? Formulate it for solution by the sequential unconstrained minimization method. Starting with a feasible point, show how the constrained minimum is approached as the parameter $r \to 0$. Draw a contour sketch to illustrate the process. Are the conditions for a constrained minimum satisfied?

6.10 For the linear function

$$F(\boldsymbol{\phi}, \psi) = \sum_{i=1}^{k} \phi_i f_i(\psi),$$

(a) Formulate the discrete minimax approximation of $S(\psi)$ by $F(\boldsymbol{\phi}, \psi)$ as a linear programming problem, assuming $\boldsymbol{\phi}$ to be unconstrained.
(b) Assuming an objective function of the form of (6.102), derive ∇U and $\mathbf{H}$ (Note that a polynomial is a special case).

6.11 Verify (6.137).

6.12 Formulate the design of a notch filter in terms of inequality constraints, given the following requirements. The attenuation should not exceed A_1 dB over the frequency range 0 to ω_1, and A_2 dB over the range ω_2 to ω_3, with $0 < \omega_1 < \omega_2 < \omega_3$. At ω_0, where $\omega_1 < \omega_0 < \omega_2$, the attenuation must exceed A_0 dB.

6.13 Devise an algorithm for finding the extrema of a well-behaved multimodal function of one variable (Figure 6-2), such as the passband response of a filter.

6.14 Discuss the scaling effects of the transformation $\phi_i = \exp \phi_i'$ (Table 6-1).

6.15 (a) Are the necessary conditions for a constrained *minimum* satisfied anywhere along the boundary of the feasible region in Figure 6-1?
(b) What about the conditions for a constrained *maximum*?

6.16 Suppose we have to minimize

$$U = \sum_{\omega_i \in \Omega_d} [L(\omega_i) - S(\omega_i)]^p$$

where $L(\omega_i)$ is the insertion loss in dB of a filter between R_g and R_L, $S(\omega_i)$ is the desired insertion loss between R_g and R_L, Ω_d is a set of discrete frequencies ω_i, and p is an even positive integer. Obtain an expression relating ∇U to $\mathbf{G}(j\omega_i)$ where the elements of $\mathbf{G}$ might be as in Table 6-2. Assume convenient values for the excitations of the original and adjoint networks.

REFERENCES

[1] R. Fletcher, "A review of methods for unconstrained optimization," *Optimization*, R. Fletcher, Ed., Academic Press, New York, 1969.

[2] E. M. L. Beale, "Nonlinear programming," *Digital Computer User's Handbook*, M. Klerer and G. A. Korn, Ed., McGraw-Hill, New York, 1967.

[3] P. Wolfe, "Methods of nonlinear programming," *Nonlinear Programming*, J. Abadie, Ed., John Wiley, New York, 1967.

[4] W. I. Zangwill, *Nonlinear Programming*, Prentice-Hall, Englewood Cliffs, New Jersey, 1969.

[5] M. J. Box, "A comparison of several current optimization methods, and the use of transformations in constrained problems," *Computer J.*, **9**, 67–77 (May, 1966).

[6] J. W. Bandler, "Optimization methods for computer-aided design," *IEEE Trans. Microwave Theory and Techniques*, **MTT-17**, 533–552 (Aug., 1969).

[7] A. V. Fiacco and G. P. McCormick, *Nonlinear Programming: Sequential Unconstrained Minimization Techniques*, John Wiley, New York, 1968.

[8] J. Kowalik and M. R. Osborne, *Methods for Unconstrained Optimization Problems*, Elsevier, New York, 1968.

[9] F. A. Lootsma, "Logarithmic programming: a method of solving nonlinear-programming problems," *Philips Res. Repts.*, **22**, 329–344 (June, 1967).

[10] J. Bracken and G. P. McCormick, *Selected Applications of Nonlinear Programming*, John Wiley, New York, 1968.

[11] H. W. Kuhn and A. W. Tucker, "Non-linear programming," *Proc. 2nd Symp. on Math. Statistics and Probability. Berkeley, Calif.*, University of California Press, 481–493, 1951.

[12] D. C. Handscomb, Ed., *Methods of Numerical Approximation*, Pergamon, Oxford, 1966.

[13] G. C. Temes and J. A. C. Bingham, "Iterative Chebyshev approximation technique for network synthesis," *IEEE Trans. Circuit Theory*, **CT-14**, 31–37 (March, 1967).

[14] G. C. Temes and D. A. Calahan, "Computer-aided network optimization the state-of-the-art," *Proc. IEEE*, **55**, 1832–1863 (Nov., 1967).

[15] G. C. Temes, "Optimization methods in circuit design," *Computer Oriented Circuit Design*, F. F. Kuo and W. G. Magnuson, Jr., Ed., Prentice-Hall, Englewood Cliffs, New Jersey, 1969.

[16] L. S. Lasdon and A. D. Waren, "Optimal design of filters with bounded, lossy elements," *IEEE Trans. Circuit Theory*, **CT-13**, 175–187 (June, 1966).

[17] A. D. Waren, L. S. Lasdon, and D. F. Suchman, "Optimization in engineering design," *Proc. IEEE*, **55**, 1885–1897 (Nov., 1967).

[18] Y. Ishizaki and H. Watanabe, "An iterative Chebyshev approximation method for network design," *IEEE Trans. Circuit Theory*, **CT-15**, 326–336 (Dec., 1968).

[19] M. R. Osborne and G. A. Watson, "An algorithm for minimax approximation in the nonlinear case," *Computer J.*, **12**, 63–68 (Feb., 1969).

[20] J. W. Bandler, T. V. Srinivasan and C. Charalambous, "Minimax optimization of networks by grazor search," *IEEE Trans. Microwave Theory and Techniques*, **MTT-20**, 596–604 (Sept., 1972).

[21] J. W. Bandler, "Computer optimization of inhomogeneous waveguide transformers," *IEEE Trans. Microwave Theory and Techniques*, **MTT-17**, 563–571 (Aug., 1969).

[22] J. W. Bandler and P. A. Macdonald, "Optimization of microwave networks by razor search," *IEEE Trans. Microwave Theory and Techniques*, **MTT-17**, 552–562 (Aug., 1969).

[23] H. J. Carlin and O. P. Gupta, "Computer design of filters with lumped-distributed elements or frequency variable terminations," *IEEE Trans. Microwave Theory and Techniques*, **MTT-17**, 598-604 (Aug., 1969).

[24] J. W. Bandler and A. G. Lee-Chan, "Gradient razor search method for optimization," *1971 International Microwave Symp., Digest of Technical Papers*, 118–119 (May, 1971).

[25] J. W. Bandler, "Conditions for a minimax optimum," *IEEE Trans. Circuit Theory*, **CT-18**, 476–479 (July, 1971).

[26] M. J. Box, D. Davies, and W. H. Swann, *Non-linear Optimization Techniques*, Oliver and Boyd, Edinburgh, 1969.

[27] R. Fletcher and M. J. D. Powell, "A rapidly convergent descent method for minimization," *Computer J.*, **6**, 163–168 (June, 1963).

[28] M. J. D. Powell, "An efficient method for finding the minimum of a function of several variables without calculating derivatives," *Computer J.*, **7**, 155–162 (July, 1964).

[29] J. W. Bandler and P. A. Macdonald, "Cascaded noncommensurate transmission-line networks as optimization problems," *IEEE Trans. Circuit Theory*, **CT-16**, 391–394 (Aug., 1969).

[30] R. Hooke and T. A. Jeeves, "'Direct search' solution of numerical and statistical problems," *J. ACM*, **8**, 212–229 (April, 1961).

[31] J. A. Nelder and R. Mead, "A simplex method for function minimization," *Computer J.*, **7**, 308–313 (Jan., 1965).

[32] H. H. Rosenbrock, "An automatic method for finding the greatest or least value of a function," *Computer J.*, **3**, 175–184 (Oct., 1960).

[33] W. I. Zangwill, "Minimizing a function without calculating derivatives," *Computer J.*, **10**, 293–296 (Nov., 1967).

[34] R. Fletcher, "Function minimization without evaluating derivatives—a review," *Computer J.*, **8**, 33–41 (April, 1965).

[35] R. Fletcher and C. M. Reeves, "Function minimization by conjugate gradients," *Computer J.*, **7**, 149–154 (July, 1964).

[36] G. C. Temes and D. Y. F. Zai, "Least *p*th approximation," *IEEE Trans. Circuit Theory*, **CT-16**, 235–237 (May, 1969).

[37] J. W. Bandler and R. E. Seviora, "Current trends in network optimization," *IEEE Trans. Microwave Theory and Techniques*, **MTT-18**, 1159–1170 (Dec., 1970).

[38] S. W. Director and R. A. Rohrer, "The generalized adjoint network and network sensitivities," *IEEE Trans. Circuit Theory*, **CT-16**, 318–323 (Aug., 1969).

[39] S. W. Director and R. A. Rohrer, "Automated network design—the frequency-domain case," *IEEE Trans. Circuit Theory*, **CT-16**, 330–337 (Aug., 1969).

[40] J. W. Bandler and R. E. Seviora, "Computation of sensitivities for noncommensurate networks," *IEEE Trans. Circuit Theory*, **CT-18**, 174–178 (Jan., 1971).

[41] C. A. Desoer and E. S. Kuh, *Basic Circuit Theory*, McGraw-Hill, New York, 1969, Chapter 9.

[42] R. E. Seviora, M. Sablatash and J. W. Bandler, "Least *p*th and minimax objectives for automated network design," *Electronics Letters*, **6**, 14–15 (Jan., 1970).

[43] G. C. Temes, "Exact computation of group delay and its sensitivities using adjoint-network concept," *Electronics Letters*, **6**, 483–485 (July, 1970).

[44] S. W. Director, "*LU* factorization in network sensitivity calculations," *IEEE Trans. Circuit Theory*, **CT-18**, 184–185 (Jan., 1971).

[45] G. A. Richards, "Second-derivative sensitivity using the concept of the adjoint network," *Electronics Letters*, **5**, 398–399 (Aug., 1969).

[46] G. C. Temes and R. N. Gadenz, "Simple technique for the prediction of dissipation-induced loss distortion," *Electronics Letters*, **6**, 836–837 (Dec., 1970).

[47] R. N. Gadenz and G. C. Temes, "Computation of dissipation-induced loss distortion in lumped/distributed networks," *Electronics Letters*, **7**, 258–260 (May, 1971).
[48] J. W. Bandler and R. E. Seviora, "Sensitivities in terms of wave variables," *Proc. 8th Annual Allerton Conf. on Circuit and System Theory*, 379–387 (Oct., 1970).
[49] J. W. Bandler and R. E. Seviora, "Wave sensitivities of networks," *IEEE Trans. Microwave Theory and Techniques*, **MTT-20**, 138–147 (Feb., 1972).
[50] J. W. Bandler and R. E. Seviora, "Computation of equivalent wave source using the adjoint network," *Electronics Letters*, **7**, 235–236 (May, 1971).

7

Microwave Filters

E. G. Cristal
Hewlett-Packard
Palo Alto, California

This chapter presents a short introduction to the design of transverse electromagnetic (TEM) microwave filters. Although the published literature on this subject is extensive, the material presented here has had to be quite limited. Consequently, we have strived to present a number of general design methods, both exact and approximate, rather than enumerate specific filter designs. On the other hand, in illustrating some of the methods, several useful filter designs have also been derived.

The physical realization of microwave filters from given electrical designs is only slightly touched upon. This topic is omitted, not because it is unimportant, but, rather, because there was not enough space available to treat the subject adequately. The difficulties in realizing practical microwave filter structures can perhaps be appreciated when one considers that, for shielded TEM structures, the range of realizable impedances is only (approximately) 40 to 1. Contrast this with lumped elements for which parameter values are available over a range of greater than 10^6.

The frequency band to which the term "microwave" applies is generally considered to lie above 1 GHz. However, the theoretical design techniques presented here are not necessarily limited to this band. Indeed, the techniques can and are used at high frequencies (HF) (3–30 MHz) as well as at 20 GHz. The limitations are essentially practical ones, for example, mechanical dimensions or tolerances in the construction of the physical hardware. It is probably fair to say that in practice the design techniques presented herein are principally used in the band 0.1 to 20 GHz.

7.1 INTRODUCTION

This chapter is specifically concerned with guided-wave systems for which the wave propagation is transverse electromagnetic (TEM), that is to say, for systems in which the field components in the direction of propagation are zero. However, the theory and techniques discussed in several topics can be extended to non-TEM guided-wave systems as well. For the most part, we will be discussing networks that consist of lumped resistors, ideal transformers, and lossless, commensurate transmission lines.* Consequently, we will use the terms "microwave filter," "microwave network" and "transmission-line filter," "transmission-line network" interchangeably. Only in a few cases will lumped inductors or capacitors be used together with transmission-line components.

TEM transmission-line networks are distinguished by having no component of electric or magnetic field in the direction of propagation. Consequently, for angular frequencies ω Maxwell's field equations simplify to

$$\frac{\partial}{\partial z}\mathbf{E}_t = -j\omega\mu\mathbf{H}_t \times \mathbf{u}_z$$

$$\frac{\partial}{\partial z}\mathbf{H}_t = -j\omega\epsilon(\mathbf{u}_z \times \mathbf{E}_t) \tag{7.1}$$

$$j\omega\epsilon E_z = \mathbf{\nabla}_t \cdot \mathbf{H}_t \times \mathbf{u}_z = 0$$

$$j\omega\mu H_z = \mathbf{\nabla}_t \cdot \mathbf{u}_z \times \mathbf{E}_t = 0, \tag{7.2}$$

where:

- $\mathbf{E}_t$ = the vector electric field transverse to the z-axis (the direction of propagation);
- $\mathbf{H}_t$ = the vector magnetic field transverse to the z-axis;
- E_z, H_z = the z components of the electric and magnetic fields;
- ϵ = the permittivity of the medium;
- μ = the permeability of the medium;
- $\mathbf{\nabla}_t$ = the gradient operator transverse to the z-axis, $(\mathbf{\nabla} - \mathbf{u}_z(\partial/\partial z))$;
- $\mathbf{u}_z$ = a unit vector in the direction of the z-axis.

* Commensurate: of proportionate measure. The lengths of commensurate transmission lines have a common divisor.

By eliminating $\mathbf{H}_t$ in (7.1) and $\mathbf{E}_t$ in (7.2), the following wave equations are obtained for the transverse electric and magnetic fields.

$$\frac{\partial^2}{\partial z^2}\begin{Bmatrix}\mathbf{E}_t\\ \mathbf{H}_t\end{Bmatrix} = \gamma^2\begin{Bmatrix}\mathbf{E}_t\\ \mathbf{H}_t\end{Bmatrix}, \tag{7.3}$$

with

$$\gamma = j\omega\sqrt{\mu\epsilon}.$$

Solutions to (7.3) are proportional to $\exp[\pm\gamma z]$. Thus, TEM waves propagate along the transmission line with a phase velocity equal to $1/\sqrt{\mu\epsilon}$, which for free space equals the velocity of light.

Reference to (7.1) reveals that

$$\mathbf{E}_t = \sqrt{\mu/\epsilon}(\mathbf{H}_t \times \mathbf{u}_z).$$

The quantity $\sqrt{\mu/\epsilon}$ has dimensions of ohms and is called the wave impedance of the medium.

Another consequence of the TEM assumption is that a unique voltage between transmission-line conductors may be defined. Since $H_z \equiv 0$, it follows that

$$\oint \mathbf{E}_t \cdot \mathbf{ds} \equiv 0$$

for all closed paths in planes transverse to the z-direction. Hence, the line integral is independent of path, and the voltage between conductors may be given by

$$V = -\int_a^b \mathbf{E}_t \cdot \mathbf{ds},$$

where a and b represent arbitrary points on the two conductor surfaces. Similarly, since $E_z \equiv 0$ the current in either of the transmission lines is given by

$$I = \oint (\mathbf{H}_t \times \mathbf{u}_z) \cdot \mathbf{ds},$$

where the path of integration is any closed curve surrounding the conductor and lying in a plane transverse to the z-axis. The *characteristic impedance* of the transmission line is defined as

$$Z_0 = \frac{V}{I}.$$

It can be proved that an additional consequence of TEM propagation is that the low-frequency concepts of capacitance and inductance remain valid, and they may be analytically or experimentally determined by conventional methods. Furthermore, the phase velocity and characteristic impedance are given by

$$v = \frac{1}{\sqrt{LC}}$$

$$Z_0 = \sqrt{\frac{L}{C}}$$

where L and C are the inductance and capacitance per unit length of the transmission line conductors.

Richards' Transformation

Richards' transformation applies to networks consisting of lumped resistors, ideal transformers, and commensurate-length lossless transmission lines [1]. These conditions ensure that driving-point immittances of such networks, $Z(s)$, will be rational functions of $\exp[\tau s]$, where:

$\tau = 2l/v$, the round-trip delay time for the shortest commensurate-length line;
l = length of the shortest commensurate-length line;
v = velocity of propagation;
$s = \sigma + j\omega$, the complex frequency variable of lumped element network theory.

Richards introduced the new frequency variable

$$S = \tanh\left(\frac{\tau s}{2}\right) = \frac{e^{\tau s} - 1}{e^{\tau s} + 1} = \Sigma + j\Omega. \tag{7.4}$$

On the real frequency axis $\Sigma = 0$, and (7.4) reduces to

$$S = j\Omega = j\tan(\theta),$$

where

$$\theta = \frac{\omega l}{v} = \text{the electrical length.}$$

Solving (7.4) for $e^{\tau s}$ gives

$$e^{\tau s} = \frac{1 + S}{1 - S}.$$

Therefore,

$$Z(S) = \text{rational function of } S. \tag{7.5}$$

There will also be occasions to use the reciprocal form of Richards' transformation, which is defined by

$$W = \coth\left(\frac{\tau s}{2}\right) = \Gamma - j\mho.$$

By the previous argument

$$Z(W) = \text{rational function of } W.$$

Richards established that commensurate transmission-line immittance functions $Z(S)$ have a positive real part Σ whenever $\sigma \geq 0$ [2]. Thus, (7.4) maps the right half of the s-plane into the right half of the S-plane, and

$$\text{Re}[Z(S)] \geq 0 \qquad \text{for} \quad \Sigma \geq 0. \tag{7.6}$$

The mapping is not one to one. However, the multiple-valueness of the inverse merely corresponds to the periodicity of transmission-line network immittance functions.

Under Richards' transformation, a close correspondence exists between lumped inductors and capacitors in the s-plane and short- and open-circuited transmission lines in the S- and W-planes. This is illustrated in Table 7-1. It is seen that open-circuited and short-circuited transmission lines in the S-plane correspond to lumped capacitors and inductors, respectively, in the s-plane. Similar correspondences hold for the W-plane elements.

In discussing commensurate transmission-line networks, it is convenient to use lumped-element symbols to represent distributed elements, and to use lumped-element phraseology. Thus, we speak of S-plane capacitors and S-plane inductors (or just capacitors and inductors), meaning distributed open-circuited and short-circuited stubs. However, the distributed-line representation shown in Table 7-1 will also be used, particularly when it is necessary to distinguish between lumped and distributed elements.*

The Unit Element

One place where the correspondence between lumped and transmission-line networks breaks down is for the two-port network consisting of a

* In this and future sections of this chapter, *all* symbolic representations for circuit elements are to be understood as distributed except in the following cases: (1) the network is clearly lumped, as indicated in the figure title or discussion in the text, (2) the network consists of both lumped and distributed elements, in which case the distributed elements will be represented by the transmission line symbols given in Table 7-1, and the lumped elements will be represented by the conventional symbols, and (3) specific statements to the contrary.

TABLE 7-1 Lumped and Distributed Element Correspondence

s-Plane			S-Plane			W-Plane		
s-Plane Element	Symbol	Admittance or Impedance	S-Plane Element	Symbol	Admittance or Impedance	W-Plane Element	Symbol	Admittance or Impedance
Capacitor	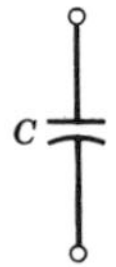 C	$Y = Cs$	Open-circuited transmission line	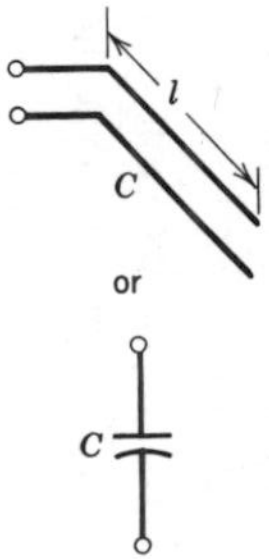 l, C or C	$Y = CS$	Short-circuited transmission line	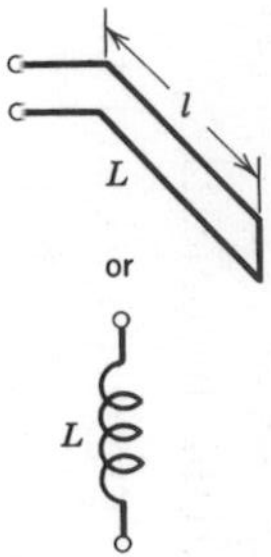 l, L or L	$Y = L^{-1} W$
Inductor	 L	$Z = Ls$	Short-circuited transmission line	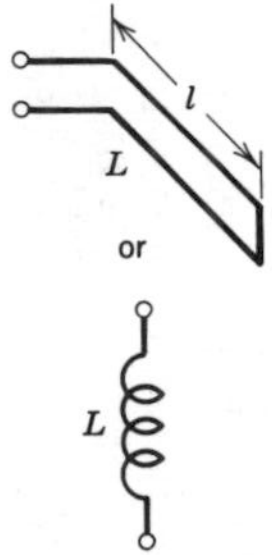 l, L or L	$Z = LS$	Open-circuited transmission line	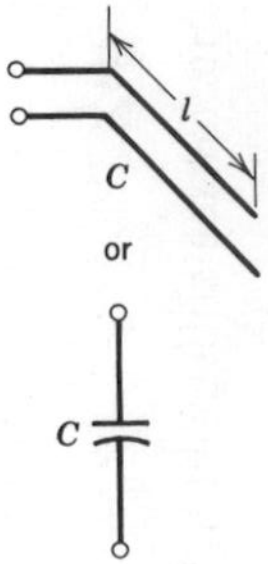 l, C or C	$Z = C^{-1} W$

commensurate-length line. A transmission line has an impedance matrix given by

$$\begin{bmatrix} V_1 \\ V_2 \end{bmatrix} = \begin{bmatrix} -jZ_0 \cot\theta & -jZ_0 \csc\theta \\ -jZ_0 \csc\theta & -jZ_0 \cot\theta \end{bmatrix} \begin{bmatrix} i_1 \\ i_2 \end{bmatrix},$$

which in terms of Richards' variable becomes

$$\begin{bmatrix} V_1 \\ V_2 \end{bmatrix} = \frac{Z_0}{S} \begin{bmatrix} 1 & \sqrt{1-S^2} \\ \sqrt{1-S^2} & 1 \end{bmatrix} \begin{bmatrix} i_1 \\ i_2 \end{bmatrix}.$$

Ozaki and Ishii referred to this network element as a "unit element," hereafter abbreviated as UE [3]. Its symbol is given in Table 7-2. Note that the unit element has a *half-order* transmission zero at $S = \pm 1$. Consequently, it has no lumped-element counterpart. However, it is interesting to note that two unit elements in cascade create a first-order transmission zero at $S = \pm 1$. Thus, two unit elements in cascade correspond to a *C*-section of lumped-element theory.

TABLE 7-2 Unit Element Nomenclature

Element	Symbol	Admittance or Impedance Matrices: S-plane	Admittance or Impedance Matrices: W-plane
Unit-Element	① —— Z —— ② (two parallel lines)	$[Z] = \frac{1}{S}\begin{bmatrix} Z_0 & Z_0 T \\ Z_0 T & Z_0 \end{bmatrix}$	$[Z] = \begin{bmatrix} Z_0 W & Z_0 T' \\ Z_0 T' & Z_0 W \end{bmatrix}$
or	or	$[Y] = \frac{1}{S}\begin{bmatrix} Y_0 & -Y_0 T \\ -Y_0 T & Y_0 \end{bmatrix}$	$[Y] = \begin{bmatrix} Y_0 W & -Y_0 T' \\ -Y_0 T' & Y_0 W \end{bmatrix}$
UE	① [Z UE] ②	$T = \sqrt{1-S^2}$	$T' = \sqrt{W^2-1}$

Richards' Theorem

If $Z(S)$ is a positive real function and $Z(S)/Z(1)$ is not identically equal to S or $1/S$, then

$$Z'(S) = \frac{Z(S) - SZ(1)}{Z(1) - SZ(S)} \tag{7.7}$$

is also a positive real function. Further, if $Z(S)$ is rational in S (the only case considered here), then a factor $S - 1$ cancels in the numerator and denominator of (7.7), and $Z'(S)$ is the same degree as $Z(S)$. If $Z(1) + Z(-1) = 0$ as well, a second factor $S + 1$ cancels in the numerator and denominator, and $Z'(S)$ is of lower degree than $Z(S)$.

This theorem is singularly important in transmission-line network theory. An immediate consequence of Richards' theorem is the following corollary:

Corollary 7-1. Let $Z(S)$ be a positive real impedance function satisfying $Z(S)/Z(1) \neq S$ or S^{-1}. Then, a unit element of value $Z(1)$ may always be extracted from $Z(S)$, leaving a positive real impedance $Z'(S)$ in cascade with the unit element. The value of $Z'(S)$ is

$$Z'(S) = \frac{Z(S) - SZ(1)}{Z(1) - SZ(S)} Z(1). \tag{7.8}$$

The impedance $Z'(S)$ is also positive real and of degree equal to $Z(S)$ if $Z(1) + Z(-1) \neq 0$. It is of degree 1 less than $Z(S)$ if $Z(1) + Z(-1) = 0$. This corollary is illustrated schematically in Figure 7-1.

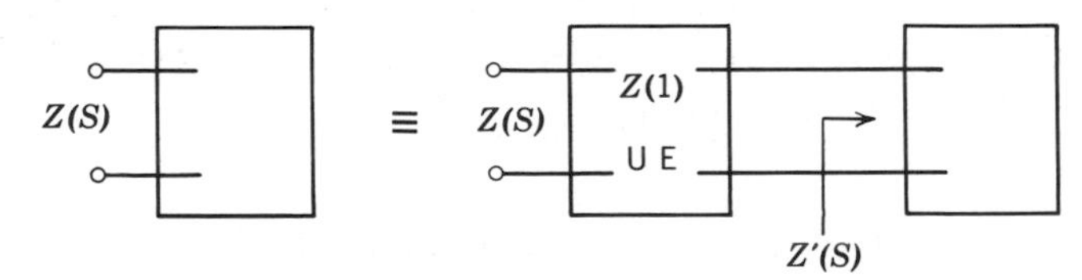

FIGURE 7-1 Schematic illustration of Corollary 1.

A second corollary is the following.

Corollary 7-2. An arbitrary reactance function is realizable by a cascade of unit elements terminated in an open or short circuit. The proof follows directly from Corollary 7-1, since $Z(1) + Z(-1) = 0$ for reactance functions.

Synthesis of Driving Point Functions

Consider Richards' variable S as the independent frequency variable. Then (7.5) and (7.6) have been shown to be necessary and sufficient for $Z(S)$ to be a realizable lumped-constant impedance [4]. Using the correspondences between lumped and distributed elements given in Table 7-1, the same conclusions carry over to commensurate transmission-line networks. Consequently, the analytical methods and impedance synthesis procedures of lumped-element theory can be extended to transmission-line networks. Techniques, such as those given by Bode [5], Darlington [6], Guillemin [7], Youla [8], and others, can be immediately taken over. With the inclusion of Richards' theorem and corollaries, these procedures may be generalized and also used for transfer function synthesis. A number of examples follow to illustrate several of the various methods.

Reactance Function Synthesis

Let us synthesize a network to realize the reactance function

$$Z(S) = \frac{8S + S^3}{2 + 7S^2}.$$

Expanding $Z(S)$ in a continued fraction expansion gives the network shown in Figure 7-2(a).

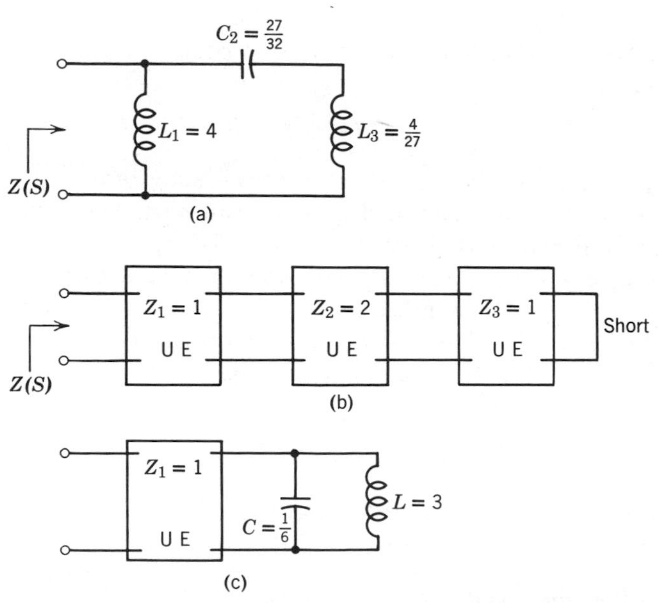

FIGURE 7-2 ***S*-plane reactance synthesis examples.**

According to Corollary 7-2, $Z(S)$ can also be developed in a cascade of unit elements. The value of the first unit element is obtained by setting $S = 1$. This gives

$$Z_1 = Z(1) = \frac{8 + 1}{2 + 7} = 1.$$

The unit element is next extracted from the impedance function $Z(S)$ via Corollary 7-1, leaving a reduced impedance function $Z'(S)$ given by

$$Z'(S) = \frac{\left[\dfrac{8S + S^3}{2 + 7S^2}\right] - S}{1 - S\left[\dfrac{8S + S^3}{2 + 7S^2}\right]} = \frac{6S[1 - S^2]}{2 - S^2 - S^4}.$$

Dividing both the numerator and the denominator of the above by $1 - S^2$ yields

$$Z'(S) = \frac{6S}{2 + S^2}. \tag{7.9}$$

The value of the second unit element is next obtained by the same procedure.

$$Z_2 = Z'(1) = \frac{6}{2 + 1} = 2.$$

Extracting the second unit element from the impedance function $Z'(S)$ leaves the impedance function

$$Z''(S) = S.$$

$Z''(S)$ may be interpreted either as an S-plane inductor of value 1 or as a unit element of value 1 short circuited at its output port. The network corresponding to this realization is given in Figure 7-2(b).

Of course, either pole extraction or unit-element extraction is always possible at any step in the synthesis, so that many canonical realizations are possible. For example, at step (7.9) the function $Z'(S)$ could be realized in Foster form by a LC antiresonant circuit with

$$L = 3,$$
$$C = 1/6.$$

This realization is given in Figure 7-2(c).

Impedance Function Synthesis

Next, let us synthesize a network to realize the impedance

$$Z(S) = \frac{1 + 2S}{1 + S + S^2} \qquad \text{or} \qquad Y(S) = \frac{1 + S + S^2}{1 + 2S}. \tag{7.10}$$

It is seen that $Y(S)$ has a pole at infinity, corresponding to a shunt S-plane capacitor of value

$$C = \lim_{S \to \infty} \frac{Y(S)}{S} = \frac{1}{2}.$$

Extraction of C from $Y(S)$ gives

$$Y'(S) = Y(S) - \frac{S}{2} = \frac{1}{2}\left[\frac{2 + S}{1 + 2S}\right].$$

$Y'(S)$ has no poles or zeros at the origin or infinity. Consequently, no capacitors or inductors can be removed. However, it is not a minimum function. Calculation of its real part gives

$$\mathrm{Re}[Y'(j\Omega)] = \frac{1 + \Omega^2}{1 + 4\Omega^2},$$

which has a minimum value of 1/4 at $\Omega = \infty$. Therefore, a conductance of 1/4 mho may be removed. Subtraction of 1/4 from $Y'(S)$ leaves

$$Y''(S) = \frac{3}{4(1 + 2S)},$$

and

$$Z''(S) = \frac{4}{3} + \frac{8S}{3}.$$

$Z''(S)$ is recognized as a resistance of 4/3 in series with an inductance of 8/3. The final network is given in Figure 7-3(a).

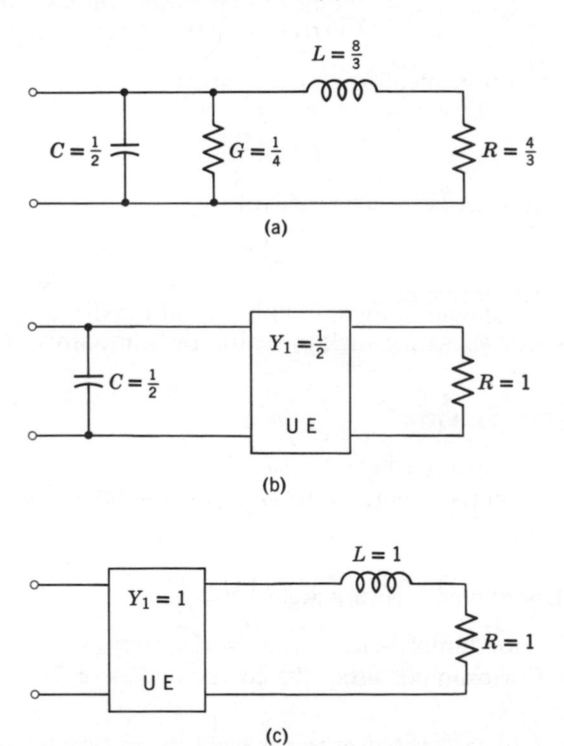

FIGURE 7-3 ***S*-plane impedance synthesis examples.**

The same function, (7.10), may be synthesized by a lossless two-port network terminated in a resistance via the Darlington method. The even part of $Z(S)$,

$$\mathrm{Ev}[Z(S)] = \frac{1 - S^2}{1 + S^2 + S^4},$$

has zeros corresponding to a 1/2-order transmission zero at $S = \pm 1$ and a first-order transmission zero at $S = \pm\infty$. The latter zero requires an A- or B-section, while the former zero requires a unit element.*

The B-section may be realized by removing a shunt capacitor from $Z(S)$, as was previously done. The admittance then remaining is

$$Y'(S) = \frac{1}{2}\left[\frac{2 + S}{1 + 2S}\right].$$

The last step is to extract a unit element from $Y'(S)$. Applying Richards' theorem gives

$$Y_1 = Y'(S)|_{S=1} = \tfrac{1}{2}(\tfrac{3}{3}) = \tfrac{1}{2}.$$

Subtracting the unit element from $Y'(S)$ leaves

$$Y''(S) = \frac{1}{2}\left\{\frac{\dfrac{1}{2}\left[\dfrac{2 + S}{1 + 2S}\right] - \dfrac{S}{2}}{\dfrac{1}{2} - \dfrac{S}{2}\left[\dfrac{2 + S}{1 + 2S}\right]}\right\} = 1.$$

This realization is shown in Figure 7-3(b). A third realization is achieved by extracting the unit element first. This is shown in Figure 7-3(c).

Transfer Function Synthesis

The methods of lumped-element, transfer-function synthesis are also readily extended to transmission-line networks. An example is postponed until Section 7.4.

Lumped and Distributed Network Relationships

Many theorems for lumped-element networks carry over to transmission-line networks. For example, using the correspondences

* It is interesting to note that synthesis of the same function in a lumped network would require augmenting (7.10) by the factor $(S + 1)$.

s-plane	S-plane
$s = \infty$	$S = \infty$
$s = 0$	$S = 0$
$s = j\omega$	$S = j\Omega = j \tan \omega l/v$
$d\omega$	$d\Omega = l/v \sec^2 \omega l/v \, d\omega$

the well-known reactance area law for minimum reactance functions $Z(\omega) = R(\omega) + jX(\omega)$,

$$\frac{2}{\pi}\int_0^\infty \frac{X(\omega)}{\omega}\, d\omega = R(\infty) - R(0)$$

becomes

$$\frac{2}{\pi}\int_0^\infty \frac{X(\Omega)}{\Omega}\, d\Omega = R(\infty) - R(0),$$

which reduces to

$$\frac{2}{\pi}\int_0^{\pi/2} \frac{X(\theta)\, d\theta}{\sin\theta\cos\theta} = R(\pi/2) - R(0).$$

Similarly, for reactive circuits, the relationship

$$\frac{dX}{d\omega} \geq \frac{X}{\omega}$$

becomes

$$\frac{dX(\Omega)}{d\Omega} \geq \frac{X}{\Omega},$$

which reduces to

$$\frac{dX(\theta)}{d\theta} \geq \frac{X(\theta)}{\sin\theta\cos\theta}.$$

Other relationships may be similarly transformed.

7.2 TRANSMISSION-LINE NETWORK IDENTITIES

In the design of transmission-line filters and other microwave components, various network identities are used to obtain networks that are electrically equivalent, but that differ in form or in component values. Such transformations provide designers with flexibility, but, more important, they are often essential to obtain networks that are physically realizable with practical dimensions. The Kuroda identities [3], shown in Table 7-3, give equations

TABLE 7-3 Transmission-Line Network Transformations

Network	Equivalent Network	Relationships	
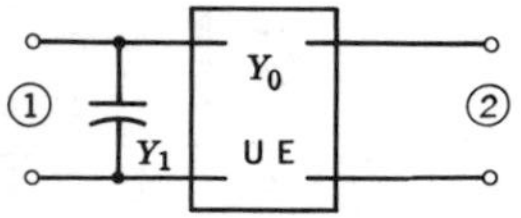 ① Y_1 Y_0 U E ②	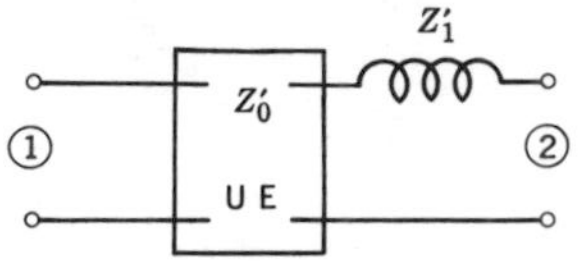 ① Z_0' U E Z_1' ②	$Z_0' = \dfrac{1}{Y_0 + Y_1}$ $Z_1' = \dfrac{Y_1/Y_0}{Y_0 + Y_1}$	
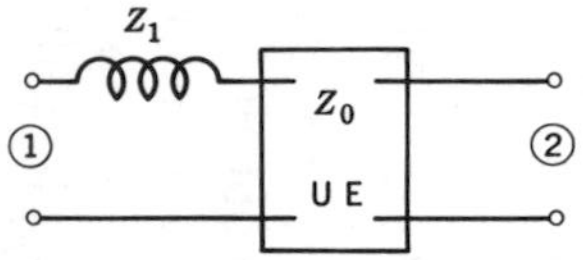 ① Z_1 Z_0 U E ②	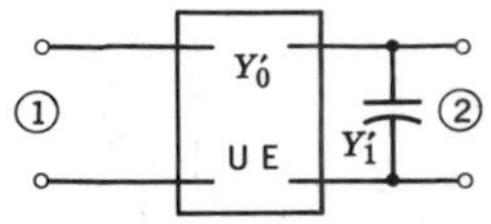 ① Y_0' U E Y_1' ②	$Y_0' = \dfrac{1}{Z_0 + Z_1}$ $Y_1' = \dfrac{Z_1/Z_0}{Z_0 + Z_1}$	
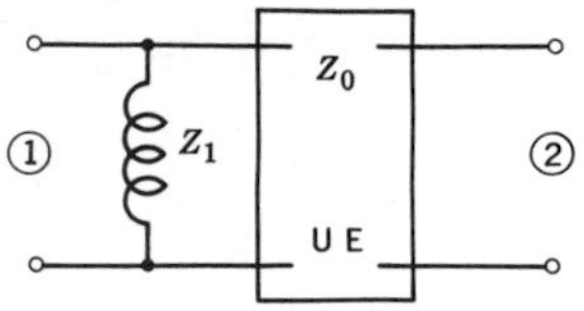 ① Z_1 Z_0 U E ②	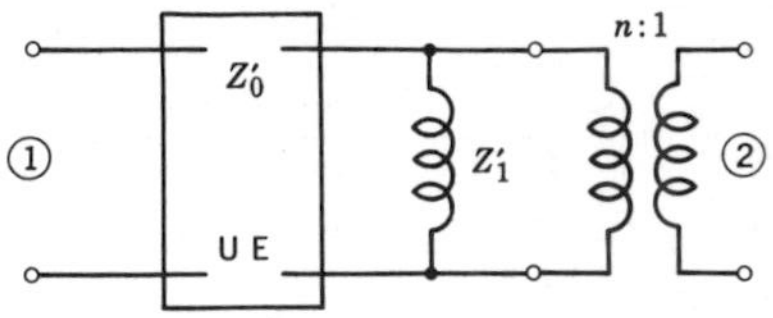 ① Z_0' U E Z_1' $n:1$ ②	$Z_0' = Z_0 n$ $\quad Z_1' = Z_1 n$	$n = \dfrac{1}{1 + \dfrac{Z_0}{Z_1}}$
	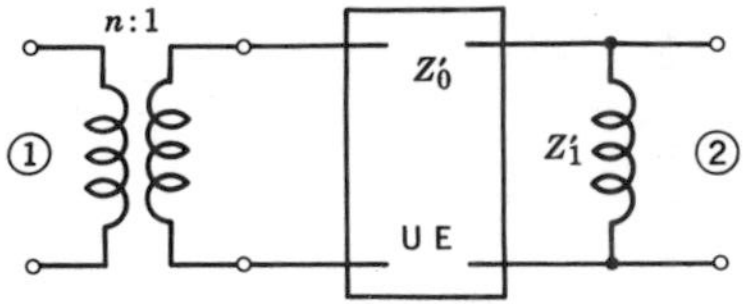 ① $n:1$ Z_0' U E Z_1' ②	$Z_0' = \dfrac{Z_0}{n}$ $\quad Z_1' = \dfrac{Z_1}{n}$	

① Y_1 Y_0 UE ②	① Y'_0 UE Y'_1 $1:n$ ②	$Y'_0 = Y_0 n \qquad Y'_1 = Y_1 n$	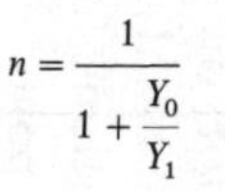 $n = \dfrac{1}{1 + \dfrac{Y_0}{Y_1}}$
	① $1:n$ Y'_0 UE Y'_1 ②	$Y'_0 = \dfrac{Y_0}{n} \qquad Y'_1 = \dfrac{Y_1}{n}$	

for interchanging a unit element with a series inductor or capacitor, and a unit element with a shunt inductor or capacitor. These identities are frequently used in designing all-pole bandstop and bandpass filters.

Levy has presented general equations for interchanging a unit element and an arbitrary positive real two-port network [9]. This is illustrated in Figure 7-4. If the unprimed symbols are used to denote network parameters

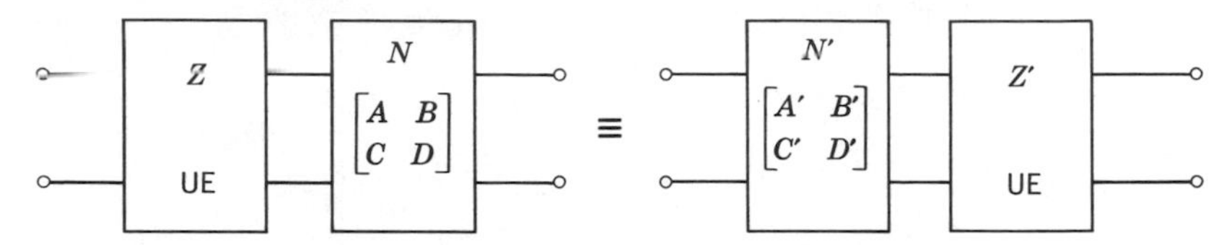

FIGURE 7-4 Generalized Kuroda's identity.

on the left side of Figure 7-4 and primed symbols for network parameters on the right side, then Z' and N' must satisfy

$$Z' = \frac{B(1) + ZD(1)}{A(1) + ZC(1)}. \tag{7.11}$$

$$\begin{aligned}(1 - S^2)A' &= A + S(CZ - BY') - S^2DZY' \\ (1 - S^2)B' &= B + S(DZ - AZ') - S^2CZZ' \\ (1 - S^2)C' &= C + S(AY - DY') - S^2BYY' \\ (1 - S^2)D' &= D + S(BY - CY') - S^2AYZ'.\end{aligned} \tag{7.12}$$

To illustrate the use of these equations, Kuroda's first identity (Table 7-3, Number 1) will be derived. For this example, the network N consists of a unit element and shunt capacitor of values Y_0 and Y_1, respectively. The $[F]$ matrix for the shunt capacitor is

$$[F] = \begin{bmatrix} 1 & 0 \\ SY_1 & 1 \end{bmatrix}.$$

The unit element Z'_0 is obtained by substituting $A = D = 1$, $B = 0$, $C = Y_1$ into (7.11). This gives

$$Z'_0 = \frac{Z_0}{1 + Z_0 Y_1}.$$

The A', B', C', D' parameters for network N' are obtained by substituting into (7.12) and dividing by $(1 - S^2)$.

$$A' = 1$$

$$B' = \frac{SZ_0^2 Y_1}{1 + Z_0 Y_1}$$

$$C' = 0$$

$$D' = 1.$$

These equations are recognized as the $[F]$ matrix for a series inductor of value

$$Z' = \frac{Z_0^2 Y_1}{1 + Z_0 Y_1}.$$

7.3 COUPLED-TRANSMISSION-LINE NETWORKS

Coupled-transmission-line networks are utilized in numerous microwave components: filters, directional couplers, matching networks, and equalizers, to name just a few. An arbitrary network of n-coupled transmission lines above a ground plane,* as shown in Figure 7-5, can be described by the admittance matrix equation [10]

$$\begin{bmatrix} i_1 \\ i_2 \\ \vdots \\ i_n \\ \vdots \\ i_{2n} \end{bmatrix} = \frac{v}{S} \begin{bmatrix} \overset{n \times n}{[C]} & \overset{n \times n}{-T[C]} \\ \overset{n \times n}{-T[C]} & \overset{n \times n}{[C]} \end{bmatrix} \begin{bmatrix} v_1 \\ v_2 \\ \vdots \\ v_n \\ \vdots \\ v_{2n} \end{bmatrix} \tag{7.13}$$

where v = velocity of light in the medium of propagation,

$$T = \sqrt{1 - S^2},$$

and the matrix $[C]$ is the Maxwellian static capacitance distribution for the coupled line network:

$$[C] = \begin{bmatrix} C_{11} & -C_{12} & -C_{13} & \cdots & -C_{1n} \\ -C_{21} & C_{22} & -C_{23} & \cdots & -C_{2n} \\ \vdots & & & & \\ -C_{n1} & -C_{n2} & \cdot & \cdot \quad \cdot & C_{nn} \end{bmatrix}$$

* The ground plane represents the common return for all lines. An equivalent representation is n pairs of transmission lines.

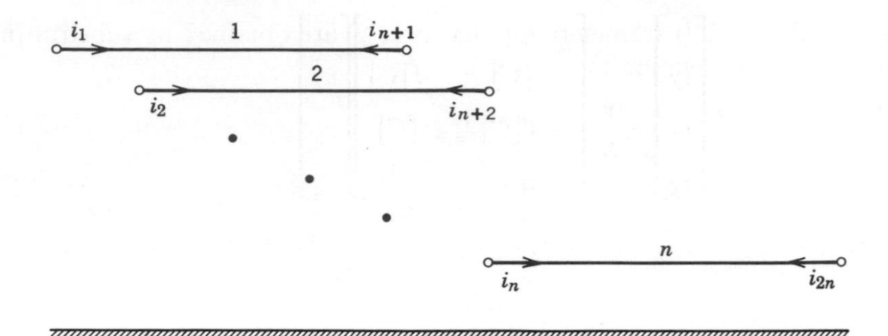

FIGURE 7-5 General *n*-wire line above a ground plane.

where

$$C_{ij} = C_{ji}$$
$$C_{ij} \geq 0$$
$$C_{ii} \geq \sum_{j \neq i} C_{ij} \qquad \text{for} \quad i = 1, 2, \ldots, n. \tag{7.14}$$

The dimensions of C_{ij} are farads/unit length. Alternatively, the impedance matrix may be used:

$$\begin{bmatrix} v_1 \\ v_2 \\ \vdots \\ v_n \\ \vdots \\ v_{2n} \end{bmatrix} = \frac{v}{S} \begin{bmatrix} \begin{matrix} n \times n \\ [L] \end{matrix} & \begin{matrix} n \times n \\ T[L] \end{matrix} \\ & \\ \begin{matrix} n \times n \\ T[L] \end{matrix} & \begin{matrix} n \times n \\ [L] \end{matrix} \end{bmatrix} \begin{bmatrix} i_1 \\ i_2 \\ \vdots \\ i_n \\ \vdots \\ i_{2n} \end{bmatrix}$$

where $[L]$ is the inductance distribution for the network, also satisfying (7.14). The dimensions of L_{ij} are henries/unit length. $[L]$ and $[C]$ together satisfy

$$[L][C] = [C][L] = \frac{1}{v^2}[I],$$

where $[I]$ is the identity matrix.

Equivalent Circuits for Coupled Two-Wire Lines

Pairs of coupled transmission lines form an important class of networks in microwave filter theory. Two-port equivalent circuits can be derived from coupled two-wire lines by utilizing the general four-port equations:

$$\begin{bmatrix} i_1 \\ i_2 \\ i_3 \\ i_4 \end{bmatrix} = \frac{v}{S} \begin{bmatrix} [C] & -T[C] \\ -T[C] & [C] \end{bmatrix} \begin{bmatrix} v_1 \\ v_2 \\ v_3 \\ v_4 \end{bmatrix} \tag{7.15}$$

where

$$[C] = \begin{bmatrix} C_{11} & -C_{12} \\ -C_{12} & C_{22} \end{bmatrix}.$$

First, boundary conditions are imposed on two of the ports to reduce the four-port to a two-port. Then the two-port network is synthesized. To illustrate the procedure, consider the two-port network obtained by short circuiting to ground ports 3 and 4. Equation (7.15) reduces to

$$\begin{bmatrix} i_1 \\ i_2 \end{bmatrix} = \frac{v}{S} \begin{bmatrix} C_{11} & -C_{12} \\ -C_{12} & C_{22} \end{bmatrix} \begin{bmatrix} v_1 \\ v_2 \end{bmatrix},$$

whose equivalent circuit is readily recognized as a π network of inductors. The equivalence is illustrated in Figure 7-6. The descriptions of coupled transmission lines by the impedance matrix, $[Z]$, and the hybrid matrices, $[G]$, $[H]$, and $[F]$ are also often useful in obtaining equivalent circuits.

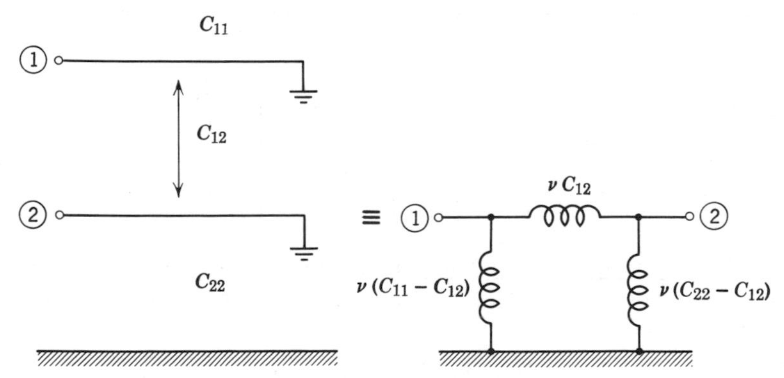

FIGURE 7-6 Two-wire line equivalent circuit example.

For cases in which the equivalent circuit is complex, classical synthesis procedures, such as Cauer's or Darlington's method, may be used.

For two-wire-line networks, in which the imposed terminal conditions are relatively simple, the above methods are straight forward and not too difficult. However, for multiwire-line networks, the reduction from multiport to two-port is usually too difficult to attempt. An alternate analysis method, which often is simpler, is the *graph-transformation method* [11]. This method sometimes yields the equivalent circuit in a few steps and does not require synthesis. For multiport networks, the graph-transformation method is almost always easier to use.

The Graph-Transformation Method

In the graph-transformation method, a schematic representation of the coupled-line network, called its graph, is drawn. The graph consists only of inductors, capacitors, transformers, and uncoupled unit elements. Next, appropriate boundary conditions are applied to the terminals of the graph. Then, various network transformations are applied until a satisfactory equivalent circuit is obtained. A graph representation of an n-wire coupled-line network is presented in Figure 7-7 for the important special case in which coupling between nonadjacent lines is negligible. The y_{ij} values shown are the unnormalized admittances of each line, which are related to the capacitance parameters of (7.13) by

$$y_{ij} = vC_{ij}.$$

In the graph representation, all lines are uncoupled and all nodal points are in one-to-one correspondence with the physical network. The dual graph on an

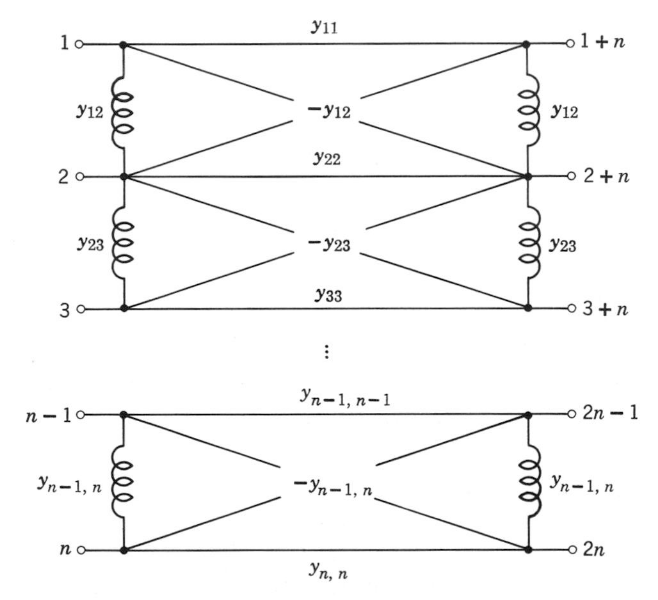

FIGURE 7-7 Graph representation of multi-wire-line network in which coupling beyond nearest neighbors is negligible.

impedance basis is useful in some cases. However, it is more complicated than that of Figure 7-7 and requires ideal transformers, thus making it more crowded and more tedious to manipulate.

Numerous coupled-line network equivalent circuits can be derived by using the graph of Figure 7-7. For example, the network of Figure 7-6 can be obtained by the steps shown in Figure 7-8. The network nodal points are in

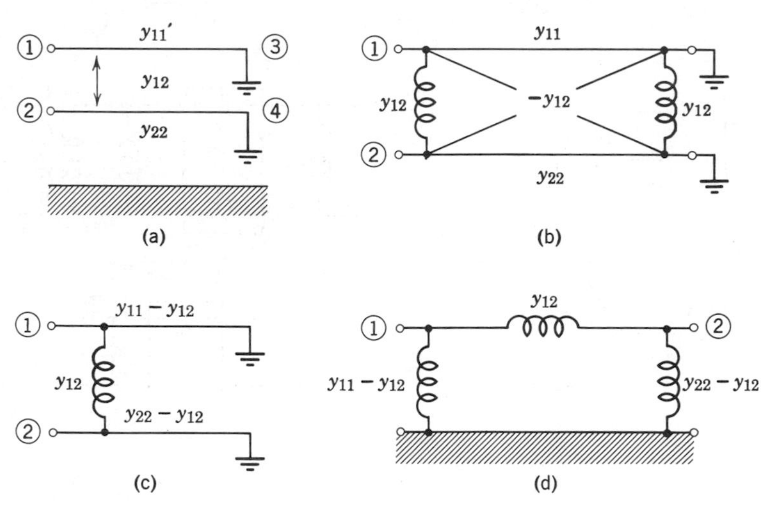

FIGURE 7-8 (a) Network of coupled lines grounded at ports three and four, (b) graph equivalent circuit, (c) reduced graph of Figure 7-8(b), and (d) final equivalent circuit.

one-to-one correspondence with the graph nodal points; hence the graph of Figure 7-8(a) takes the form shown in Figure 7-8(b).* Since the inductor between ports 3 and 4 is short circuited at both ends, it can be removed from the circuit. The unit elements of $-y_{12}$ admittances are in parallel with y_{11} and y_{22}, so the graph of Figure 7-8(b) reduces to that of Figure 7-8(c). The unit elements $(y_{11} - y_{12})$ and $(y_{22} - y_{12})$ are short circuited at one end, hence they are equivalent to inductors. The final equivalent circuit of Figure 7-8(d) is thus obtained. In the general case, or when the imposed boundary conditions are not short circuits, the transformations of Sato [12] and Pang [13] are often helpful in obtaining a useful equivalent circuit.

Table 7-4 presents equivalent circuits for a few of the many 2- and 3-wire coupled-line networks commonly used in microwave component design.

* The ground plane, being incidental to the graph, is dispensed with.

TABLE 7-4 Coupled-Transmission-Line Equivalent Circuits

	Transmission Line Network	Equivalent Circuit
1.	① 1, 2 ②	Y_1, Y_2, Z_0, UE, ①, ②
2.	① 1, 2 ②	Y_0, UE, Z_1, Z_2, $-1:1$, ①, ②
3.	① 1 ②, 2	Z_0, UE, Y_1, ①, ②
4.	① 1 ②, 2	Z_1, Y_0, UE, ①, ②

Analysis Equations	Synthesis Equations	Realizability Constraints
$Y_1 = \dfrac{1}{(L_{11} - L_{12})v}$ $Y_2 = \dfrac{1}{(L_{22} - L_{12})v}$ $Z_0 = vL_{12}$	$vL_{11} = Z_1 + Z_0$ $vL_{22} = Z_2 + Z_0$ $vL_{12} = Z_0$	None
$Z_1 = \dfrac{1}{v(C_{11} - C_{12})}$ $Z_2 = \dfrac{1}{v(C_{22} - C_{12})}$ $Y_0 = vC_{12}$	$vC_{11} = Y_1 + Y_0$ $vC_{22} = Y_2 + Y_0$ $vC_{12} = Y_0$	None
$Z_0 = vL_{11}$ $Y_1 = vC_{11} - \dfrac{1}{vL_{11}}$ $Y_1 = \dfrac{vC_{12}^2}{C_{22}}$	$vC_{11} = Y_1 + Y_0$ $Y_1 \le vC_{12} \le Y_1 + Y_0$ $vC_{22} = \dfrac{(vC_{12})^2}{Y_1}$	None
	Symmetrical case: $C_{11} = C_{22}$ $vC_{11} = Y_1 + Y_0$ $vC_{12} = \sqrt{Y_1(Y_1 + Y_0)}$	None
$Y_0 = vC_{11}$ $Z_1 = vL_{11} - \dfrac{1}{vC_{11}}$ $Z_1 = \dfrac{(vL_{12})^2}{vL_{22}}$	$vL_{11} = Z_1 + Z_0$ $Z_1 \le vL_{12} \le Z_1 + Z_0$ $vL_{22} = \dfrac{(vL_{12})^2}{Z_1}$	None
	Symmetrical case: $L_{11} = L_{22}$ $vL_{11} = Z_1 + Z_0$ $vL_{12} = \sqrt{Z_1(Z_1 + Z_0)}$	None

TABLE 7-4 (continued)

Transmission Line Network	Equivalent Circuit
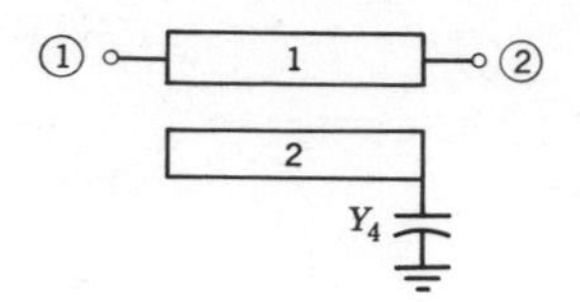	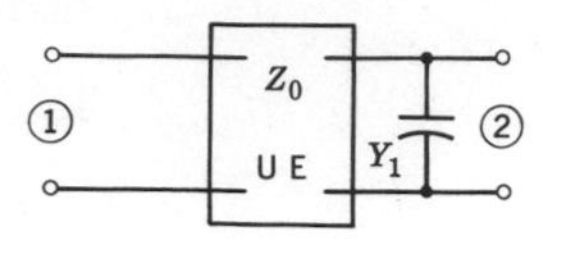
5.	
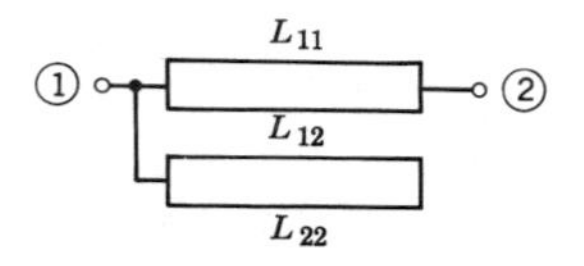	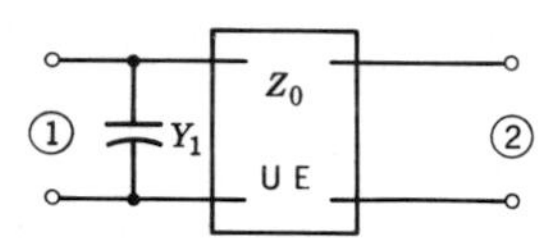
6.	

Analysis Equations	Synthesis Equations	Realizability Constraints
$Z_0 = vL_{11}$ $Y_1 = \frac{(vC_{12})^2}{vC_{22}}\left\{\frac{1}{1 + vC_{22}/Y_4}\right\}$	Define $Y'_1 = Y_1\left[1 + \frac{vC_{22}}{Y_4}\right]$ Choose: 1) $Y_4 > Y_1$ 2) $vC_{22} > \frac{Y_1}{1 - Y_1/Y_4}$ Then $vC_{11} = Y'_1 + Y_0$ $vC_{12} = \sqrt{Y'_1 vC_{22}}$	$Y_4 > Y_1$
	Symmetrical case Choose: 1) $Y_4 > Y_1$ Then $vC_{11} = Y_4\left\{\frac{Y_0 + Y_1}{Y_4 - Y_1}\right\}$ $vC_{12} = \frac{\sqrt{Y_1 Y_4(Y_0 + Y_1)(Y_0 + Y_4)}}{Y_4 - Y_1}$	$Y_4 > Y_1$
$Z_0 = vL_{11}$ $Y_1 = \frac{(vC_{22} - vC_{12})^2}{vC_{22}}$	Define: $k^2 = \frac{C_{12}^2}{C_{11}C_{22}}$ Choose: $k^2 \leq \left[\frac{Y_0 - Y_1}{Y_0 + Y_1}\right]^2$ $vC_{11} = \frac{Y_0}{1 - k^2}$ $vC_{22} = (k\sqrt{vC_{11}} + \sqrt{Y_1})^2$ $vC_{12} = k\sqrt{(vC_{11})(vC_{22})}$	$Y_0 \geq Y_1$
	Symmetrical case $C_{11} = C_{22}$ $vC_{11} = \frac{Z_1}{4}(Y_1 + Y_0)^2$ $vC_{12} = \frac{Z_1}{4}(Y_0^2 - Y_1^2)$ $k^2 = \left[\frac{Y_0 - Y_1}{Y_0 + Y_1}\right]^2$	$Y_0 \geq Y_1$

TABLE 7-4 (continued)

	Transmission Line Network	Equivalent Circuit
7.	① 1, ② 2, Y_c; $Y_c = \dfrac{v^2(L_{11} - L_{12})(L_{22} - L_{12})}{vL_{12}\lvert vL\rvert}$; $\lvert vL\rvert = \begin{vmatrix} vL_{11} & vL_{12} \\ vL_{12} & vL_{22} \end{vmatrix}$	Z_1, Z_2, Y_3
8.	① 1, ② 2	① Y_1, Y_2, Y_3 ②; ① Y_1', Y_2', Y_3' ②
9.	① 1, ② 2	① Z_1, Z_2, Z_3 ②; ① Z_1', Z_2', Z_3' ②
10.	① 1, 2, ② 3	① Y_1, Z_2, Y_3 ②, $n:1$, $1:m$

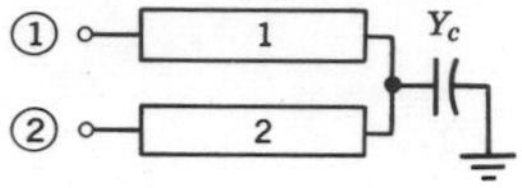

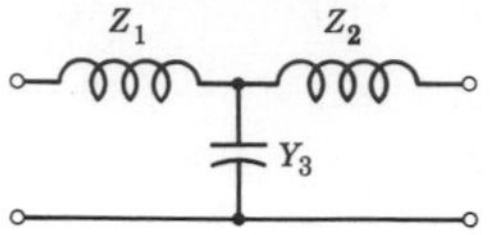

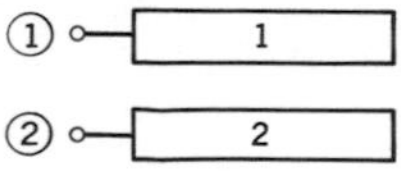

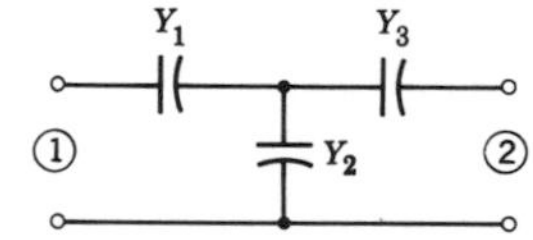

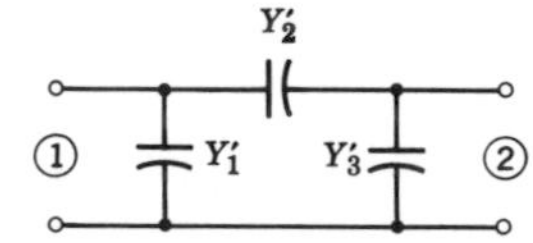

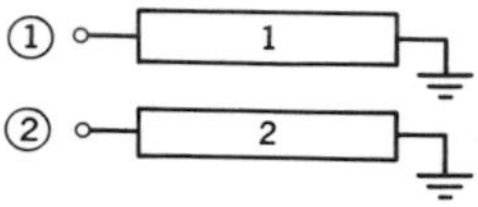

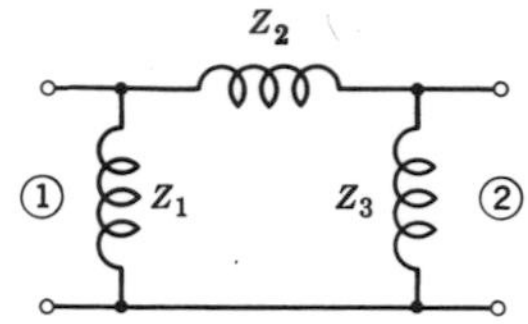

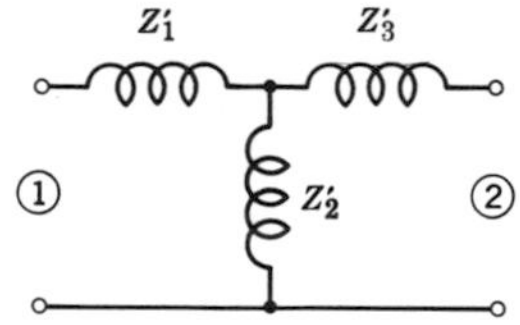

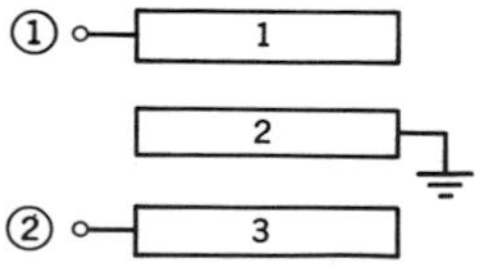

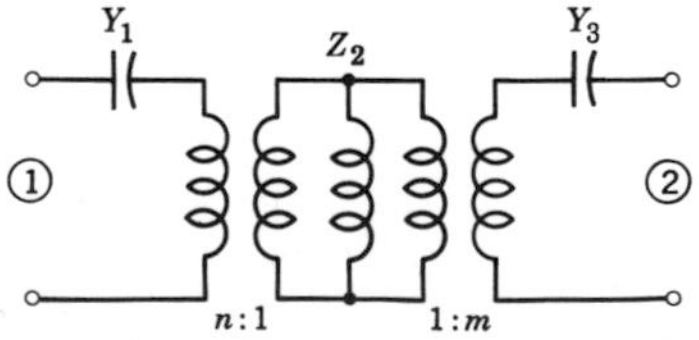

Analysis Equations	Synthesis Equations	Realizability Constraints
$Z_1 = v(L_{11} - L_{12})$	$vL_{11} = Z_1 + Z_3$	None
$Z_2 = v(L_{22} - L_{12})$	$vL_{22} = Z_2 + Z_3$	
$Z_3 = vL_{12}$	$vL_{12} = Z_3$	
	$Y_c = \dfrac{v^2(L_{11} - L_{12})(L_{22} - L_{12})}{vL_{12}\lvert vL \rvert}$	
$Z_1 = v(L_{11} - L_{12})$	$vL_{11} = Z_1 + Z_2$	None
$Z_2 = vL_{12}$	$vL_{12} = Z_2$	
$Z_3 = v(L_{22} - L_{12})$	$vL_{22} = Z_3 + Z_2$	
$Y'_1 = v(C_{11} - C_{12})$	$vC_{11} = Y'_1 + Y'_2$	
$Y'_2 = vC_{12}$	$vC_{12} = Y'_2$	
$Y'_3 = v(C_{22} - C_{12})$	$vC_{22} = Y'_3 + Y'_2$	
$Y_1 = v(C_{11} - C_{12})$	$vC_{11} = Y_1 + Y_2$	None
$Y_2 = vC_{12}$	$vC_{12} = Y_2$	
$Y_3 = v(C_{22} - C_{12})$	$vC_{22} = Y_3 + Y_2$	
$Z'_1 = v(L_{11} - L_{12})$	$vL_{11} = Z'_1 + Z'_2$	
$Z'_2 = vL_{12}$	$vL_{12} = Z'_2$	
$Z'_3 = v(L_{22} - L_{12})$	$vL_{22} = Z'_3 + Z'_2$	
$Y_1 = vC_{11}$	$vC_{11} = Y_1$	1. $0 \le n \le 1$
$Y_3 = vC_{33}$	$vC_{33} = Y_3$	2. $0 \le m \le 1$
$Y_2 = v\left[C_{22} - \dfrac{C_{12}^2}{C_{11}} - \dfrac{C_{23}^2}{C_{33}}\right]$	$vC_{12} = nY_1$	3. $\dfrac{Y_2}{Y_1} > n(1 - n)$
$n = \dfrac{C_{12}}{C_{11}}$	$vC_{23} = mY_3$	$+ m(1 - m)\dfrac{Y_3}{Y_1}$
$m = \dfrac{C_{23}}{C_{33}}$	$vC_{22} = Y_2 + n^2 Y_1 + m^2 Y_3$	

TABLE 7-4 (continued)

	Transmission Line Network	Equivalent Circuit
		TRANSFORMER
11.	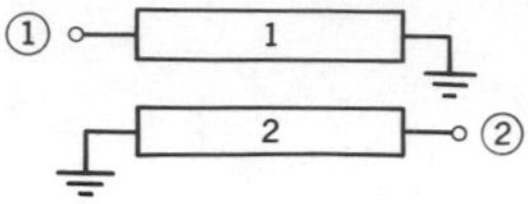 ① 1 2 ②	$-m:1$, Y_0, UE, Z_2, ① ②
12.	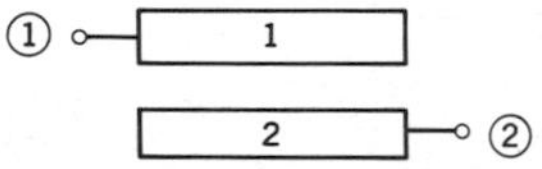 ① 1 2 ②	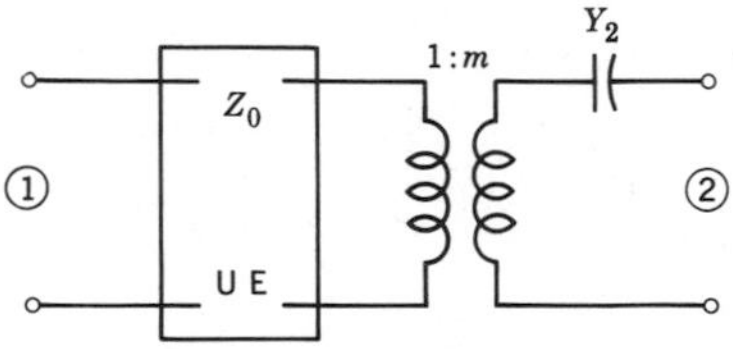 $1:m$, Z_0, UE, Y_2, ① ②
13.	① 1 2 ②	$m:1$, Y_0, UE, Z_2, ① ②
		$1:m$, Z_2, Y_0, UE, ① ②

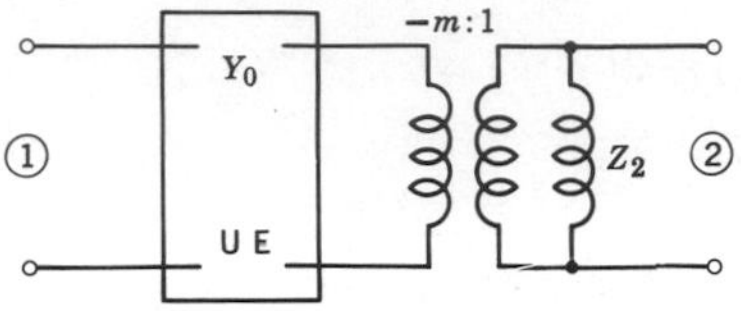

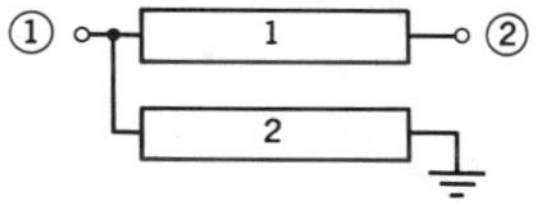

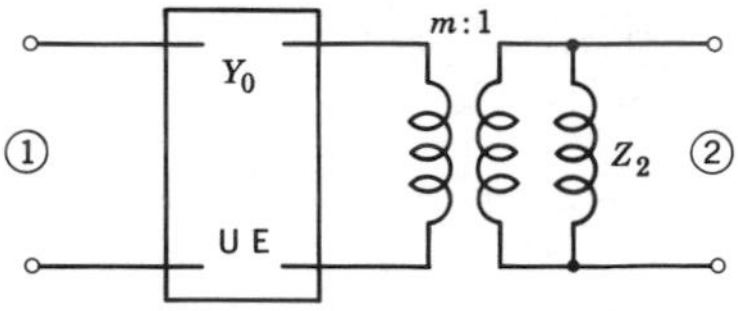

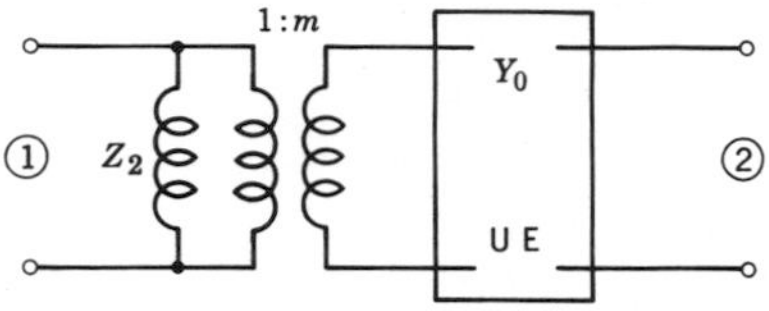

Analysis Equations	Synthesis Equations	Realizability Constraints
NETWORKS		
$m = \frac{C_{12}}{C_{11}} = \frac{L_{12}}{L_{22}}$	$vC_{11} = Y_0$	1. $0 \leq m \leq 1$
$Y_0 = vC_{11}$	$vC_{12} = mY_0$	2. $\frac{Y_2}{Y_0} \geq m(1 - m)$
$Z_2 = vL_{22}$	$vC_{22} = m^2 Y_0 + Y_2$	
$m = \frac{L_{12}}{L_{11}} = \frac{C_{12}}{C_{22}}$	$vL_{11} = Z_0$	1. $0 \leq m \leq 1$
$Z_0 = vL_{11}$	$vL_{12} = mZ_0$	2. $\frac{Z_2}{Z_0} \geq m(1 - m)$
$Y_2 = vC_{22}$	$vL_{22} = m^2 Z_0 + Z_2$	
$m = \left[1 + \frac{C_{22} - C_{12}}{C_{11} - C_{12}}\right]^{-1}$	$vC_{11} = Y_2 + Y_0 m^2$	1. $0 \leq m \leq 1$
$Y_0 = v[(C_{11} - C_{12}) + (C_{22} - C_{12})]$	$vC_{12} = Y_2 - m(1 - m)Y_0$	2. $\frac{Y_2}{Y_0} \geq m(1 - m)$
$Z_2 = v[(L_{11} - L_{12}) + (L_{22} - L_{12})]$	$vC_{22} = Y_2 + Y_0(1 - m)^2$	
$m = 1 - \frac{C_{12}}{C_{11}}$	$vC_{11} = Y_0$	1. $0 \leq m \leq 1$
$Y_0 = vC_{11}$	$vC_{12} = Y_0(1 - m)$	2. $\frac{Y_2}{Y_0} \geq m(1 - m)$
$Z_2 = vL_{22}$	$vC_{22} = Y_2 + Y_0(1 - m)^2$	

7.4 FILTER DESIGN PROCEDURES

Optimum Chebyshev and Butterworth Filters

Horton and Wenzel [14] have derived generalized Chebyshev and Butterworth transfer functions for lossless all-pole, transmission-line filters.* Filters of these kinds contain only shunt or series C's and L's, ideal transformers, and nonredundant unit elements.†

Let S_{11}, S_{22}, and S_{21} be the complex scattering parameters of the two-port filter. Then the generalized Chebyshev transfer function is

$$\left|\frac{1}{S_{21}}\right|^2 = 1 + \epsilon_p[T_m(x)T_n(y) - U_m(x)U_n(y)]^2. \tag{7.16}$$

The generalized Butterworth transfer function is

$$\left|\frac{1}{S_{21}}\right|^2 = 1 + (x)^{2m}(y)^{2n}. \tag{7.17}$$

The symbols in (7.16) and (7.17) are defined as:

$$T_m(z) = \begin{cases} \cos(m \cos^{-1} z), & \text{for } z \le 1 \\ \cosh(m \cosh^{-1} z), & \text{for } z > 1 \\ \text{Chebyshev function of the first kind} \end{cases}$$

$$U_n(z) = \begin{cases} \sin(n \cos^{-1} z) & \text{for } z \le 1 \\ \sinh(n \cosh^{-1} z) & \text{for } z > 1 \\ \text{unnormalized Chebyshev function of the second kind} \end{cases}$$

$$m = \begin{cases} \text{the number of transmission zeros at } \pm\infty \\ \text{the number of nonredundant } L\text{'s and } C\text{'s} \end{cases}$$

$$n = \begin{cases} \text{the number of transmission zeros at } \pm 1 \\ \text{the number of nonredundant unit elements} \end{cases}$$

ϵ_p = passband ripple tolerance‡

$$\theta_c = \begin{cases} \text{equiripple bandedge for Chebyshev filters} \\ \text{3 dB bandedge for Butterworth filters} \end{cases}$$

$\theta_c = \dfrac{\pi}{2}\left\{1 - \dfrac{w_p}{2}\right\}$ or $\dfrac{\pi}{2}\left\{1 - \dfrac{w_s}{2}\right\}$, where w_p, w_s are the fractional pass and stop bandwidths, respectively.

* The term all-pole as used here allows half-order transmission zeros at $S = \pm 1$.

† A redundant unit element contributes the factor $[(1 - S)/(1 + S)]^{1/2}$ to the overall transfer function. Since this is an allpass function, a redundant unit element does not contribute to the amplitude response.

‡ Often, the passband ripple is stated in dB. Then $\epsilon_p = 10^{A_p/10} - 1$ where A_p is the passband ripple in dB.

For bandstop filters

$$x = \frac{S}{S_c} = \frac{\tan(\theta)}{\tan(\theta_c)} = \frac{\Omega}{\Omega_c},$$

$$y = \left[\frac{S}{\sqrt{1 - S^2}}\right] \Bigg/ \left[\frac{S_c}{\sqrt{1 - S_c^2}}\right] = \frac{\sin(\theta)}{\sin(\theta_c)}. \tag{7.18}$$

And for bandpass filters

$$x = \frac{W}{W_c} = \frac{\cot(\theta)}{\cot(\theta_c)} = \frac{\mho}{\mho_c},$$

$$y = \left[\frac{W}{\sqrt{1 - W^2}}\right] \Bigg/ \left[\frac{W_c}{\sqrt{1 - W_c^2}}\right] = \frac{\cos(\theta)}{\cos(\theta_c)}. \tag{7.19}$$

Typical filter responses corresponding to these functions are depicted in Figure 7-9.* Note in (7.16) and (7.17) that for a given kth order filter any

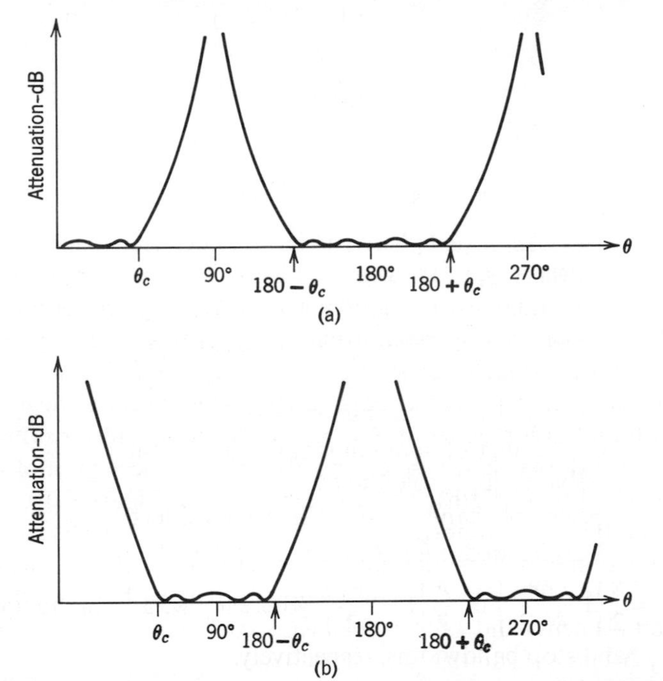

FIGURE 7-9 Typical commensurate transmission-line filter responses: (a) bandstop, (b) bandpass.

* Since commensurate transmission-line filters are periodic with period π, any filter can be considered as either bandpass or bandstop. However, usually the filter is classified according to its low-frequency behavior. Also, depending on the application and bandwidth, the bandstop filter might be considered lowpass, and the bandpass filter might be considered highpass.

combination of m and n satisfying $m + n = k$ yields a response having essentially the same passband behavior. The selectivity, input impedance, and transfer phase, however, will differ depending on the specific choices for m, n. To show more clearly the dependence of the selectivity on m and n consider a Chebyshev bandpass filter for which m and n are permitted to vary, but their sum is held to a constant value k. The attenuation in the stopband is

$$\left|\frac{1}{S_{21}}\right|^2 = 1 + \epsilon_p \cosh^2[m \cosh^{-1} x + n \cosh^{-1} y],$$

where x and y are given by (7.19). This may be rewritten as

$$\left|\frac{1}{S_{21}}\right|^2 = 1 + \epsilon_p \cosh^2[m' \cosh^{-1} x], \tag{7.20}$$

which accentuates the more familiar Chebyshev form. Here,

$$m' = m + n\alpha, \tag{7.21}$$

and

$$\alpha = \frac{\cosh^{-1} y}{\cosh^{-1} x}. \tag{7.22}$$

Since $m + n = k$, (7.21) is equivalent to

$$m' = k - (1 - \alpha)n.$$

Examination of (7.22) reveals that α is monotonic, decreasing with increasing $\mho$, and is always ≤ 1. Thus, m' is always less than or equal to k and greater than or equal to m, signifying that for a given order filter, the selectivity is maximized by maximizing m, the number of LC type elements. This is not a particularly surprising result since the LC type element produces transmission zeros on the real frequency axis, whereas the unit element produces transmission zeros at $S = \pm 1$. However, a close examination of the function α shows that it is very nearly unity in the region of cut-off for small fractional bandwidths. This is clearly seen in Figure 7-10, in which α has been plotted against normalized frequency, $\mho/\mho_c$, with fractional bandwidth as a parameter. Note that for fractional bandwidths (w_p) less than 0.2, α is greater than 0.95 for $\mho/\mho_c$ up to 3. Thus, for narrow-bandwidth passband filters, unit elements and LC type elements contribute about equally to the selectivity over a substantial part of the band.

For bandstop filters, the same graph may be used, except that now the abscissa is to be regarded as Ω/Ω_c rather than $\mho/\mho_c$, and the fractional stop bandwidth, w_s, is to be used rather than w_p. Thus, a similar conclusion is reached for bandstop filters, except that it is wideband-bandstop filters for which unit elements and LC type elements contribute about equally to the

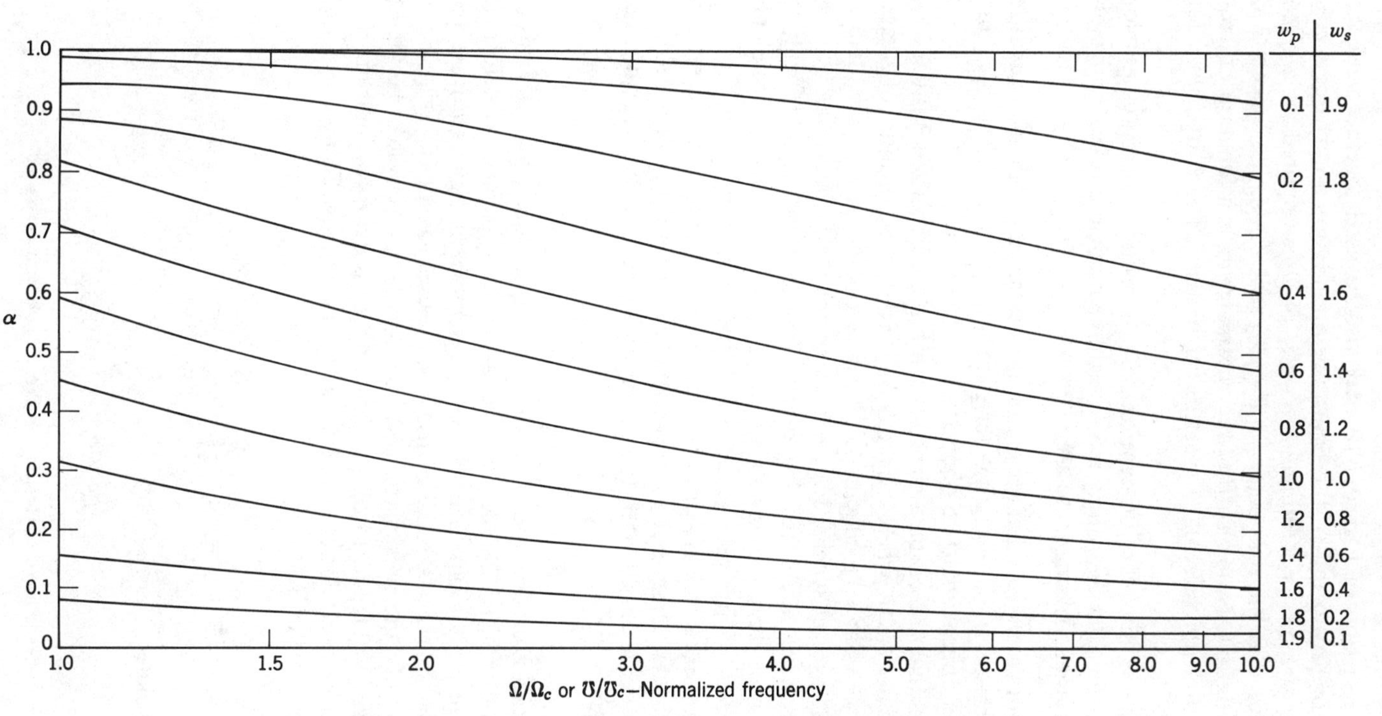

FIGURE 7-10 α versus Ω/Ω_c or $\mho/\mho_c$ with fractional bandwidth as a parameter.

selectivity. The graph of Figure 7-10 may also be used together with the nomographs of Kawakami [15] to determine attenuation at any frequency, or, in inverse fashion, to determine the order of filter required for a given selectivity. For this, the parameter m' of (7.20) replaces the parameter n in Kawakami's figures. Note also that the parameter m' is a function of frequency, so that a new value of m' must be calculated for each frequency for precise results.

In practice, the number of LC type elements and unit elements chosen for a given kth-order filter is strongly influenced by practical physical realizability considerations. Among these are:

1. In a completely shielded structure, it is extremely difficult to realize characteristic impedances greater than approximately 200 ohms and less than approximately 5 ohms.
2. In filters that use coupled transmission lines, normalized mutual coupling values C_{12}/ϵ (where ϵ is the permittivity of the medium) of greater than 6 to 8 are very difficult to obtain, and they are very sensitive to small changes in spacing (e.g., manufacturing tolerances).
3. In microwave filters utilizing direct coupled stubs, it is difficult to realize more than one L and C at a single junction. Thus, unit elements are needed to physically separate LC type resonators.
4. In many cases, the maximum value of m is limited by the type of physical realization. For example, in one common type of microwave filter to be discussed later, the maximum value of m is 1 regardless of the value of k.

Exact Synthesis Example As is the case in lumped networks, synthesis of transmission-line filters should be carried out in a suitable transformed variable in order to minimize the loss of significant figures during the synthesis process. However, the example below uses Richards' variable S in order to simplify the presentation. A seven-section Chebyshev bandpass filter, having $w_p = 1.0$ and 0.1 dB passband ripple, is to be synthesized.* The filter contains 2 unit elements and 5 LC type elements. Thus, $m = 5$ and $n = 2$. The parameter S_c is

$$S_c = j \tan \theta_c = j \tan \left[\frac{\pi}{2} \left(1 - \frac{w_p}{2} \right) \right] = j \tan(45°) = j$$

and the parameter ϵ_p is given by

$$\epsilon_p = 10^{0.1/10} - 1 = 0.0233.$$

* The example is from the paper of Horton and Wenzel [14] with permission.

The transmission function is, therefore,

$$|S_{21}|^2 = \frac{1}{1 + 0.0233\left[T_5\left(\frac{S_c}{S}\right)T_2\left(\frac{S\sqrt{1-S_c^2}}{S_c\sqrt{1-S^2}}\right) - U_5\left(\frac{S_c}{S}\right)U_2\left(\frac{S\sqrt{1-S_c^2}}{S_c\sqrt{1-S^2}}\right)\right]^2}.$$

Substitution of j for S_c, use of the identity $|S_{11}|^2 + |S_{21}|^2 = 1$ which holds for lossless networks, and simplification of the polynomials yields

$$|S_{11}|^2 = \frac{[0.98S^6 + 8.15S^4 + 17.64S^2 + 10.78]^2}{-S^{14} + 2.96S^{12} + 14.96S^{10} + 100.96S^8 + 308.69S^6 + 487.06S^4 + 380.46S^2 + 116.23}.$$

The next step is to determine the function $S_{11}(S)$ from its squared magnitude. For this the denominator is factored numerically using a computer. Since the denominator of $S_{11}(S)$ must be Hurwitz, the left hand roots are assigned to $S_{11}(S)$. On the other hand, the numerator need not be Hurwitz. Its zeros may be distributed arbitrarily so long as the assignation yields a function symmetrical about the real axis. In the present example, the numerator was factored by inspection and the left half plane zeros were selected.* The resulting $S_{11}(S)$ is given by

$$S_{11}(S) = -\frac{0.98S^6 + 8.15S^4 + 17.64S^2 + 10.78}{S^7 + 6.16S^6 + 17.47S^5 + 32.36S^4 + 39.13S^3 + 37.57S^2 + 20.73S + 10.78}.$$

From $S_{11}(S)$ the input impedance is obtained via

$$Z_{\text{in}} = \frac{1 + S_{11}}{1 - S_{11}}$$

yielding

$$Z_{\text{in}} = \frac{S^7 + 5.18S^6 + 17.47S^5 + 24.21S^4 + 39.13S^3 + 19.93S^2 + 20.73S}{S^7 + 7.13S^6 + 17.47S^5 + 40.51S^4 + 39.13S^3 + 55.21S^2 + 20.73S + 21.56}.$$

Since the odd parts of the numerator and denominator of Z_{in} are identical, it is known that the network representing Z_{in} can be symmetrical. Thus, only half the network need be synthesized. Although the unit elements may be removed at any time during the synthesis cycle, in the present example a unit element was removed first. Thus, via Richards' theorem,

$$Z_{\text{UE}} = Z_{\text{in}}(1) = 0.63.$$

* Selection of the right half plane zero would yield the dual circuit.

The impedance remaining after subtracting the unit element is

$$Z'_{\text{in}}(S) = \frac{1.84S^6 + 10.20S^5 + 18.84S^4 + 33.65S^3 + 20.08S^2 + 20.98S}{4.64S^6 + 21.09S^5 + 64.83S^4 + 82.37S^3 + 128.05S^2 + 60.52S + 62.96}$$

Z'_{in} is seen to have a zero at the origin, and therefore Y'_{in} has a pole there. The pole corresponds to a shunt inductor whose value is

$$L_1 = \left[\lim_{S\to 0} SY'_{\text{in}}(S)\right]^{-1} = 0.33.$$

After removing the inductor L_1, the remaining admittance is

$$Y''_{\text{in}}(S) = \frac{13.98S^5 + 46.87S^4 + 102.8S^3 + 77.15S^2 + 80.25S}{5.54S^5 + 30.73S^4 + 56.77S^3 + 101.4S^2 + 60.52S + 62.96}.$$

Y''_{in} has a zero at the origin, and therefore $Z''_{\text{in}}(S)$ has a pole there. The pole corresponds to a series capacitor whose value is given by

$$C_2 = \left[\lim_{S\to 0} SZ''_{\text{in}}(S)\right]^{-1} = 1.27.$$

In a similar manner, a second shunt inductor of value $L_3 = 0.26$ is next obtained. At this point, the synthesis may be terminated, since it is known that the filter is symmetrical. The final network is shown in Figure 7-11.

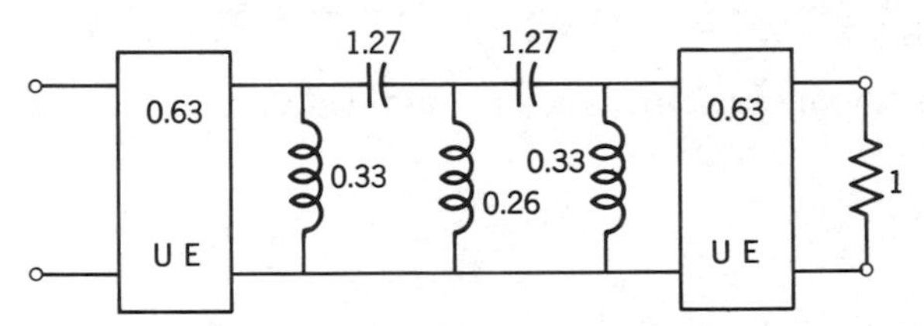

FIGURE 7-11 Chebyshev filter having optimum responses for 5 LC elements and 2 unit elements.

Direct Mapping Techniques

Several design methods are based on the use of a doubly-terminated lumped-lowpass ladder filter, shown schematically in Figure 7.12.*

The g_i parameters are capacitance (or conductance) for shunt elements and inductance (or resistance) for series elements. Tables of element values are available that give prescribed Chebyshev, Butterworth, and other filter

* The singly-terminated filter in which the source immittance is zero is also sometimes used, as are lumped-element elliptic-function filters.

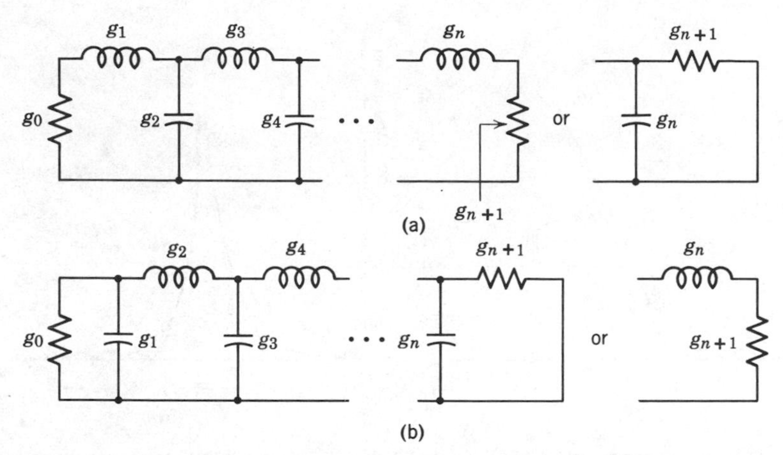

FIGURE 7-12 Dual forms of a lowpass prototype ladder filter.

characteristics [16]. The element values are typically given in normalized form such that for Chebyshev filters the equi-ripple cut-off frequency occurs at $\omega' = 1$; while in Butterworth filters the 3 dB cut-off frequency occurs at $\omega' = 1$. The impedance level is usually set such that g_0 equals 1.

Often a suitable transmission-line filter can be designed by applying a scaled version of Richards' transformation directly to a lumped-element prototype filter of this kind. For example, a transformation that converts a lumped-element lowpass-ladder filter into a bandstop transmission line filter is

$$\omega' = \left\{\omega_x' \cot\left(\frac{\pi}{2}\frac{\omega_x}{\omega_0}\right)\right\}\tan\left(\frac{\pi}{2}\frac{\omega}{\omega_0}\right) = \Lambda_s \tan\left(\frac{\pi}{2}\frac{\omega}{\omega_0}\right). \tag{7.23}$$

Similarly, a bandpass transformation is

$$\omega' = -\left\{\omega_x' \tan\left(\frac{\pi}{2}\frac{\omega_x}{\omega_0}\right)\right\}\cot\left(\frac{\pi}{2}\frac{\omega}{\omega_0}\right) = -\Lambda_p \cot\left(\frac{\pi}{2}\frac{\omega}{\omega_0}\right), \tag{7.24}$$

where the definitions of Λ_s and Λ_p are evident from (7.23) and (7.24), and

- ω' is the frequency variable for the lumped-element lowpass-prototype filter,
- ω is the frequency variable for the transmission-line filter,
- ω_0 is the center frequency of the transmission-line filter,
- ω_x is any arbitrary frequency for the transmission-line filter,
- ω_x' is the corresponding frequency for the lumped-element filter.

The frequency ω of the transmission-line filter maps into its corresponding frequency ω' of the prototype, as illustrated in Figure 7-13 for the case of the bandstop transformation. Once the parameters ω_x', ω_x, and ω_0 are selected,

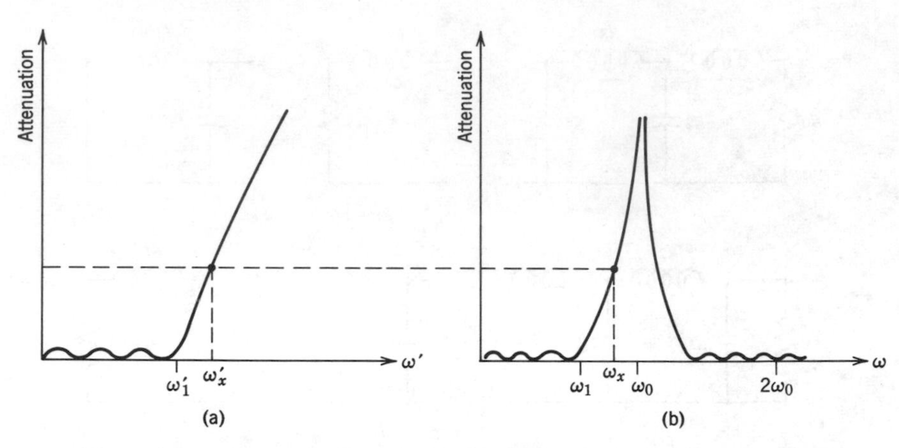

FIGURE 7-13 Tangent transformation applied to a Chebyshev lowpass filter: (a) prototype response, (b) mapped response.

the frequency characteristics of the transmission-line filter are uniquely determined for all ω. Inversely, the response of the prototype filter at frequency ω' is identical to that of the bandstop filter at

$$\omega = \frac{2}{\pi}\omega_0 \tan^{-1}\left[\frac{\omega'}{\Lambda_s}\right],$$

and for bandpass filters at

$$\omega = -\frac{2}{\pi}\omega_0 \cot^{-1}\left[\frac{\omega'}{\Lambda_p}\right].$$

Under the mappings of (7.23) and (7.24), the prototype filter shown in Figure 7-12 transforms into the transmission-line filters shown in Figures 7-14(a) and (b), respectively. If the order of the filter is 3 or less, it is sometimes possible to realize the filters by connecting the transmission-line stubs at a single reference plane. However, the success of this form of realization depends on the impedance values of the stubs, which in turn are related to the bandwidth and impedance level of the filter. In practice it is found that wide bandwidths are usually necessary for direct realization.

Another method of realization when n is 3 or more is to partition the filters into T sections of (LCL), (LCL) ... for bandstop filters, and (CLC), (CLC) ... for the bandpass filters. Then, for bandstop filters one may use the coupled-line identity of Table 7-4, Number 7 to give a filter realization of

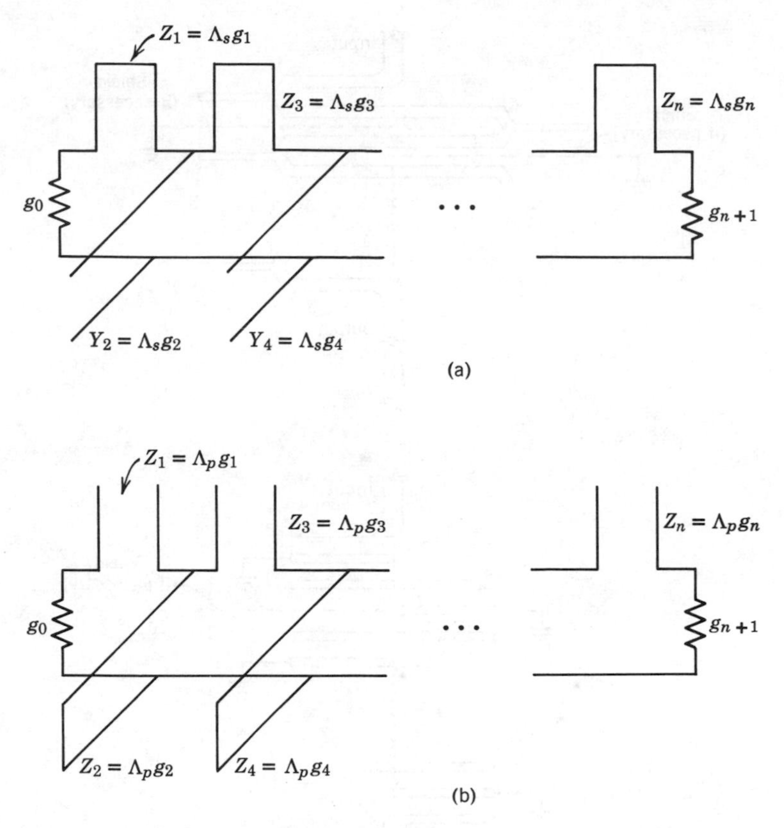

FIGURE 7-14 Transformed prototype networks by tangent and cotangent mappings: (a) tangent mapping, (b) cotangent mapping.

the type shown in Figure 7-15(a). This technique has been reported by Sato [17]. For the bandpass filter one may use the coupled-line identity of Table 7-4, Number 10 to give the physical realization shown in Figure 7-15(b). This technique has been reported by Levy [18]. In the latter case, ideal transformers must be inserted into the network, but this is always theoretically possible. However, in practice, it is found that these realizations are possible only over restricted bandwidths because of physical realizability limitations.

Other microwave filters that use tangent and cotangent mappings for their design are the digital-elliptic bandpass and bandstop filters [19], and the stepped-digital elliptic filter [20].

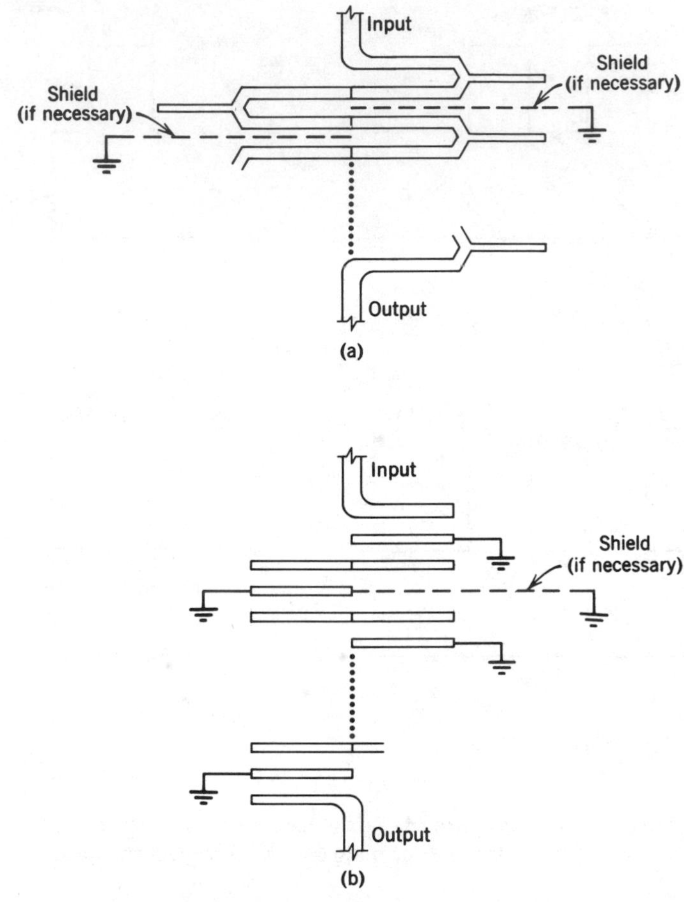

FIGURE 7-15 Coupled-line realizations of bandstop and bandpass filters: (a) bandstop filter, (b) bandpass filter.

An Exact Design Method of Ozaki and Ishii

In practice, the previous direct mapping design techniques are limited in their range of applications. Wide bandwidths are usually required, and physical realizations above (approximately) 4 GHz can be difficult to achieve. A variation of the direct mapping technique particularly useful for bandstop filters was described by Ozaki and Ishii [3]. Their method consists of introducing redundant unit elements into the transmission line filter in a way that physically separates the L's and C's by one or more unit elements. The method turns out to be practical for an extensive range of bandwidths, while at the same time physical realizations are made much easier. Let us illustrate the method with the design of a 4-resonator bandstop filter. The first step is to perform the transformation of (7.23) on a suitable lowpass prototype. Next, a number of unit elements are inserted between the source resistance and the first resonator of the transformed filter, and likewise between the load resistance and the last resonator. These steps are illustrated in Figure 7-16(a) and (b). The characteristic impedances of the unit elements

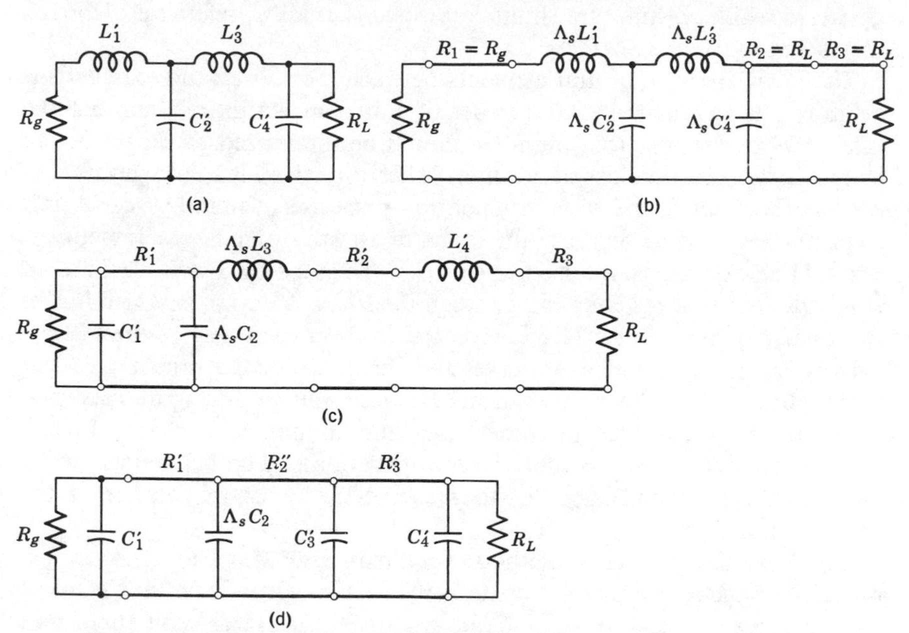

FIGURE 7-16 Example of a design method of Ozaki and Ishii: (a) prototype, (b) transformed prototype with 3 unit elements added, (c) Kuroda's identity applied once at each end of filter, and (d) Kuroda's identity applied to $\Lambda_s L_3$ and R'_2 ; L'_4 and R_3.

are chosen equal to their respective terminations, so as not to alter the attenuation of the filter. Specifically, if the transfer function of the lowpass lumped-element prototype is

$$S_{21}(\omega'),$$

then the response of the transmission line filter before inserting the unit elements is

$$S_{21}(\Lambda_s \tan\theta),$$

and the response after inserting k unit elements is

$$\left(\frac{1 - j\tan\theta}{1 + j\tan\theta}\right)^{k/2} S_{21}(\Lambda_s \tan\theta).$$

The last step in the design procedure is to utilize Kuroda's identities, Table 7-3, Numbers 1 and 2, to transpose the unit elements and stubs in a way that shifts the unit elements into the interior of the filter. The transpositions are effected in a way that separates the stubs by at least 1 unit element. The process is illustrated in Figure 7-16(c) and (d). Since series stubs are usually more difficult to realize than shunt stubs in a shielded structure, a design in which all stubs are shunt connected is usually preferable, although this is not so in all cases.

The initial division of unit elements between the source and load ends is arbitrary, as well as the total number of unit elements introduced into the filter. (Of course, the total number should be minimized to minimize the length and dissipation loss of the filter.) Therefore, there is a large number of designs that realize the same attenuation response. Generally, $n - 1$ unit elements are used to separate the stubs of an nth-order filter. Sometimes, $3(n - 1)$ unit elements are used to separate the stubs by 3 unit elements to eliminate troublesome coupling between the stubs. A design in which half of the total number of unit elements is fed in from each end of the filter is usually satisfactory, and it preserves the symmetry of the prototype filter. Other divisions of the unit elements between source and load may give better impedance values in some cases and should be considered on a case-to-case basis. Often in multiplexer applications, all unit elements are fed in from the load end in order to preserve the input impedance of the prototype.

Based on the preceding method, Schiffman and Matthaei derived formulas for bandstop filters for up to 5 resonators with single unit element spacing between resonators, and for up to 3 resonators with three unit element spacing between resonators [21]. Their formulas are based on equal, or near equal, division of the unit elements between the source and load ends of the filter.

A numerical procedure based on the preceding design method has been formulated as a computational algorithm [22]. The algorithm can be used in conjunction with a desk calculator or programmed on a computer. Arbitrary division of the unit elements between source and load ends of the filter can be easily accommodated.

For bandstop filters having wide stop bandwidths, physical realizations are usually accomplished by connecting the shunt stubs directly to the unit elements. For filters of narrow stop bandwidths, the required stub impedances are much too large for practical realizations. However, the physical realizations shown in Figure 7-17(a) and (b) are practical for air-line and printed circuits, respectively. The transformation from direct-coupled stub design to coupled-line design is readily accomplished with the identities of Table 7-4, Numbers 3 and 5.

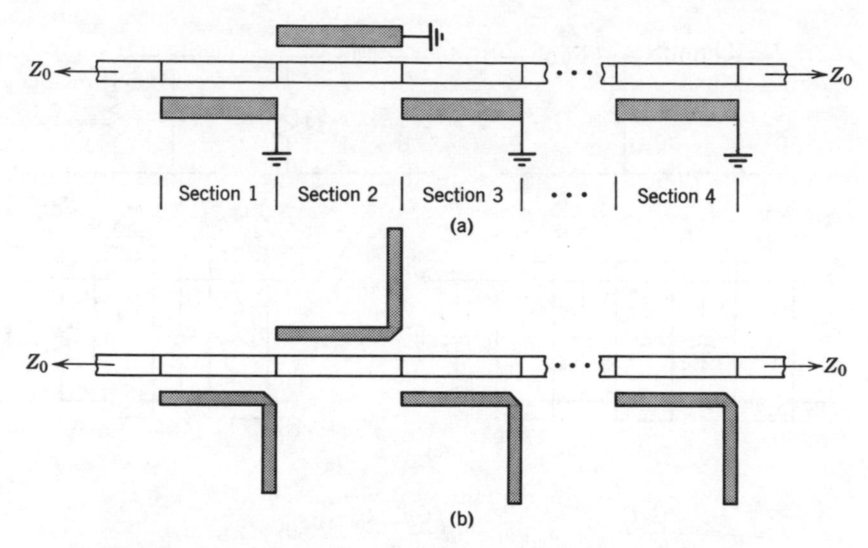

FIGURE 7-17 Coupled-line realizations for narrow to moderate stop bandwidth bandstop filters.

An Approximate Design Method of Cohn

A design method due to Cohn makes use of lowpass lumped-element prototype filters together with circuit approximations to ideal immittance inverters [23, 24]. The method works satisfactoraly for fractional bandwidths up to about 15%.

An ideal impedance inverter is a two-port having the *ABCD* matrix

$$\begin{bmatrix} 0 & jK \\ j/K & 0 \end{bmatrix}.$$

It has the property that the impedance seen at one pair of terminals is the reciprocal of the impedance at the second pair of terminals scaled by the inverter constant squared:

$$Z_{\text{in}} = K^2/Z_{\text{out}}.$$

A similar result holds on a dual basis for ideal admittance inverters.

A ladder network of series and shunt resonators can be replaced by an equivalent network (on an immittance and insertion loss basis), consisting of either all series or all shunt resonators separated by ideal inverters. The equivalences are shown in Figure 7-18. The choice for C_i, L_i, and terminating immittances in the equivalent networks of Figure 7-18(b) and (c) are completely arbitrary. The relationships given in this figure can be derived by

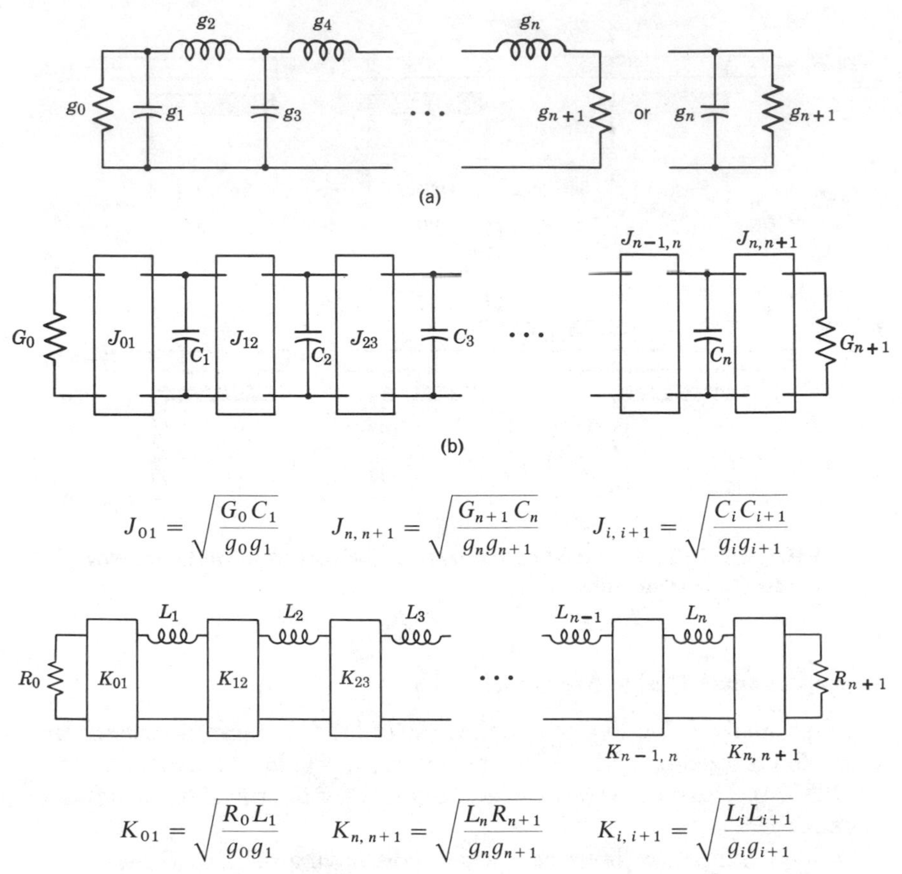

$$K_{01} = \sqrt{\frac{R_0 L_1}{g_0 g_1}} \qquad K_{n,n+1} = \sqrt{\frac{L_n R_{n+1}}{g_n g_{n+1}}} \qquad K_{i,i+1} = \sqrt{\frac{L_i L_{i+1}}{g_i g_{i+1}}}$$

FIGURE 7-18 Equivalent lowpass networks using inverters: (a) prototype, (b) equivalent network using admittance inverters, and (c) equivalent network using impedance inverters.

expanding the input immittances of the prototype network and the equivalent networks in continued fractions and by equating corresponding terms. The same technique is applicable to bandpass filters, as shown in Figure 7-19.

Two important generalizations, shown in Figure 7-20 and 7-21, are obtained by replacing the lumped *L-C* resonators by distributed circuits [25]. These can be microwave cavities, quarter-wavelength lines, or any other suitable resonant devices. In theory, the reactances of the distributed circuits should equal those of the lumped resonators at all frequencies. In practice, they approximate the reactances of the lumped resonators only near resonance. However, this is sufficient for narrow band filters. For convenience, the

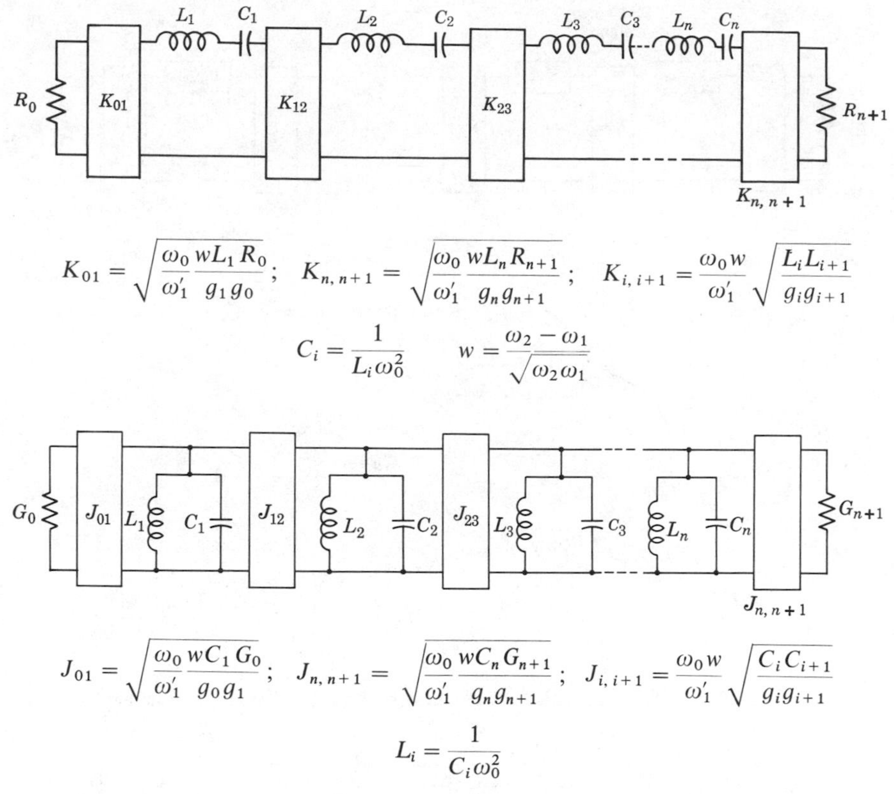

FIGURE 7-19 Bandpass filters using impedance and admittance inverters.

distributed resonator reactance and reactance slope are made equal to their corresponding lumped-resonator values at band center. For this, a quantity denoted as the slope parameter is introduced. The slope parameter for resonators having zero reactance at band center is denoted by χ and is defined by

$$\chi = \frac{\omega_0}{2} \frac{dX(\omega)}{d\omega}\bigg|_{\omega = \omega_0},$$

where $X(\omega)$ is the reactance of the distributed resonator. This and the dual circuit relationships are shown in Figure 7-20 and 7-21.

Ideal immittance inverters are not realizable with passive elements alone. However, they can be approximated over narrow bandwidths by several circuits, a few of which are shown in Figure 7-22. Although these circuits require negative elements or negative line lengths, in practice the negative

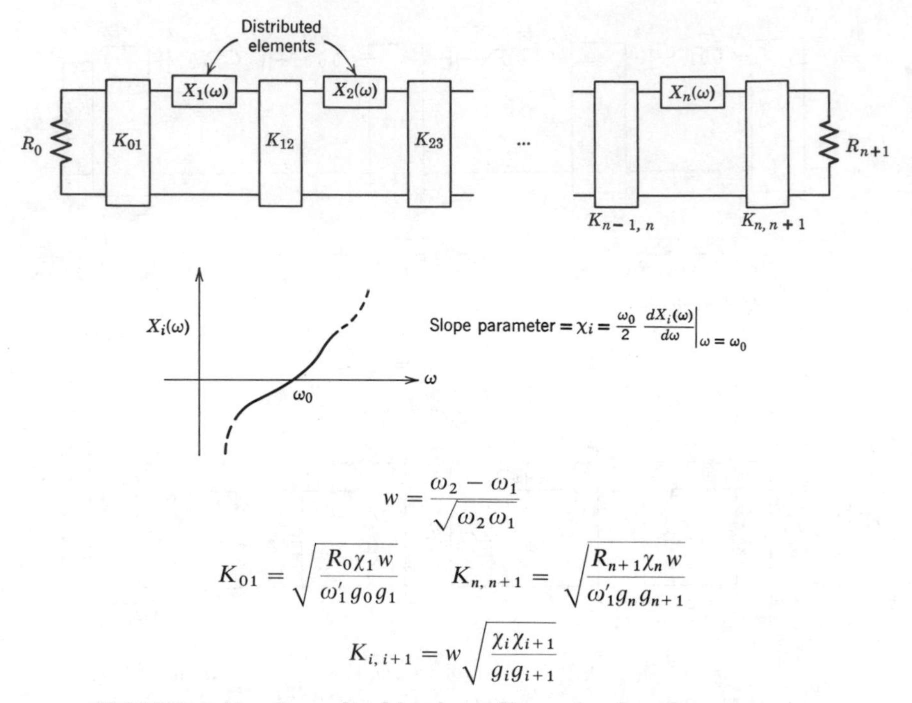

$$w = \frac{\omega_2 - \omega_1}{\sqrt{\omega_2 \omega_1}}$$

$$K_{01} = \sqrt{\frac{R_0 \chi_1 w}{\omega_1' g_0 g_1}} \qquad K_{n,\,n+1} = \sqrt{\frac{R_{n+1} \chi_n w}{\omega_1' g_n g_{n+1}}}$$

$$K_{i,\,i+1} = w \sqrt{\frac{\chi_i \chi_{i+1}}{g_i g_{i+1}}}$$

FIGURE 7-20 Generalized bandpass filter using impedance inverters.

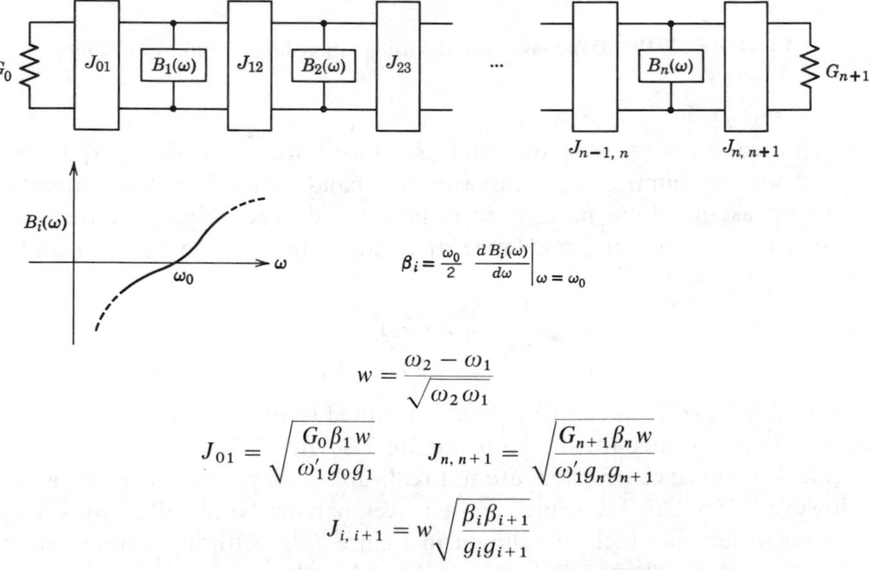

$$w = \frac{\omega_2 - \omega_1}{\sqrt{\omega_2 \omega_1}}$$

$$J_{01} = \sqrt{\frac{G_0 \beta_1 w}{\omega_1' g_0 g_1}} \qquad J_{n,\,n+1} = \sqrt{\frac{G_{n+1} \beta_n w}{\omega_1' g_n g_{n+1}}}$$

$$J_{i,\,i+1} = w \sqrt{\frac{\beta_i \beta_{i+1}}{g_i g_{i+1}}}$$

FIGURE 7-21 Generalized bandpass filter using admittance inverters.

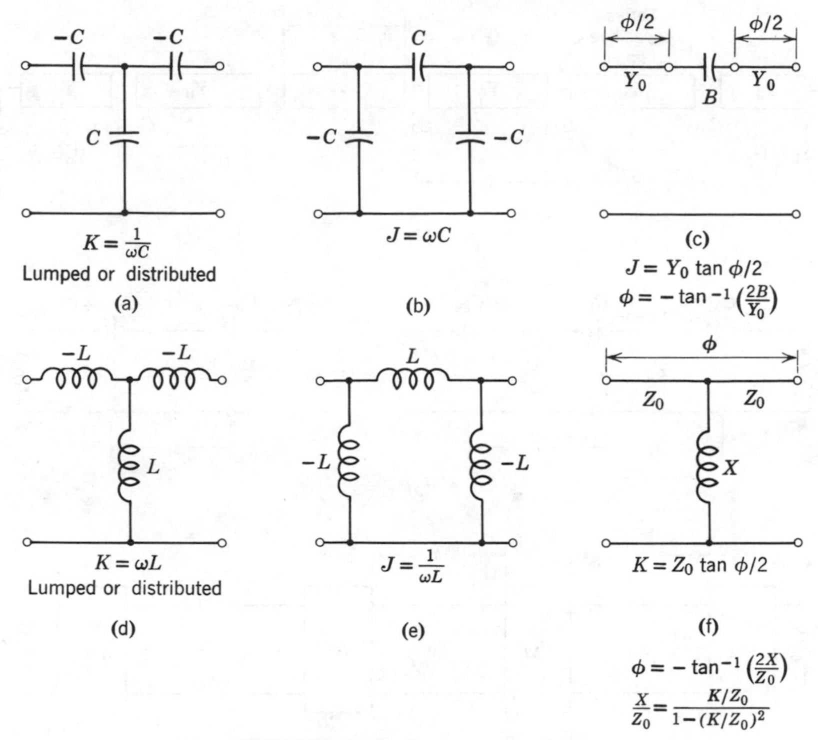

FIGURE 7-22 Inverter circuits.

quantities can be absorbed in adjacent elements. Note that the inverter circuits are frequency sensitive, which is one reason why filters utilizing these circuits are restricted to narrow bandwidths. Another useful inverter is the unit element in the vicinity of $\theta = 90°$.

Example Derivation

A derivation of the design equations for the end-coupled half-wave resonator-bandpass filter will serve to illustrate the latter synthesis method. The filter under consideration is illustrated schematically in Figure 7-23(a), and its equivalent circuit is given in Figure 7-23(b). The design task is to determine the electrical lengths, θ_i, for $i = 1$ through n, and the values of *lumped* capacitances, $C_{i,\,i+1}$, for $i = 0$ through n, that will achieve a specified filter performance. One method of deriving the design equations is to first transform the circuit of Figure 7-23(b) into the general form given in Figure 7-21. Then, relationships can be written between corresponding quantities of the actual circuit and idealized circuit with inverters. From these relationships the design equations can be obtained.

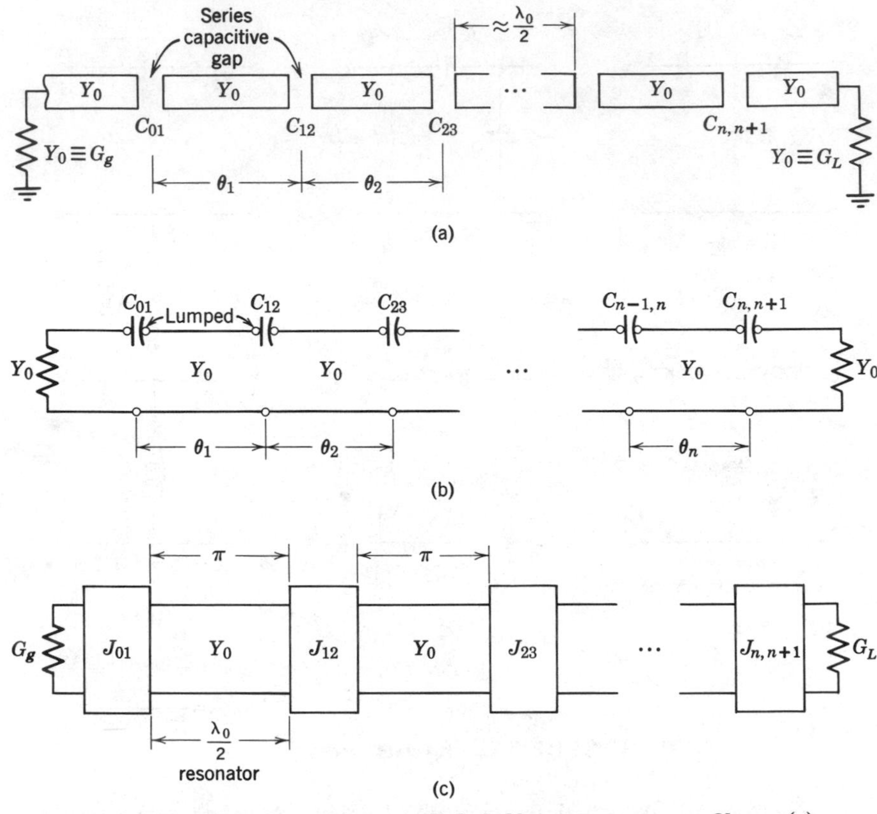

FIGURE 7-23 Capacitive coupled half-wave resonator filter: (a) schematic representation of the end-coupled half-wave filter (The ground plane is omitted), (b) equivalent circuit, (c) equivalent circuit with inverters.

A preliminary step before beginning the design is to note that a unit element in the vicinity of $\theta = \pi$ behaves approximately like a shunt susceptance. According to Table 7-2, the impedance matrix of a unit element for $\theta \approx \pi$ is

$$[Z] \approx \begin{bmatrix} \dfrac{Z_0}{j \tan \theta} & \dfrac{Z_0}{j \tan \theta} \\ \dfrac{Z_0}{j \tan \theta} & \dfrac{Z_0}{j \tan \theta} \end{bmatrix},$$

which corresponds to a shunt susceptance $B(\omega)$ of value

$$B(\omega) = Y_0 \cot \theta \approx Y_0(\theta - \pi).$$

The slope parameter of $B(\omega)$ is

$$\beta = \frac{\pi}{2} Y_0. \tag{7.25}$$

Negative line lengths sufficient to achieve the inverter circuit of Figure 7-22(c) are inserted on both sides of each coupling capacitor, $C_{i,\,i+1}$ in Figure 7-23(b). The required line lengths [as given in Figure 7-22(c)] are

$$\phi_{i,\,i+1} = -\tan^{-1}\left[\frac{2B_{i,\,i+1}}{Y_0}\right], \qquad i = 0, 1, \ldots, n \tag{7.26}$$

where $B_{i,\,i+1}/Y_0 = \omega_0 C_{i,\,i+1}/Y_0$ are the normalized coupling susceptances, and ω_0 is the center frequency of the filter. Next, the line length θ_i is adjusted so that the net electrical length between inverters is π. Thus, θ_i must satisfy

$$\theta_i = \pi + \tfrac{1}{2}[\phi_{i-1,\,i} + \phi_{i,\,i+1}], \qquad i = 1, 2, \ldots, n. \tag{7.27}$$

Substituting (7.26) into (7.27) gives

$$\theta_i = \pi - \frac{1}{2}\left[\tan^{-1}\left(\frac{2B_{i-1,\,i}}{Y_0}\right) + \tan^{-1}\left(\frac{2B_{i,\,i+1}}{Y_0}\right)\right], \qquad i = 1, 2, \ldots, n. \tag{7.28}$$

Using (7.28) for the electrical line lengths, the circuit of Figure 7-23(b) now takes the form shown in Figure 7-23(c), which by the previous discussion is equivalent to Figure 7-11. All that remains now is to relate the inverter values of Figure 7-23(c) to the prototype filter and nominal-design fractional bandwidth, from which the coupling capacitances $C_{i,\,i+1}$ are easily solved.

The inverter values required to achieve a specified prototype filter response were given in Figure 7-21. Substitution of (7.25) into those equations gives

$$J_{01} = \sqrt{\frac{G_g(\pi/2)Y_0 w}{\omega'_1 g_n g_{n+1}}}$$

$$J_{n,\,n+1} = \sqrt{\frac{G_L(\pi/2)Y_0 w}{\omega'_1 g_n g_{n+1}}}$$

$$J_{i,\,i+1} = \frac{Y_0(\pi/2)w}{\omega'_1}\sqrt{\frac{1}{g_i g_{i+1}}}. \tag{7.29}$$

Substituting these into the equations in Figure 7-22(c) and solving for the coupling susceptances yields

$$B_{i,\,i+1}/Y_0 = \frac{J_{i,\,i+1}/Y_0}{1 - (J_{i,\,i+1}/Y_0)^2}, \qquad \text{for} \quad i = 0, 1, \ldots, n. \tag{7.30}$$

Equations (7.29), (7.30), and (7.28) are the required design equations.

In summary, the design steps are:

(1) Based on a given set of filter specifications, choose a suitable lowpass prototype and fractional bandwidth.
(2) Determine $J_{i,i+1}/Y_0$ and $B_{i,i+1}/Y_0$ for $i = 0, 1, \ldots, n$ from (7.29) and (7.30), respectively.
(3) Determine θ_i for $i = 1, 2, \ldots, n$ from (7.28).

The dual case, which uses inductive coupled half-wave resonators, is useful in coaxial line and also waveguide. This geometry is shown symbolically in Figure 7-24. The corresponding design equations are identical to

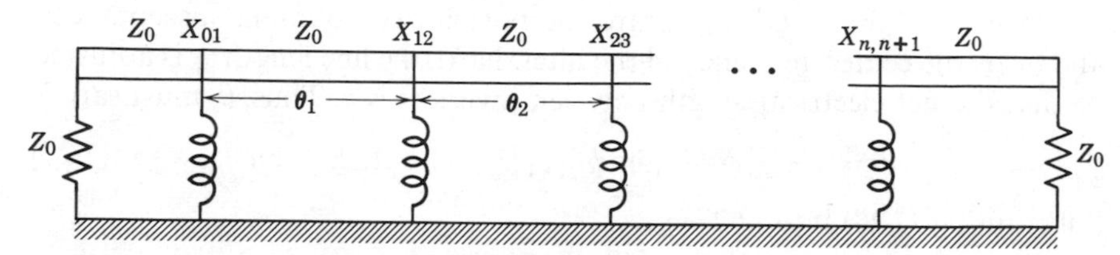

FIGURE 7-24 Dual circuit to Figure 7-23(a).

(7.29), (7.30), and (7.28), with impedance inverters $K_{i,i+1}/Z_0$ replacing admittance inverters $J_{i,i+1}/Y_0$ and reactances $X_{i,i+1}/Z_0$ and replacing susceptances $B_{i,i+1}/Y_0$.

An Approximate Design Method of Matthaei

An approximate design procedure for bandpass filters, due to Matthaei, uses a lowpass lumped-element prototype filter, inverters, and image-parameter techniques [25, 26]. The method gives satisfactory results for bandwidths of from very narrow to approximately 3 : 1. Note in Figure 7-18 that the parameters L_i in the all series, and C_i in the all shunt equivalent circuits, are arbitrary. If they are all chosen equal, the equivalent filter is readily partitioned into symmetrical sections, as shown in Figure 7-25 for the shunt C case. The image admittances for all but the end sections are readily calculated from the well-known formula

$$Y_I = \sqrt{y_{11}^2 - y_{12}^2},$$

giving

$$Y_I = \sqrt{J_{i-1,i}^2 - \left(\frac{\omega' C}{2}\right)^2}, \tag{7.31}$$

where the admittance inverters $J_{i-1,i}$ are related to the prototype filter by the equations in Figure 7-18(b).

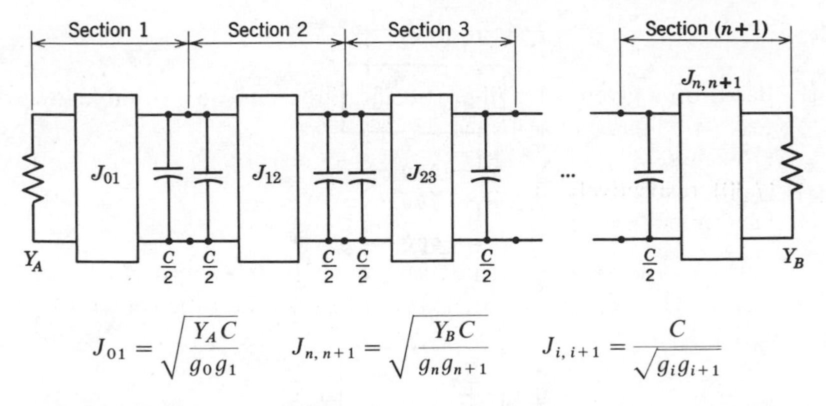

FIGURE 7-25 Example of symmetrical partitioning of Figure 7-18(b).

A microwave filter structure that may similarly be divided into cascaded symmetrical sections is the half-wave parallel-coupled resonator circuit shown in Figure 7-26. It is seen to consist of $n + 1$ cascaded pairs of short-circuited coupled lines of the type given in Table 7-4, Number 2. In Matthaei's design procedure, the interior-coupled-line sections are made symmetrical, while the end-coupled-line sections are made asymmetrical in

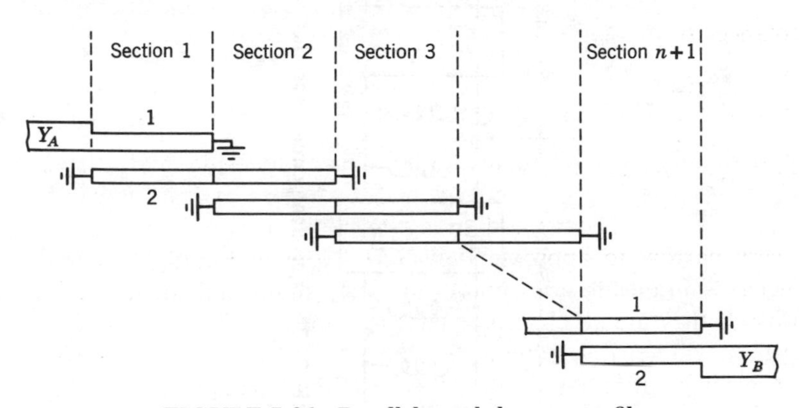

FIGURE 7-26 Parallel-coupled-resonator filter.

order to explicitly incorporate transformers into the equivalent circuit. For the end sections, the equivalent circuit of Table 7-4, Number 11 is appropriate. Hence, the equivalent circuit for the parallel-coupled resonator filter is that shown in Figure 7-27.* Note in the representation that the unit

* The $-1:1$transformers of each coupled section have been eliminated in Figure 7-27 since they do not affect the attenuation response.

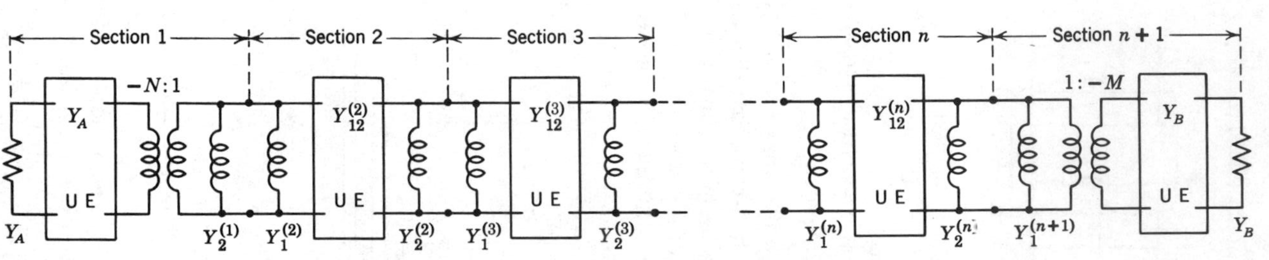

FIGURE 7-27 Equivalent circuit for the parallel-coupled resonator filter of Figure 7-26.

elements of each end section have been set equal to their respective terminating admittances.

The image impedance for the ith-interior parallel-coupled section in Figure 7-27 is

$$Y_i = Y_{12}^{(i)}\sqrt{1 - \left[2\frac{Y_1^{(i)}}{Y_{12}^{(i)}} + \left(\frac{Y_1^{(i)}}{Y_{12}^{(i)}}\right)^2\right]\cot^2\theta}. \tag{7.32}$$

In the present design procedure, the image admittance of each interior section of the prototype is required to equal the image admittance of the corresponding section of the microwave filter. Of course, this is not possible for all frequencies, but since there are two degrees of freedom, (7.31) and (7.32) can be equated and satisfied at two points. In order to obtain good performance over the entire passband, the points of match usually chosen are band center, $\omega' = 0$ corresponding to $\theta = 90°$, and band edge, $\omega' = -\omega_1'$ corresponding to $\theta = \theta_1$.* Equating (7.31) and (7.32) at $\omega' = 0$ and $\theta = 90°$ gives

$$Y_{12}^{(i)} = J_{i-1,\,i} \tag{7.33}$$

for the ith section. Equating them at $\omega' = -\omega_1'$ and $\theta = \theta_1$ gives

$$(Y_{12}^{(i)})^2\{1 - [2x + x^2]\cot^2\theta_1\} = J_{i-1,\,i}^2 - \left(\frac{\omega' C}{2}\right)^2, \tag{7.34}$$

where $x = Y_1^{(i)}/Y_{12}^{(i)}$. Substituting (7.33) into (7.34), solving for $Y_1^{(i)}$, and simplifying gives

$$Y_1^{(i)} = \omega_1' C\left\{\sqrt{\left(\frac{J_{i-1,\,i}}{\omega_1' C}\right)^2 + \left(\frac{\tan\theta_1}{2}\right)^2} - \frac{J_{i-1,\,i}}{\omega_1' C}\right\}. \tag{7.35}$$

The derivation of the equations for the end sections differs from that of the interior sections. We consider only section 1, since the results are directly applicable to section $n + 1$. The equivalent circuit for the coupled line circuit constituting section 1 is shown in Figure 7-28. The input admittance from port 2 is

$$Y = -jY_2^{(1)}\cot(\theta) + N^2 Y_A. \tag{7.36}$$

The admittance seen from the corresponding reference plane in the circuit of Figure 7-25 is

$$Y' = j\frac{\omega' C}{2} + \frac{J_{01}^2}{Y_A}. \tag{7.37}$$

*For an alternate approach, see reference [27].

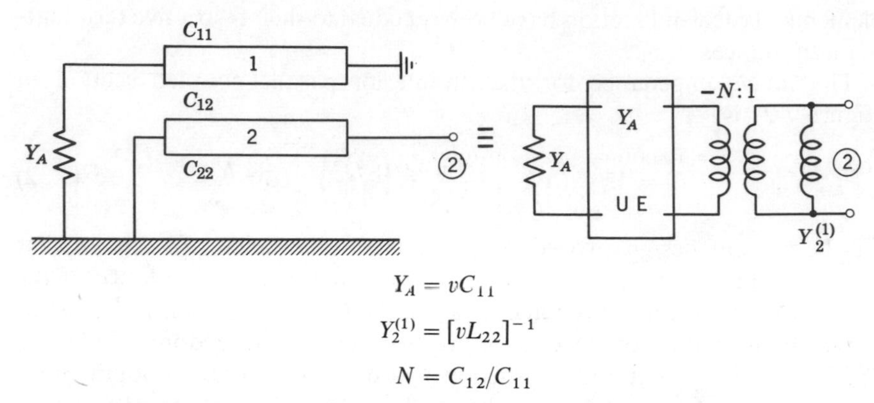

$$Y_A = vC_{11}$$

$$Y_2^{(1)} = [vL_{22}]^{-1}$$

$$N = C_{12}/C_{11}$$

FIGURE 7-28 End section of parallel-coupled-resonator filter and its equivalent circuit.

Equating the real parts of (7.36) and (7.37) gives

$$N = \frac{J_{01}}{Y_A}. \tag{7.38}$$

The imaginary parts of (7.37) and (7.38) are automatically equal at $\omega' = 0$, $\theta = 90°$. They may also be required to be equal at bandedge. This gives

$$Y_2^{(1)} \cot \theta_1 = \frac{\omega_1' C}{2},$$

or

$$Y_2^{(1)} = \omega_1' C \frac{\tan \theta_1}{2}. \tag{7.39}$$

Extending these results to section $n + 1$ gives

$$M = \frac{J_{n,n+1}}{Y_B}, \tag{7.40}$$

$$Y_1^{(n+1)} = \omega_1' C \frac{\tan \theta_1}{2}. \tag{7.41}$$

To obtain the coupled line parameters $vC_{11}^{(i)}$ and $vC_{12}^{(i)}$ for the interior sections, (7.34) and (7.35) are substituted into the synthesis equations given in Table 7-4, Number 2. Similarly, (7.38) and (7.39) may be substituted into the synthesis equations in Table 7-4, Number 11, to obtain the coupled line parameters for the first section, $vC_{11}^{(1)}$, $vC_{12}^{(1)}$, and $vC_{22}^{(1)}$. The parameters for the $n + 1$ section may be similarly obtained using (7.40) and (7.41). The results are summarized in Table 7-5.

TABLE 7-5 Design Equations for Half-Wave Parallel-Coupled Filters of the Form Given in Figure 7-26*

Define:

$$w = \text{fractional bandwidth} = \frac{2(f_2 - f_1)}{f_2 + f_1}$$

$$\theta_1 = \frac{\pi}{2}\left(1 - \frac{w}{2}\right)$$

$$h = \frac{\omega_1' C}{Y},$$ a dimensionless scale factor used to obtain a convenient admittance level within the filter

$$\tau = \frac{\tan\theta_1}{2}$$

<u>Section 1</u>

$$J_1 = (\omega_1' g_0 g_1)^{-1/2}$$

$$\frac{\nu C_{11}^{(1)}}{Y_A} = 1$$

$$\frac{\nu C_{12}^{(1)}}{Y_A} = \sqrt{h}\, J_1$$

$$\frac{\nu C_{22}^{(1)}}{Y_A} = h(J_1^2 + \tau)$$

<u>Section $n+1$</u>

$$J_{n+1} = (\omega_1' g_n g_{n+1})^{-1/2}$$

$$\frac{\nu C_{11}^{(n+1)}}{Y_B} = 1$$

$$\frac{\nu C_{12}^{(n+1)}}{Y_B} = \sqrt{h}\, J_{n+1}$$

$$\frac{\nu C_{22}^{(n+1)}}{Y_B} = h(J_{n+1}^2 + \tau)$$

<u>Sections $i = 2, 3, \ldots, n$</u>

$$J_i = \frac{1}{\omega_1' \sqrt{g_{i-1} g_i}}$$

$$\frac{\nu C_{11}^{(i)}}{Y_A} = h\sqrt{J_i^2 + \tau^2} \qquad \frac{\nu C_{12}^{(i)}}{Y_A} = hJ_i$$

* An n-resonator prototype yields $n+1$ coupled sections

The dimensionless scale factor h in Table 7-5 should be selected to obtain a satisfactory impedance level within the filter, and to maximize the unloaded resonator Q's. However, often a satisfactory choicc for h is that which makes one of the end sections symmetrical. To achieve this requires (for the generator side)

$$\nu C_{11}^{(1)} = \nu C_{22}^{(1)}.$$

Placing the above constraint on the equations for section 1 in Table 7-5 gives the result

$$h = \frac{1}{J_1^2 + \tau}. \tag{7.42}$$

In most filter applications the terminations Y_A and Y_B are equal, usually 0.02 mho. Furthermore, for Chebyshev and maximally flat prototypes, $J_1 = J_{n+1}$, so that the value for h given in (7.42) makes the $(n + 1)$ section symmetrical as well.

The dual filter having open-circuited ends is shown in Figure 7-29. This filter form is well suited for printed-circuit construction, where grounding of the resonators is difficult. The design equations are identical to those in Table 7-5, with C replaced by L, and Y_A and Y_B replaced by Z_A and Z_B, respectively.

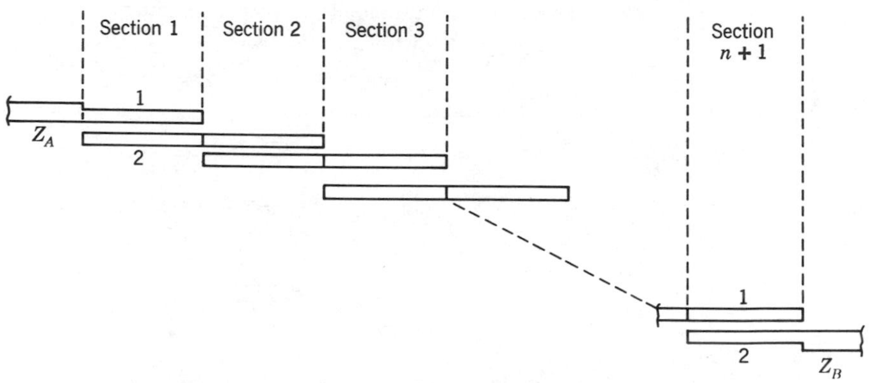

FIGURE 7-29 Half-wave parallel-coupled-resonator filter.

The equivalent circuit for the filter of the previous example reveals that two unit elements do not contribute to the attenuation response, while $n - 1$ unit elements and effectively one stub do contribute. The generic form of the transmission-line transfer function corresponding to a kth-order prototype is therefore of the form

$$S_{21}(S) \triangleq \frac{S[\sqrt{1 - S^2}]^{k-1}}{H_k(S)} \left[\frac{1 - S}{1 + S}\right],$$

where $H_k(S)$ is a Hurwitz polynomial of degree k. Thus, if the filter design were based on a Chebyshev or Butterworth lowpass prototype, a very good estimate of the attenuation response could be obtained by letting $m = 1$, and $n = k - 1$ in (7.16) or (7.17). Alternatively, one could compute m' from (7.21) and Figure 7-10, and then obtain the attenuation using the nomographs of Kawakami.

A variation of the previous design procedure yields a class of filters having a transmission characteristic of the form

$$S_{21}(S) = \frac{S^3[\sqrt{1 - S^2}]^{n-3}}{H_n(S)},$$

which is seen to have a 3rd-order zero at the origin, and $n - 3$ contributing unit elements. These filters are more suitable for wide bandwidths with regard to both ease of physical realization and selectivity. Details are given in references [25] and [26].

7.5 SUMMARY

TEM microwave filters can be viewed theoretically as consisting of couplings and interconnections of transmission lines along which the wave propagation is transverse-electromagnetic. Consequences of the TEM assumption are that individual and multicoupled transmission lines can be characterized by their static capacitance and inductance matrices, and that the characteristic impedance and propagation velocity of lines are independent of frequency.

If the lengths of the transmission lines comprising a given network are commensurate, application of Richards' transformation allows the use of many of the analytical methods of lumped-element filter theory. The application of Richards' transformation to a length of transmission line yields a canonical network element, termed the unit element, that has no counterpart in lumped element theory. The unit element has half-order transmission zeros at $S = \pm 1$.

The unit element appears naturally in the equivalent circuits of many coupled transmission line networks, and can contribute to a filter's selectivity. Additionally, it may be used advantageously to physically separate stubs, or along with various transformations to obtain equivalent filters having differing physical realizations. Several design procedures, both exact and approximate, were presented that utilized these concepts in obtaining practical microwave bandpass and bandstop filters.

PROBLEMS

7.1 Develop the impedance function

$$Z = \frac{4S^4 + 6S^2 + 1}{4S^3 + 4S}$$

in the first Cauer form of reactive network.

7.2 For the impedance function of Problem 1, extract first a unit element and then develop the remainder in the first Cauer form.

7.3 For the impedance function of Problem 1, utilize Richards' procedure to develop the function into a network of all unit elements.

7.4 Utilize Kuroda's identities of Table 7-3, Numbers 1 and 2, to transform the Cauer network in Problem 1 to the all unit element network obtained in Problem 3.

7.5 Using the Darlington procedure, synthesize a network for the impedance function

$$Z(S) = \frac{1 + 2S + 2S^2}{1 + 2S + 2S^2 + 2S^3}.$$

7.6 Develop the impedance function of Problem 5 in the form of three shunt capacitors separated by unit elements.

7.7 Determine coupled-line realizations for the network of Problem 6 using
(1) Table 7-4, Number 3
(2) Table 7-4, Number 5
(3) Table 7-4, Number 6.

7.8 Derive the identity of Table 7-4, Number 2 using the C-matrix description. Rederive the same circuit using the graph-transformation procedure.

7.9 Using Cohn's procedure derive design equations for the filter of Figure 7-26.

REFERENCES

[1] P. I. Richards, "Resistor-transmission-line circuits," *Proc. IRE*, **36**, 217–220 (Feb., 1948).
[2] P. I. Richards, "A special class of functions with positive real parts in half plane," *Duke Math. J.*, **14**, 777–788 (Sept., 1947).
[3] H. Ozaki and J. Ishii, "Synthesis of a class of strip-line filters," *IRE Trans. Circuit Theory*, **CT-5**, 104–109 (June, 1958).
[4] O. Brune, "Synthesis of a finite two-terminal network whose driving-point impedance is a prescribed function of frequency," *J. Math. Phys.*, **10**, 191–236 (Oct., 1931).
[5] H. Bode, *Network Analysis and Feedback Amplifier Design*, D. Van Nostrand, Princeton, New Jersey, 1945.
[6] S. Darlington, "Synthesis of reactance four-poles which produce prescribed insertion loss characteristics," *J. Math. Phys.*, **18**, 257–353 (Sept., 1939).
[7] E. A. Guillemin, *Synthesis of Passive Networks*, John Wiley, New York, 1957.
[8] D. C. Youla, "A new theory of cascade synthesis," *IEEE Trans. Circuit Theory*, **CT-9**, 244–260 (Sept., 1961).
[9] R. Levy, "A general equivalent circuit transformation for distributed networks," *IEEE Trans. Circuit Theory*, **CT-12**, 457–458 (Sept., 1965).
[10] H. Uchida, *Fundamentals of Coupled Lines and Multiwire Antennas*, Sasaki, Sendai, Japan, 1967.
[11] R. Sato and E. G. Cristal, "Simplified analysis of coupled transmission-line networks," *IEEE Trans. Microwave Theory and Techniques*, **MTT-18**, 122–131 (March, 1970).

[12] R. Sato, "Two universal transformations for distributed commensurate transmission line networks," *Electronic Letters*, **6**, 173–174 (March, 1970).

[13] K. K. Pang, "A new transformation for microwave network synthesis," *IEEE Trans. Circuit Theory*, **CT-13**, 235–238 (June, 1966).

[14] M. C. Horton and R. J. Wenzel, "General theory and design of optimum quarter-wave TEM-Filters," *IEEE Trans. Microwave Theory and Techniques*, **MTT-13**, 316–327 (May, 1965).

[15] M. Kawakami, "Nomographs for Butterworth and Chebyshev filters," *IRE Trans. Circuit Theory*, **CT-10**, 288–291 (June, 1963).

[16] L. Weinberg, *Network Analysis and Synthesis*, McGraw-Hill, New York, 1962.

[17] R. Sato, *Transmission-Line Circuits*, Korona, Tokyo, Japan, 1963, Chap. 9.

[18] R. Levy, "Three-wire-line interdigital filters of Chebyshev and elliptic-function characteristic for broad bandwidths," *Electronic Letters*, **2**, 455–456 (Dec., 1966).

[19] M. C. Horton and R. J. Wenzel, "The digital elliptic filter—a compact sharp-cutoff design for wide bandstop or bandpass requirements," *IEEE Trans. Microwave Theory Techniques*, **MTT-15**, 307–314 (May, 1967).

[20] J. D. Rhodes, "The stepped digital elliptic filter," *IEEE Trans. Microwave Theory and Techniques*, **MTT-17**, 178–184 (April, 1969).

[21] B. M. Schiffman and G. L. Matthaei, "Exact design of microwave band-stop filters," *IEEE Trans. Microwave Theory and Techniques*, **MTT-12**, 6–15 (Jan., 1964).

[22] E. G. Cristal, "Addendum to 'An exact method for synthesis of microwave bandstop filters'," *IEEE Trans. Microwave Theory and Techniques*, **MTT-12**, 369–382 (May, 1964).

[23] S. B. Cohn, "Direct-coupled-resonator filters," *Proc. IRE*, **45**, 187–196 (Feb., 1957).

[24] S. B. Cohn, "Parallel-coupled transmission-line-resonator filters," *IRE Trans. Microwave Theory and Techniques*, **MTT-6**, 223–231 (April, 1958).

[25] G. L. Matthaei, L. Young, and E. M. T. Jones, *Microwave Filters, Impedance Matching Networks, and Coupling Structures*, McGraw-Hill, New York, 1964.

[26] G. L. Matthaei, "Design of wide-band (and narrow-band) band-pass microwave filters on the insertion loss basis," *IEEE Trans. Microwave Theory and Techniques*, **MTT-8**, 580–593 (Nov., 1960).

[27] E. G. Cristal, "New design equations for a class of microwave filters," *IEEE Trans Microwave Theory and Techniques*, **MTT-19**, 486–490 (May, 1971).

8

Active Filters with Lumped RC Networks

S. K. Mitra
University of California
Davis, California

The type of filters we have considered so far in this book are passive filters. Even though these filters do provide satisfactory performances for many applications, there are cases where they do not provide an attractive solution to the system designer because of their cost, size, and other reasons. In these cases, active networks provide an alternative solution to the filtering problem.

There are basically four types of active filters: (1) active filters with lumped R-C elements, (2) active filters with distributed R-C elements, (3) active N-path filters, and (4) digital filters. The first type of active filters, more commonly known as *active RC filters*, is the subject of discussion of this chapter. The remaining three types are discussed in later chapters of this book.

The commercial availability of inexpensive silicon monolithic operational amplifiers has influenced significantly the design and development of active RC filters in recent years. Good quality active filters are now being produced using these amplifiers and thin- or thick-film R-C components, and thus providing the well-known benefits of hybrid IC fabrication. The useful frequency range of active filters is determined primarily by the active elements being used, and with the present state of the art in the design of IC amplifiers, the range is roughly from dc to 1 MHz. Within this frequency range, the active filters are being used in increasing numbers in communication systems, control systems, and instrumentation. The basic purpose of this chapter is to outline several design methods of such filters which make use of high gain operational amplifiers as the active elements.

8.1 PRELIMINARY CONSIDERATIONS

Active filters can be designed either in *direct form* or in *cascade form* [1]. In the direct form, the prescribed transfer function is realized directly as one filter, whereas, in the cascade form, it is realized as a cascade of lower-order sections. A number of direct realization methods have been reported [1]. We shall describe briefly in this chapter two such design approaches, namely, the simulated-inductance approach and the frequency-dependent-negative-resistance approach. Cascade realization methods normally make use of active second-order filter sections and passive first-order filter sections. The active biquadratic sections are usually amplifier based realizations. A large portion of this chapter will be devoted to this second design approach.

We next review some essential definitions and concepts.

The Operational Amplifier

An ideal operational amplifier is a two-input, single-output voltage-controlled voltage source whose output voltage is proportional to the difference of the input voltages. With reference to the model shown in Figure 8-1,

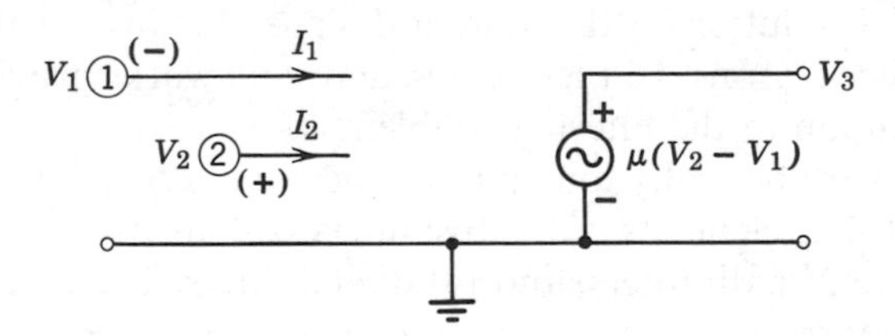

FIGURE 8-1 Controlled-source model of an operational amplifier.

we can define the idealized device as:

$$\begin{aligned} V_3 &= \mu(V_2 - V_1) \\ \mu &\to \infty \\ I_1 &= I_2 = 0. \end{aligned} \tag{8.1}$$

The open-loop gain μ ideally is independent of frequency, and also of temperature and voltage levels at the input terminals. It should be noted that the output voltage is always of opposite polarity to that of terminal 1 (the inverting input terminal) and is of the same polarity as that of the voltage at terminal 2 (the noninverting input terminal). From the definition, it is clear that the input impedances and common-mode impedance are infinite, and the output impedance is zero. An additional requirement implied in the idealized model is the so-called "zero offset" requirement:

$$V_3 \to 0 \quad \text{if} \quad (V_2 - V_1) \to 0. \tag{8.2}$$

The ideal operational amplifier is symbolically represented by Figure 8-2.

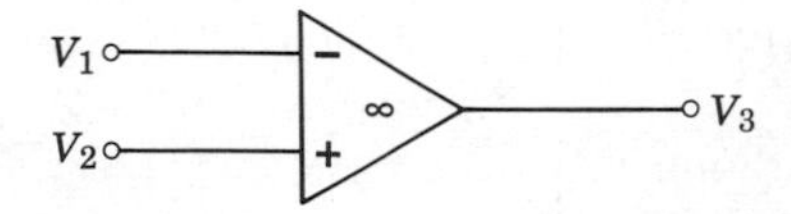

FIGURE 8-2 Schematic representation of an operational amplifier.

A practical operational amplifier is, however, a nonideal device having an open-loop gain which decreases monotonically with frequency from a high value at dc. Its input and output impedances are finite. For linear operation, input and output signal levels are required to stay within certain limits. In addition, the common-mode rejection error and finite unequal common-mode impedance of the two input terminals, and input offset voltage and current, are also sources of nonideal behavior. In practice, these high gain amplifiers are used with negative feedback and can be made to provide satisfactory performance.

A number of practical operational amplifiers have a gain characteristic which rolls off faster than 12 dB/octave and as a result will oscillate when feedback is applied [2]. Consequently, external compensation should be provided for stable operation. Figure 8-3 illustrates the compensating arrangement for Fairchild μA 715.

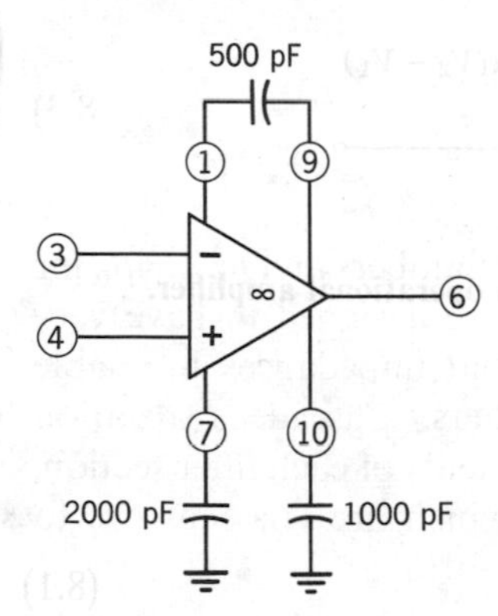

FIGURE 8-3 Compensation arrangement for μA 715 for unity closed-loop gain.

The operational amplifier can be used to realize various kinds of linear active elements [2]. For example, Figure 8-4 shows the realization of a noninverting type voltage amplifier (voltage-controlled voltage source with positive gain) and an inverting type voltage amplifier (voltage controlled voltage source with negative gain).

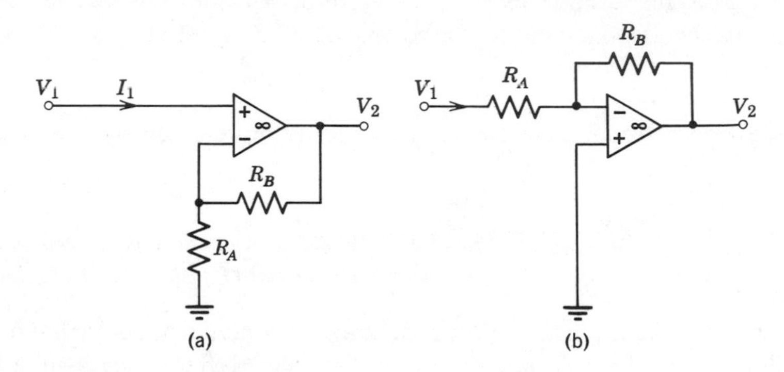

FIGURE 8-4 Realization of finite gain voltage amplifiers: (a) noninverting type and (b) inverting type.

Biquadratic Filter Section

As indicated earlier, a higher-order active filter is commonly designed by cascading second-order and first-order filter sections with proper isolation between stages. The first-order filter section is realizable by passive RC networks. Hence, most design methods concentrate on realizing second-order transfer functions of the form:

$$T(s) = \frac{N(s)}{D(s)} = \frac{\alpha_2 s^2 + \alpha_1 s + \alpha_0}{s^2 + \beta_1 s + \beta_0} \tag{8.3}$$

where, of course, the coefficients α_i and β_i are real numbers and $\beta_1^2 < 4\beta_0$ for complex poles. The second-order filter sections are designed to have very high input impedances and extremely low output impedances to enable cascading without additional isolation amplifiers. Cascade realization allows independent post design corrective adjustments of each filter section, and it almost always leads to filter structures which are less sensitive to parameter changes than those designed in direct form.

Pole Pair Q and Pole Frequency

A convenient description of the filter's performance at frequencies of interest is by means of the following two system parameters:

$$Q = \frac{\sqrt{\beta_0}}{\beta_1} \tag{8.4}$$

$$\omega_n = \sqrt{\beta_0} \tag{8.5}$$

Since the pole pair Q and pole frequency ω_n are directly related to the coefficients of the denominator polynomial, these can be readily expressed in terms of the network parameters. It should be noted that for filter sections exhibiting resonance, Q and ω_n as defined above are approximately equal to the Q-factor of the resonance characteristic and the resonant frequency respectively.

Commonly used criteria in comparing various active filter sections are the range of realizable Q values and the sensitivities of Q and ω_n with respect to the active and passive parameters describing the filter sections. These criteria will also be employed in this chapter.

We now turn our attention to a very useful network transformation technique.

RC : CR Transformation

The RC : CR transformation [3] of an active RC network* is the process of generating another active RC network by replacing each resistor R_i and each capacitor C_j in the original network by a capacitor of value $1/R_i$ F and a resistor of value $1/C_j$ ohms, respectively. The network functions of the original and the derived network are related. For example, if the original network is characterized by an impedance function $Z_A(s)$, the corresponding impedance function $Z_B(s)$ of the derived network is given as:

$$Z_B(s) = \frac{1}{s} Z_A\left(\frac{1}{s}\right). \tag{8.6}$$

On the other hand, if the original network is described by a transfer voltage (current) ratio $t_A(s)$, then the corresponding network function of the derived network is expressed by

$$t_B(s) = t_A\left(\frac{1}{s}\right). \tag{8.7}$$

*It is tacitly assumed here that the active RC network being considered consists of only voltage amplifiers and R-C elements.

The active filters of Figure 8-5(a) and (b) are related by RC : CR transformation. Analysis of each network separately leads to:

$$t_A(s) = \frac{V_{2A}}{V_{1A}} = \frac{6}{s^2 + 2s + 3} \tag{8.8}$$

$$t_B(s) = \frac{V_{2B}}{V_{1B}} = \frac{2s^2}{s^2 + \frac{2}{3}s + \frac{1}{3}}. \tag{8.9}$$

It is seen that the above transfer voltage ratios indeed satisfy relation (8.7).

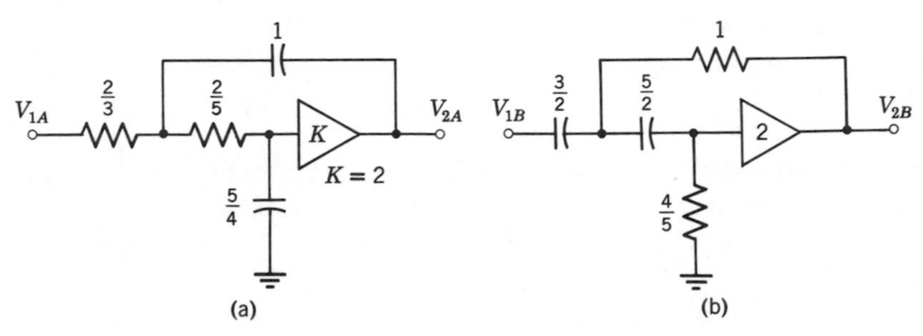

FIGURE 8-5 Illustration of RC-CR transformation: (a) lowpass filter and (b) highpass filter. Element values in ohms and farads.

It should be pointed out here that the transformation process only affects the R-C elements and not the amplifiers. In addition, the transformation process is identical to the commonly used lowpass-to-highpass transformation and can be used to construct a highpass active filter from a lowpass active filter and vice-versa.

8.2 SENSITIVITY

The active and passive network parameters, e.g., the voltage gain of an amplifier, resistance value, etc., vary with changes in temperature, aging, and many other external and internal causes. These parameter variations may cause an active filter to behave significantly differently from its desired idealized performance.

An estimate of the effect of parameter variations on the network characteristic is given by sensitivity figures. There are various types of sensitivities, each type depending on the particular application of the network in question. Since for the most part in this book we are concerned with the design of frequency-selective filters, we review here those definitions which are more pertinent to our objectives.

Q-Sensitivity and Pole Frequency Sensitivity

The sensitivity of the pole pair Q with respect to a network parameter x is defined as:

$$S_x^Q = \frac{d(\ln Q)}{d(\ln x)} = \frac{x}{Q}\frac{dQ}{dx}. \tag{8.10a}$$

Thus, S_x^Q gives the incremental change in Q due to an incremental change in x. For example, if S_x^Q is equal to 0.5, then it implies that a 2% change in x will cause a 1% change in Q.

The sensitivity of the pole frequency ω_n is similarly defined as:

$$S_x^{\omega_n} = \frac{x}{\omega_n}\frac{d\omega_n}{dx}. \tag{8.10b}$$

Example 8-1. Analyze the lowpass passive filter of Figure 8-6 [4].

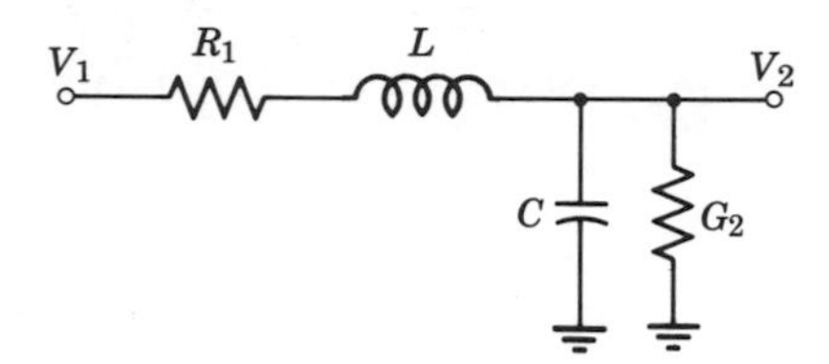

FIGURE 8-6 A passive RLC lowpass filter.

Solution: By elementary calculation,

$$\frac{V_2}{V_1} = \frac{\dfrac{1}{LC}}{s^2 + \left(\dfrac{G_2}{C} + \dfrac{R_1}{L}\right)s + \left(\dfrac{R_1 G_2 + 1}{LC}\right)}. \tag{8.11}$$

The pole pair Q of the filter is

$$Q = \frac{\sqrt{LC(R_1 G_2 + 1)}}{LG_2 + R_1 C}.$$

The various Q-sensitivities are easily computed as:

$$S_L^Q = -S_C^Q = \frac{R_1 C - LG_2}{2(LG_2 + R_1 C)} < \frac{1}{2}$$

$$|S_{R_1}^Q| = \frac{1}{2}\left|\frac{R_1 G_2}{R_1 G_2 + 1} - \frac{2R_1 C}{LG_2 + R_1 C}\right| < 1$$

$$|S_{G_2}^Q| = \frac{1}{2}\left|\frac{R_1 G_2}{R_1 G_2 + 1} - \frac{2LG_2}{LG_2 + R_1 C}\right| < 1. \tag{8.12}$$

The pole frequency ω_n is

$$\omega_n = \sqrt{\frac{R_1 G_2 + 1}{LC}}.$$

The corresponding sensitivities are

$$S_L^{\omega_n} = S_C^{\omega_n} = -\frac{1}{2}$$

$$S_{R_1}^{\omega_n} = S_{G_2}^{\omega_n} = \frac{R_1 G_2}{2(R_1 G_2 + 1)} < \frac{1}{2}. \tag{8.13}$$

Note from (8.12) and (8.13) that irrespective of the actual element values, both the Q- and ω_n-sensitivities are always very small.

Root Sensitivity

If $s = p_i$ is a root of the polynomial $D(s)$, then the root sensitivity of $D(s)$ with respect to a parameter x is defined as:

$$\mathcal{S}_x^{p_i} = \frac{dp_i}{dx/x} \tag{8.14}$$

The root sensitivity in general will be a complex number, whereas the Q-sensitivity is a real number. It follows readily from the definition that

$$\mathcal{S}_x^{p_i^*} = (\mathcal{S}_x^{p_i})^*. \tag{8.15}$$

If we write $p_i = \sigma_i + j\omega_i$, then we can express

$$\mathcal{S}_x^{p_i} = \mathcal{S}_x^{\sigma_i} + j\mathcal{S}_x^{\omega_i} \tag{8.16}$$

assuming the parameter x is real. In (8.16), $\mathcal{S}_x^{\sigma_i} = d\sigma_i/d(\ln x)$ and $\mathcal{S}_x^{\omega_i} = d\omega_i/d(\ln x)$.

Expression for calculating the root sensitivity is based on the fact that the pole polynomial $D(s)$ can be written as [5]:

$$D(s) = D_1(s) + xD_2(s) \tag{8.17}$$

where $D_1(s)$ and $D_2(s)$ are polynomials independent of x. If $s = p_i$ is a simple zero of $D(s)$, then it can be shown [6] that

$$\mathcal{S}_x^{p_i} = \left.\frac{(s - p_i)D_1(s)}{D(s)}\right|_{s=p_i} \tag{8.18a}$$

$$= \left.\frac{-x(s - p_i)D_2(s)}{D(s)}\right|_{s=p_i}. \tag{8.18b}$$

Example 8-2. Consider the network of Figure 8-5(a). We wish to calculate the pole sensitivities with respect to the gain K of the amplifier, whose nominal value is 2.

Solution: The transfer voltage ratio of the filter is

$$t_A(s) = \frac{3K}{s^2 + (5 - \frac{3}{2}K)s + 3}.$$

The pole polynomial for nominal values of the parameters is

$$D(s) = (s^2 + 5s + 3) - K(\tfrac{3}{2}s) = s^2 + 2s + 3.$$

Hence $D_1(s) = (s^2 + 5s + 3)$ and $D_2(s) = -\frac{3}{2}s$. The poles are at $p_1 = -1 + j\sqrt{2}$. and $p_1^* = -1 - j\sqrt{2}$. Using (8.18b) we obtain

$$\mathscr{S}_K^{p_1} = \left.\frac{2(s + 1 - j\sqrt{2})(\frac{3}{2}s)}{s^2 + 2s + 3}\right|_{s=-1+j\sqrt{2}} = 1.5 + j\frac{1.5}{1.414}.$$

It can be shown that the angle of the tangent to the root locus of $D(s, k)$ at $s = p_i$ is equal to the argument of $\mathscr{S}_x^{p_i}$ [6]. Hence, the aim of the designer should be not only to minimize $|\mathscr{S}_x^{p_i}|$ but also to bring arg $\mathscr{S}_x^{p_i}$ close to $\pm 90°$ for better stability margin.

Coefficient Sensitivity

If we write the polynomial $D(s)$ as

$$D(s) = \sum_{i=0}^{n} \beta_i s^i, \tag{8.19}$$

then the sensitivity of the coefficient β_i with respect to a network parameter x is defined as

$$\hat{S}_x^{\beta_i} \triangleq \frac{d(\ln \beta_i)}{d(\ln x)} = \frac{x}{\beta_i}\frac{d\beta_i}{dx} \tag{8.20}$$

Unlike the Q-sensitivity and the root sensitivity, the coefficient sensitivity is not a physically measurable (or meaningful) quantity. However, it can be used to calculate other sensitivities as illustrated next.

Relation Between Sensitivities

From (8.3), the denominator polynomial of a second-order filter is

$$D(s) = s^2 + \beta_1 s + \beta_0, \tag{8.21}$$

which alternately can be written as

$$D(s) = (s + \sigma_0 + j\omega_0)(s + \sigma_0 - j\omega_0) = s^2 + 2\sigma_0 s + (\sigma_0^2 + \omega_0^2). \quad (8.22)$$

From (8.21) and (8.22) one readily obtains [7]

$$\mathscr{S}_x^{\sigma_0} = \frac{\beta_1}{2} \hat{S}_x^{\beta_1} \quad (8.23)$$

$$\mathscr{S}_x^{\omega_0} = \frac{1}{\sqrt{4\beta_0 - \beta_1^2}} \left[\beta_0 \hat{S}_x^{\beta_0} - \frac{\beta_1^2}{2} \hat{S}_x^{\beta_1} \right], \quad (8.24)$$

where of course $p_0 = \sigma_0 + j\omega_0$.

Similarly, the following relations can be derived from (8.4) and (8.5):

$$S_x^Q = \tfrac{1}{2}\hat{S}_x^{\beta_0} - \hat{S}_x^{\beta_1} \quad (8.25)$$

$$S_x^{\omega_n} = \tfrac{1}{2}\hat{S}_x^{\beta_0}. \quad (8.26)$$

The relations between pole sensitivities and S_x^Q and $S_x^{\omega_n}$ are also of interest. These are

$$\mathscr{S}_x^{\sigma_0} = \frac{\omega_n}{2Q} (S_x^{\omega_n} - S_x^Q) \quad (8.27)$$

$$\mathscr{S}_x^{\omega_0} = \frac{\omega_n}{2Q} \left[\sqrt{4Q^2 - 1}\, S_x^{\omega_n} + \frac{S_x^Q}{\sqrt{4Q^2 - 1}} \right]. \quad (8.28)$$

A number of second-order active RC filters are essentially third-order systems reduced by pole-zero cancellation. It is thus desirable to derive analytical expressions relating the Q-sensitivities and pole sensitivities to the readily calculable coefficient sensitivities for the third-order filters. It is apparent that the denominator polynomial of such filters can be expressed in either of the following forms:

$$D(s) = (s + \gamma)\left(s^2 + \frac{\omega_n}{Q} s + \omega_n^2 \right) = s^3 + \beta_2 s^2 + \beta_1 s + \beta_0$$

$$= (s + \gamma)(s^2 + 2\sigma_0 s + \sigma_0^2 + \omega_0^2) \quad (8.29)$$

where $s = -\sigma_0 \pm j\omega_0$ is the complex pole pair. It can be easily shown that the desired relations are [7]:

$$S_x^{\omega_n} = \frac{(Q\gamma - \omega_n)\beta_0 \hat{S}_x^{\beta_0} + Q\omega_n^2\beta_1 \hat{S}_x^{\beta_1} - Q\omega_n^2\gamma\beta_2 \hat{S}_x^{\beta_2}}{2\omega_n^2[Q(\omega_n^2 + \gamma^2) - \omega_n\gamma]}$$

$$S_x^Q = \frac{\beta_0(2\omega_n Q^2 + \gamma Q - \omega_n)\hat{S}_x^{\beta_0} + \beta_1(\omega_n^2 Q - 2\omega_n\gamma Q^2)\hat{S}_x^{\beta_1} + \beta_2\omega_n^2(Q\gamma - 2\omega_n Q^2)\hat{S}_x^{\beta_2}}{2\omega_n^2[(\omega_n^2 + \gamma^2)Q - \omega_n\gamma]}$$

$$S_x^\gamma = \frac{Q\left[\dfrac{\beta_0}{\gamma}\hat{S}_x^{\beta_0} - \beta_1 \hat{S}_x^{\beta_1} + \beta_2\gamma\hat{S}_x^{\beta_2}\right]}{(\omega_n^2 + \gamma^2)Q - \omega_n\gamma}. \tag{8.30}$$

Frequency Response Sensitivity

For bandpass filters, the sensitivity of the magnitude characteristic is directly related to the Q- and ω_n-sensitivities. As stated earlier, for high Q values the pole pair Q and the pole frequency ω_n are almost equal to the quality factor Q_0 of the resonance characteristic and the resonant frequency ω_0. Consider a second-order bandpass transfer function

$$T(s) = \frac{s}{s^2 + \dfrac{\omega_n}{Q}s + \omega_n^2}. \tag{8.31}$$

The magnitude function is then

$$|T(j\omega)|^2 = \frac{\omega^2}{(\omega_n^2 - \omega^2)^2 + \left(\dfrac{\omega_n}{Q}\cdot\omega\right)^2}. \tag{8.32}$$

The sensitivity of the magnitude is defined as:

$$S_x^{|T(j\omega)|} = \frac{x}{|T(j\omega)|}\cdot\frac{d|T(j\omega)|}{dx}, \tag{8.33}$$

which is seen to be a function of frequency. It can be shown that at $\omega = \omega_n$

$$S_x^{|T(j\omega)|} = S_x^Q - S_x^{\omega_n}, \tag{8.34}$$

and at the 3 dB cut-off frequencies

$$S_x^{|T(j\omega)|} \cong -Q \cdot S_x^{\omega_n} + \tfrac{1}{2}(S_x^Q - S_x^{\omega_n}). \tag{8.35}$$

Equations (8.34) and (8.35) indicate that both S_x^Q and $S_x^{\omega_n}$ are important in determining the variation of the frequency response around resonance. As we shall show later, it is possible to design active filters with very low values for these sensitivities. But as (8.35) points out, around the 3 dB points the deviation of the magnitude characteristic may be quite significant for very high Q filters.

With these preliminaries, we next consider several typical active RC filter synthesis procedures. The following two sections describe a number of biquadratic filter sections employing finite gain voltage amplifiers as the active elements. Sections 8.5 and 8.6 outline two methods of direct realization of a higher-order transfer function.

8.3 SINGLE-LOOP FEEDBACK APPROACH

The single-loop feedback configurations are basically single amplifier structures which employ either positive feedback or negative feedback to realize complex poles. The amplifier output terminal is also chosen to be the filter output, thus providing a low-impedance output. Filter input impedance can be made high by appropriate impedance scaling. As a result these filter sections can be cascaded without additional isolation amplifiers. For convenience, these filter sections realize second-order transfer functions. (Note that first order transfer functions are realizable with passive R-C elements only.)

These filter structures make use of either finite-gain or infinite-gain voltage amplifiers, either of which can be realized by monolithic operational amplifiers in practice. A noninverting type (positive-gain) voltage-amplifier realization is shown in Figure 8-4(a). Analysis yields for this circuit

$$V_2 = \left(1 + \frac{R_B}{R_A}\right) V_1$$

$$I_1 = 0. \tag{8.36}$$

Effects of finite input and output impedances of the operational amplifier are very small provided the open-loop gain of the amplifier is reasonably large (10^4 or more). The circuit of Figure 8-4(b) realizes an inverting type (negative-gain) voltage amplifier characterized by the following input-output relations:

$$V_2 = -\frac{R_B}{R_A} V_1$$

$$I_1 = \frac{1}{R_A} V_1 . \tag{8.37}$$

It is seen from the above that the input impedance of this inverting type voltage amplifier is essentially R_A, and hence any filter configuration employing this negative gain amplifier must have a resistance at the input of the amplifier to account for this input resistance.

We now present several design techniques for single-loop feedback type active filters. For convenience, the filters will be designed to have a pole frequency ω_n normalized to unity. Any other value can be easily obtained by appropriate frequency scaling.

Lowpass and Highpass Filter Sections

A lowpass second-order filter section is characterized by a transfer function of the form:

$$t_v = \frac{H}{s^2 + \frac{1}{Q}s + 1}. \tag{8.38}$$

An example of this type of filter is the one of Figure 8-7 [8], which has a voltage-transfer ratio given by:

$$\frac{V_2}{V_1} = \frac{\frac{KG_1G_2}{C_1C_2}}{s^2 + \left[\frac{(G_1+G_2)}{C_1} + \frac{(1-K)G_2}{C_2}\right]s + \frac{G_1G_2}{C_1C_2}}. \tag{8.39}$$

It follows from (8.4) that the pole pair Q of this filter is given as:

$$Q = \frac{\sqrt{C_1C_2G_1G_2}}{C_2(G_1+G_2) + (1-K)C_1G_2}. \tag{8.40}$$

Design equations for $\omega_n = 1$ can be obtained by comparing (8.39) with (8.38). Thus

$$\frac{G_1+G_2}{C_1} + \frac{(1-K)G_2}{C_2} = \frac{1}{Q}$$

$$\frac{G_1G_2}{C_1C_2} = 1. \tag{8.41}$$

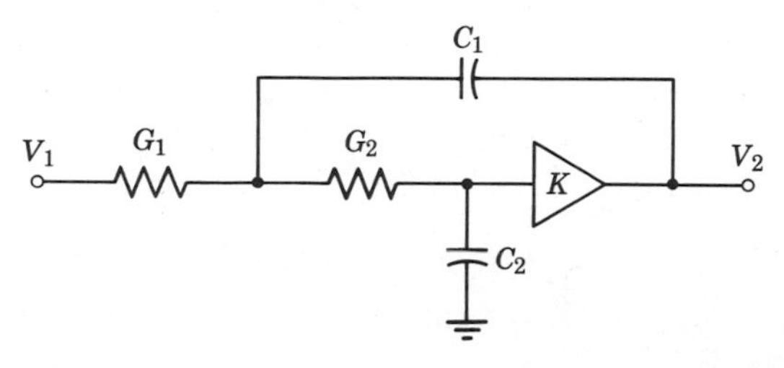

FIGURE 8-7 A positive feedback lowpass filter section.

Note that the number of unknown quantities is five, whereas the number of equations is two. Hence there is no unique solution. One approach to solving (8.41) would be to preselect some element values and then solve for the rest. One such solution is

$$G_1 = G_2 = C_1 = C_2 = 1$$
$$K = 3 - \frac{1}{Q}. \tag{8.42}$$

To calculate the sensitivities, we observe that

$$\beta_1 = \frac{G_1 + G_2}{C_1} + \frac{(1-K)G_2}{C_2}$$
$$\beta_0 = \frac{G_1 G_2}{C_1 C_2}. \tag{8.43}$$

Consider the effect of the amplifier gain K first. Then nominally, $\beta_1 = 3 - K$ and $\beta_0 = 1$. Hence

$$\hat{S}_K^{\beta_1} = \frac{-K}{3-K} = 1 - 3Q$$
$$\hat{S}_K^{\beta_0} = 0. \tag{8.44}$$

Substituting the above in (8.23), (8.24), and (8.25), we obtain

$$\mathscr{S}_K^{\sigma_0} = -\frac{1}{2}\left(3 - \frac{1}{Q}\right)$$
$$\mathscr{S}_K^{\omega_0} = \frac{3 - \dfrac{1}{Q}}{2\sqrt{4Q^2 - 1}}$$
$$S_K^Q = 3Q - 1. \tag{8.45}$$

Consider a Q-value of ten. Then $\mathscr{S}_K^{\sigma_0} = -1.45$ and $\mathscr{S}_K^{\omega_0} = 0.0725$. This indicates that the movement of the pole position is mainly parallel to the real axis. In addition, $\mathscr{S}_K^{\sigma_0}$ being negative indicates that an increase in gain brings the pole closer to the $j\omega$-axis and hence closer to oscillation. This high sensitivity feature is clearly demonstrated by the high Q-sensitivity of the circuit. Other Q-sensitivities can similarly be derived and are as follows:

$$S_{G_1}^Q = -S_{G_2}^Q = -Q + \tfrac{1}{2}$$
$$S_{C_1}^Q = -S_{C_2}^Q = 2Q - \tfrac{1}{2}. \tag{8.46}$$

Note that these are also very high. As a matter of fact, high Q-sensitivities are always the feature of all single-loop positive-feedback active filters [4]. However, this type of filter requires very low gain values, and as a result it can be used over a wide frequency range.

Unity gain amplifiers (voltage followers) have very high gain stability, making them attractive in filter design. A unity gain amplifier can be used in the lowpass filter of Figure 8-7 at the expense of element value spread. If we set $K = 1$ in (8.39) and then solve for element values, we get as one possible solution:

$$G_1 = G_2 = 1$$

$$C_1 = 2Q \quad , \quad C_2 = \frac{1}{2Q}. \tag{8.47}$$

The Q-sensitivities for this case are [9]:

$$S^Q_{G_1} = S^Q_{G_2} = 0$$

$$S^Q_{C_1} = -S^Q_{C_2} = \tfrac{1}{2}$$

$$S^Q_{\mu} \cong \frac{2Q^2}{\mu},$$

where μ is the open-loop gain of the operational amplifier realizing the unity gain amplifier. The passive parameter sensitivities are very low. The active parameter sensitivity is low provided $Q^2 \ll \mu$. This restricts the usefulness of this structure to low Q filter realizations.

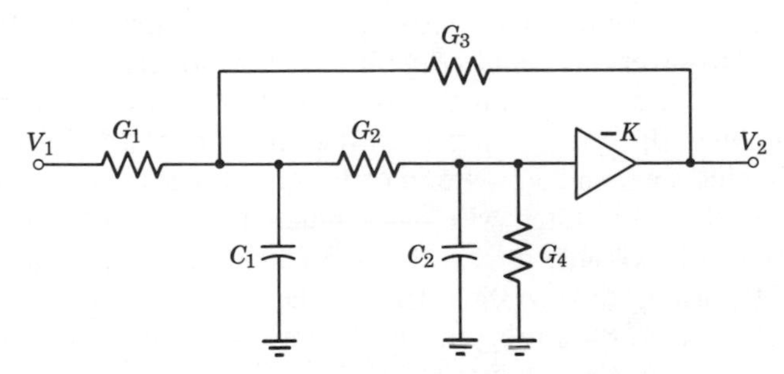

FIGURE 8-8 A negative feedback lowpass filter section.

We now consider a lowpass filter employing negative feedback. The circuit is shown in Figure 8-8 [8, 10]. Analysis leads to:

$$\frac{V_2}{V_1} = \frac{\dfrac{-KG_1G_2}{C_1C_2}}{s^2 + \left[\dfrac{G_1 + G_2 + G_3}{C_1} + \dfrac{G_2 + G_4}{C_2}\right]s + \dfrac{G_2(G_1 + G_3 + KG_3) + G_4(G_1 + G_2 + G_3)}{C_1C_2}}. \tag{8.48}$$

Comparing the above with (8.38), we can obtain the required design equations. One solution of these is:

$$\begin{aligned} G_1 = G_2 = G_3 = G_4 &= 1 \\ C_1 = C_2 &= 5Q \\ K &= 25Q^2 - 5. \end{aligned} \tag{8.49}$$

The pole pair Q of this filter is

$$Q = \frac{\sqrt{C_1 C_2[G_2(G_1 + G_3 + KG_3) + G_4(G_1 + G_2 + G_3)]}}{C_2(G_1 + G_2 + G_3) + C_1(G_2 + G_4)}. \tag{8.50}$$

The corresponding Q-sensitivity figures are:

$$S_K^Q = \frac{K}{2(5 + K)} < \frac{1}{2}$$

$$S_{C_1}^Q = -S_{C_2}^Q = \frac{1}{10}$$

$$S_{G_1}^Q = \frac{-2K}{10(5 + K)} \cong -\frac{1}{5} \qquad S_{G_2}^Q = \frac{K - 5}{10(5 + K)} \cong \frac{1}{10}$$

$$S_{G_3}^Q = \frac{3K}{10(5 + K)} \cong \frac{3}{10} \qquad S_{G_4}^Q = \frac{5 - 2K}{10(5 + K)} \cong -\frac{1}{5}. \tag{8.51}$$

Examination of these results indicates that the filter has very low Q-sensitivities. However, the required gain even for moderate Q values is very high. Hence, if the circuit of Figure 8-4(b) is used to realize the inverting-type voltage amplifier, the open-loop gain of the operational amplifier should be considerably higher than $25Q^2 - 5$ in order to stabilize the closed loop gain. This implies that this filter will have much lower bandwidth than the previous one. It is possible to realize a Q of 50 at very low frequencies, less than 100 Hz, using this configuration. These low-sensitivity and high-amplifier-gain features are inherent in all single-loop-feedback active filters employing negative-gain amplifiers [4].

Highpass filters are realized by making an RC : CR transformation on the lowpass filter section. The following example illustrates the procedure.

Example 8-3. Let us realize a fourth-order highpass filter having a maximally flat magnitude characteristic in the passband. The transfer function is given as:

$$t(s) = \frac{Hs^4}{(s^2 + 0.7653s + 1)(s^2 + 1.8477s + 1)}.$$

Solution: We first realize the corresponding lowpass version:

$$t'(s) = t\left(\frac{1}{s}\right) = \frac{H}{(s^2 + 0.7653s + 1)(s^2 + 1.8477s + 1)},$$

as a cascade of two second-order lowpass structures of the form of Figure 8-7. Figure 8-9(a) realizes $t_1(s) = H/(s^2 + 0.7653s + 1)$ and Figure 8-9(b) realizes $t_2(s) = H_2/(s^2 + 1.8477s + 1)$. Note that in the realization of these two transfer functions we have used the design equations given by (8.47). Connecting these two networks in cascade and using the RC : CR transformation described earlier, we obtain the desired highpass filter realization shown in Figure 8-9(c) where the operational amplifier realizations of unity gain amplifiers have been used.

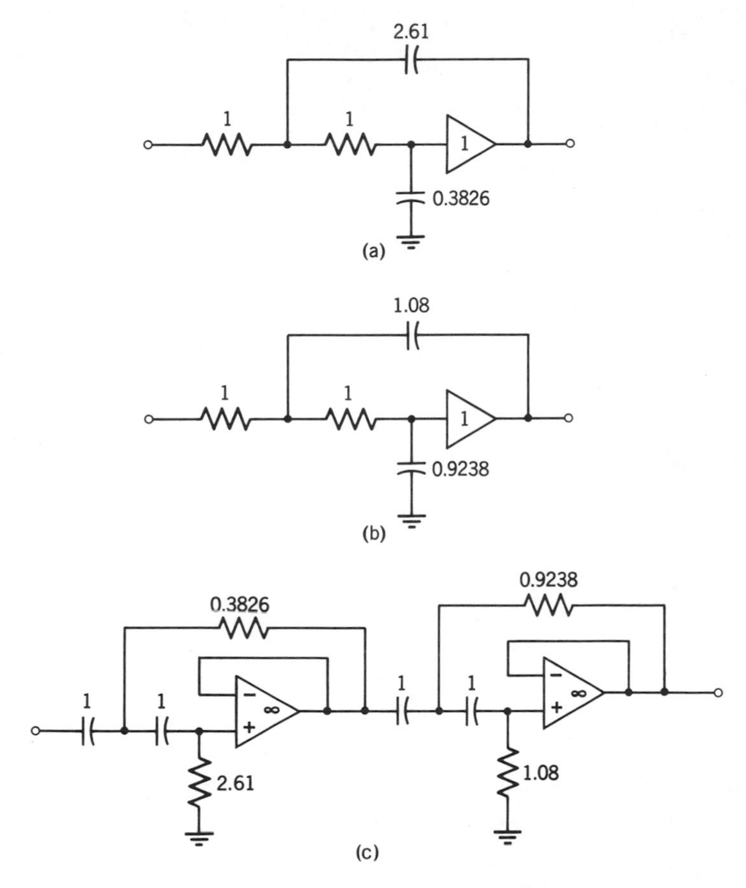

FIGURE 8-9 Steps in the realization of a 4th-order highpass Butterworth filter. Element values in ohms and farads.

Bandpass Filter Section

A second-order bandpass filter is described by a transfer function of the form:

$$t(s) = \frac{Hs}{s^2 + \frac{1}{Q}s + 1}. \tag{8.52}$$

This type of transfer function can also be realized by both negative feedback and positive feedback structures. An example of the former is the one shown in Figure 8-10 [8]. Its transfer-voltage ratio is given as:

$$\frac{V_2}{V_1} = \frac{\frac{-KG_1 s}{(1+K)C_1}}{s^2 + \frac{(C_2G_1 + C_1G_2 + C_2G_2)}{(1+K)C_1C_2}s + \frac{G_1G_2}{(1+K)C_1C_2}}. \tag{8.53}$$

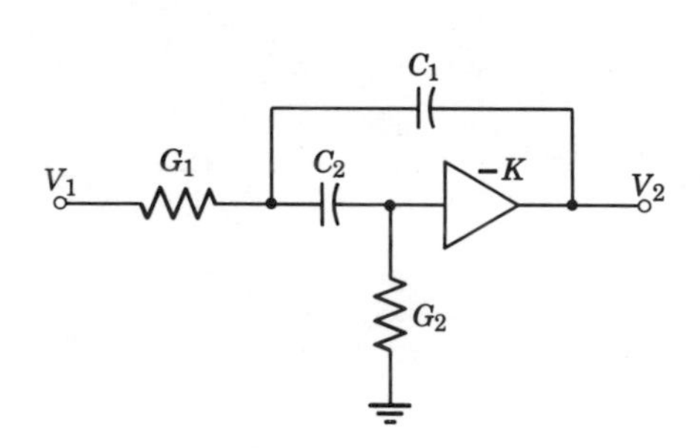

FIGURE 8-10 A negative feedback bandpass filter section.

As before, the design equations are not unique. One set of solutions [11] is as follows:

$$G_1 = G_2 = 1$$

$$C_1 = C_2 = \frac{1}{3Q}$$

$$K = 9Q^2 - 1. \tag{8.54}$$

The corresponding Q-sensitivities are

$$S_K^Q = \frac{K}{2(1+K)} < \frac{1}{2}$$

$$S_{G_2}^Q = S_{G_1}^Q = \frac{1}{6}$$

$$S_{C_1}^Q = -S_{C_2}^Q = \frac{1}{6}. \tag{8.55}$$

The high-gain value can be reduced somewhat by allowing some element spread. One such solution [12] is:

$$\frac{C_1}{C_2} = \frac{G_1}{G_2} = 100$$

$$K \cong 4Q^2 - 1 \tag{8.56}$$

and the resulting Q-sensitivities are:

$$S^Q_{C_1} = -S^Q_{C_2} = S^Q_{G_1} = -S^Q_{G_2} \cong 0.0025. \tag{8.57}$$

The reduction in sensitivities is almost 1,000 to 15!

A bandpass filter section using a positive-gain amplifier [13] is shown in Figure 8-11. Analyzing we obtain

$$\frac{V_2}{V_1} = \frac{K\dfrac{G_1}{C_1}s}{s^2 + \left[\dfrac{G_1 + G_2 + G_3 - KG_2}{C_1} + \dfrac{G_3}{C_2}\right]s + \dfrac{G_3(G_1 + G_2)}{C_1 C_2}}. \tag{8.58}$$

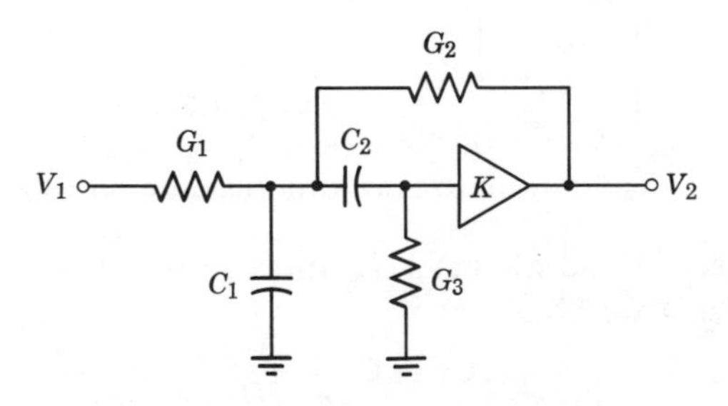

FIGURE 8-11 A positive feedback bandpass filter section.

One set of solutions for $\omega_n = 1$ would be as follows:

$$G_1 = G_2 = 1 \quad , \quad G_3 = \frac{1}{2}$$

$$C_1 = C_2 = 1$$

$$K = 3 - \frac{1}{Q}. \tag{8.59}$$

The pertinent Q-sensitivities are:

$$S_K^Q = 3Q - 1$$

$$S_{C_1}^Q = -S_{C_2}^Q = -\frac{Q}{2} + \frac{1}{2}$$

$$S_{G_1}^Q = -Q + \frac{1}{4}, \qquad S_{G_2}^Q = 2Q - \frac{3}{4}, \qquad S_{G_3}^Q = -Q + \frac{1}{2}. \tag{8.60}$$

Observe again the very high sensitivity feature of single-amplifier structures employing positive feedback.

We have noted that the negative-feedback type active filters discussed so far have very low sensitivity figures. However, the required amplifier gain is very high, and this limits the operation of this type of filter to low frequencies. It is possible to reduce the required amplifier gain at the expense of slightly higher sensitivities by appropriately choosing the transmission zeros of the feedback path (the so-called *phantom zeros*). One such network is shown in Figure 8-12 [7]. This structure is a modified version of a network

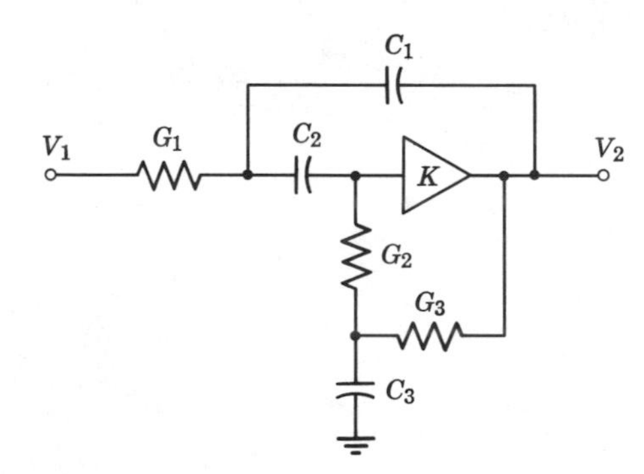

FIGURE 8-12 A positive feedback bandpass biquadratic section obtained by pole-zero cancellation.

proposed by Kerwin and Shaffer [14] and can realize a bandpass transfer function. Analyzing we obtain

$$\frac{V_2}{V_1} = \frac{\dfrac{KG_1}{C_1 C_3(1-K)}(C_3 s + G_2 + G_3)s}{s^3 + \beta_2 s^2 + \beta_1 s + \beta_0} \tag{8.61}$$

where

$$\beta_2 = \frac{G_2 + G_3}{C_3} + \frac{(C_1 + C_2)G_2 + C_2 G_1}{C_1 C_2(1-K)}$$

$$\beta_1 = \frac{(C_1 + C_2)G_2 G_3(1-K) + C_2 G_1(G_2 + G_3) + C_3 G_1 G_2}{C_1 C_2 C_3(1-K)}$$

$$\beta_0 = \frac{G_1 G_2 G_3}{C_1 C_2 C_3}. \tag{8.62}$$

One set of design equations could be

$$C_1 = C_2 = 1$$
$$G_2 = G_3 = 1$$
$$G_1 = C_3 = 2$$
$$K = -(4Q - 1). \tag{8.63}$$

Substituting this in (8.61) and (8.62) we obtain

$$\frac{V_2}{V_1} = \frac{-\dfrac{4Q-1}{2Q}s(s+1)}{s^3 + \left(1 + \dfrac{1}{Q}\right)s^2 + \left(1 + \dfrac{1}{Q}\right)s + 1} = \frac{-\dfrac{4Q-1}{2Q}s}{s^2 + \dfrac{1}{Q}s + 1}. \tag{8.64}$$

It follows that $\omega_n = \gamma = 1$. Also $\beta_2 = \beta_1 = 1 + 1/Q$ and $\beta_0 = 1$. This simplifies the sensitivity expressions of (8.30) as:

$$S_x^Q = \frac{(Q+1)(\hat{S}_x^{\beta_0} - \hat{S}_x^{\beta_1} - \hat{S}_x^{\beta_2})}{2} \tag{8.65}$$

and similarly for $S_x^{\omega_n}$ and S_x^{γ}. From above and (8.62) the various Q-sensitivities are easily calculated as:

$$S_K^Q = 1 - \frac{1}{4Q}$$

$$S_{G_1}^Q = \frac{1}{4}(2Q-1), \qquad S_{G_2}^Q = -\frac{1}{8}(2Q+1), \qquad S_{G_3}^Q = -\frac{1}{8}(2Q-3)$$

$$S_{C_1}^Q = -\frac{1}{8}(2Q-3), \qquad S_{C_2}^Q = -\frac{1}{8}(2Q+1), \qquad S_{C_3}^Q = \frac{1}{4}(2Q-1). \tag{8.66}$$

Note from (8.63), the gain value is now proportional to Q rather than Q^2. But the Q-sensitivities are now correspondingly higher than those values for the circuit of Figure 8-10 for example (see (8.55)).

Filter Section with $j\omega$-Axis Transmission Zeros

Another very useful second-order transfer function is the one with a pair of $j\omega$-axis zeros:

$$t_v = \frac{H(s^2 + \alpha_0)}{s^2 + \dfrac{1}{Q}s + 1} \tag{8.67}$$

Here three cases have to be examined separately: (1) $\alpha_0 < 1$, (2) $\alpha_0 > 1$, and (3) $\alpha_0 = 1$. One proposed circuit for the case $\alpha_0 < 1$ is the one shown in Figure 8-13 [15] which has a voltage-transfer ratio given by:

$$\frac{V_2}{V_1} = \frac{K(s^2 + a^2)}{s^2 + 2a(G + 2 - K)s + (1 + 2G)a^2}. \tag{8.68}$$

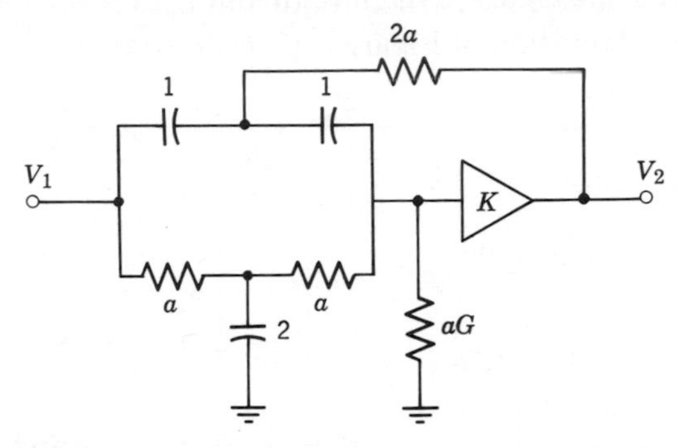

FIGURE 8-13 A filter section having $j\omega$-axis transmission zeros. Element values in ohms and farads.

Comparing (8.68) with (8.67) we identify

$$\begin{aligned} a^2 &= \alpha_0 \\ 2a(G + 2 - K) &= \frac{1}{Q} \\ (1 + 2G)a^2 &= 1 \end{aligned} \tag{8.69}$$

from which we obtain one design equation as:

$$\begin{aligned} a &= \sqrt{\alpha_0} \\ G &= \frac{1}{2\alpha_0} - \frac{1}{2} \\ K &= 2 - \left[\frac{1}{2aQ} - G\right]. \end{aligned} \tag{8.70}$$

Observe that the pole pair Q is given by

$$Q = \frac{\sqrt{1 + 2G}}{2(G + 2 - K)}, \tag{8.71}$$

and the Q-sensitivity with respect to K is thus

$$S_K^Q = \left(\frac{3\alpha_0 + 1}{2\alpha_0}\right)Q - 1. \tag{8.72}$$

The second case, $\alpha_0 > 1$, is related to the previous case by RC : CR transformation and will thus be left as an exercise. We now consider the last case, $\alpha_0 = 1$. A suitable filter structure for this case is obtained by slightly modifying the configuration of Figure 8-13 as illustrated in Figure 8-14 [16].

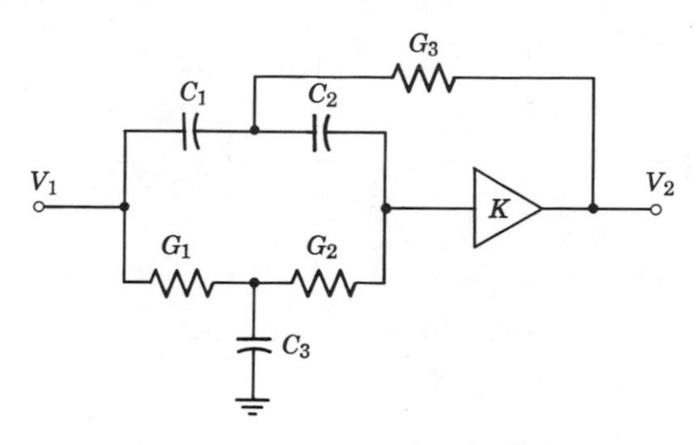

FIGURE 8-14 Another realization of a filter section with $j\omega$-axis transmission zeros.

Here we have

$$\frac{V_2}{V_1} = \frac{K(C_1 C_2 s^2 + G_1 G_2)}{C_1 C_2 s^2 + [C_2(G_1 + G_2)(1 - K) + (C_1 + C_2)G_2]s + G_1 G_2}, \tag{8.73}$$

provided $C_3 = C_1 + C_2$ and $G_3 = G_1 + G_2$. Comparing (8.73) with (8.67), one set of design equation is obtained as

$$K = 1$$

$$C_1 = C_2 = 1, \qquad C_3 = 2$$

$$G_1 = 2Q, \qquad G_2 = \frac{1}{2Q}$$

$$G_3 = 2Q + \frac{1}{2Q}. \tag{8.74}$$

The Q-sensitivity with respect to K can be readily calculated from (8.73) as:

$$S_K^Q = 2Q^2 + \tfrac{1}{2}. \tag{8.75}$$

Even though the sensitivity is numerically very high, unity gain amplifiers have very good gain stability and hence are more practical to use.

Q-Invariant Filter Sections

It is possible to design single-amplifier filter sections with either negative or positive feedback exhibiting zero Q-sensitivity with respect to the amplifier gain [17]. Two such filter structures are shown in Figure 8-15.

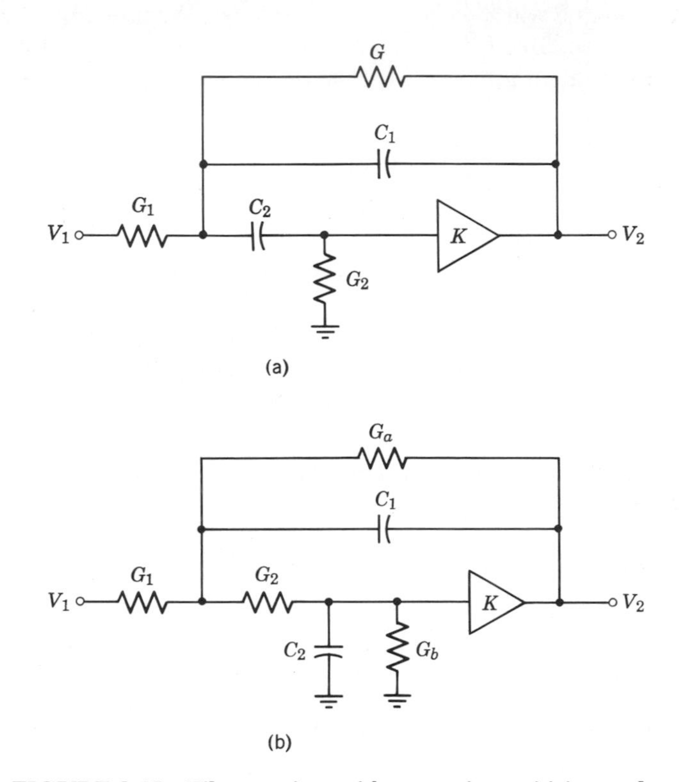

FIGURE 8-15 Filter sections with zero gain sensitivity products.

The circuit of Figure 8-15(a) employs negative feedback and realizes a bandpass transfer function. It can be analyzed to yield the following voltage-transfer ratio:

$$\frac{V_2}{V_1} = \frac{G_1 C_2 s}{C_1 C_2(1 - K)s^2 + [C_2(G_1 + G + G_2 - KG) + C_1 G_2] + (G_1 + G)G_2}. \tag{8.76}$$

Comparing (8.76) with (8.52) and equating like coefficients, the values of the passive components and the amplifier gain can be obtained. There are many possible sets of solutions of these design equations. One interesting set for $\omega_n = 1$ is as follows:

$$C_1 = C_2 = \frac{1}{6Q}$$

$$G_1 = \frac{36Q^2 - 4}{36Q^2 - 1}; \quad G_2 = 1; \quad G = \frac{3}{36Q^2 - 1}$$

$$K = -(36Q^2 - 1).$$

The corresponding sensitivities are:

$$S_K^Q = 0; \quad S_K^{\omega_n} = -\frac{1}{2}\left(1 - \frac{1}{36Q^2}\right)$$

$$S_{C_1}^Q = -S_{C_2}^Q = S_{G_1}^Q = \frac{1}{3}$$

$$S_G^Q = -\frac{1}{2}; \quad S_{G_2}^Q = \frac{1}{6}$$

$$S_{G_1}^{\omega_n} = S_{G_2}^{\omega_n} = -S_{C_1}^{\omega_n} = -S_{C_2}^{\omega_n} = \frac{1}{2}; \quad S_G^{\omega_n} = 0.$$

A positive-feedback lowpass filter section exhibiting zero Q-sensitivity with respect to the amplifier gain is shown in Figure 8-15(b). For the following set of element values:

$$G_1 = G_2 = 1 \qquad C_1 = C_2 = \frac{1}{\sqrt{4Q^2 - 1}}$$

$$G_a = G_b = \frac{2Q}{\sqrt{4Q^2 - 1}} \qquad K = 3 + \frac{\sqrt{4Q^2 - 1}}{Q} \tag{8.77}$$

the filter has zero Q-sensitivity with respect to K. In (8.77), ω_n is assumed to have unity value. The pertinent sensitivities are

$$S_K^Q = 0 \qquad S_K^{\omega_n} = -2\beta(3Q + \beta)$$

$$S_{G_1}^Q = S_{G_2}^Q = -S_{G_a}^Q = -S_{G_b}^Q = \frac{\beta^2}{2}$$

$$S_{C_1}^Q = -S_{C_2}^Q = \frac{\beta}{2}(4Q + \beta)$$

$$S_{G_1}^{\omega_n} = -\frac{\beta}{2}(2Q + \beta) - 1$$

$$S_{G_2}^{\omega_n} = -S_{G_a}^{\omega_n} = -S_{G_b}^{\omega_n} = -\frac{\beta}{2}(2Q + \beta)$$

where $\beta = \sqrt{4Q^2 - 1}$.

In both of the circuits, the amplifier gain is finite. Hence the circuits exhibit also zero-gain sensitivity product. It should be noted that the circuit of Figure 8-15(a) being a negative feedback realization is associated with very low Q- and ω_n-sensitivities with respect to all parameters at a cost of very high amplifier gain. On the other hand, the circuit of Figure 8-15(b) requires a low-gain amplifier, but has very high sensitivities (proportional to Q^2) with respect to other parameters, which is a characteristic of positive-feedback realization.

The general method of realization of filters with zero-gain sensitivity product is outlined in reference [17].

8.4 MULTILOOP FEEDBACK APPROACH

In multiloop active filters, two or more amplifiers are normally used. As in the case of single-loop feedback-type active filters, here also the finite-gain voltage amplifiers can be realized by monolithic operational amplifiers. We plan to outline several design techniques in this section.

State-Variable Realization

The most popular multiloop feedback circuit is the so-called state-variable realization scheme [18], which is essentially an analog computer simulation of a second-order transfer function [19]. One such realization is shown in Figure 8-16. Analyzing, one obtains

$$\frac{V_2}{V_1} = \frac{G_3(G_1 + G_4)}{G_1(G_2 + G_3)} \left[\frac{\dfrac{G_5 G_6}{C_1 C_2}}{s^2 + \dfrac{G_2 G_5(G_1 + G_4)}{G_1(G_2 + G_3)C_1} s + \dfrac{G_4 G_5 G_6}{G_1 C_1 C_2}} \right]. \tag{8.78}$$

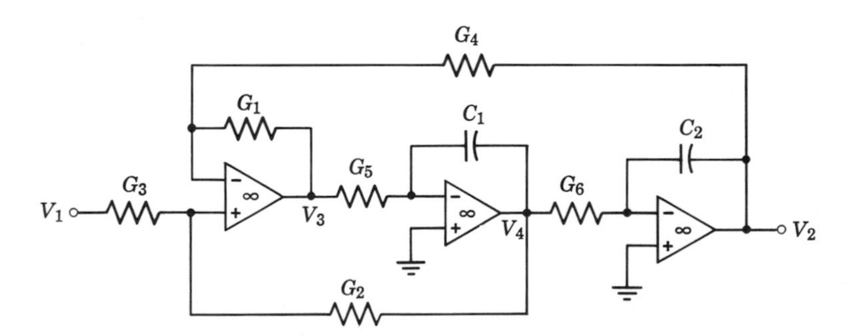

FIGURE 8-16 State-variable realization.

Design equations can be easily derived. For normalized value of ω_n equal to unity, we can set

$$G_1 = G_4 = G_5 = G_6 = 1$$
$$\frac{G_3}{G_2} = 2Q - 1$$
$$C_1 = C_2 = 1. \tag{8.79}$$

For this set of values, (8.78) reduces to:

$$\frac{V_2}{V_1} = \frac{2Q-1}{Q}\left[\frac{1}{s^2 + \frac{1}{Q}s + 1}\right]. \tag{8.80}$$

Since $V_4 = -(C_2/G_6)sV_2 = -sV_2$, and $V_3 = -(C_1/G_5)sV_4 = s^2V_2$, it follows that

$$\frac{V_3}{V_1} = \frac{2Q-1}{Q}\left[\frac{s^2}{s^2 + \frac{1}{Q}s + 1}\right] \tag{8.81}$$

$$\frac{V_4}{V_1} = \frac{2Q-1}{Q}\left[\frac{-s}{s^2 + \frac{1}{Q}s + 1}\right]. \tag{8.82}$$

Consequently, the filter structure of Figure 8-16 simultaneously realizes a bandpass, a highpass, and a lowpass transfer function depending on the choice of the output terminal. Transfer functions having arbitrary transmission zeros can be realized by appropriately summing the three voltages V_2, V_3 and V_4 by means of an additional summing amplifier shown in Figure 8-17. The input-output relations given by (8.83) can be used to get

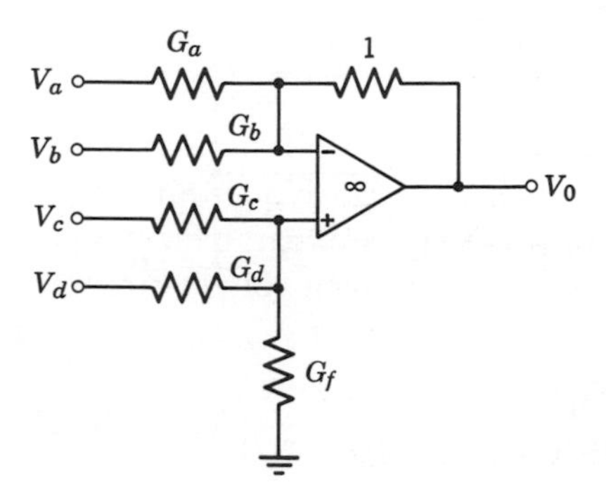

FIGURE 8-17 A summing amplifier.

the appropriate numerator:

$$V_0 = \frac{G_a + G_b + 1}{G_c + G_d + G_f}(G_c V_c + G_d V_d) - G_a V_a - G_b V_b. \tag{8.83}$$

The pole frequency ω_n and pole pair Q are given by

$$\omega_n = \sqrt{\frac{G_4 G_5 G_6}{G_1 C_1 C_2}} \tag{8.84}$$

$$Q = \sqrt{\frac{G_1 G_4 G_6 C_1}{G_5 C_2}} \cdot \frac{G_2 + G_3}{G_2(G_1 + G_4)}. \tag{8.85}$$

The above expressions suggest a method for post-design adjustments of ω_n and Q. First, ω_n (which is approximately the resonant frequency for high Q values) is adjusted by trimming G_1. Then Q can be adjusted by trimming either G_2 or G_3.

The Q-sensitivities of this structure are:

$$S^Q_{C_1} = -S^Q_{C_2} = \frac{1}{2}$$

$$S^Q_{G_1} = -S^Q_{G_4} = \frac{G_1 - G_4}{2(G_1 + G_4)} = 0$$

$$S^Q_{G_3} = -S^Q_{G_2} = \frac{G_3}{G_2 + G_3} = 1 - \frac{1}{2Q}$$

$$S^Q_{G_6} = -S^Q_{G_5} = \frac{1}{2}. \tag{8.86}$$

The Q-sensitivities with respect to the gains of the operational amplifiers are also low, provided the gain values are larger than the Q of the pole pair. There are a number of variations of the state-variable realizations [20, 21]. The effect of a nonideal operational amplifier on the performance of these filters has been examined by Thomas [22].

Q-Invariant Filter Sections

With the aid of two voltage amplifiers, one can design biquadratic filter sections having zero Q-sensitivities with respect to passive parameters. One such circuit is shown in Figure 8-18 [23, 24]. The pertinent voltage-transfer ratio is given as:

$$\frac{V_2}{V_1} = \frac{K_1 K_2 G_1 C_2 s}{(1 - K_1 K_2)C_1 C_2 s^2 + (C_2 G_1 + C_1 G_2)s + G_1 G_2}. \tag{8.87}$$

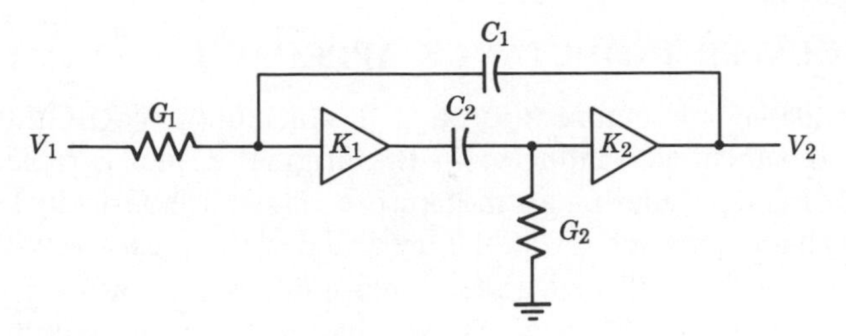

FIGURE 8-18 A bandpass filter section having zero passive Q-sensitivities.

Comparing above with (8.52) we obtain for $\omega_n = 1$ one set of design values as given below:

$$G_1 = G_2 = 1$$

$$C_1 = C_2 = \frac{1}{2Q}$$

$$K_1 K_2 = -(4Q^2 - 1). \tag{8.88}$$

The corresponding sensitivities are:

$$S_{K_1}^{Q} = S_{K_2}^{Q} = \frac{-K_1 K_2}{2(1 - K_1 K_2)} < \frac{1}{2}$$

$$S_{G_1}^{Q} = S_{G_2}^{Q} = S_{C_1}^{Q} = S_{C_2}^{Q} = 0$$

$$S_{K_1}^{\omega_n} = S_{K_2}^{\omega_n} = -\frac{1}{2}\left(1 - \frac{1}{4Q^2}\right)$$

$$S_{G_1}^{\omega_n} = S_{G_2}^{\omega_n} = -S_{C_1}^{\omega_n} = -S_{C_2}^{\omega_n} = -\frac{1}{2}.$$

Note that the passive Q-sensitivities are zero and all other sensitivities are very low. From (8.88) it is evident that one of the amplifiers must be a noninverting type, and the other must be an inverting type. If operational amplifiers are used to implement the finite-gain amplifiers, then K_1 is made positive and is realized by means of the circuit of Figure 8-4(a). K_2 is now negative. If G_2 is considered to be its input conductance, then the composite circuit can be implemented as shown in Figure 8-4(b).

A general biquadratic active filter section having similar sensitivity properties will be found in reference [25]. Other filter sections having zero Q-sensitivities with respect to the passive parameters have also been reported [26, 27].

8.5 SIMULATED INDUCTANCE APPROACH

In the simulated inductance approach, a conventional RLC filter is first designed, and then each inductor in the original design is replaced by an active RC circuit simulating an inductance. There are basically two advantages of such an approach. First, theory and design of resistively-terminated LC filters are well established, and extensive tables are now available on the design of such filters. Second, as we shall show, most commonly used doubly-terminated LC filter structures have inherently extremely low parameter sensitivities making them very suitable for the design of filters with stringent requirements, such as telephone channel filters with very sharp cut-off characteristics.

By means of a simple example, we showed earlier the low sensitivity feature of an RLC resonator. To prove this property for a general doubly-terminated LC filter [28], consider such a filter as shown in Figure 8-19. If

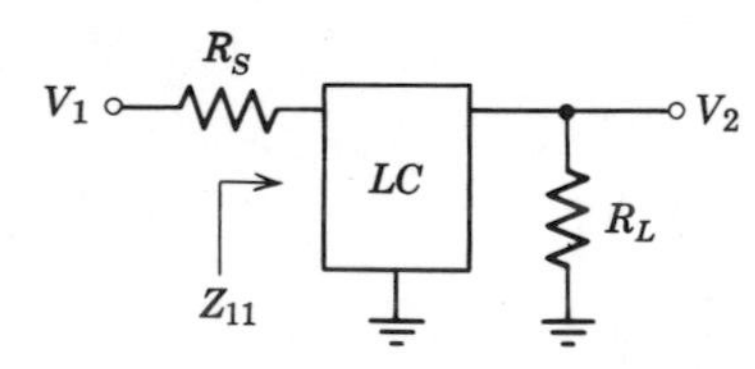

FIGURE 8-19 A doubly terminated LC filter.

the input impedance of the terminated filter at any frequency $s = j\omega$ is $Z_{11}(j\omega) = R_{11} + jX_{11}$, then the power P_{in} into the filter is

$$P_{\text{in}} = \frac{V_1^2 R_{11}}{(R_s + R_{11})^2 + X_{11}^2} \tag{8.89}$$

which is also equal to the power P_L delivered to the load resistor R_L because the filter network is lossless. Let us now compute the incremental change in P_L due to an incremental change in the value of a component x. Now

$$\frac{\partial P_L}{\partial x} = \frac{\partial P_L}{\partial R_{11}} \cdot \frac{\partial R_{11}}{\partial x} + \frac{\partial P_L}{\partial X_{11}} \cdot \frac{\partial X_{11}}{\partial x}.$$

From (8.89), one easily computes

$$\frac{\partial P_L}{\partial R_{11}} = \frac{R_s^2 - R_{11}^2 + X_{11}^2}{\{(R_{11} + R_s)^2 + X_{11}^2\}^2} V_1^2$$

$$\frac{\partial P_L}{\partial X_{11}} = -\frac{2R_{11}X_{11}}{\{(R_{11} + R_s)^2 + X_{11}^2\}^2} V_1^2.$$

Normally these filters are designed to have maximum power transfer at frequencies of minimum loss in the passband. This implies that at these frequencies

$$R_{11} = R_s$$
$$X_{11} = 0,$$

and hence it follows that in the passband

$$\frac{\partial P_L}{\partial R_{11}} = 0, \qquad \frac{\partial P_L}{\partial X_{11}} = 0.$$

As a result the sensitivity $S_x^{P_L}$ to any component is thus zero. Similar reasoning shows that the filter's output impedance must also be matched to the load resistor for zero sensitivity in the passband.

There are many ways by which a simulated inductance can be obtained. Some of these are reviewed next.

Capacitor-Terminated Active Gyrator

An ideal active gyrator is defined by the following short-circuit admittance parameters:

$$\begin{aligned} y_{11} &= 0 \qquad & y_{12} &= g_a \\ y_{21} &= -g_b \qquad & y_{22} &= 0. \end{aligned} \tag{8.90}$$

The input admittance of the gyrator, when terminated at port 2 by a capacitor C, is

$$Y_{\text{in}} = y_{11} - \frac{y_{12}y_{21}}{y_{22} + sC} = \frac{g_a g_b}{sC}, \tag{8.91}$$

which is the admittance of an inductor. A practical gyrator is however a nonideal device. One source of nonidealness is due to nonzero y_{11} and y_{22}. The effect of this is to make the simulated inductor appear like a lossy inductor [29] having an equivalent inductance L_{eq} of value:

$$L_{\text{eq}} \cong \frac{Cg_a g_b}{(y_{11}y_{22} + g_a g_b)^2 + \omega^2 y_{11}^2 C^2} \tag{8.92}$$

and a Q-factor of value,

$$Q = \frac{\omega C g_a g_b}{y_{11}y_{22}^2 + g_a g_b y_{22} + \omega^2 y_{11} C^2}. \tag{8.93}$$

Hence, to increase the Q of the simulated inductor, one must design the active gyrator to have very small y_{11} and y_{22}. Another source of nonidealness is the presence of phase shifts due to charge storage in the active elements. The effect of this is to make y_{12} and y_{21} complex numbers instead of completely real quantities. If we write $y_{12} = g_a e^{-j\phi_a}$ and $y_{21} = -g_b e^{-j\phi_b}$, then letting $\phi = \phi_a + \phi_b$, the input impedance of the capacitor-terminated nonideal gyrator is

$$j\frac{\omega C}{g_a g_b} e^{j\phi} \cong -\frac{\omega C \phi}{g_a g_b} + j\frac{\omega C}{g_a g_b},$$

where we have used $e^{j\phi} \cong 1 + j\phi$. It is thus seen that the effect of a negative phase angle in the gyration admittances is to produce a negative resistance in series with the inductor [29]. This negative resistance by proper adjustment can be made to cancel the positive series resistance produced by nonzero y_{11} and y_{22} to design very high Q simulated inductances [29, 30].

Realization of the active gyrator is best achieved by connecting two voltage-controlled current sources of opposite transconductances in parallel, with the input port of one controlled source connected to the output port of the other (Figure 8-20). To make y_{11} and y_{22} very small, the voltage-

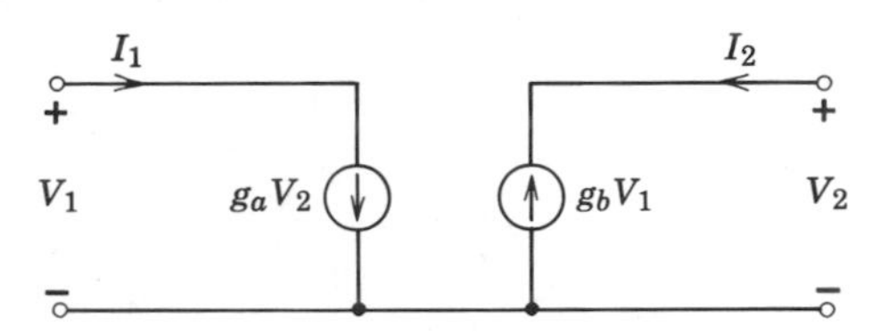

FIGURE 8-20 Controlled-source representation of an active gyrator.

controlled current sources should each have very high input and output impedances. Silicon monolithic gyrator circuits based on this approach have been reported [30–32].

Various other gyrator circuit designs suitable for monolithic construction have been reported, but their discussion is beyond the scope of this chapter. An excellent review of gyrator circuits will be found in reference [29].

Resistor-Terminated Generalized Impedance Converters

An alternate approach to simulating an inductor is by means of a generalized impedance converter (GIC), which is characterized by the following h-parameters:

$$\begin{aligned} h_{11} &= 0 \qquad & h_{12} &= a(s) \\ h_{21} &= -b(s) \qquad & h_{22} &= 0. \end{aligned} \tag{8.94}$$

The input impedance seen at port 1 when port 2 is terminated by a resistance R_L is given as:

$$Z_{11} = h_{11} - \frac{h_{12}h_{21}}{h_{22} + \dfrac{1}{R_L}} = a(s)b(s)R_L. \tag{8.95}$$

It is thus evident that if the product $a(s)b(s)$ is proportional to s, then the resistor-terminated GIC simulates an inductor. One realization of such an GIC is sketched in Figure 8-21 [33]. For this device $h_{12} = 1$ and

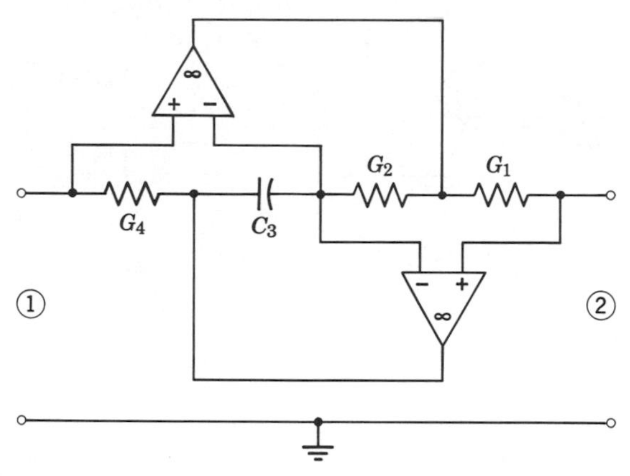

FIGURE 8-21 A generalized impedance converter suitable for inductance simulation.

$h_{21} = -sC_3 G_1/G_2 G_4$. This implies that when port 2 is terminated by a resistance R_L, the input impedance at port 1 is equal to $sC_3 G_1 R_L/G_2 G_4$, i.e., the one-port behaves as an inductance of $C_3 G_1 R_L/G_2 G_4$ henries! There are several variations of this basic circuit. However, for the purpose of inductance simulation, the GIC of Figure 8-21, which is a modification of Riordan's circuit [34], is considered to be the best circuit.

Using this circuit, it is possible to simulate inductors with a maximum Q value over 1000 at low frequencies provided the capacitor C_3 is of very high quality. The quality of this capacitor incidentally is an important factor affecting the Q of the simulated inductor at very low frequencies.

Floating-Inductor Simulation

The methods described earlier lead to simulated inductors having a grounded terminal. However, floating inductors are often needed in filter design, for example, in lowpass filters and some bandpass filter structures. One simple solution to this problem is to cascade two gyrators with a capacitor connected across the common port [35], as shown in Figure 8-22(a). The equivalent circuit of this arrangement is sketched in Figure 8-22(b), which reduces to a floating inductor (Figure 8-22(c)), provided

$$g_a = g_d \qquad g_b = g_c. \tag{8.96}$$

Thus, for a perfect simulation the two gyrators must be identical. Any mismatch would result in the network of Figure 8-22(b), which has undesirable effects in the case of sharp cut-off filters like the telephone channel filters [36]. The effect of mismatch may not be serious in the case of simple filters.

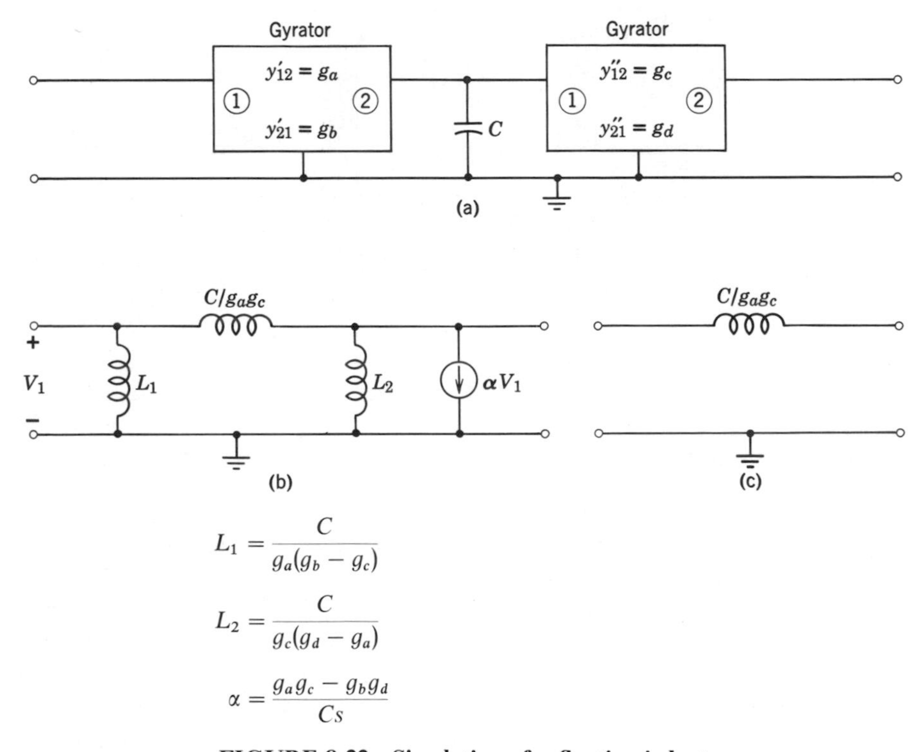

FIGURE 8-22 Simulation of a floating inductor.

Alternative more practical solutions to the realization of a floating inductor appear to be the use of private floating power supply for each gyrator [36] or the use of semifloating gyrator [37].

Two alternative solutions to the floating-inductor problem were recently proposed [38]. In certain types of bandpass filters, one encounters a floating inductor along with additional capacitive elements as shown in Figure 8-23(a). An equivalent representation of this two-port is Figure 8-23(b), which employs a grounded inductor and a grounded negative capacitor. A

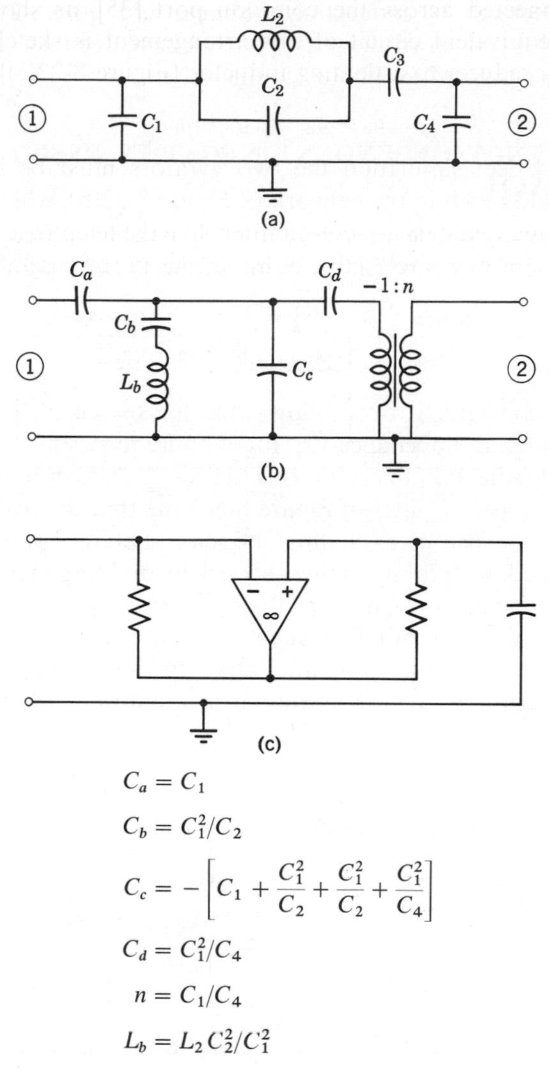

$$C_a = C_1$$

$$C_b = C_1^2/C_2$$

$$C_c = -\left[C_1 + \frac{C_1^2}{C_2} + \frac{C_1^2}{C_2} + \frac{C_1^2}{C_4}\right]$$

$$C_d = C_1^2/C_4$$

$$n = C_1/C_4$$

$$L_b = L_2 C_2^2/C_1^2$$

FIGURE 8-23 An alternate solution to the problem of the realization of a floating inductor.

negative capacitor is readly realized by the circuit of Figure 8-23(c). Note that the ideal transformer shown in the equivalent circuit of Figure 8-23(b) is eliminated later by impedance scaling all elements to the right of the transformer.

In many filter realizations, floating inductors also appear across two grounded inductors in the form of a Π-circuit. This type of Π-network can be realized effectively by connecting the circuit of Figure 8-21 in a back-to-back fashion [38]. Both of these approaches have been successfully implemented in the design of telephone channel filters.

8.6 FREQUENCY-DEPENDENT NEGATIVE RESISTANCE APPROACH

A frequency-dependent negative resistance (to be abbreviated as FDNR) is a one-port whose impedance $Z(s)$ is either of the following forms:

$$Z(s) = K/s^2 \tag{8.97}$$

$$Z(s) = Ks^2 \tag{8.98}$$

where K is a constant. It thus follows that for $s = j\omega$, $Z(j\omega)$ is a negative resistance having an impedance varying with frequency.

To illustrate the usefulness of this device [39], consider the doubly-terminated lowpass LC filter of Figure 8-6. Note that the realization of this filter requires the use of a floating inductor. Let us apply a frequency-dependent impedance scaling to this network by dividing each impedance by s. This implies that to obtain the transformed network, we replace each resistor R_i by a capacitor of $1/R_i$ F, each inductor L_i by a resistor of value L_i ohms, and each capacitor C_i by an FDNR having an impedance $Z_i(s) = 1/C_i s^2$. The resultant network is shown in Figure 8-24. Observe that the

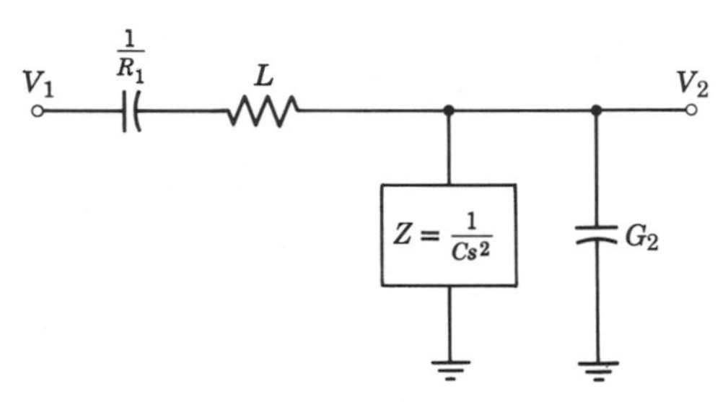

FIGURE 8-24 Realization of a lowpass filter section using a frequency-dependent negative resistance.

voltage-transfer ratio of the network of Figure 8-24 is identical to that of Figure 8-6 and is given by (8.11).

Realization of an FDNR can be achieved by means of a generalized impedance converter. For example, if in the circuit of Figure 8-21 we replace the resistor G_1 by a capacitor C_1, and terminate port 2 by a resistor R_L, then the input impedance seen at port 1 is $(C_1 C_3 R_L/G_2 G_4)s^2$. On the other hand, if port 1 is terminated by a resistor R_L, the input impedance seen at port 2 is

$G_2 G_4 R_L/C_1 C_3 s^2$. An extensive analysis of various types of generalized impedance converters will be found in reference [40].

8.7 SUMMARY

Several typical active RC filter design methods employing operational amplifiers as the active elements have been outlined in this chapter. Sections 8.3 and 8.4 discussed the design of RC amplifier type filter structures. For convenience and sensitivity reasons, this type of active filter is designed as a cascade of second-order filter sections. The most useful class of such second-order sections, discussed here, are the lowpass, highpass, bandpass sections, and those with $j\omega$-axis zeros. An excellent extensive discussion on this type of filter will be found in reference [41].

Two very simple but direct approaches to inductorless filter design are the simulated inductance approach (Section 8.5) and the frequency-dependent negative-resistance approach (Section 8.6). These methods begin by realizing a specified transfer function by terminated LC filters. Next, in the simulated inductance approach, each inductor in the realization is replaced by an active RC one-port simulating an inductance, whereas in the second approach, a frequency-dependent impedance scaling is used to convert the RLC filter into a network composed of resistors, capacitors, and frequency-dependent negative resistances.

Because of limited space, only a sampling of typical design methods has been included in this chapter. For a more complete discussion we refer the reader to several books on the subject [1, 2, 42–44].

ACKNOWLEDGEMENT

Thanks are due to Professor S. C. Dutta Roy and Miss B. A. Shanta Pai for their constructive criticisms.

PROBLEMS

8.1 For each of the following bandpass transfer functions,

(a) $$\frac{3s}{2s^2 + s + 6}$$

(b) $$\frac{s}{s^2 + s + 10}$$

(c) $$\frac{s}{s^2 + 3s + 2},$$

determine the pole pair Q and pole frequency ω_n. Compare these figures with the Q of the actual frequency response and the actual resonant frequency.

8.2 Determine the active filter configurations which are RC : CR transformations of the networks of Figures 8-7, 8-8, 8-10, 8-11, 8-12, 8-13, and 8-14.

8.3 Analyze the active filters of Figure 8-5 and compute the pole and Q-sensitivities of each circuit.

8.4 A number of possible decompositions of the polynomial $(s^2 + 2s + 10)$ are given below:
(a) $(s + 1)^2 + 9K$
(b) $(s + \sqrt{10})^2 - 4.3246Ks$
(c) $(s^2 + s + 8) + K(s + 2)$
where the nominal value of the parameter K is unity. Compute the root sensitivities with respect to K for each decomposition and compare the values.

8.5 Determine the coefficient sensitivities of the denominator polynomial of (8.11) with respect to R_1, G_2, L, and C. Using these results and (8.25), compute the Q-sensitivities. Check your answer with those given by (8.12).

8.6 Derive (8.25) and (8.26).

8.7 Derive (8.27) and (8.28).

8.8 Derive the sensitivity expressions of (8.30).

8.9 Solve (8.36) for the element values by setting the gain of amplifier as two $(K = 2)$.

8.10 Verify the expressions for the sensitivities given by (8.45).

8.11 Derive the Q-sensitivities of (8.46).

8.12 Realize the following transfer functions using single-amplifier structures:

(a) $\dfrac{H}{s^2 + \sqrt{2}\,s + 1}$ (b) $\dfrac{H}{s^3 + 2s^2 + 2s + 1}$

(c) $\dfrac{H}{s^2 + 3s + 3}$ (d) $\dfrac{Hs^2}{s^2 + \sqrt{2}\,s + 1}$

(e) $\dfrac{Hs}{s^3 + 2s^2 + 2s + 1}$ (f) $\dfrac{Hs^2}{s^3 + 2s^2 + 2s + 1}$.

8.13 Verify the expressions of (8.51), (8.55) and (8.57).

8.14 Compute the ω_n- and γ-sensitivities of the filter of Figure 8-12.

8.15 Realize the following transfer functions using single-amplifier structures per pole pair:

(a) $\dfrac{Hs}{s^2 + \sqrt{2}\,s + 1}$ (b) $\dfrac{Hs}{s^2 + 3s + 3}$

(c) $\dfrac{Hs}{(s^2 + 0.7653s + 1)(s^2 + 1.8477s + 1)}$

(d) $\dfrac{Hs^2}{(s^2 + 0.7653s + 1)(s^2 + 1.8477s + 1)}$

(e) $\dfrac{H(s^2 + 2)}{s^2 + s + 10}$ (f) $\dfrac{H(s^2 + 2)}{s^2 + s + 1}$ (g) $\dfrac{H(s^2 + 2)}{s^2 + s + 2}$.

8.16 Realize the transfer functions of Problem 8.15 using the state-variable approach.

REFERENCES

[1] S. K. Mitra, Ed., *Active Inductorless Filters*, IEEE Press, New York, 1971.

[2] S. K. Mitra, *Analysis and Synthesis of Linear Active Networks*, John Wiley, New York, pp. 447–469.

[3] S. K. Mitra, "A network transformation for active RC networks," *Proc. IEEE*, **55**, 2021–2022 (Nov., 1967).

[4] S. K. Mitra, "Filter design using integrated operational amplifiers," *1969 WESCON Technical Papers Session 4*, San Francisco, Calif. (Aug., 1969).

[5] H. W. Bode, *Network Analysis and Feedback Amplifier Design*, Van Nostrand, New York, 1945, p. 10.

[6] S. K. Mitra, *Analysis and Synthesis of Linear Active Networks*, John Wiley, New York, 1969, pp. 175–177.

[7] M. A. Soderstrand and S. K. Mitra, "Sensitivity analysis of third-order filters," *Mexico 1971 International IEEE Conference on Circuits, Systems and Computers* (Jan., 1971).

[8] R. P. Sallen and E. L. Key, "A practical method of designing RC active ffilters," *IRE Trans. Circuit Theory*, **CT-2**, 74–85 (March, 1955).

[9] L. Scultety, "Active RC network using IC operational amplifiers," *Proc. 4th Colloquium on Microwave Communication, Budapest, Hungary*, **II**, Paper CT-24 (April, 1970).

[10] P. L. Taylor, "Flexible design method for active RC two-ports," *Proc. IEEE* (*London*), **110**, 1607–1616 (Sept., 1963).

[11] P. R. Geffe, "RC-amplifier resonators for active filters," *IEEE Trans. Circuit Theory*, **CT-15**, 415–419 (Dec., 1968).

[12] M. A. Soderstrand, "Sensitivity Studies of Multi-loop Feedback Active Filters," M. S. Thesis, University of California, Davis, Calif., (Sept., 1969).

[13] W. J. Kerwin and L. P.. Huelsman," "The design of high performance active RC band-pass filters," *IEEE Intl. Conv. Record*, **14**, Part 10, 74–80 (1960).

[14] W. J. Kerwin and C. V. Shaffer, "An integrable if amplifier of high stability," *Proc. Eleventh Midwest Symp. on Circuit Theory*, Univ. of Notre Dame (May, 1968).

[15] W. J. Kerwin, "An active RC elliptic function filter," *1966 Region VI Conference Record, Tucson, Arizona*, 647–654 (April, 1966).

[16] R. M. Inigo, "Active filter realization using finite gain voltage amplifiers," *IEEE Trans. Circuit Theory*, **CT-17**, 445–448 (Aug., 1970).

[17] M. A. Soderstrand and S. K. Mitra, "Active RC filters with zero gain-sensitivity-product," *Symp. Digest, 1972 IEEE International Symposium on Circuit Theory*, North Hollywood, Calif., (April, 1972). pp. 340–344.

[18] W. J. Kerwin, L. P. Huelsman and R. W. Newcomb, "State variable synthesis for insensitive integrated circuit transfer functions," *IEEE J. Solid State Circuits*, **SC-2**, 87–92 (Sept., 1967).

[19] W. H. Schüssler, *On the Representation of Transfer Functions and Networks on Analog Computers*, Westdeutscher Verlag, Cologne, 1961.

[20] H. Sutcliffe, "Tunable filter for low frequencies using operational amplifiers," *Electronic Eng.*, **36**, 399–403 (June, 1964)

[21] J. Tow, "Active RC filters—A state-space realization," *Proc. IEEE.*, **56**, 1137–1139 (June, 1968).

[22] L. C. Thomas, "The Biquad: Part 1—Some practical design considerations," *IEEE Trans. Circuit Theory*, **CT-18**, 358–361 (May, 1971).

[23] P. R. Geffe, "A Q-invariant active resonator," *Proc. IEEE* (*Lett.*), **57**, 1442 (Aug., 1969).

[24] M. A. Soderstrand and S. K. Mitra, "Extremely low sensitivity active RC filter," *Proc. IEEE* (*Lett.*), **57**, 2175 (Dec., 1969).

[25] M. A. Soderstrand and S. K. Mitra, "Design of active filters with zero passive Q-sensitivity," *IEEE Trans. Circuit Theory*, **CT-20**, 289–294, (May, 1973).

[26] T. Deliyannis, "A low pass filter with extremely low sensitivity," *Proc. IEEE* (*Lett.*), **58**, 1366–1367 (Sept., 1970).

[27] A. E. Sanderson, "A Q-invariant, ω_0—invariant active resonator," *Proc. IEEE* (*Lett.*), **60**, 908–909 (July, 1972).

[28] D. F. Sheahan, "Inductorless Filters," *Tech. Rept. No. 6560-15, Systems Theory Lab.*, Stanford University, Stanford, Calif., (Sept., 1967).

[29] H. J. Orchard, "Gyrator circuits," *Active Filters*, L. P. Huelsman, Ed., McGraw-Hill, New York, 1970, pp. 90–127.

[30] H. T. Chua and R. W. Newcomb, "Integrated direct-coupled gyrator," *Electronics Letters*, **3**, 182–184 (May, 1967).

[31] R. G. Hove and C. A. Kleingarter, "Silicon monolithic gyrator using FET's," *1969 WESCON Technical Papers Session 4*, San Francisco, Calif. (Aug., 1969).

[32] S. S. Haykim, S. Kramer, J. Schewchun, and D. H. Treleaven, "Integrated-circuit implementation of direct-coupled gyrator," *IEEE J. Solid-State Circuits*, **SC-4**, 164–66 (June, 1969).

[33] A. Antoniou, "Realization of gyrators using operational amplifiers and their use in RC-active network synthesis," *Proc. IEE* (*London*), **116**, 1838–1850 (Nov., 1969).

[34] R. H. S. Riordan, "Simulated inductors using differential amplifiers," *Electronics Letters*, **3**, 50–51 (Feb., 1967).

[35] A. G. J. Holt and J. Taylor, "Method of replacing ungrounded inductances by grounded gyrators," *Electronics Letters*, **1**, 105 (1965).

[36] D. F. Sheahan, "Gyrator-floatation circuit," *Electronics Letters*, **3**, 39–40 (1967).

[37] W. H. Holmes, S. Gruetzmann, and W. E. Heinlein, "Direct-coupled gyrators with floating ports," *Electronics Letters*, **3**, 46–47 (1967).

[38] H. J. Orchard and D. F. Sheahan, "Inductorless band-pass filters," *IEEE J. Solid-State Circuits*, **SC-5**, 108–118 (June, 1970).
[39] L. T. Bruton, "Network transfer functions using the concept of frequency-dependent negative resistance," *IEEE Trans. Circuit Theory*, **CT-16**, 406–408 (Aug., 1969).
[40] L. T. Bruton, "Non-ideal performance of a class of two-amplifier positive impedance converters," *IEEE Trans. Circuit Theory*, **CT-16**, 541–549 (Nov., 1969).
[41] W. J. Kerwin, "Active RC network synthesis using voltage amplifier," *Active Filters*, L. P. Huelsman, Ed., McGraw-Hill, New York, 1970, Chapter 2.
[42] K. L. Su, *Active Network Synthesis*, McGraw-Hill, New York, 1965.
[43] L. P. Huelsman, *Theory and Design of Active RC Circuits*, McGraw-Hill, New York, 1968.
[44] R. W. Newcomb, *Active Integrated Circuit Synthesis*, Prentice-Hall, Englewood Cliffs, New Jersey, 1968.

9

Active Distributed $\overline{\text{RC}}$ Networks

Ralph W. Wyndrum, Jr.
Bell Telephone Laboratories
Whippany, New Jersey

During the past decade there has been considerable interest in the characterization, analysis, and synthesis of distributed RC ($\overline{\text{RC}}$) networks. The basic motivation has been that thin film integrated circuits most readily afford $\overline{\text{RC}}$ network realizations. In addition, some silicon integrated circuits, and various relatively large area semiconductor devices, can be accurately characterized only as distributed RC networks.

The mid-1960's provided us with several new approaches to the synthesis of passive distributed RC networks [1, 2, 3, 4]. With these basic approaches, various researchers evolved methods to synthesize active distributed RC (A$\overline{\text{RC}}$) networks. "A$\overline{\text{RC}}$ networks" refer to topologies including passive $\overline{\text{RC}}$ structures and discrete active components. If ever a truly active *and* distributed structure becomes practical (including distributed controlled sources, for example), it will be a simple matter to evolve appropriate mathematical synthesis procedures based on what is now known.

Like their discrete active RC counterparts discussed earlier in Chapter 8, A$\overline{\text{RC}}$ networks are capable of realizing lowpass, bandpass, and highpass functions with rapid cut-off characteristics and controlled out-of-band performance. $\overline{ARC}$ networks are often inherently more integrable than their discrete counterparts; interconnections, for example, can be largely eliminated.

This chapter reviews pertinent passive distributed $\overline{RC}$ analysis and synthesis techniques as necessary. Several active $\overline{RC}$ network synthesis approaches are indicated for the realization of practical families of network characteristics. Those using operational amplifiers are stressed, to the exclusion of NIC (Negative Impedance Converter) approaches. The chapter closes with considerations of the fabrication, component tolerances, and sensitivities of $\overline{ARC}$ networks.

9.1 PRELIMINARY CONSIDERATIONS

Passive $\overline{RC}$ Networks

The understanding of the behavior of $\overline{RC}$ transmission lines is based on the differential equations for the voltage and current along a finite transmission line segment. An incremental section of such a network is shown in Figure 9-1. The indicated approximation holds for very small Δx, so that the

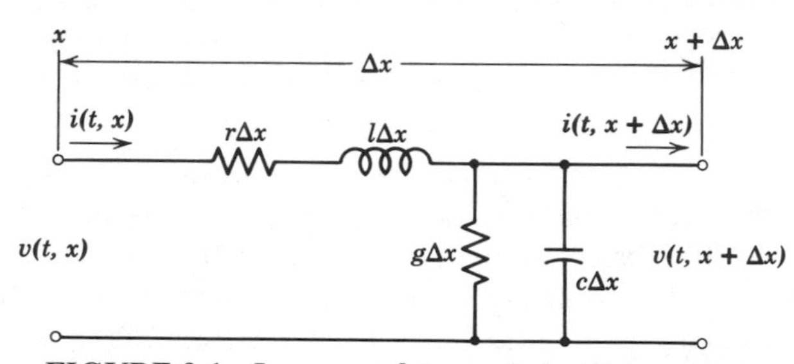

FIGURE 9-1 Incremental transmission line segment.

general parameters are given as r ohms-per-meter, l henries-per-meter, g mhos-per-meter, and c farads per meter. Clearly, only the fundamental transmission mode is considered; higher-order modes are neglected. For most transmission lines, those without rapidly varying cross-sections, this assumption is valid. The resulting transmission line equations, therefore, are simply [4]

$$\frac{\partial v(t, x)}{\partial x} = -l\frac{\partial i(t, x)}{\partial t} - ri(t, x) \tag{9.1}$$

$$\frac{\partial i(t, x)}{\partial x} = -c\frac{\partial v(t, x)}{\partial t} - gv(t, x). \tag{9.2}$$

Progressing in the usual fashion through the analysis of (9.1) and (9.2), and assuming that both v and i possess Laplace transforms with respect to time, the first-order homogeneous linear differential equations (9.3) and (9.4) evolve, which assume zero initial conditions:

$$\frac{d}{dx} V(s, x) = (r + sl)I(s, x) \tag{9.3}$$

$$\frac{d}{dx} I(s, x) = -(g + sc)V(s, x). \tag{9.4}$$

At this point we have not said anything concerning the nature of the network line parameters r, c, l, and g. Our concern is with distributed RC lines; i.e., we assume g and l to be zero. If r and c are independent of x, the line is said to be a uniform $\overline{\text{RC}}$ ($\text{U}\overline{\text{RC}}$) line [1]. Correspondingly, if r and c are functions of x, the line is said to be nonuniform. "Uniformity" thus refers to the cross-section of the line as a function of the distance x. Although a wide variety of functions $r(x)$ and $c(x)$ are mathematically conceivable, both r and c, for any practical homogeneous line, are intimately related to one another by geometry (for instance c is directly related, and r inversely related, to the width of the line). This recognition simplifies the analysis of the situation. A line for which $r(x)$ and $c(x)$ vary monotonically with x is defined as a "tapered" line, which will be considered in the next section.

If we first consider the $\text{U}\overline{\text{RC}}$ line we may derive its $ABCD$ parameters as well as the open-circuit impedance and short-circuit admittance parameters from the solutions of (9.3) and (9.4):

$$\begin{aligned} V(s, x) &= A_1 \cosh \Gamma x + A_2 \sinh \Gamma x \\ I(s, x) &= B_1 \cosh \Gamma x + B_2 \sinh \Gamma x \end{aligned} \tag{9.5}$$

where the general propagation function Γ (including nonzero l and g) is given by

$$\Gamma = \sqrt{(r + sl)(g + sc)}.$$

and for our purposes, with $l = g = 0$, this may be simplified to

$$\Gamma = \sqrt{src}.$$

Thus, the parameters cited above for a line of length L, total resistance $R = (rL)$, and total capacitance $C = (cL)$ become

$$\begin{bmatrix} A & B \\ C & D \end{bmatrix} = \begin{bmatrix} \cosh \sqrt{sRC} & \sqrt{\dfrac{R}{sC}} \sinh \sqrt{sRC} \\ \sqrt{\dfrac{sC}{R}} \sinh \sqrt{sRC} & \cosh \sqrt{sRC} \end{bmatrix}, \tag{9.6}$$

$$\begin{bmatrix} z_{11} & z_{12} \\ z_{21} & z_{22} \end{bmatrix} = \sqrt{\frac{R}{sC}} \begin{bmatrix} \coth\sqrt{sRC} & \operatorname{csch}\sqrt{sRC} \\ \operatorname{csch}\sqrt{sRC} & \coth\sqrt{sRC} \end{bmatrix}, \tag{9.7}$$

and

$$\begin{bmatrix} y_{11} & y_{12} \\ y_{21} & y_{22} \end{bmatrix} = \sqrt{\frac{sC}{R}} \begin{bmatrix} \coth\sqrt{sRC} & -\operatorname{csch}\sqrt{sRC} \\ -\operatorname{csch}\sqrt{sRC} & \coth\sqrt{sRC} \end{bmatrix}. \tag{9.8}$$

Note that the network functions are transcendental and therefore significantly different from those of discrete RC networks. Clearly some modification or transformation is required to make use of these frequency characteristics, and basically two such approaches have been reported in the literature [3, 4]. The first is to approximate the hyperbolic functions by the first several terms of their series or product expansion. This is equivalent, from another point of view, to identifying a small network of discrete resistors and capacitors which together will provide an approximation to the $\overline{\text{RC}}$ line in question. A second approach is to modify the network theory used for microwave circuits* and then bring Richard's Theorem [1] to bear on $\overline{\text{RC}}$ networks. Ultimately this allows one to work in a transformed frequency plane where the immittance functions are no longer transcendental, but rather rational fractions of the independent variable. This has a strong advantage: it leads naturally to an exact synthesis technique. Since $\overline{\text{URC}}$ transfer and driving-point functions are represented by hyperbolic functions, only the artifice of carrying out the synthesis in another complex variable permits us to work with a finite number of poles and zeros. The disadvantage, of course, is the heuristic insight lost by working in an arbitrary and rather meaningless transformed-frequency plane. The reader is referred to the literature for a detailed description of these techniques [1, 2, 4]. Their salient features will be included in the following sections where they are necessary to understand the active $\overline{\text{RC}}$ synthesis technique in question. Figure 9-2 indicates the synthesis procedures involved for the realization of an $\overline{\text{RC}}$ driving point or transfer impedance.

9.2 EARLY OPERATIONAL AMPLIFIER $\overline{\text{RC}}$ FILTER DESIGN

Initial active $\overline{\text{RC}}$ filter designs [5–8] employed simple $\overline{\text{RC}}$ networks in the feedback loop of a high-gain amplifier. The intent was to invert the characteristics of an $\overline{\text{RC}}$ network so as to provide highly selective active filters. The $\overline{\text{RC}}$ network was a selective rejection network, achieved by combining a simple $\overline{\text{RC}}$ segment such as discussed in the last section, with one or more external lumped impedances, to produce a real frequency

* See Chapter 7.

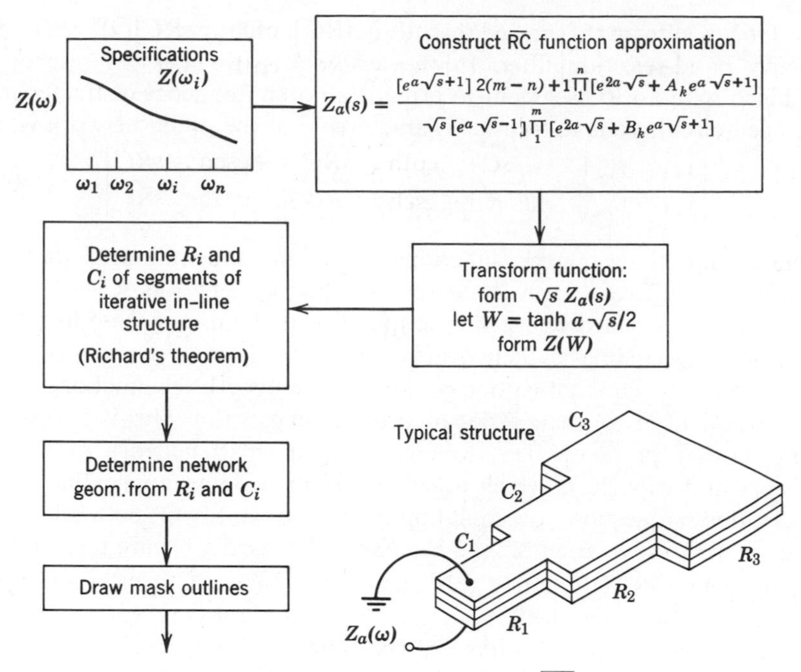

FIGURE 9-2 Computer generation of $\overline{RC}$ networks.

transmission zero. Two networks achieving this characteristic are shown in Figure 9-3. The network shown in Figure 9-3(b) combines a distributed-augmenting network in series* with the basic RC section. To generate the real zero of transmission, for the given networks Z_B, a series network Z_A is found such that the Z_{12} term of the composite network is equal to 0 at some frequency, by the appropriate series combination of a basic $\overline{RC}$ network with a series-augmenting lumped network.

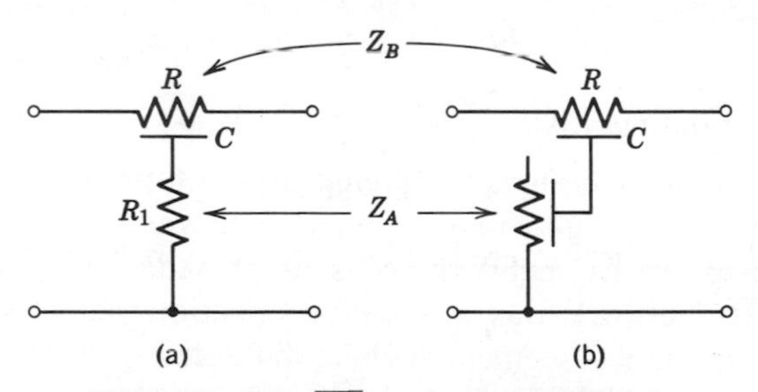

FIGURE 9-3 $\overline{RC}$ notch rejection filters.

126 * In the sense of a series connection of two ports.

In 1962, Kaufman [5] showed that the network of Figure 9-3(a)* realized a selective band-rejection filter. In the feedback path, a series resistor is combined with an $\overline{\text{RC}}$ segment to provide a notch frequency characteristic. The open-circuit voltage-transfer characteristic of the notch network is

$$t_v(\omega) = \frac{\alpha + u \sinh u}{\alpha \cosh u + u \sinh u} \tag{9.9}$$

where

$$u = (1 + j)\left(\frac{\omega}{\omega_1}\right)^{1/2}$$

$$\alpha = \frac{R}{R_1} \tag{9.10}$$

and

$$\omega_1 = \frac{2}{RC}.$$

Note that

$$t_v(\infty) = 1 = t_v(0).$$

A perfect notch is obtained when

$$\alpha + u \sinh u = 0.$$

There exists an infinite but countable set of discrete values for α which will allow such a zero. The first zero occurs at ω_{01}, where

$$\frac{\omega_{01}}{\omega_1} = 5.5951 \cdots. \tag{9.11}$$

or

$$\omega_{01} = \frac{(11.1902 \cdots)}{RC} \tag{9.12}$$

and the corresponding α is

$$\alpha_1 = 17.786 \cdots. \tag{9.13}$$

The third and all higher-ordered zeros occur very nearly at

$$\frac{\omega_{0,n}}{\omega_1} = \left(n\pi - \frac{\pi}{4}\right)^2. \tag{9.14}$$

* Figures 9-3 through 9-7 adapted from reference [5] with permission of the Institute of Electrical and Electronic Engineers.

Each of these requires a different value of α. There are no zeros of the denominator for real frequency. The transmission zero arises because Z_{12} of the $\overline{RC}$ network is exactly $-R_1$ at ω_{01}, and R_1 is in series with the $\overline{RC}$ network.

The behavior of the $t_v(\omega)$ for α near 17.8 was investigated thoroughly and is shown by the polar plot of $t_v(\omega)$ in Figure 9-4. Corresponding to the polar plot, both amplitude and phase are plotted in Figures 9-5 and 9-6. The

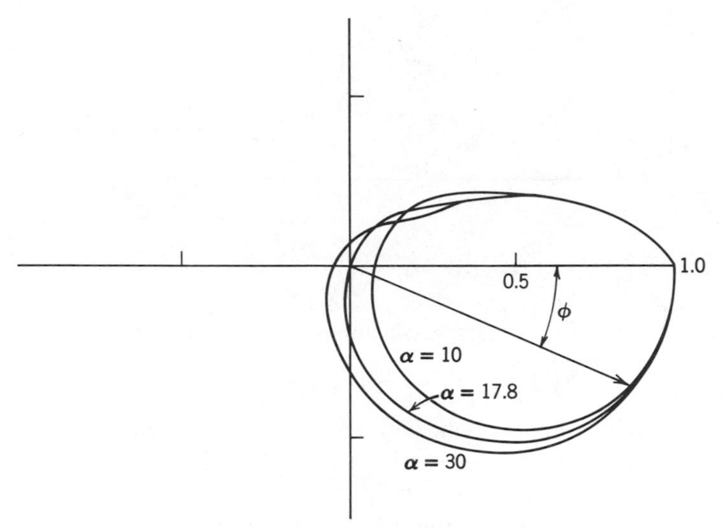

FIGURE 9-4 Polar plots of the complex gain for several α.

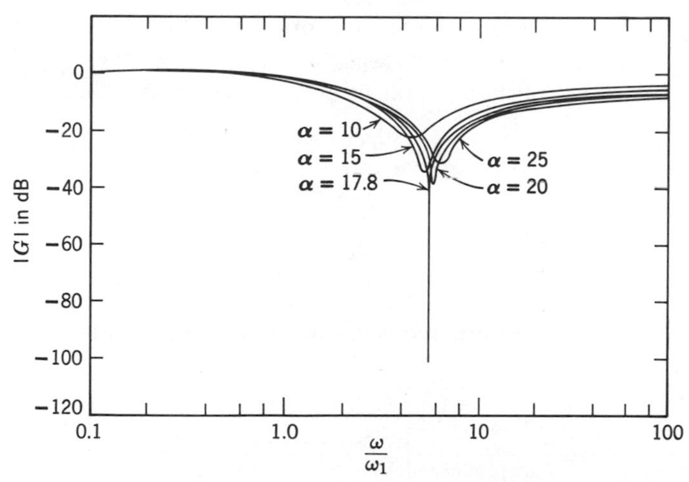

FIGURE 9-5 $|G(\omega)|$ **versus** ω/ω_1.

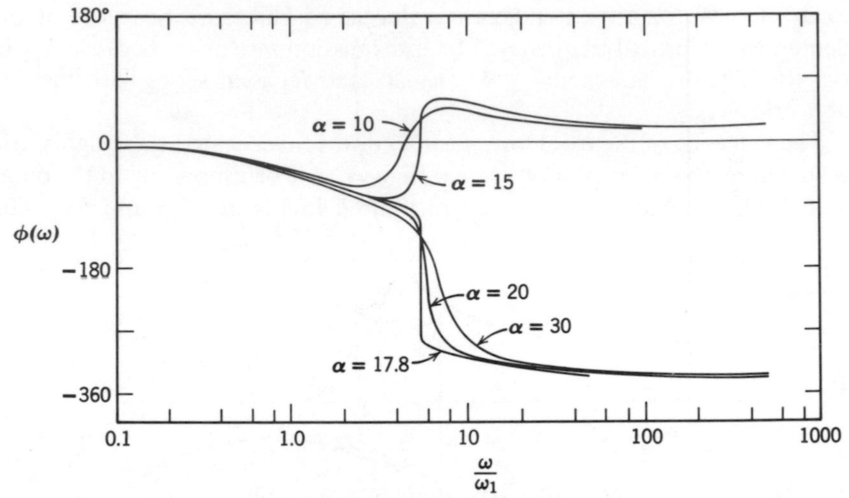

FIGURE 9-6 $\phi(\omega)$ versus (ω/ω_1).

second (and further) transmission zeros correspond to far less selective rejection characteristics.

The earliest and most important application of the $\overline{RC}$ notch network in an active configuration was in conjunction with a high-gain amplifier to provide a narrow band tuned amplifier (Figure 9-7). The notch network rejection characteristic is essentially inverted when the forward gain is sufficiently high, and the selectivity, Q, becomes proportional directly to the amplifier gain. Stable filters with Q less than 15 or 20, however, can be constructed in this manner. Oscillators can also be readily designed using this configuration, though appropriate limiting circuitry must be provided.

Kaufman's $\overline{RC}$ notch network affords a suitable replacement for the "twin-T" network in many applications such as in the bandpass filter of

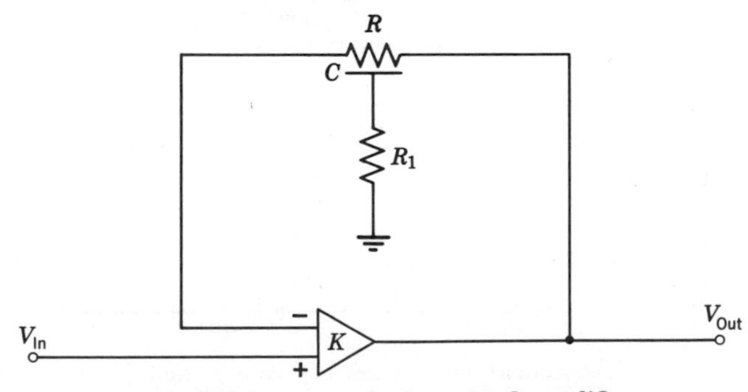

FIGURE 9-7 A high-Q tuned amplifier.

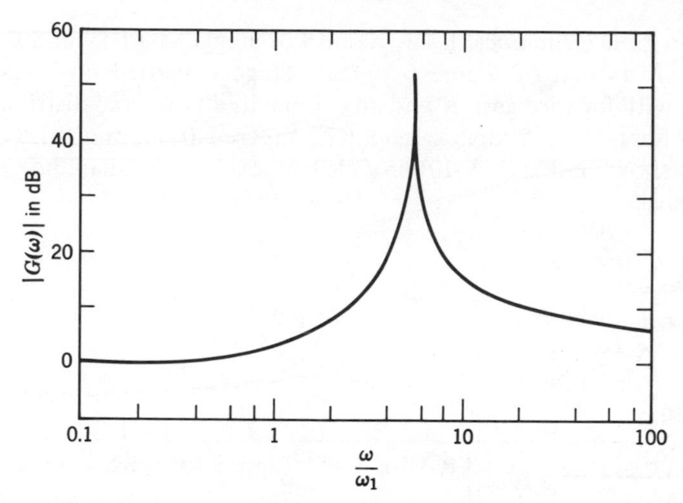

FIGURE 9-8 Response of an amplifier employing selective $\overline{RC}$ feedback ($K = 500$, $\alpha = 17.8$).

Figure 9-7. Its selectivity, however, is poorer than that of the twin T, particularly at frequencies above the notch, as shown in Figures 9-8 and 9-9. Trimming of $\overline{RC}$ notch networks for prescribed notch depth and frequency is detailed elsewhere [7].

Following Kaufman's work, Herskowitz and Wyndrum [8] described an approximately maximally flat narrow band amplifier that might be realized,

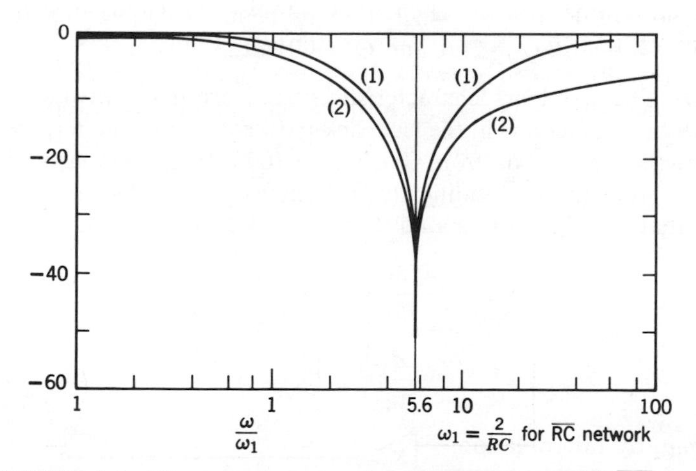

FIGURE 9-9 Theoretical curves for (1) twin-T and (2) $\overline{RC}$ notch networks.

using thin film techniques, by a cascade of stagger-tuned bandpass filter stages such as that of Figure 9-7. Each stage consisted of a wide band amplifier with forward gain K and an exponentially-tapered distributed RC rejection filter in its feedback loop. The tapered (nonuniform) $\overline{RC}$ notch filters (as shown in Figure 9-10) provide better selectivity than their uniform counterparts.

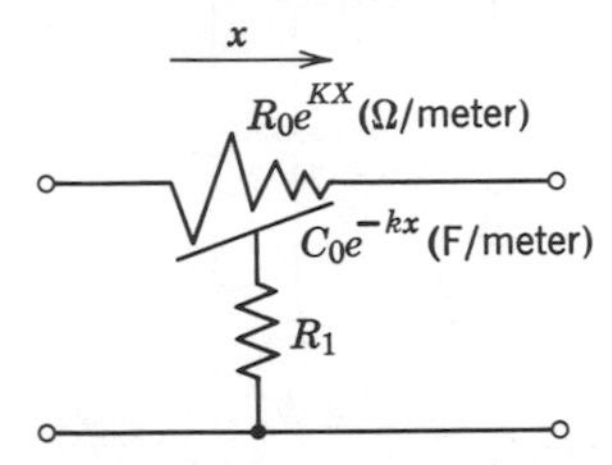

FIGURE 9-10 Tapered $\overline{RC}$ notch network.

Computer solutions for the zero locations of the hyperbolic-voltage transfer function of the exponentially-tapered distributed rejection filters were determined as a function of the distributed-network parameters and geometry. Since the loci of the closed loop poles of the amplifier tend toward the open-loop (feedback network) zeros for very high forward gain, these zeros approximate the closed loop poles. Maximally flat or equi-ripple magnitude responses are obtained by cascading appropriate feedback amplifier stages, to realize the proper pole configurations.

Example 9-1. Design a bandpass filter using a uniform $\overline{RC}$ network in the configuration of Figure 9-7, such that its center frequency is 100 KHz, $t_v(100\text{ kHz}) = 60$ dB, $Q = 21.6$, and $C = .001$ uF.

Solution: The bandpass characteristic results from the inversion of the band-reject characteristic in the feedback path by the high gain (here 60 dB) in the forward path. Thus, from 60 dB, $K = 1000$. Next we interpolate along the $\beta = 0$ curve (corresponding to uniform $\overline{RC}$ networks) in Figure 9-11 realizing that $Q = 21.6$ corresponds to $\omega/\sigma = 43.2$. This leads to the solution,

$$\frac{\omega}{\omega_1} = 10.8$$

$$\frac{\sigma}{\omega_1} = 0.25.$$

Thus, again by interpolation,

$$\alpha = 16.4 = \frac{R}{R_1}$$

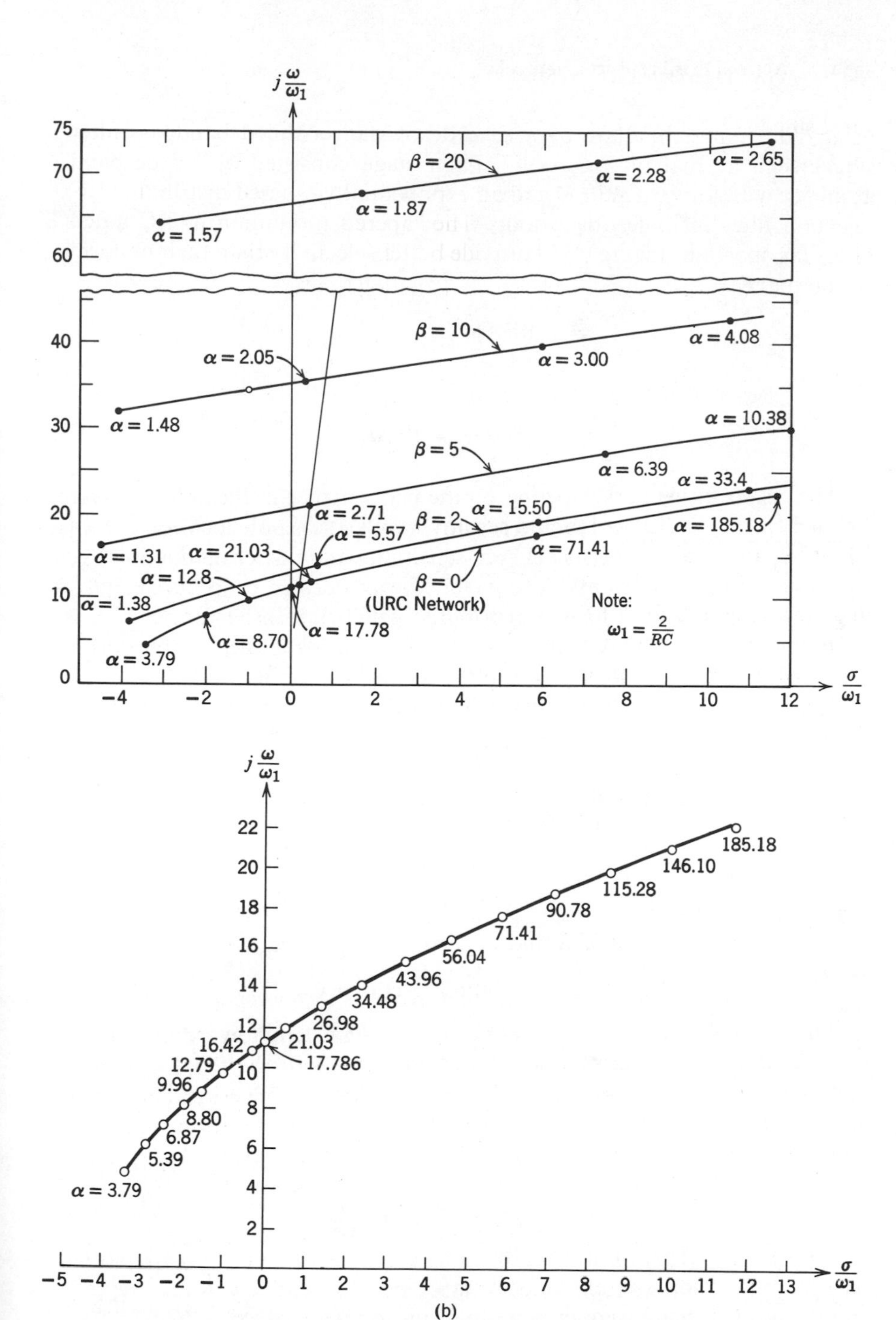

FIGURE 9-11 (a) Locations of filter transfer function zeros. (b) Detail for Figure 9-11(a) for $\beta = 0$.

and since

$$\omega_1 = \frac{2}{RC},$$

$$R = \frac{1}{\pi f_1 C} = \frac{1}{(3.14)(10^5)(10^{-9})}$$

$$= 3.18\ K\Omega.$$

Thus

$$R_1 = \frac{R}{16.4} = 194\Omega.$$

The effects of network tapering on the performance of the amplifier were related to power dissipation and sensitivity of critical pole locations. It was shown that a trade off exists between sensitivity and power dissipation over the range of taper constants. Also included was a design procedure employing computer solutions for the distributed-network transfer-function zeros. Tapering may be achieved simply in the thin-film structure by appropriate masking and photolithography applied to the vacuum deposited films.

The open-circuit voltage-transfer function of the tapered $\overline{\text{RC}}$ network in Figure 9-10* is

$$t_v(s) = \frac{\gamma_0 e^{kL} + sC_0 R_1 \sinh \gamma_0 L}{\gamma_0 \cosh \gamma_0 L + (k + sC_0 R_1) \sinh \gamma_0 L} \tag{9.15}$$

where

$$\gamma_0 = \sqrt{k^2 + sR_0 C_0}$$

and the x-dependencies of R and C per unit length are:

$$r(x) = R_0 e^{2kx} \qquad \text{and} \qquad c(x) = C_0 e^{-2kx}.$$

The poles of a high gain feedback amplifier stage tend toward the zeros of the RC network placed in the feedback loop. Therefore, the zero locations of the feedback networks near the real frequency axis become critical in specifying the bandpass properties of the stagger-tuned configuration. Again, an infinite array of zeros exists for the hyperbolic transfer function $t_v(s)$. These zeros are the roots of

$$\gamma_0 e^{kL} + sC_0 R_1 \sinh \gamma_0 L = 0. \tag{9.16}$$

The poles of $t_v(s)$ are on the negative real axis. The zeros of $t_v(s)$ are plotted in Figure 9-11 for several values of taper $\beta = kL$, with $\alpha = R_0 L e^{\beta}/R_1$ as a parameter. Further design curves are given in reference [8].

* Figures 9-10 and 9-11 adapted from [8] by permission.

Sensitivity of Q, defined as in (8.10), to α was found to increase rapidly with taper β, for $\beta > 10$. Experimental results verified the theory for both the uniform $\overline{\text{RC}}$ and tapered $\overline{\text{RC}}$ bandpass filters described above.

At this point in time there was a retrenchment of effort on the part of those involved with passive RC network research. They tried to understand better the more sophisticated $\overline{\text{RC}}$ networks placed in active network topologies, and to acquire broader classes of filter and equalization characteristics. Few additional active $\overline{\text{RC}}$ network results were evolved until about 1966. By that time passive $\overline{\text{RC}}$ networks were well understood both from an analytical and a synthetical point of view. From an analytical viewpoint virtually all nonuniform lines which could be analyzed in a tractable fashion (and many others, too!) had been studied. In addition, several passive synthesis procedures had been proposed, based on both uniform and nonuniform $\overline{\text{RC}}$ networks, which provided for the synthesis of prescribed and quite general frequency characteristics or transient responses. The next sections discuss each of several broad approaches to active $\overline{\text{RC}}$ network synthesis, bringing in the relevant passive synthesis technique as it is pertinent.

9.3 THE DESIGN OF RC-$\overline{\text{RC}}$ ACTIVE FILTERS

There are fundamentally three approaches to active RC network synthesis. Each of these approaches is distinguished on the basis of the ideal active-gain element employed. One approach uses negative impedance converters (NIC's), the second, operational amplifiers, and the third, gyrators. Approaches using negative impedance converters were stressed during the 1950's (they provide the most tractable mathematical solutions), but very little study was made of their sensitivity to component tolerances. Ultimately, when the design of active filters proceeded beyond the purely academic stage into the realm of practical realizations, it was found that the sensitivities associated with NIC-type synthesis procedures were generally excessive.

A great deal of work has been done to realize a gyrator, and this element shows substantial promise. However, since its fundamental objective in active synthesis is to convert the capacitor into a structure which behaves like an inductor for use in LC synthesis, it has yet to be proposed as a serious candidate for $\overline{\text{RC}}$ network synthesis.

Operational amplifiers then take on the dominant role in providing active $\overline{\text{RC}}$ filter characteristics. Certain operational amplifier active $\overline{\text{RC}}$ configurations have acceptably low sensitivities, and perhaps just as importantly, make use of inexpensive commercial integrated operational amplifiers with near ideal characteristics. Essentially, operational amplifiers provide a voltage-controlled voltage source (VCVS) capability as was first used in a broad sense by Sallen and Key [10].

In most active RC configurations, one is struck by the large number of passive elements which far outstrip both the cost and the size of the active element, i.e., the operational amplifier. A potential advantage of using $\overline{\text{RC}}$ networks in active configurations is the reduction of the number of passive elements and the substantial reduction which may be expected in the substrate area taken by the passive elements. For example, $\overline{\text{RC}}$ elements combine both the capacitor and resistor functions on the same substrate area. A simple active $\overline{\text{RC}}$ network which provides sharp cut-off lowpass characteristics [11, 12] is shown in Figure 9-12. The $[ABCD]$ transfer matrix of the

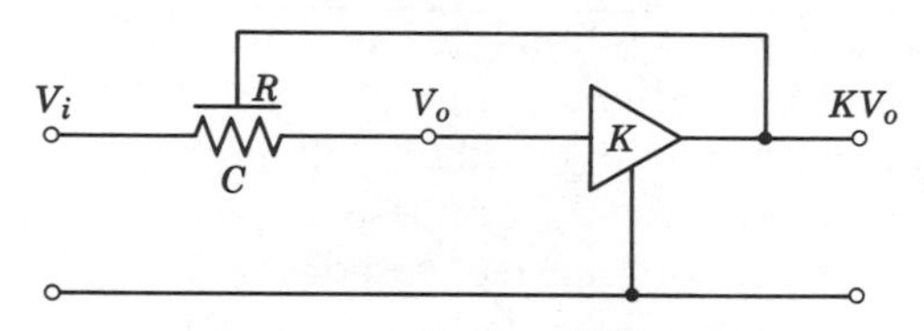

FIGURE 9-12 Active $\overline{\text{RC}}$ filter.

passive-distributed RC network, with its capacitive terminal grounded, is given by

$$F(s) = \begin{bmatrix} \cosh\sqrt{s\tau} & \dfrac{R\sinh\sqrt{s\tau}}{\sqrt{s\tau}} \\ \dfrac{\sqrt{s\tau}\sinh\sqrt{s\tau}}{R} & \cosh\sqrt{s\tau} \end{bmatrix} \tag{9.17}$$

where $\tau = RC$.

It is simple to show that the overall transmission matrix of the active network is given by

$$F(s) = \cosh\sqrt{s\tau}\begin{bmatrix} (1-k) + k\,\text{sech}\sqrt{s\tau} & \dfrac{R\tanh\sqrt{s\tau}}{\sqrt{s\tau}} \\ \dfrac{(1-k)\sqrt{s\tau}\tanh\sqrt{s\tau}}{R} & 1 \end{bmatrix}. \tag{9.18}$$

The voltage transfer function of this network

$$t_v(s) = \frac{V_0}{V_i}(s) = \frac{1}{k + (1-k)\cosh\sqrt{sRC}} \tag{9.19}$$

permits complex pole pairs to be realized rather readily. The first (dominant) poles are given by

$$sRC = (\log(M + \sqrt{M^2 - 1}))^2 \tag{9.20}$$

where

$$M = -k/(1-k). \tag{9.21}$$

These dominant pole solutions, shown in Figure 9-13,* lead to real frequency responses as shown in Figure 9-14, for several values of k from $k = 0$ (i.e., the passive structure) to $k \cong .921$. Values of $k > .921$ result in oscillatory response. A third-order Butterworth filter characteristic is included in this figure for comparison of amplitude cut-off with frequency.

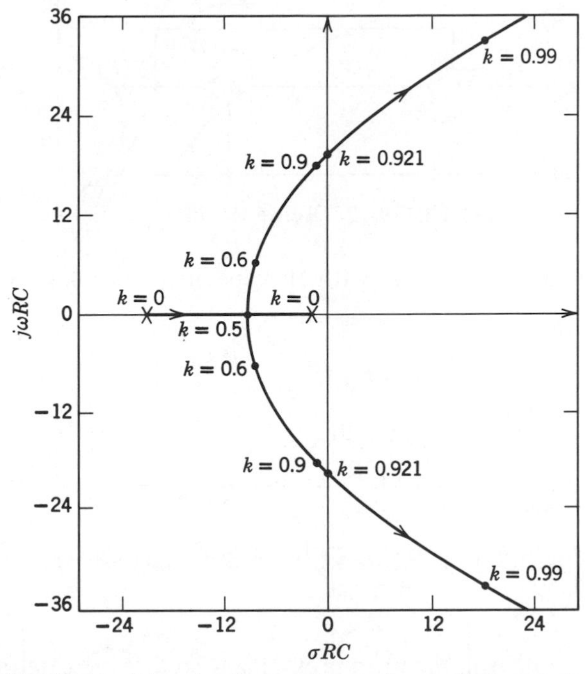

FIGURE 9-13 First dominant pole root locus of Figure 9-12 as function of k.

The value of k for which the dominant poles cross the $j\omega$-axis is about .921. Because the sensitivity of $t_v(s)$ to k,

$$S_k^{t_v} = \frac{dt_v/t_v}{dk/K} = \frac{1}{(1-k)} t_v(s) \tag{9.22}$$

* Figures 9-13 through 9-18 adapted from [12], with permission of the Institute of Electrical and Electronic Engineers.

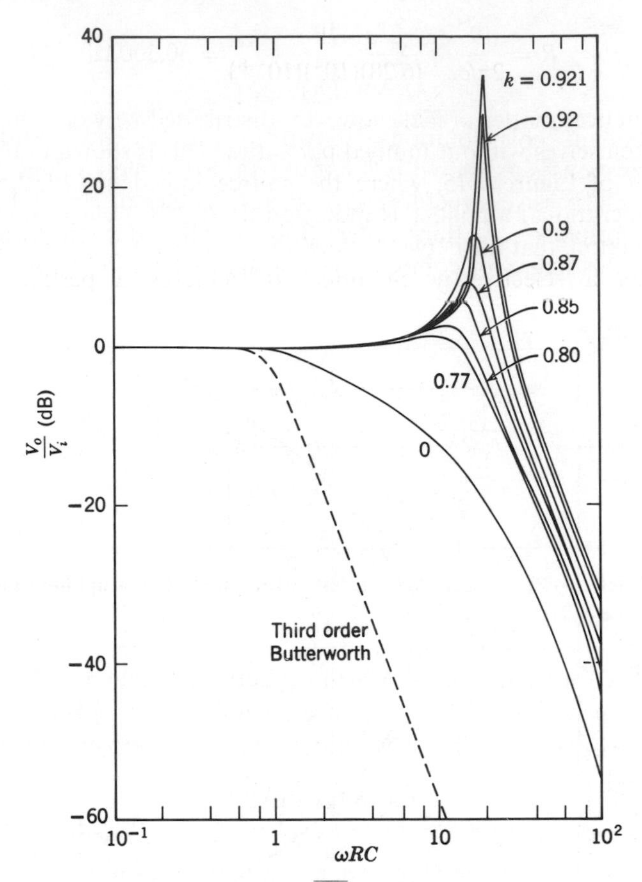

FIGURE 9-14 Active $\overline{\text{RC}}$ frequency response curves.

involves $(1 - k)$ in its denominator, preliminary studies indicate that active $\overline{\text{RC}}$ structures have lower sensitivities to k than their active lumped RC counterparts, such as Sallen and Key circuits, where values of k approaching .99 are required in analogous configurations to achieve similar selectivities and lowpass characteristics [12].

Example 9-2. Design an A$\overline{\text{RC}}$ network using the configuration of Figure 9-12 for which the 3 dB cut-off frequency is 10 kHz, the frequency domain overshoot is 3 dB, and the total capacitance of the distributed network is 10,000 pF.

Solution: From Figure 9-14, to achieve 3 dB overshoot, $k = 0.80$. For $k = 0.80$, the normalized 3 dB cut-off frequency $\omega RC = 19$. Thus

$$R = \frac{19}{2\pi fc} = \frac{19}{(6.28)(10^4)(10^{-8})} = 30{,}300\ \Omega.$$

As a practical matter of fact, however, distributed networks never really exist by themselves without lumped parasitics. This is shown in the equivalent circuit of Figure 9-15, where the source impedance of the voltage-follower operational amplifier is indicated as R_1. It is clear, following the previous section, that the network shown in this figure, even with $k = 0$, can perform as a selective notch filter. It achieves a perfect notch at

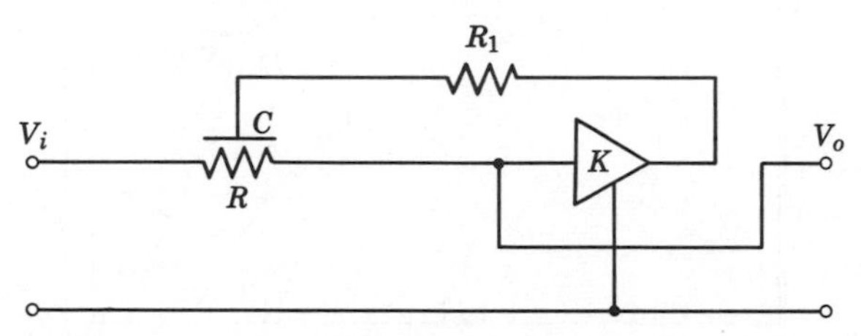

FIGURE 9-15 Active $\overline{\text{RC}}$ filter with parasitic amplifier output impedance R_1.

$\omega_0 \cong 11.19/RC$ if $R_1/R \cong .0562$. When the active RC network of Figure 9-12 is altered to include a source impedance, a similar response is to be expected letting $\alpha = R_1/r$ and $\tau = RC$. The voltage transfer function is given by

$$t_v(s) = \frac{1 + \alpha\sqrt{s\tau}\sinh\sqrt{s\tau}}{k + (1-k)\cosh\sqrt{s\tau} + \alpha\sqrt{s\tau}\sinh\sqrt{s\tau}}. \tag{9.23}$$

For $k = 0$, this reduces to the passive notch network; for $\alpha = 0$ it reduces to the network of Figure 9-12. For $\alpha = 0$ and $k = 0$, it reduces to the case of the passive uniform $\overline{\text{RC}}$ segment.

The frequency response of the expected lowpass characteristic of the network of Figure 9-12 is seriously degraded when the VCVS is attended by even a small source resistance R_1. In fact, for $\alpha = R_1/R$ as small as 10^{-4}, the lowpass response is degraded as shown in Figures 9-16 and 9-17 for $k = .8$, and .9, with α as a parameter. Clearly some caution is warranted when building such a lowpass filter using practical voltage amplifiers.

Example 9-3. Assume an active $\overline{\text{RC}}$ network of the configuration given in Figure 9-15 is designed for audio application with $R = 36K\Omega$ and $k = 0.8$. What maximum output impedance R_1 can be tolerated to insure that the out-of-band response is down more than 40 dB up to a decade above the 3 dB cut-off frequency?

Solution: From Figure 9-16, $\alpha \leq .0001$. Thus $R_1 \leq 3.6\Omega$.

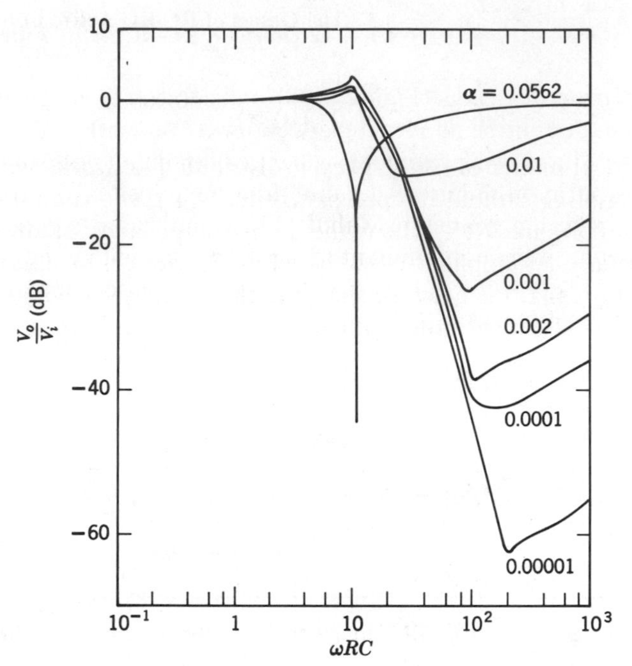

FIGURE 9-16 Lowpass response for $k = 0.8$.

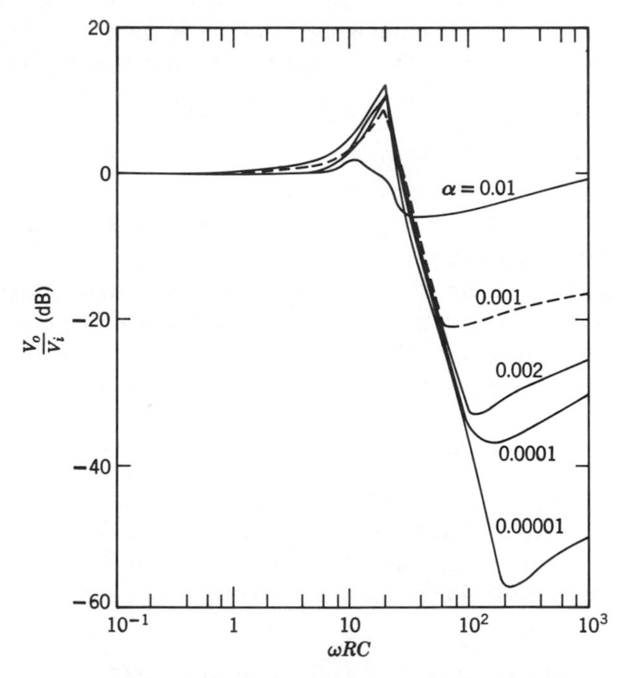

FIGURE 9-17 Lowpass response for $k = 0.9$.

An $\overline{\text{RC}}$ network cascade (Figure 9-18) overcomes the undesired high-frequency response of the above single-section $\overline{\text{RC}}$ network.

Design procedures using computer-generated dominant pole solutions for filters of the form of Figure 9-12, but using exponentially tapered $\overline{\text{RC}}$ networks, have been presented by Mahdi [13]. Mahdi has also demonstrated that the presence of the nondominant poles associated with either the U$\overline{\text{RC}}$ or E$\overline{\text{RC}}$ active filters tends to linearize the phase response, compared to the lumped ARC counterpart.

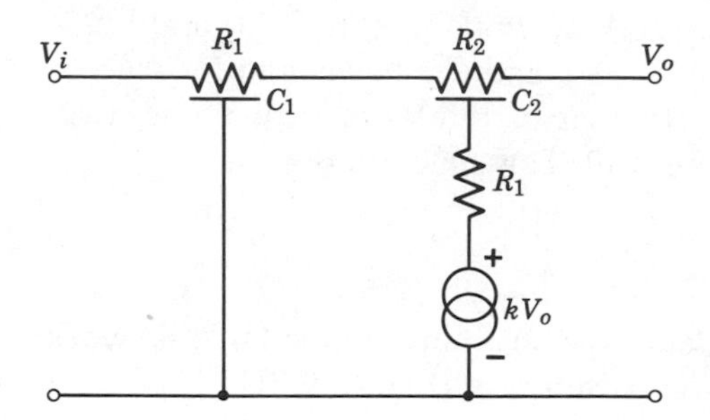

FIGURE 9-18 Two segment active $\overline{\text{RC}}$ section.

Kerwin and Huelsman have taken an approach making use of both distributed and lumped RC networks combined with active elements [11, 14, 15]. This synthesis method for a prescribed amplitude response, corresponding to any combination of left half plane poles and $j\omega$-axis zeros, requires only a cascade of a few simple distributed-lumped-active (DLA) RC networks.

Figure 9-19 shows a lumped active RC network which may realize two complex poles and two real frequency zeros. Figure 9-20 shows a DLA network which produces a very similar amplitude response, but it is quite clearly much simpler and requires a smaller voltage-controlled-voltage-source gain ($A_2 < A_1$). Kerwin's synthesis procedure is based on the concept of effective dominant pole (discussed by Ghausi and Kelly) [4]. One can clearly see that the network topology of Figure 9-20 is identical to the previously noted notch network of Figure 9-15 except for the lumped capacitor.

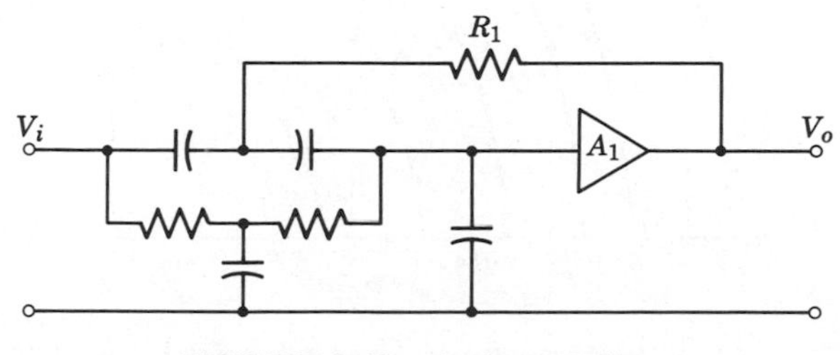

FIGURE 9-19 Active RC filter.

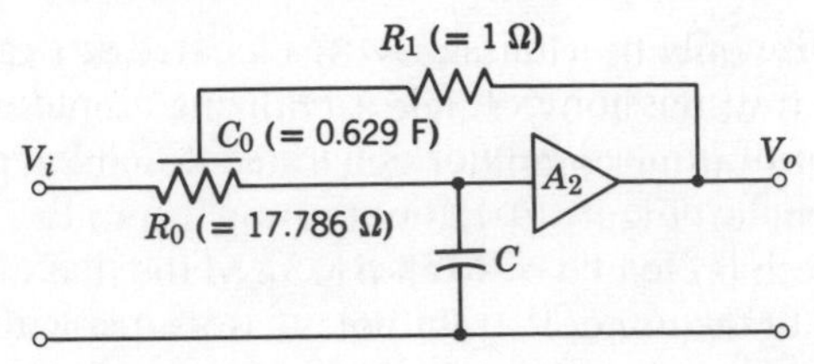

FIGURE 9-20 Active $\overline{RC}$ network to produce two complex zeros and two complex poles.

Kerwin is essentially matching the DLA network (Figure 9-20) amplitude response (with the specific values of R_0, R_1, and C_0 shown, and for various values of gain A_2 and capacity C) to that of the rational two-pole, two-zero function associated with Figure 9.19. This function is given by

$$t_v(s) = \left.\frac{A_2 \varepsilon_1 (s^2 + 1)}{s^2 + \delta_1 s + \varepsilon_1}\right|_{\varepsilon_1 < 1} .$$

Kerwin has plotted the "equivalent pole positions" of the DLA network as a function of the DLA network parameters (Figure 9-21). This DLA

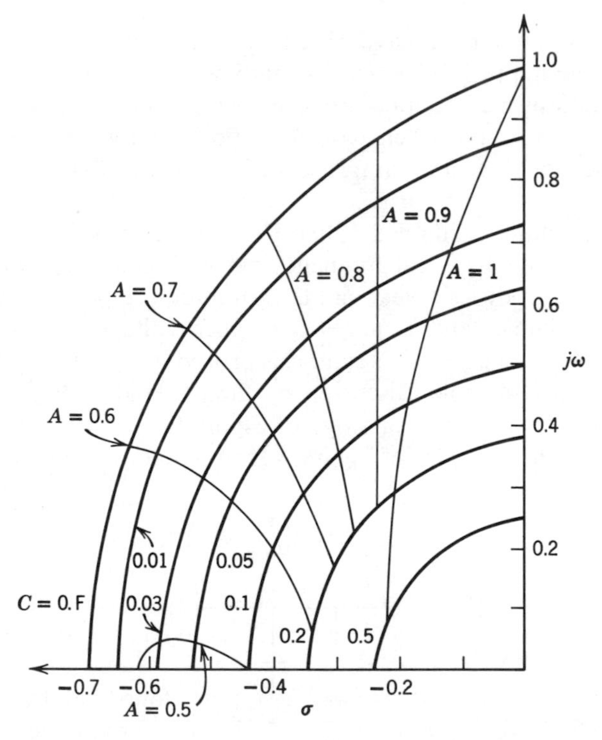

FIGURE 9-21 Equivalent pole positions for Figure 9-20.

network will realize only transmission zeros located at a radius greater than the radius to the transmission poles. For realizing amplitude characteristics matched to those of lumped networks with specified lowpass performance resulting from specific pole-pair positions, one chooses the values of A and C (shown in Figure 9-21) for the DLA network. This procedure provides the network values in Figure 9-20 resulting in transmission zeros (for sharp cut-off) normalized to $\omega = 1$, as well as the specified transmission poles.

If the poles of the voltage-transfer function are intended to lie at a radius from the origin greater than the zeros, and in fact anywhere outside of the region circumscribed by $C = 0$, then the building block shown in Figure

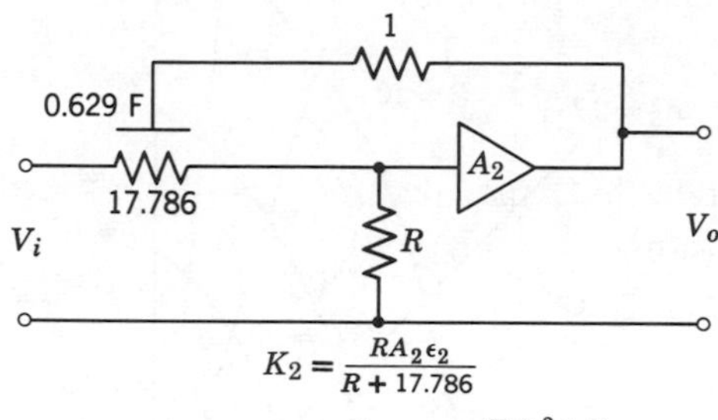

FIGURE 9-22

9-22 with the specific values shown may be used. Again a set of design curves in reference [16] permits one to locate the dominant pole locations and to read off corresponding values of R and A_2. Essentially then we have replaced the capacitor with the resistor in very much the same fashion as Sallen and Key [10] have done.

To realize a simple single pole pair, the active discrete element RC network shown in Figure 9-23(a) may be used. The resulting transfer function is of the form $V_0/V_i = K/(s^2 + \delta s + \varepsilon)$. However, as indicated previously, the network of Figure 9-23(b) will provide essentially the same characteristic to about a decade above the resonant frequency, and for equivalent performance one need only choose equivalent pole positions for the $\overline{ARC}$ network and refer to Figure 9-23(c) to identify appropriate values of A_3 and C.

Kerwin has shown that a given five-pole, four-(real frequency) zero lowpass elliptic filter function can be very closely approximated by a cascade of the appropriate $\overline{ARC}$ building blocks.

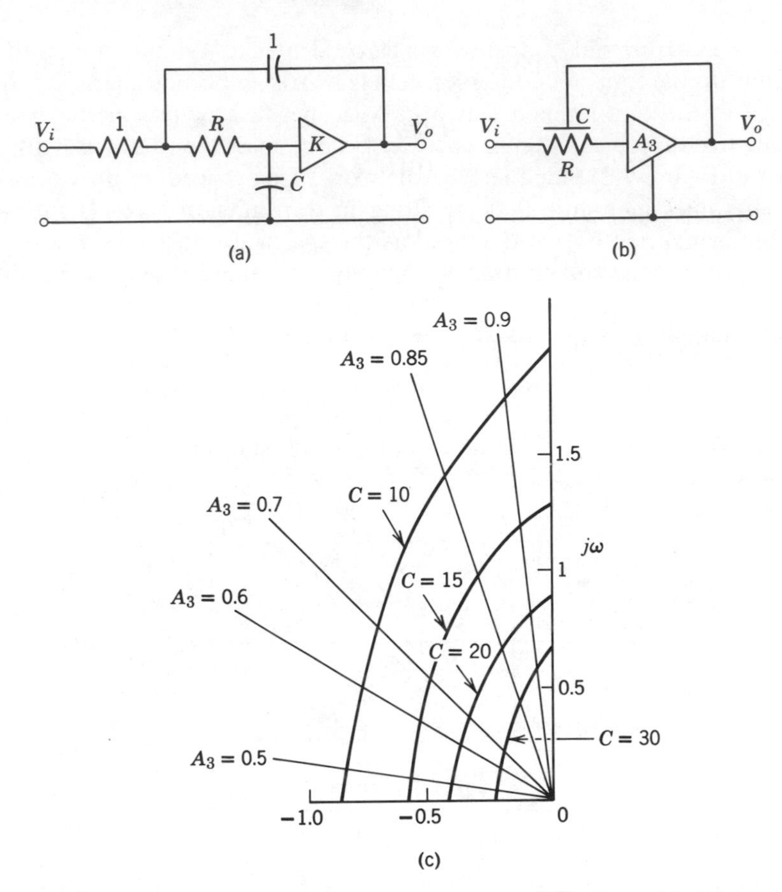

FIGURE 9-23 (a) Active RC filter. (b) Active $\overline{RC}$ filter. (c) DLA design chart for network of Figure 9-23(a).

Example 9-4. Design a network [16] which achieves an overall voltage-transfer function given by

$$t_v(s) = \frac{0.168(0.359s^2 + 1)(0.738s^2 + 1)}{(s^2 + 0.488s + .501)(s^2 + 0.114s + .808)(s + 0.416)} \tag{9.24}$$

corresponding to an equi-ripple passband with a tolerance of 0.5 dB, and an equi-ripple stopband with a minimum attenuation of 40 dB).

Solution: We may partition the overall transfer characteristic into the product of the three transfer characteristics, t_a, t_b, and t_c:

$$t_a(s) = \frac{0.359s^2 + 1}{s^2 + 0.488s + .501} \tag{9.25}$$

$$t_b(s) = \frac{0.738s^2 + 1}{s^2 + 0.144s + .808} \tag{9.26}$$

$$t_c(s) = \frac{0.416}{s + 0.416}. \tag{9.27}$$

Substituting $\rho^2 = 0.359s^2$ in $t_a(s)$, so that the zeros are normalized to $\omega = \pm 1$, we obtain

$$t_a'(\rho) = \frac{\rho^2 + 1}{2.785\rho^2 + 0.814\rho + 0.501}. \tag{9.28}$$

We must perform the same normalization in all cases so that the design chart in Figure 9-21 may be used. The resulting primed networks are shown in Figure 9-24, and the overall network characteristic is provided in Figure 9-25.

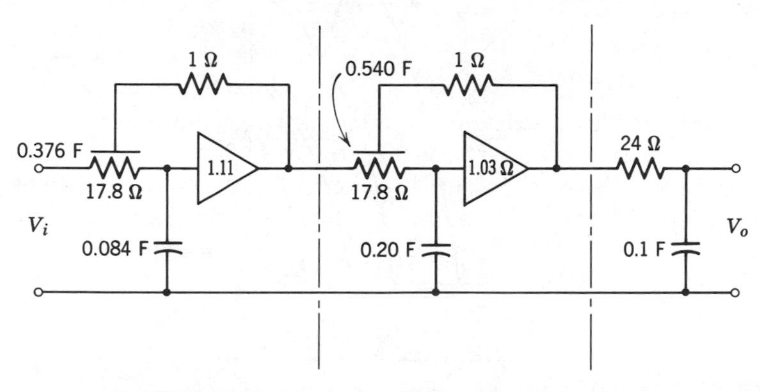

FIGURE 9-24 Normalized DLA network example.

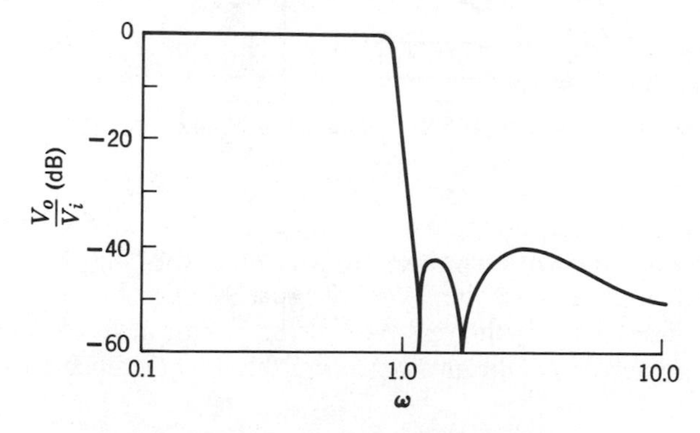

FIGURE 9-25 DLA network response.

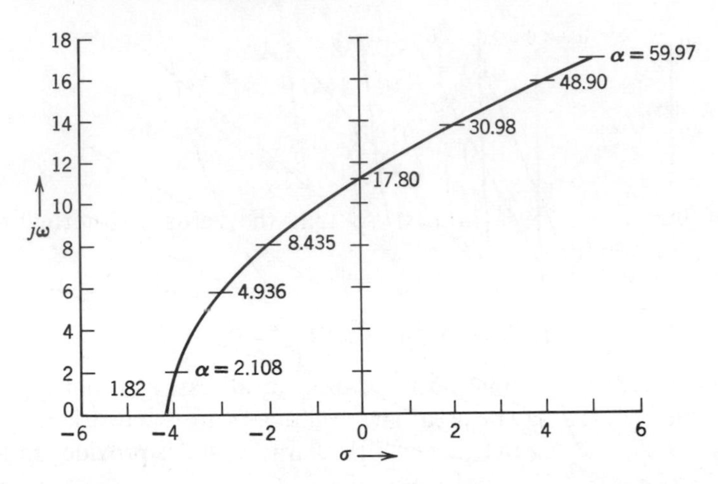

FIGURE 9-26 Locus of the zeros of the network of Figure 9-3 as a function of $1/R_1$.

As indicated earlier, the augmented $\overline{URC}$ passive segment of Figure 9-3 can provide complex or imaginary zeros, including those in the right half plane. These complex zeros have the positions shown in Figure 9-26 as functions of $\alpha = R/R_1$. A passive D-L network may be used in the active configuration of Figure 9-27 to obtain a bandpass characteristic with

$$t_v(0) = t_v(\infty) = 0. \tag{9.29}$$

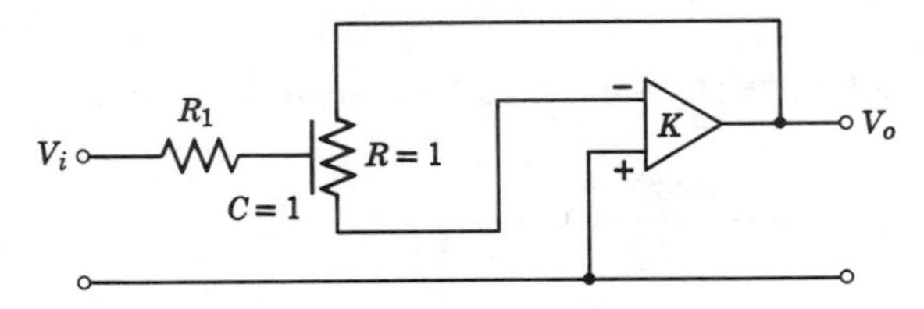

FIGURE 9-27 Bandpass DLA network with phantom right half plane zeros.

For this network, a computer program [18] has provided Q and Q-sensitivity (Figure 9-28), and peak frequency ω_0 (Figure 9-29), as functions of amplifier gain K. In these cases, σ (the real part of the phantom zero* position) is the design parameter. σ is chosen approximately by reference to Figure 9-28.

* See Section 8.3.

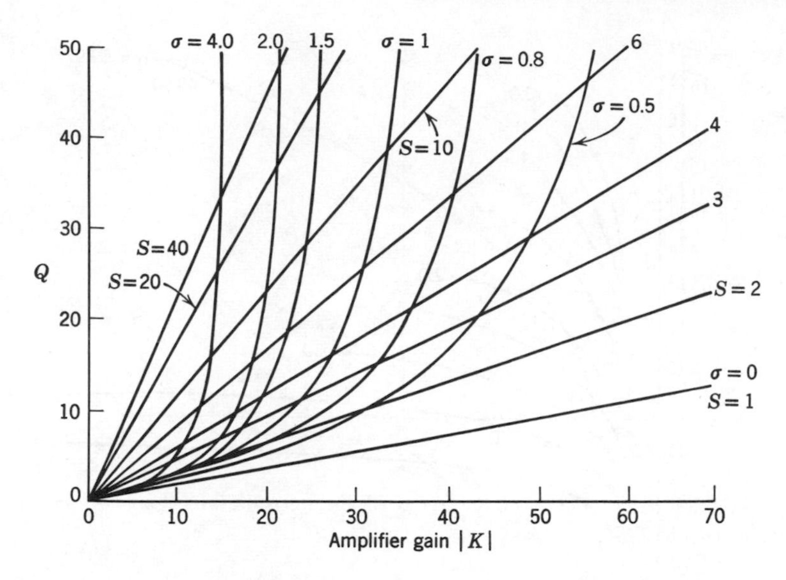

FIGURE 9-28 Q **versus** $|K|$ **as a function of the right half plane phantom zero position** (σ) **for the network of Figure 9-27.**

Then [18]

$$R_1 \simeq \frac{1}{17.80 + 5.40\sigma + 0.529\sigma^2} \tag{9.30}$$

within $\pm .4\%$ from $\sigma = 0$ to $\sigma = 5.0$. As the phantom zeros go further into the right half plane, the gain K required for a particular system gain is lessened.

Example 9-5. Assume we require a bandpass filter [18] with $Q = 50$ and $S_K^Q < 10$.

Solution: Referring to Figure 9-28, we find $K \simeq 42$ and $\sigma \simeq 0.8$. From (9.30), $R_1 \simeq .045\Omega$. Then from Figure 9-29, one solution is that $\omega_0 = 12$ rps, assuming $|K| = 42$ dB. Sensitivity to passive-element parameters is very small, and $S_C^Q = 0$. Finally, if R_1 tracks R (as it would in an IC realization),

$$S_R^Q + S_{R_1}^Q = 0.$$

All that remains is sensitivity of Q to K, given above, and to excess phase variations.

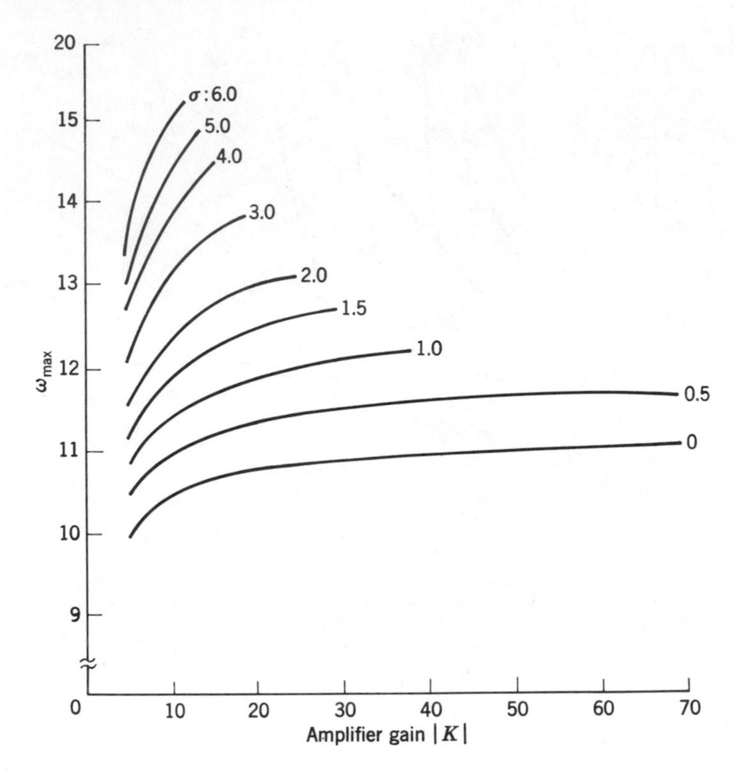

FIGURE 9-29 Frequency of maximum response versus amplifier gain and σ.

9.4 STABILITY OF ACTIVE $\overline{RC}$ NETWORKS

In 1969, Woronka and Barclow applied Wyndrum's exact synthesis procedure [1] for $\overline{RC}$ network synthesis and drawing on the methodology of microwave-filter theory, an active filter configuration using the Linvill approach, and no lumped elements resulted. This structure (Figure 9-30) required only commonground-cascaded $\overline{RC}$ segments with shunt $\overline{RC}$ stubs for the passive components. It expanded on an earlier approach [19] which utilized a negative impedance converter (NIC) cascaded with $\overline{RC}$ network sections.

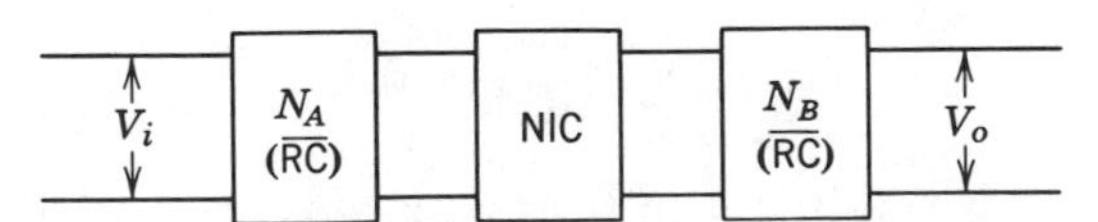

FIGURE 9-30 Active $\overline{RC}$ network configuration.

Perhaps more importantly, Woronka and Barclow [20] described the conditions necessary to realize an active uniform $\overline{\text{RC}}$ network with a z_{21} achieving complex poles and complex finite transmission zeros. First, let us review the pertinent background material. A U$\overline{\text{RC}}$ element (Figure 9-31)*

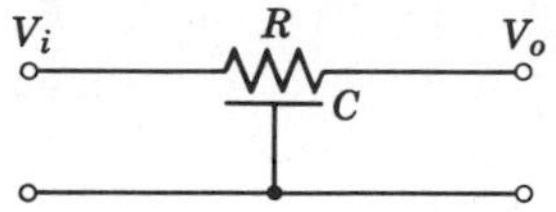

V_i R V_o C

$$Z_{11} = Z_{22} = \frac{R \coth \sqrt{sRC}}{\sqrt{sRC}} \Rightarrow \frac{R \coth \sqrt{sRC}}{\sqrt{RC}} \Rightarrow \frac{\alpha}{W}$$

$$Z_{12} = Z_{21} = \underbrace{\frac{R \operatorname{csch} \sqrt{sRC}}{\sqrt{sRC}}}_{s \text{ plane } (\overline{\text{RC}})} \Rightarrow \underbrace{\frac{R \operatorname{csch} \sqrt{sRC}}{\sqrt{RC}}}_{\sqrt{s} \text{ plane } (\overline{\text{LC}} \text{ plane})} \Rightarrow \underbrace{\frac{\alpha(1 - W^2)^{1/2}}{W}}_{W \text{ plane (L-C plane)}}$$

FIGURE 9-31 Single segment characterization.

has a hyperbolic immittance matrix with an infinity of negative real poles and zeros. In the initial development of exact passive U$\overline{\text{RC}}$ synthesis procedures [1], cascade connections of common-ground U$\overline{\text{RC}}$ elements were introduced with shunt stubs to achieve finite transmission zeros. Parallel interconnection of such networks would permit complex transmission zeros. A positive real conformal transformation converts the transcendental expressions to rational functions of a new variable, W. The $\overline{\text{RC}}$ functions become lumped LC functions, i.e., polynomial ratios. Richards' theorem is used to complete the synthesis. Necessary and sufficient conditions were determined for realizability of the driving-point impedance functions, and sufficient conditions for transfer impedance functions [1]. Detailed stability conditions are derived in the W plane, which are, of course, independent of the topological realization [20, 21].

The U$\overline{\text{RC}}$ single and double segments to be used in this procedure are characterized in Figures 9-31 and 9-32. Cascaded $\overline{\text{RC}}$ elements of course, produce only negative real poles, though they generally possess an infinite number of distinct poles and zeros which are difficult to manipulate in a synthesis procedure. Fortunately, it is possible to map these functions into

*Figure 9-31 through 9-36 adapted from reference [20] with permission of the Institute of Electrical and Electronic Engineers.

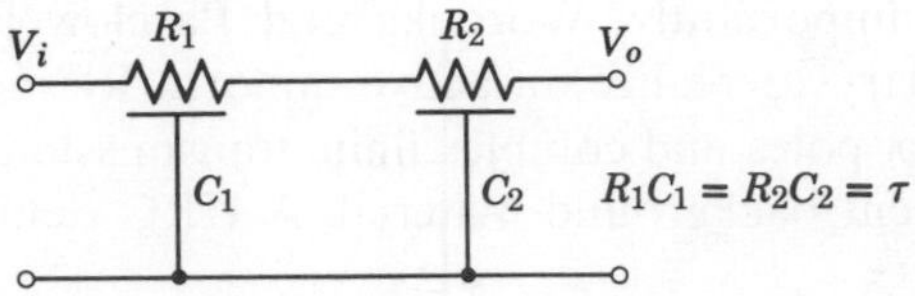

V_i R_1 R_2 V_o

C_1 C_2 $R_1C_1 = R_2C_2 = \tau$

$$\frac{Z_{11}}{R_1 R_2} = \frac{[1 + R_1/R_2 \tanh \sqrt{s\tau}]}{\sqrt{s\tau} \tanh \sqrt{s\tau}\,(R_1 + R_2)} \Rightarrow \frac{[1 + R_1/R_2 \tanh \sqrt{s\tau}]}{\sqrt{\tau} \tanh \sqrt{s\tau}\,(R_1 + R_2)}$$

$$\Rightarrow \frac{(1 + W^2)\alpha_1\alpha_2}{W(\alpha_1 + \alpha_2)}$$

$$\frac{Z_{21}}{R_1 R_2} = \frac{1}{\sqrt{s\tau} \cosh \sqrt{s\tau} \sinh \sqrt{s\tau}\,(R_1 + R_2)}$$

$$\Rightarrow \frac{R_1 R_2}{\sqrt{\tau} \cosh \sqrt{s\tau} \sinh \sqrt{s\tau}\,(R_1 + R_2)} \Rightarrow \frac{\alpha_1\alpha_2}{\left(\dfrac{W}{1 - W^2}\right)(\alpha_1 + \alpha_2)}$$

where $\alpha_1 = \dfrac{\sqrt{\tau}}{R_1}$

and $\alpha_2 = \dfrac{\sqrt{\tau}}{R_2}$

FIGURE 9-32 Double segment characterization.

other complex frequency planes where they become simple ratios of rational polynomials:

$$Z_{L\bar{C}}(\sqrt{s}) = \sqrt{s}\, Z_{\overline{RC}}(s) \tag{9.31}$$

which leaves an infinity of poles and/or zeros shifted from the negative real axis in the s-plane to the imaginary axis in the $\sqrt{s}$-plane. A second transformation:

$$W = \tanh \sqrt{s\tau} \tag{9.32}$$

provides lumped LC functions, shown in Figures 9-31 and 9-32.

The use of these basic $\overline{RC}$ elements in cascade and shunt (stub) configurations within the structure of Figure 9-30 permits the realization of lowpass, highpass, and high-Q bandpass structures.

The W $(= \Sigma + j\Omega)$ plane coordinates in terms of the $\sqrt{s}\,(= U + jV)$ plane are

$$\Sigma = \frac{\sinh \sqrt{\tau}\, U \cosh \sqrt{\tau}\, U}{\cos^2 \sqrt{\tau}\, V + \sinh^2 \sqrt{\tau}\, U} \quad \text{(Real Axis)}$$

$$\Omega = \frac{\sin \sqrt{\tau}\, V \cos \sqrt{\tau}\, V}{\cos^2 \sqrt{\tau}\, V + \sinh^2 \sqrt{\tau}\, U} \quad \text{(Imaginary Axis)}\ .$$

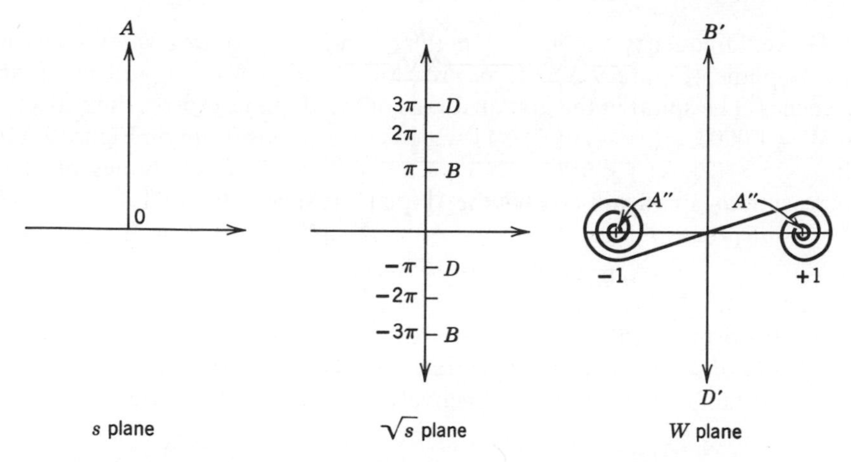

FIGURE 9-33 Identification of regions after transformation.

The entire s-plane will map into the closed upper half (or lower half) of the $\sqrt{s}$-plane (Figure 9-33). The positive real frequency axis in the s-plane corresponds to the 45° (and 225°) line in the $\sqrt{s}$-plane. The semi-infinite strip in the $\sqrt{s}$-plane (Figure 9-34) with range

$$-\frac{\pi}{2\sqrt{RC}} < \mathrm{Im}(\sqrt{s}) < \frac{\pi}{2\sqrt{RC}}$$
$$-\infty < \mathrm{Re}(\sqrt{s}) < +\infty$$

will map into the entire W-plane.

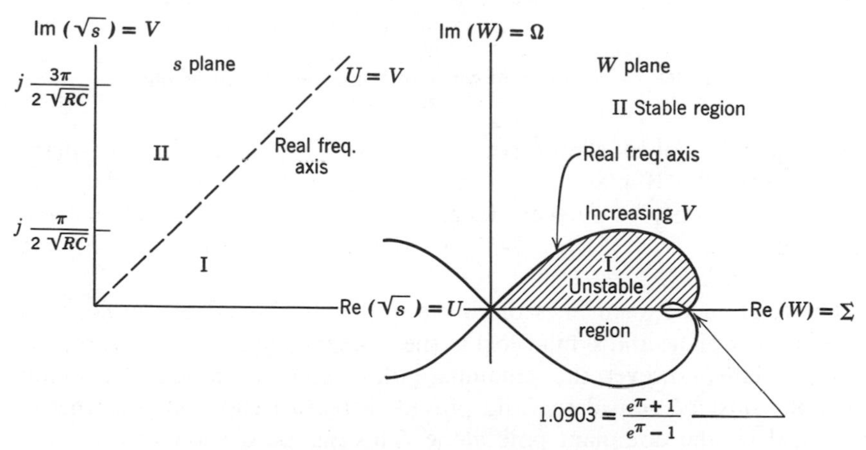

FIGURE 9-34 $\sqrt{s} \rightarrow W$-plane mapping.

The real frequency axis ($U = V$ in the $\sqrt{s}$-plane) becomes a spiral locus in the W-plane (Figure 9-33). It begins, for $s = 0$, at $W = 0$, and has four branches. The spiral in the first quadrant of the W-plane crosses the real axis at $W = 1.0903 = (e^\pi + 1)/(e^\pi - 1)$, which corresponds (note on Figure 9-33) to $\sqrt{s} = \pi/(2\sqrt{RC}) + j\pi/(2\sqrt{RC})$ or $s = j\pi^2/(2RC)$. It continues on, for increasing ω, and again crosses the W-plane real axis at

$$s = j\frac{\pi^2}{RC}, \text{ i.e., } W = 0.996272\cdots = \frac{e^{2\pi} - 1}{e^{2\pi} + 1}.$$

The transformed real frequency axis divides the W-plane into regions of stable and unstable operation (Figure 9-34) which correspond to the left half and right half of the s-plane, respectively. Figure 9-35 illustrates a detailed

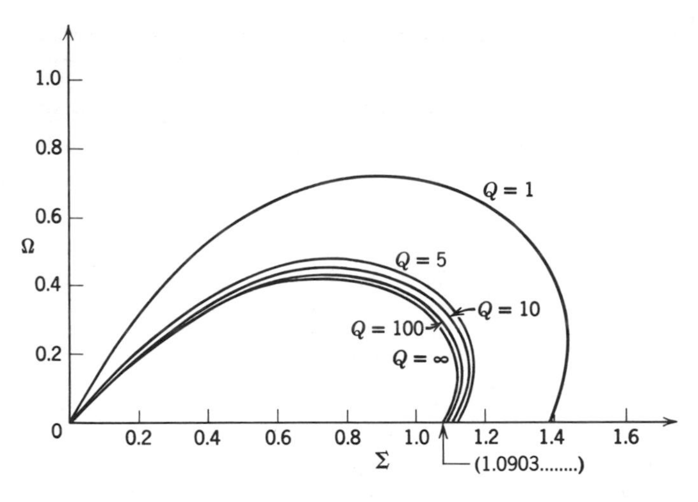

FIGURE 9-35 Lines of constant Q mapped into W plane.

mapping of lines of constant Q [i.e., selectivity, $Q = (\omega_0)/(2\sigma)$] in the s-plane, mapped into the W-plane. Their intersections with the real W-plane axis provide the basis for further design of s-plane transfer functions with prescribed selectivity (Figure 9-35), particularly those requiring only real W-plane poles.

The relation between lines of constant $Q[= (\omega_0)/(2\sigma)]$ of the *dominant* pole in the s-plane and similar loci in the W-plane are given in Figure 9-35. In the s-plane, however, the remaining poles which accompany the dominant pole must be considered and provide a perturbation on performance predicted by the dominant pole alone. This has been shown to be small within a decade of the dominant pole frequency. Design curves, which relate

Q to the X of the $(\tanh^2 \sqrt{s\tau} - |X|^2)$ term for the case where only positive real axis W-plane poles are realizable, are given in Figure 9-36. Poles in the

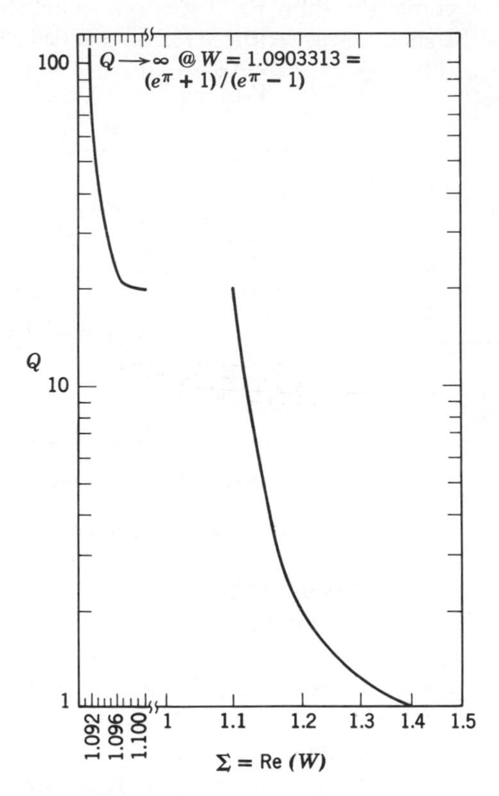

FIGURE 9-36 Design curve for choice of $W = \Sigma + j0$ as a function of $Q = \omega_0/2\sigma$ of dominant pole.

W-plane must lie outside the spirals labelled $U = V$ to insure stability. All this leads to:

Theorem: A necessary and sufficient condition for a stable-active-distributed RC network function rational in $W = \tanh(a\sqrt{s})/2$, and whose numerator (in W) is of degree less than or equal to that of the denominator, is that all of the W-plane poles map outside the region (shown in Figure 9-35 for the first quadrant) formed by

$$\frac{e^{|\phi|}e^{j\phi} - 1}{e^{|\phi|}e^{j\phi} + 1}$$

where $-\pi \leq \phi = \arg(1 + W)/(1 - W) \leq \pi$.

9.5 TRANSVERSAL $\overline{RC}$ FILTERS

In 1966 and 1967 W. I. H. Chen and R. C. Levine considered the possibility of using wholly integrable thin-film $\overline{RC}$ networks to achieve a prescribed time-domain response such as was required for transversal equalizer characteristics. Transversal filters (as we can see in Figure 9-37) have a special

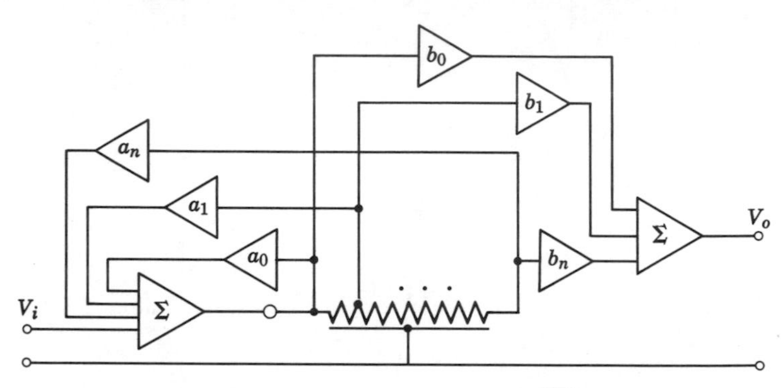

FIGURE 9-37 Recursive tapped $\overline{RC}$ line.

distribution of taps which may be interconnected in any fashion, and whose outputs are generally summed to form the desired output signal. Traditionally, transversal filters have used distributed or lumped LC (lossless) delay lines. Except at very high frequencies, the sheer volume of the hardware required to implement this approach has severely limited the use of transversal filters. Of course, the $\overline{RC}$ transmission lines which may be used in their stead lead to a very large reduction in the filter size, and they open up the possibility that the $\overline{RC}$ structure (which obeys the diffusion equation) may provide an adequate replacement for the pure $\overline{LC}$ delay network. The synthesis procedure, then, essentially requires that one determine a process to compute the contribution of each individual tap waveform, and a process by which these are suitably summed to provide the overall signal. Transversal $\overline{RC}$ filters may be either of the recursive or nonrecursive types; i.e., with or without feedback. Chen and Levine provided a digital computer technique for identifying the coefficients required to synthesize various overall filter specifications. Working on the basis of synthesizing a prescribed *time-domain* impulse response, they provided an algorithm to determine the coefficients α_i, resulting in the least mean square error approximation to the desired impulse response.

The output signal at a tap in the absence of feedback in response to an impulse excitation is shown in Figure 9-38* This, of course, is the exact

* Figures 9-38 through 9-40 adapted from reference [22], with permission of the Institute of Electrical and Electronics Engineers, Inc.

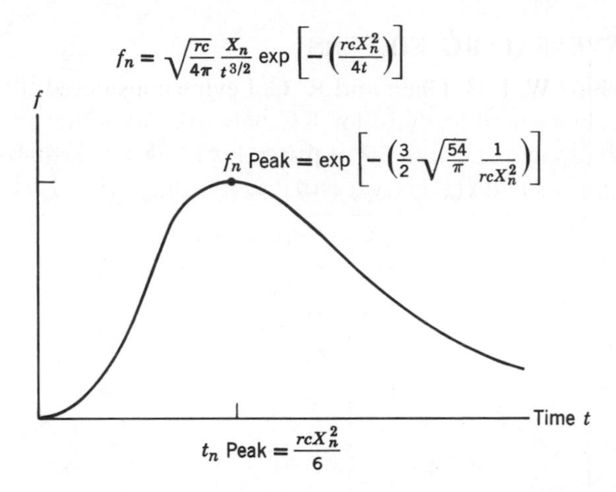

FIGURE 9-38 Impulse response of $\overline{RC}$ transversal filter at tap distance X_n from input.

solution assuming an infinite structure, or one which is terminated in another $\overline{RC}$ line so that no impulse reflection exists at the output. Note several features of the tap functions. First of all, the time at which the peak occurs is proportional to the square of the distance of the tap from the input. The peak amplitude of the signal is inversely proportional to the square of this distance. We are not seeing delay in the usual sense but rather dispersion (Part of the signal appears immediately at all taps when the impulse is applied at the input). The family of responses associated with finite line is shown in Figure 9-39, assuming taps are spaced according to the rule, $X_n = kV_n$.* These tap signals include both the original signal and the reflected signals arising from the two unmatched ends of the line. An open-circuit termination is assumed at the output, and a voltage-source excitation is assumed at the input. Because the characteristic impedance, z_0, of an $\overline{RC}$ line must be terminated in an infinite $\overline{RC}$ line for zero reflection, it is impossible to construct any network which provides a perfect impedance match at the output terminal. The waveforms at those taps near the input of the $\overline{RC}$ line differ very little from what they would be if the line were infinite. Chen and Levine found it necessary to space the taps at points x_n according to a rule: $x_n = k_n$ where k_n is the scaling factor which assures that the peaks of the various waveforms have an equal-time spacing. In his doctoral thesis, Chen showed that this spacing guarantees the completeness of the family of the individual tap functions. That is to say that the functions which are available

* W. I. H. Chen [22] has shown that such spacing guarantees the completeness of the family of tap functions.

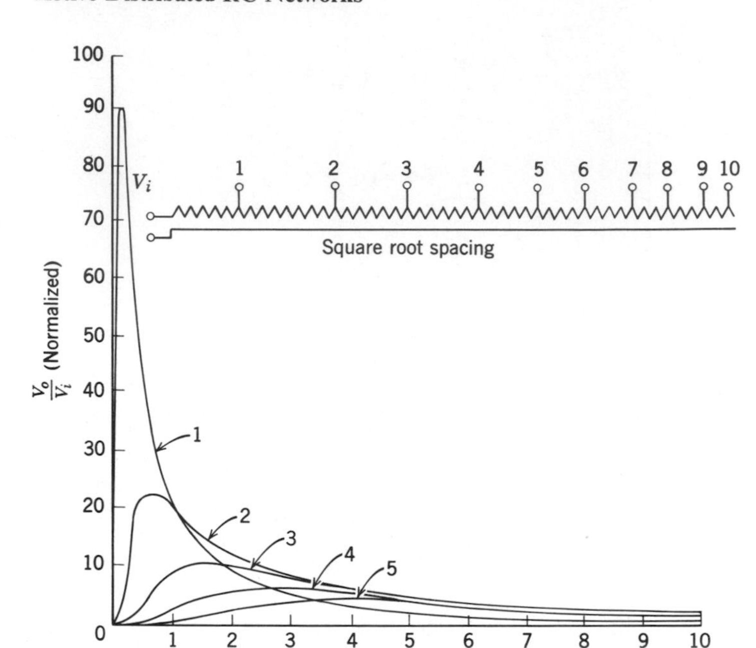

FIGURE 9-39 Impulse response of finite $\overline{RC}$ line with taps spaced according to the rule $X_n = k\sqrt{n}$.

at the taps spaced according to the above rule can be combined to form any finite energy signal such as those that would be derived for lowpass, bandpass, or equalizer filters; this is the most important conclusion in Chen's thesis. Chen and Levine restricted their synthesis procedure to the combination of waveforms which realized an impulse response in the time domain corresponding to a lowpass filter structure in the frequency domain. Their approximation was based on the method of least squares, where the error was measured in the time domain. Thus the error was indicated as

$$I = \int_0^\infty \left[h(t) - \sum_{n=1}^{N} \alpha_n h_n(t) \right]^2 dt. \tag{9.33}$$

Here h is the required impulse response, h_n is the impulse response of the nth tap, and N is the number of the last tap. Of course it was assumed that the taps were spaced according to the rule given above, resulting in N simultaneous algebraic equations. Because the tap functions were shown to be linearly independent, the nonsingular solution of these equations for each of the proper values for x_n was assured. Computations were complex and had to be carried out on a digital computer. Both positive and negative tap values were required. The earlier examples by Chen and Levine tried to

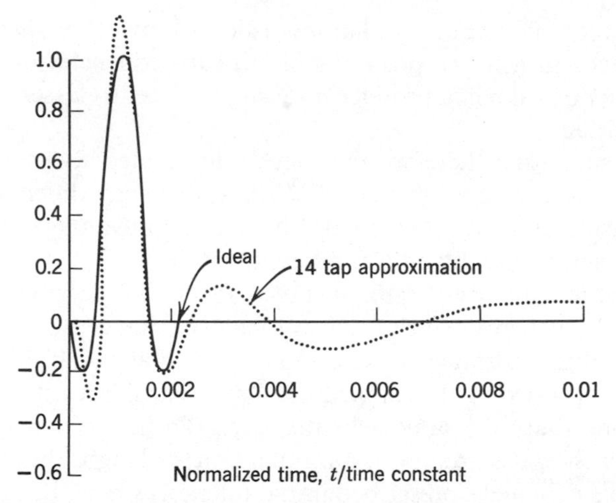

FIGURE 9-40 14 tap approximation to desired time-domain function.

approximate the response shown in Figure 9-40, using a 14 tap $\overline{\text{RC}}$ line. This was, of course, a nonrecursive lowpass filter, and the time-domain approximation that was achieved is shown in dotted format in Figure 9-40.

One of the broad conclusions one can reach concerning $\overline{\text{RC}}$ transversal filters is that it is difficult to realize practical lowpass or bandpass filter characteristics using a small number of taps with a nonrecursive structure. As the approximation to simple functions corresponding to lowpass filters is required to become better, hundreds of taps are necessary to implement the desired approximation with, say, .1 to .3 dB accuracy.

Alternatively, however, there is a possibility of using a recursive structure which has not been thoroughly studied. As our recursive-digital-filter experience indicates, the number of taps required might be a great many less—perhaps a total of 3 to 10 taps. Results by D. A. Spaulding [23] have shown this to be the case, although substantially further effort needs to be applied to achieve a true synthesis procedure.

9.6 LUMPED VERSUS DISTRIBUTED ACTIVE RC FILTERS

When one looks at the poles and zeros which characterize both lumped and distributed RC networks, one is struck by their basic similarity. In fact, the only substantial difference between the poles and zeros of lumped networks and those of distributed networks lies in the difference of their numbers. That is, the distributed networks are characterized by an infinity of poles and zeros. Nonetheless, the frequency performance of the two network classes are fundamentally similar. For the driving-point impedance functions, the poles and zeros alternate along the negative real axis, and they result in monotonic frequency characteristics. Many other similarities may also be noted. In general, positions of critical frequencies of RC and $\overline{RC}$ transfer functions do not produce radically different classes of frequency responses either.

It is not surprising therefore that active distributed RC networks using NIC's or operational amplifiers do not show strikingly different characteristics from similar active-lumped RC filters. The advantages to be gained from using active-distributed RC filters must lie in the simplicity of their realization. Thus, in any specific instance, it may be possible to achieve a distributed RC network which has less total capacitance (thus total area) or perhaps less total resistance. It is also to the advantage of the distributed RC network (as opposed to the lumped network) that the resistive and capacitive functions share the same substrate area. To insure the practicality of these approaches, the broad question of a practical algorithm for trimming $\overline{RC}$ networks to achieve prescribed initial tolerances must be answered. The corresponding problem in lumped RC networks has a vastly simpler solution; one trims discrete resistor values. How does one trim a distributed resistor which is intimately associated with (and cannot be physically divorced from) a distributed capacitor? This major problem has been largely ignored and severely limits the use of active $\overline{RC}$ filters.

The ultimate in fabrication technology that may be possible with distributed RC networks would be to realize both the distributed network and the active element within a single silicon integrated chip. Although this has been suggested by Kaufman, it has never been practically executed.

9.7 FABRICATION CONSIDERATIONS

To date the term active $\overline{RC}$ network synthesis has meant that the passive part of the network realization is an $\overline{RC}$ structure, and the active part of the realization consists of discrete controlled sources. There have been no practical realizations of distributed active elements. It is possible in the near future, that thin-film transistors or large area PN devices, for example, may be found which are capable of providing active distributed gain.

Nonetheless, the combination of silicon integrated circuits as essentially wideband-lumped-gain elements and thin-film $\overline{RC}$ networks has provided the basis for the synthesis techniques described in this chapter. The distributed RC passive networks are most reasonably constructed in thin film. The actual distributed network may be either tapered or uniform. The $\overline{RC}$ network may also provide highpass characteristics. The transfer functions of each of these network configurations have been indicated previously.

In general, either one of two approaches can be used for thin film deposition: vacuum evaporation, or cathode sputtering. These approaches have been thoroughly documented in the literature. If cathode sputtering is used (assuming tantalum is the metal being sputtered), then the tantalum is sputtered on a suitable glass or ceramic substrate where portions of the tantalum that are not wanted are photolithographically removed. Following this, the remaining tantalum is partially oxidized to form a dielectric Ta_2O_5 layer. Finally, gold is deposited as the counterelectrode. Using this approach, the passive distributed RC network may have a total capacitance which is accurate to within a few percent, and the resistance layer may be trimmed to tenths of a percent. The active devices may be provided by beam-leaded, silicon-integrated circuit chips, bonded to interconnecting pads on the thin film substrate. Operational amplifiers have been available as SIC chips for some time; gyrators are just now becoming available as individual chips; and negative impedance converters are quite clearly integrable.

If the distributed RC networks are to be made in silicon technology, then we may use the geometry associated with large area PN junctions that are used to realize small capacitors in SICs. The PN junction to the substrate is strongly back-biased, and the uppermost P type layer is strongly conducting. Evaporated contacts will result in an appropriate $\overline{RC}$ network. Note that this is the same geometry that is used for integrated capacitors. It is unlikely, however, for practical distributed networks, that the silicon technology can provide adequate precision. There is one saving grace here, and that is the capacitance of PN junction itself may be continuously electronically tuned, perhaps by a feedback error voltage, to provide the desired transfer function characteristics. Such a procedure was suggested by Kaufmann in his early paper on distributed RC notch networks [5].

Finally, tapering of $\overline{RC}$ networks, in thin film and certainly in silicon technology, is no small feat for significant taper factors. It is generally not possible to take advantage of large taper factors ($\beta > 10$, where $r(x) = R_0 e^{\beta x}$) because of photographic limitations.

ACKNOWLEDGEMENTS

The author wishes first to clearly acknowledge the contributions which have been referenced and interrelated in this chapter. In addition, J. S. Fu, Y. Fu, A. Grabel, J. J. Golembeski, W. J. Kerwin, and W. Woronka reviewed the text and offered valuable comments.

PROBLEMS

9.1 It is required to design a high Q selective amplifier using the configuration of Figure 9-7. Assume that K is to be achieved by an operational amplifier with an open loop gain of 1,000, a 3 dB bandwidth of 10 KHz and for which $Z_{in} > 10^6$

and $Z_{out} < 10\ \Omega$. Provide design values for R, C, and R_1 so that a bandpass amplifier with $Q = 10$ and $\omega_0 = 400$ Hz results. Choose values of R, C, and R_1 so that loading effects of the amplifier-network interface are negligible.

9.2 Provide a rough sketch of the layout of a thin film network which would realize R_1, R, and C. Assume a resistance film density of 50 $\Omega/\square$ and capacitance density of .1 $\mu F/cm^2$.

9.3 Now, assume the same conditions as stated in Problem 1, except that a tapered RC network is to be used, and the Q of the bandpass structure is to be 20.

Using the design curves of Figure 9-11, and observing the caution in Problem 1 concerning network loading, find values of α and thus R_1, R, and C when $\beta = 2$, 5, and 10.

9.4 Sketch the thin film layout for the solution to Problem 3, with $\beta = 2$.

9.5 Design a low pass filter of the topology of Figure 9-12, with a peak 3 dB > $G(0) = 0$ dB before cutoff.
Assume that $G(440\text{ Hz}) = -10$ dB. Neglect Z_{out} for the amplifier.

9.6 Repeat Problem 5, except that the output impedance of the op amp must now be assumed to be 1.0 Ω; out of band rejection must be greater than -40 dB.

9.7 Design an active filter using the topology of Figure 9-20 with "equivalent transmission poles" at $S = -100 \pm j10^3$, and transmission zeros at $S = \pm j3 \times 10^3$. To calculate R_1, recall that the transmission zero arises in the same fashion as for the notch network of Kaufman (Figure 9-3a).

9.8 Assume the filter of Figure 9-27 is to be used to realize a bandpass amplifier with $Q = 25$ at $W = 1$ and a sensitivity $S_K^Q \leq 10$. Using the design charts in the following figures, determine R_1 and K.

REFERENCES

[1] R. W. Wyndrum, Jr., "The exact synthesis of distributed RC networks," Technical Report 400-76, Department of Electrical Engineering, New York University, May 1963; *IEEE Convention Record 1965.*

[2] R. P. O'Shea, "Synthesis using distributed RC networks," *IEEE International Convention Record*, Part 7, 18–29 (1965).

[3] P. M. Chirlian, *Integrated and Active Network Analysis and Synthesis,* Prentice-Hall, Englewood Cliffs, New Jersey, 1967.

[4] M. S. Ghausi and J. J. Kelly, *Introduction to Distributed Parameter Networks,* Holt, Rinehart and Winston, New York, 1968.

[5] W. M. Kaufman, "Theory of a monolithic null device and some novel circuit applications," *Proc. IRE*, **48**, 1540–1545 (Dec., 1960).

[6] W. M. Kaufman and S. J. Garret, "Tapered distributed filters," *IRE Trans. Circuit Theory*, **CT-9**, No. 4, 329–336 (1962).

[7] J. J. Golembeski, "Distributed RC network tuning," *IEEE J. Solid State Circuits*, **SE4**, 425–427 (Dec., 1969).

[8] G. J. Herskowitz and R. W. Wyndrum, Jr., "Design of distributed RC feedback networks for bandpass amplifiers," *Semiconductor Products*, **7**, 13–19 (1964). Also G. J. Herskowitz and R. Wyndrum, "Distributed RC networks for approximately maximally flat narrow band amplifiers," *Proceedings of the 9th I.R.E. East Coast Conference on Aerospace and Navigational Electronics*, Baltimore, Maryland, (Oct., 1962) pgs. 1.1.2-1 through 1.1.2-7.

[9] M. J. Gay, "The design of feedback tuned amplifiers using distributed bridge-*T* networks," *Microelectronics and Reliability*; **3**, 93–107 (1964).

[10] R. P. Sallen and E. L. Key, "A practical method of designing RC active amplifiers," *IRE Trans. Circuit Theory*, **CT-2**, 74–85 (March, 1955).

[11] W. J. Kerwin, "Analysis and Synthesis of Active RC Networks Containing Distributed Lumped Elements," Ph. D. Dissertation, Stanford University, (Aug., 1966).

[12] R. W. Wyndrum, Jr., "Active distributed RC networks" *IEEE J. Solid State Circuits*, **SC-3**, 308–310 (Sept., 1968).

[13] H. Mahdi, "Synthesis of Active Distributed RC Circuits Using Uniform and Exponentially Tapered $\overline{RC}$ Networks," Ph.D. Thesis, Northeastern University (March, 1970).

[14] W. Kerwin and L. P. Huelsman, "Digital computer analysis of distributed lumped active RC networks" *IEEE J. of Solid State Circuits*, **3**, No. 1, 26–29 (March, 1968).

[15] L. P. Huelsman, "The distributed-lumped active network," *IEEE Spectrum*, 55 (Aug., 1969).

[16] W. J. Kerwin, "Synthesis of active RC networks containing distributed and lumped elements," *Proceedings of the Asilomar Conference on Circuits and Systems*, Pacific Grove, California, November 1, -3, 1967.

[17] S. P. Johnson and L. P. Huelsman, "A high-*Q* distributed-lumped-active network configuration with zero real part sensitivity," *Proc. IEEE*, **58**, 491–492 (March, 1970).

[18] W. J. Kerwin and C. V. Shaffer, "An integrable IF amplifier of high stability," *Proceedings, Eleven Midwest Symposium on Circuit Theory* (May, 1968).

[19] R. W. Wyndrum, Jr., "Microelectronic active filters," *New York University Technical Report*, 400–459 (May, 1962).

[20] R. W. Wyndrum, Jr., R. Barcklow, and W. Woronka, "Active distributed RC networks synthesis, *Proceedings Twelfth IEEE Midwest Circuit Symposium*, University of Texas (1969), Section XIII, pp. 7.1–7.12.

[21] J. Bourquin and T. Trick, "Stability of distributed RC networks," *Proceedings of the Eleventh Midwest Symposium on Circuit Theory*, Purdue University (May, 1968).

[22] W. I. H. Chen and R. C. Levine, "Computer design transversal filters using thin film RC distributed networks," *Proceedings of the Tenth Midwest Symposium on Circuit Theory of the IEEE* (May, 1967).

[23] D. A. Spaulding, "On the use of a tapped distributed RC transmission line in zero forcing adaptive equalizer," *Bell Telephone Laboratories*, unpublished memorandum (June, 1967).

10

Application, Characterization, and Design of Switched Filters

M. R. Aaron
Bell Telephone Laboratories
Holmdel, New Jersey

10.1 PRELIMINARIES

Switched filters are used in many areas of communication, control, and computing where information is processed or transmitted in pulse form. Here we concentrate on the design of switched filters as applied to communications systems. Such filters are used as a bridge from the world of the continuous time signal to the domain of discrete time and possibly discrete amplitude, as in Pulse Amplitude Modulation (PAM) and Pulse Code Modulation (PCM) Systems. Before we go on to delineate these areas of application more clearly, let us ask and answer a few simple questions. How do we categorize such a filter? What is its schematic representation? What assumptions are appropriate for analysis and design? For the bulk of this chapter we will concentrate on filters like the one shown in Figure 10-1. To make matters concrete

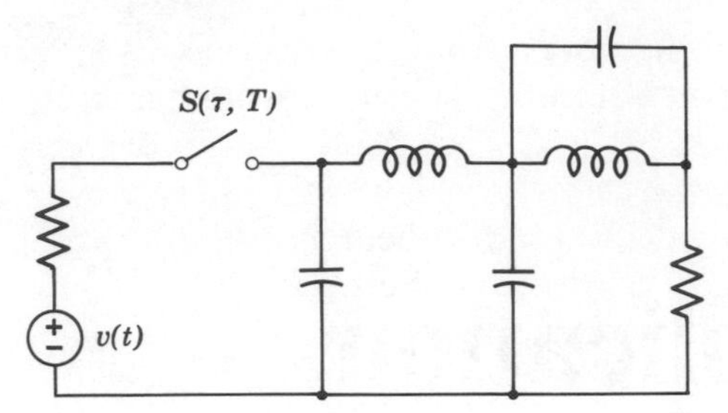

FIGURE 10-1 Typical switched filter.

we show lumped passive network elements, though active networks can and will be considered. Thus we see that this switched filter is like any other passive filter with the exception that we have imbedded a switch in one branch. Hardly surprising! Further, if the switch satisfies the properties given below, it will be clear that the switched filter fits into the category of linear, time-varying networks. It is assumed that the switch may be characterized as follows:

1. The switch has a low resistance when it is conducting (on, closed).
2. When the switch is in its so called "nonconducting" state (off, open), it is represented by a high impedance, usually infinite.
3. Transitions between the on and off states are instantaneous.
4. The switch is closed for a time interval τ every T seconds.
5. The above characteristics are signal independent.

Clearly the terms low and high impedances are relative terms and are assumed to be satisfied with respect to the elements of the passive network. Similarly, the transitions are instantaneous when compared with the time constants of the network in which the switch is imbedded.

When these assumptions are satisfied, the switched filter degenerates to the special class of linear, *periodically* time-varying networks—the network is piecewise stationary. Stated another way, in each regime, on or off, the network is linear and time invariant. The fact that the specific topology may be different for each of the two regions is no handicap in analysis. That the network has distinct regions where it is constant makes it amenable to the more conventional analytical techniques used for linear time-invariant differential equations. This was recognized early in the switched filter era by Bennett [1], who developed a general, exact method for determining the steady state response of the networks we have characterized thus far. Desoer [2] formalized and simplified Bennett's work by the use of the Bashkow [3] A-matrix, and he was able to obtain both the steady state and transient response in one fell swoop. Sandberg [4] took a different tack and considered the analysis of networks from the viewpoint of periodically-variable piecewise-constant elements. The network in Figure 10-1 may be considered a particular case of those considered by Sandberg when the switch is replaced by a periodically-varying resistor. All of these exact techniques fall into what the modern network theorist calls state-space techniques [5, 6]. Unlike the above approaches, Fettweis [7] used pole-zero and Fourier methods to attain the exact steady state response of networks containing a single periodically-operated switch. Until quite recently, the exact methods did not have a strong impact on design. General results were quite complicated, and they did not present the designer with easily understandable design alternatives. Approximate analytical techniques due initially to Cattermole [8], Crowley [9], Desoer [10], and others [11] served to simplify the analysis and thus point the way toward useful design approaches. This is the area that we emphasize in this chapter. Modern computer-aided design with its reliance on successive approximation techniques has enhanced the approximate approaches [12] and increased the utility of the exact analytical techniques. With this procedure, initial designs generated from approximate analysis, and synthesis techniques can be tested analytically using exact subroutines [13]. Elements (or singularities) can be altered in successive runs to converge on the desired response. The early view [14] of networks "belching from the computer" does indeed describe the current scene!

Our objectives in this chapter are several-fold. First, we want to indicate where switched filters are used. This is done in Section 10.3. Our second and main objective is to trace the design techniques that are now in use. Stated another way, how do we get the initial approximations to feed the computer? Here we emphasize frequency domain approaches that have long been in use but never published [11]. Another objective is to indicate how well the simplifying assumptions are satisfied in practice. Our final aim is to indicate the state of the art and its current directions. Throughout, our emphasis is on the circuits and their design, not on mathematical rigor. Therefore, we are not concerned with those academic, invalidating, pathological cases that never arise in practice. As a matter of fact, our mathematical needs are modest (principally Laplace and Fourier transform), and they are for the most part covered in the next section where formulas we need are derived—thus making this chapter essentially self contained.

10.2 SAMPLING REPRESENTATIONS AND RELATIONSHIPS

Since the process of sampling is basic to both the systems and networks under discussion, it will be profitable to recall a few simple ideas underlying the conversion of continuous time waveforms to their discrete time samples.

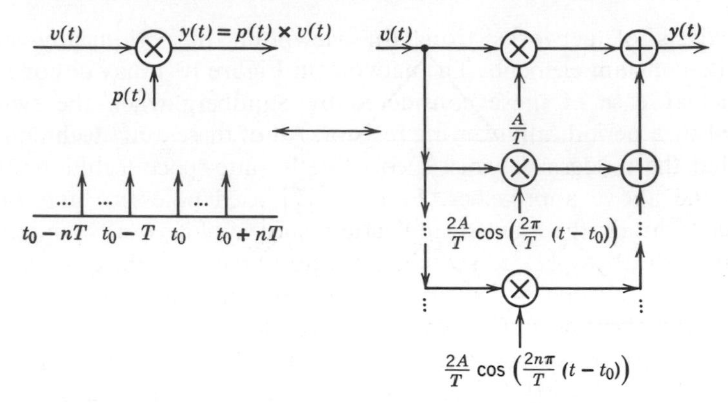

FIGURE 10-2 Sampling as pulse amplitude modulation.

Here we examine the sampling process from the standpoint of Pulse Amplitude Modulation (PAM). Within this framework, sampling is viewed as the process of multiplying a pulse carrier $p(t)$ with the signal $v(t)$ in a product modulator as depicted in Figure 10-2. If each impulse has area A, then the pulse carrier may be represented by

$$p(t) = A \sum_{n=-\infty}^{\infty} \delta(t - nT - t_0) \qquad 0 < t_0 < T \tag{10.1}$$

where $f_s = 1/T$ is the sampling frequency and t_0 is a "starting" time. Since this impulse train is periodic, it may be expanded in a Fourier series and multiplied by $v(t)$ to give the output*

$$y(t) = \frac{A}{T} \sum_{n=-\infty}^{\infty} v(t) e^{j2\pi n f_s(t - t_0)}. \tag{10.2}$$

Throughout this section it is assumed that the signal $v(t)$ is real and has no points of discontinuity except possibly at $t = 0$. Then the Fourier transform of (10.2) yields [15]

$$Y(\omega) = \frac{A}{T} \sum_{m=-\infty}^{\infty} V(\omega + m\omega_s) e^{+jm\omega_s t_0} \tag{10.3}$$

where $Y(\omega)$ is the Fourier transform of the modulator output, $V(\omega)$ is the Fourier transform (spectrum) of the input signal, $\omega = 2\pi f$ and $\omega_s = 2\pi f_s$. As shown in Figure 10-3 the magnitudes of the translates of the signal spectrum overlap when the sampling frequency is too low. This precludes recovery of

* Equation (10.2) may be interpreted as a bank (infinite) of sine wave product modulators whose carrier frequencies are at integer multiples of the sampling frequency. Each modulator is excited by the input signal, and all outputs are added together.

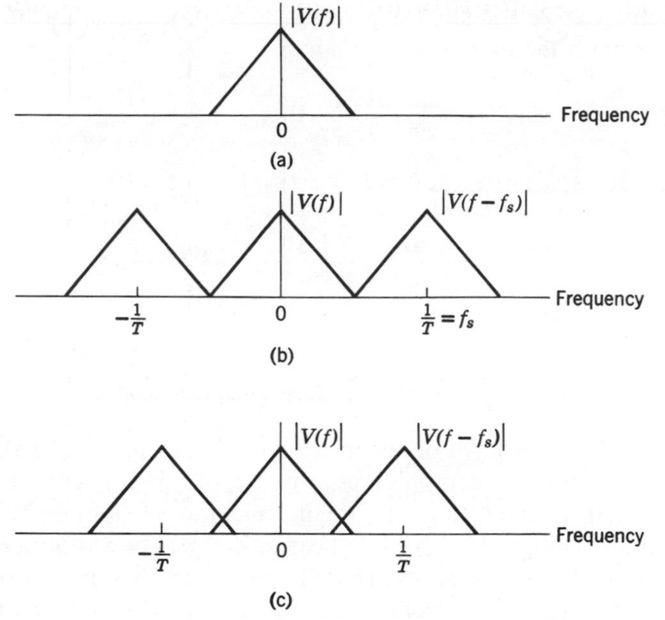

FIGURE 10-3 Amplitude spectrum of sampled signal: (a) spectrum of original signal, (b) spectrum of sampled signal, (c) spectrum with reduced sampling frequency.

the undistorted original signal from its samples by means of linear filtering. There are two remedies—increase the sampling frequency to reduce the overlap (foldover distortion, aliasing), and/or bandlimit the signal prior to sampling. As we shall discuss, these are the primary reasons for our interest in switched filters.

There is another form for the Fourier transform of the sampled output train which we derive below. This, in turn, will enable us to derive some other relationships that will prove to be useful later on. If we return to (10.1), multiply by $v(t)$ and take the Fourier transform term by term,

$$Y(f) = A \sum_{n=-\infty}^{\infty} v(nT + t_0)e^{-j2\pi f(nT+t_0)}. \tag{10.4}$$

Equating (10.3) and (10.4) gives the relationship that is at the heart of the design of switched filters; namely

$$\sum_{n=-\infty}^{\infty} v(nT + t_0)e^{-j\omega(nT+t_0)} = \frac{1}{T} \sum_{m=-\infty}^{\infty} V(\omega + m\omega_s)e^{+jm\omega_s t_0}. \tag{10.5}$$

Equation (10.5) is one form of what is known in the mathematical literature as the Poisson Summation Formula [15]. In addition to its importance here, it is a useful vehicle for summing series. If we take the limit of (10.5) as $t_0 \to 0$, we obtain the more common form

$$\sum_{n=-\infty}^{\infty} v(nT)e^{-j\omega nT} = \frac{1}{T}\sum_{m-\infty}^{\infty} V(\omega + m\omega_s) + \frac{v(0+)}{2}. \tag{10.6}$$

In terms of Laplace transforms (for $v(t) = 0$, for $t < 0$)

$$\sum_{n=0}^{\infty} v(nT)e^{-snT} = \frac{1}{T}\sum_{m=-\infty}^{\infty} V(s + jm\omega_s) + \frac{v(0+)}{2} \tag{10.7}$$

or

$$\frac{v(0+)}{2} + \sum_{n=1}^{\infty} v(nT)e^{-snT} = \frac{1}{T}\sum_{m=-\infty}^{\infty} V(s + jm\omega_s). \tag{10.8}$$

In the sequel, we will have occasion to make heavy use of (10.8) with the signal $v(t)$ replaced by the impulse response $z(t)$ of a driving point (one-port) impedance $Z(s)$. In addition it will be convenient to consider the response of $Z(s)$ to a finite width pulse $P(s)$. In all cases, $Z(s)$ will be a rational function of s. For rational $Z(s)$, the series in (10.8) may be readily summed and $v(0+)$ easily evaluated. We will collect some of the alternative forms here for ready reference. To avoid some of the mathematical distractions associated with impulses into networks whose impedance function $Z(s)$ ideally approaches a constant (for example, a pure resistor), we either excite the network with a finite width pulse or join the real world and append a parasitic capacitance or do both. In the case of a finite width pulse whose transform is $P(s)$, we thus substitute $P(s)Z(s)$ for $V(s)$ in (10.8). In this way we confine our attention to the physically interesting case where $P(s)Z(s)$ approaches $1/Cs^k$ with $k \geq 1$ as s goes to infinity. For purposes of the present section, we assume that $k = 1$ and $P(s) = 1$.

If

$$Z(s) = \frac{N(s)}{D(s)} = \frac{\sum_{n=0}^{N} a_n s^n}{\sum_{m=0}^{M} b_m s^m} \tag{10.9}$$

where $M = N + 1$, then from the initial value theorem

$$z(0+) = \lim_{s\to\infty} sZ(s) \to \frac{a_N}{b_M s^{M-N-1}} = \frac{1}{C}. \tag{10.10}$$

An alternative form for $z(0+)$ may be obtained by considering the partial fraction expansion or the inverse transform of $Z(s)$. We leave it to the reader to show that (for distinct poles)

$$z(0+) = \sum_{m=1}^{M} \text{Residues of } Z(s) \text{ at its poles.} \tag{10.11}$$

Since we have assumed $Z(s)$ to be rational, its impulse response consists of sums of exponentials which makes it easy to evaluate the expression on the right side of (10.12) as shown.*

$$\frac{1}{T}\sum Z(s + jm\omega_s) = \frac{z(0+)}{2} + \sum_{n=1}^{\infty} z(nT)e^{-snT}. \tag{10.12}$$

Clearly if

$$Z(s) = \frac{N(s)}{D(s)} = \sum_{m=1}^{M} \frac{A_m}{s + \gamma_m} \tag{10.13}$$

has distinct poles $(-\gamma_m)$ with negative real parts, whose residues are A_m, then from

$$z(t) = \sum_{m=1}^{M} A_m e^{-\gamma_m t} \tag{10.14}$$

the reader may readily verify

$$\sum_{n=0}^{\infty} z(nT)e^{-snT} = \sum_{m=1}^{M} \frac{A_m}{1 - e^{-(s+\gamma_m)T}} \tag{10.15}$$

and

$$\sum_{n=1}^{\infty} z(nT)e^{-snT} = \sum_{m=1}^{M} \frac{A_m e^{-(s+\gamma_m)T}}{1 - e^{-(s+\gamma_m)T}}. \tag{10.16}$$

Since the difference between (10.15) and (10.16) is $z(0+)$, (10.11) is proven in a rather roundabout way. From (10.11), (10.12), and (10.16) we get

$$\frac{1}{T}\sum Z(s + jn\omega_s) = \frac{1}{2}\sum_{m=1}^{M} A_m \left[\frac{1 + e^{-(s+\gamma_m)T}}{1 - e^{-(s+\gamma_m)T}}\right] \tag{10.17}$$

$$= \frac{1}{2}\sum_{=1} A_m \coth(s + \gamma_m)T/2, \tag{10.18}$$

which, incidentally, from the form of $Z(s)$ in (10.13), enables us to verify the well-known expansion (usually shown for $\omega_s = \pi$),

$$\frac{1}{\gamma} + 2\sum_{n=1}^{\infty} \frac{\gamma}{\gamma^2 + n^2\omega_s^2} = \frac{\pi}{\omega_s}\coth\left(\frac{\pi\gamma}{\omega_s}\right). \tag{10.19}$$

Before we conclude this section, we consider an idealization that is useful conceptually. Assume that $Z(s)$ is minimum reactance (no poles on the $j\omega$-axis) [34] and has a constant real part (R) over the band up to half the

* Where limits on sums are not shown, they extend from $-\infty$ to ∞.

sampling frequency and zero elsewhere. Then its response to a unit area current impulse is (Problem 10.4)

$$z(t) = \left(\frac{2R}{T}\right)\frac{\sin \pi t/T}{\pi t/T} \qquad \text{for} \quad t \geq 0$$

$$= 0 \qquad \text{for} \quad t < 0. \tag{10.20}$$

(Of course, such an impulse response can only be approximated by a finite lumped parameter network (see Problem 10.5).) Since $z(nT) = 0$ for $n \neq 0$, and $z(0+) = 2R/T$, then from (10.6) or (10.12) (with $s = j\omega$) we have

$$\sum Z(\omega + m\omega_s) = R = \frac{T}{2C} \tag{10.21}$$

where the last equality defining $C = T/2R$ follows from Bode's resistance integral theorem [34] (same as initial value theorem for impedance functions). We will return to this concept and the significance of (10.21) as we proceed.

This collects most of the formulas we need. We hasten to emphasize that the results are not new and are covered in various texts on Z-transforms and Laplace transforms [15]. They were derived here within the sampling framework to tie matters down to this unified physical view.

10.3 FIELDS OF USE

Switched Modulators

One of the earliest applications of switched networks was in modulators used for frequency translation of signals. These modulators came into importance after the invention of the wave filter and are at the cornerstone of Frequency Division Multiplex (FDM) communications. Indeed, theories and techniques developed more than thirty years ago for the design of these networks are particularly relevant to the design of the switched filters for Time Division Multiplex (TDM) that we concentrate upon. This fact does not seem to be as widely appreciated as it should be. When it is recalled that product modulation and sampling are intimately intertwined, it should send us scurrying back to the work of the pioneers on switched modulators (Peterson [16], Hussey [16], Caruthers [17], Belevitch [18], Tucker [19], and others [20]*). Toward that end consider the switched modulator block diagram of Figure 10-4. The objective of the circuit is to translate the input signal spectrum to the lower sideband (for example) about the carrier. The

* Reference [20] contains an extensive bibliography.

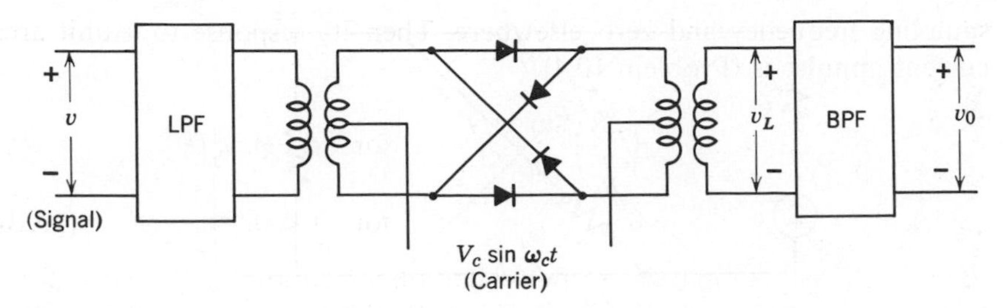

FIGURE 10-4 Block diagram of switched modulator.

input signal is bandlimited to a frequency much lower than the sinusoidal carrier by a lowpass filter whose output impedance as seen through the input transformer is assumed to be R. Similarly, the impedance viewed toward the output bandpass filters via the output transformer is also assumed to be R, and it is designed to pass the lower sideband about the carrier. The circuit containing the diodes, carrier supply, and transformers is assumed to be perfectly balanced. Furthermore, it is assumed that the following conditions are satisfied.

1. Each diode in the ring modulator has constant forward resistance r_f and constant back resistance R_b.
2. The carrier amplitude V_c is assumed to be much larger than the signal. Thus the carrier serves primarily to switch the diodes in the ring between their forward and back states. This assumption suggests that the carrier looks like a large square wave as far as the signal is concerned.
3. Each filter impedance seen by the bridge, namely R, is assumed to satisfy $R = \sqrt{r_f R_b}$.

Under the above assumptions, we can draw the equivalent lattice network for the positive half cycle of the carrier as shown in Figure 10-5(a). Using known relationships for balanced lattice networks [34] (or unfolding the bridge), we can write the voltage at the input to the bandpass filter as (denoting operations during the positive half cycle of the carrier by subscript 1)

$$v_{L_1} = \frac{v}{2}\left[\frac{1-\left(\dfrac{r_f}{R_b}\right)^{1/2}}{1+\left(\dfrac{r_f}{R}\right)^{1/2}}\right]. \tag{10.22}$$

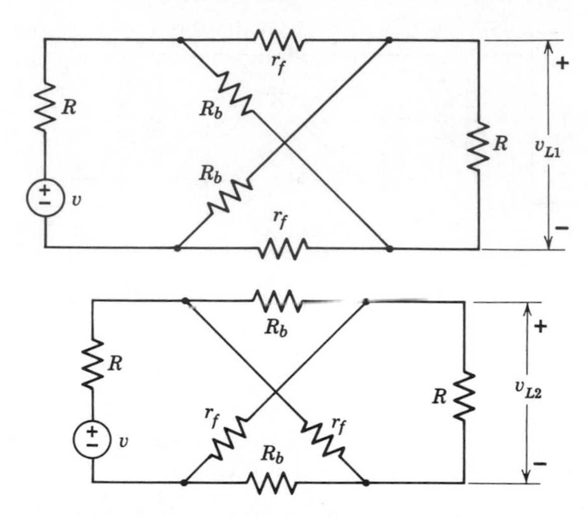

FIGURE 10-5 Idealized equivalent circuits for switched modulator: (a) positive 1/2 cycle of carrier, (b) negative 1/2 cycle of carrier.

Using Figure 10-5(b), v_L during the negative half cycle of the carrier (subscript 2) is derived as

$$v_{L_2} = -\frac{v}{2}\left[\frac{1-\left(\frac{r_f}{R_b}\right)^{1/2}}{1+\left(\frac{r_b}{R_b}\right)^{1/2}}\right]. \tag{10.23}$$

Equations (10.22) and (10.23) can be combined by writing

$$v_L = \frac{1}{2}\left[\frac{1-\left(\frac{r_f}{R_b}\right)^{1/2}}{1+\left(\frac{r_f}{R_b}\right)^{1/2}}\right] x(t)v(t) \tag{10.24}$$

with the switching function

$$x(t) = \begin{cases} 1 \text{ positive } 1/2 \text{ cycle of the carrier} \\ -1 \text{ negative } 1/2 \text{ cycle of the carrier} \end{cases} \tag{10.25}$$

which is periodic with period $2\pi/\omega_c$ and can be expanded in a Fourier series to give

$$x(t) = \frac{4}{\pi}[\cos \omega_c t - 1/3 \cos 3\omega_c t + \cdots.$$

It is apparent from (10.24) (obtained by Caruthers in 1939) that the circuit has been reduced to the product modulator considered in the section on sampling representations. In the present case, however, the carrier is a square wave rather than an impulse train. Further, its fundamental frequency is generally much higher than twice the highest frequency component in the signal.

It should be clear that we have made a number of idealizations in the analysis to reduce the circuit to such a simple form. *From the standpoint of efficient power transfer, the filter impedances should be constant in the desired bands, and either very high for the input filter or very low for the output filter (or vice versa) in the region of the undesired sidebands* [17]. *For the switched filters we consider later, similar properties will be seen to be desirable.* Other practical factors such as nonlinearity, circuit unbalances, parasitic capacitance, and realistic frequency-selective terminations have been considered by several authors. Recent work on the subject, taking into account the properties of modern semiconductor diodes, has been done by one of the pioneers, D. G. Tucker [21].

Time Division Switching

Major impetus for the analytical and design studies of the 1950's came from renewed interest in time division switching and time division multiplexing. Re-examination of both areas was stimulated by the invention of the transistor. In the rudimentary time division switching system depicted in Figure 10-6, each subscriber line is provided with a lowpass filter and a sampling

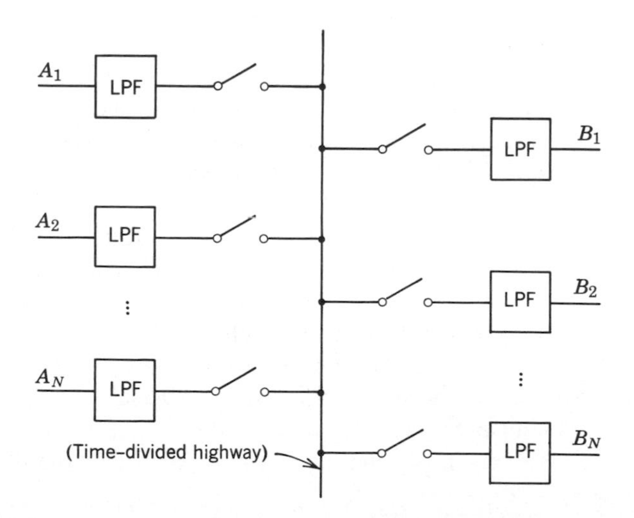

FIGURE 10-6 Rudimentary time division switching network.

gate. Simultaneous closing of the gates on two lines provides a talking path. This talking path is closed for a short period at the sampling rate, nominally 8 kHz (to transmit speech of bandwidth less than 4 kHz according to sampling theory). The time of closure in each sampling period must be short to permit other subscribers to time share the common "highway." As the number of users is increased, the time allotted per channel decreases. If it is assumed that the filter impedances as seen by the switch are resistive, then the energy transmitted between talkers is directly proportional to the ratio of on-to-off time. This leads to a considerable loss in energy when the ratio is small, as it will be when a large number of subscribers must be accommodated. Fortunately, techniques were developed to store the input signal energy during the time the connection was open and ideally transfer it all to the output in the short time interval when the connection is made. The most common approach, known as resonant transfer, depicted in Figure 10-7, was

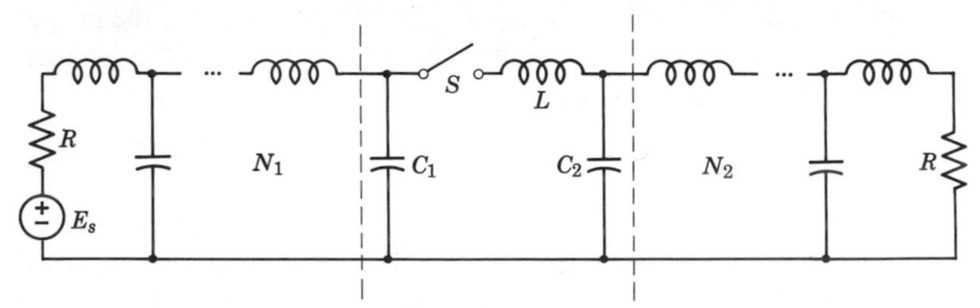

FIGURE 10-7 Resonant transfer switching network.

invented independently by several people* in the 1950's. Cattermole, in addition to contributing techniques, contributed fundamental understanding to this field. His early paper [8] on the subject is a landmark. He builds a clear physical picture, gives a logical evolution of the resonant transfer circuit, and puts forth design conditions upon which succeeding contributors were able to build. *A careful reading of his contribution in reference* [8] *is a must for a serious student in this field.* Space precludes our giving it detailed coverage here.

The ideal behavior of the circuit of Figure 10-7 is quite simple. Consider that portion of Figure 10-7 within the dashed lines and reproduced as Figure 10-8(a). While the switch is open, the last capacitor of the input filter is charged toward the source voltage. The switch is closed for a time equal to a half period of the resonant circuit. During this time, the current waveform is a half cycle sine-wave, as shown in Figure 10-8(b), and the voltage waveforms are cosinusoidal. The capacitors interchange their charges, and when the switch is opened, the charge on the capacitor of the output filter

* H. B. Haard and C. G. Svala, U.S. Patent 2,718,621 (March 11, 1953; first filed in Sweden on March 12, 1952). K. W. Cattermole, British Patent 753,645 (December 3, 1954). J. A. T. French, British Patent 841,555 (April 14, 1955). W. D. Lewis, U.S. Patent 2,936,337 (January 9, 1957).

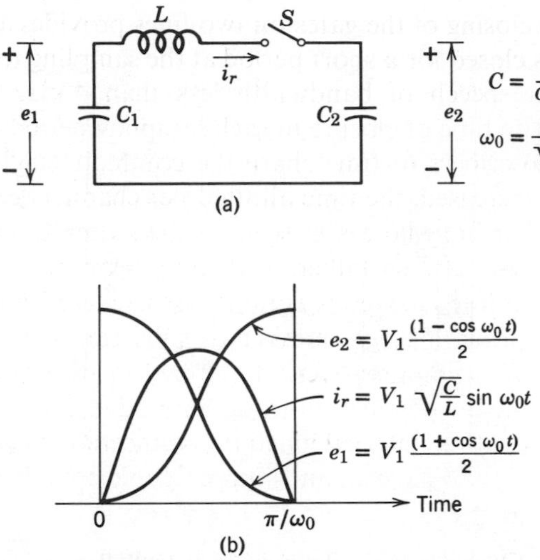

FIGURE 10-8 Resonant transfer waveforms: (a) circuit, (b) waveforms.

leaks off to the load. The clue to the approximate analysis of the circuit is the fact that during the short closure time, the currents and voltages in the circuit elements external to the central-resonant-transfer mesh remain essentially constant. This is crucial to all approximate analytical and design techniques. In Section 10.4 we consider this circuit in more detail. For a most comprehensive discussion of resonant transfer circuits and an extensive bibliography, the reader is referred to the encyclopedic paper by Fettweis [22], who has himself made significant contributions to this field. The most common use of resonant transfer switching is in PAM Time Division Switching Systems used in Private Branch Exchanges [23, 24].

Though we stress resonant transfer between lowpass filters, it is easy to show that it is possible to have resonant transfer between bandpass filters or between a lowpass and a bandpass filter. The latter case has been of interest in a combined switching and frequency division multiplex application emphasized by Thrasher [25, 26].

Another form of Time Division Switching, called digital time division switching, involves switching signals in digital form. Filters required in the digital conversion process are noted in the next section.

Time Division Multiplexing and PCM

An area closely related to Time Division Switching is Time Division Multiplexing [27]. In this case, the time division highway of Figure 10-6 is split into sending and receiving halves as shown in Figure 10-9. PAM transmission over any appreciable distance, at any significant rate, is to be avoided

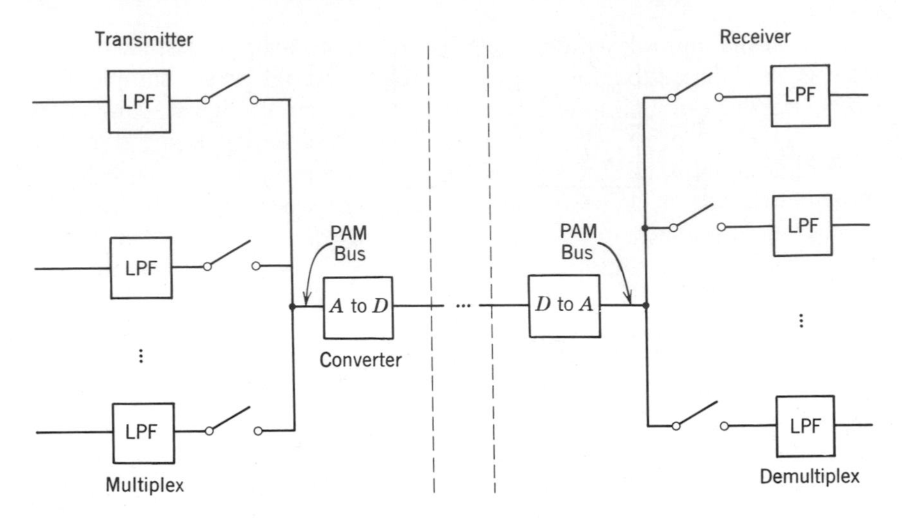

FIGURE 10-9 Time division multiplex.

because of problems attendant to crosstalk. Therefore, this multiplex is generally used as the first step on the road to PCM as shown in Figure 10-9. In a PCM system [28], the multiplexed PAM train is fed into a time shared analogue-to-digital converter to produce a digital approximation to the continuum of PAM pulses. In this case, the switched filter might take the form as shown in Figure 10-10, which looks very much like half of the resonant transfer circuit in Figure 10-7. One major exception here is that the capacitor C_L is common to the PAM bus and is connected in succession

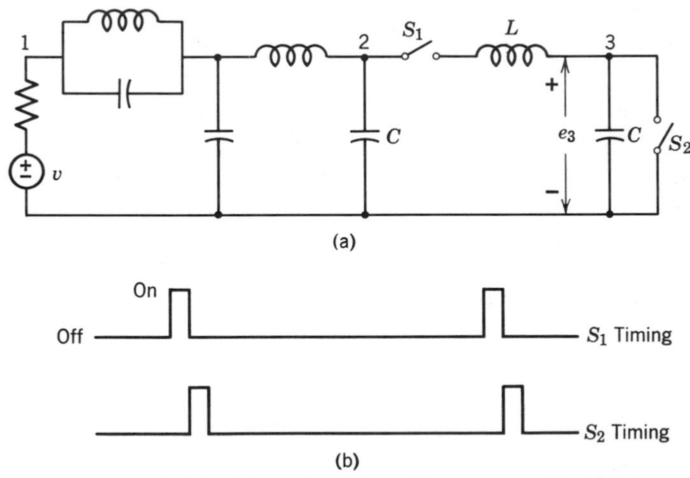

FIGURE 10-10 Multiplex filter: (a) network, (b) timing.

to each of the multiplexed channels. Therefore, to avoid crosstalk between lines, C_L is clamped to a low impedance (discharged) between samples from two successive channels as shown in the timing chart of Figure 10-10. Often, to avoid buildup of parasitic capacitance, to reduce switching speed requirements, and to minimize crosstalk, the multiplex is composed in stages [29] as shown in Figure 10-11. This means that additional switched networks called transfer gates (or resamplers) are required to access the A to D converter to each of the separate multiplex buses.

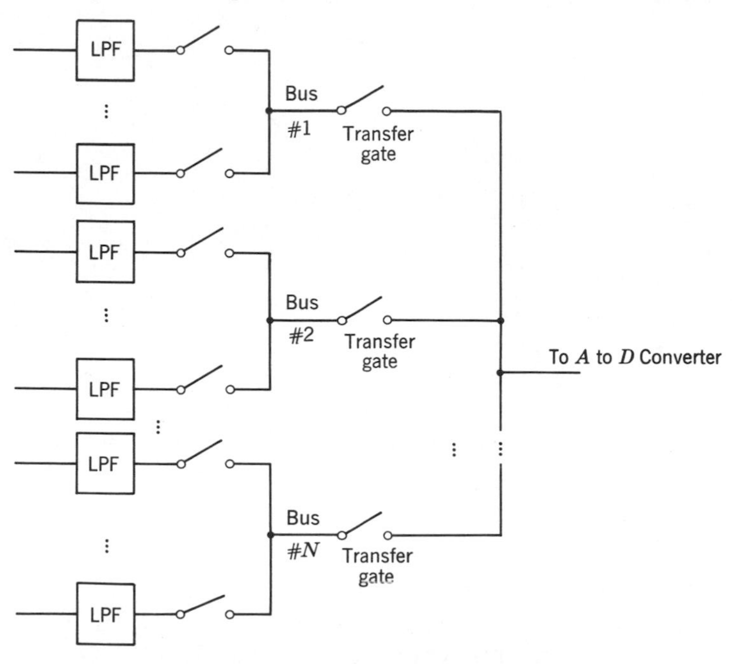

FIGURE 10-11 Multistage multiplex.

Another type of network used in a PCM terminal is the sample and hold circuit [30] as shown in Figure 10-12(a). This may be used in a single channel system to sample the output of a bandlimiting filter or in a multichannel system following filtering and sampling. In the latter case the sample and hold may be used to take a sample out of the center of pulse, where it is relatively uncorrupted by switching transients and gate control leak through. In either case the A to D converter is fed a sample ideally held flat for the conversion interval as shown in Figure 10-12(b).

At the receiving terminal, channel reconstruction filters similar to that illustrated in Figure 10-1 are required to reconstitute close replicas of the original analog signal. Our principal emphasis will be on filters for this application.

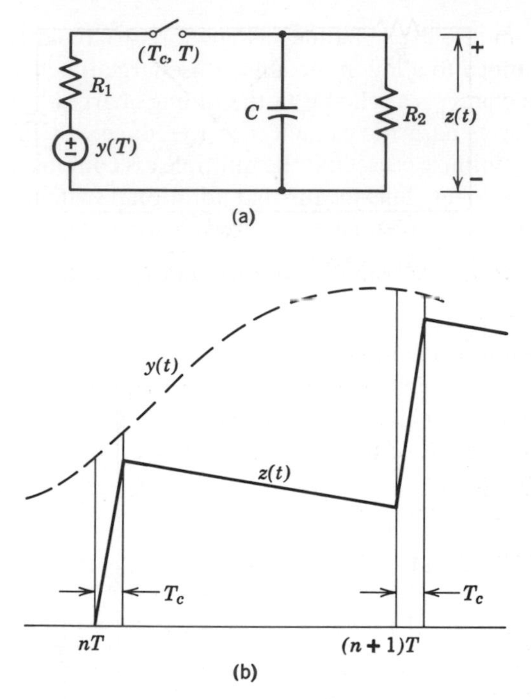

FIGURE 10-12 Ideal sample-and-hold circuit and waveforms.

Other Applications

In carrier control systems it is desirable to provide loop shaping without converting the double sideband suppressed carrier signal to baseband. Networks for providing this "envelope" shaping are called "AC compensation networks" or synchronous networks. They may be realized without inductors by employing periodically operated switches and R's and C's. In many respects they are similar to the switched modulator discussed earlier. One simple synchronous network analyzed in detail by Sun and Frisch [31] may be represented as in Figure 10-13 by an RC network connected to a capacitor through a double-pole-double-throw switch (DPDT). Clearly the DPDT switch is equivalent to the switched modulator with a periodic switching function describing the duty cycle of the switch. Sun and Frisch use the state-space approach to provide exact analysis techniques and a useful synthesis method. Further details on hardware realization and other references may be found in the book by K. A. Ivey [32].

It should be apparent that switched filters are required in sampled data control systems. Perhaps the most common network used here is the

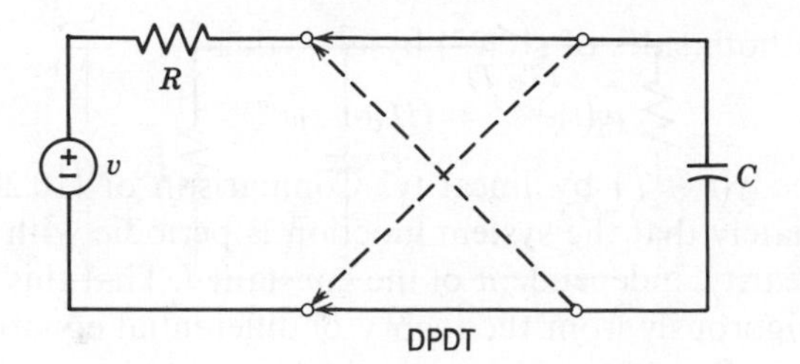

FIGURE 10-13 AC compensation network for carrier control systems.

previously mentioned sample and hold circuit. It serves to give a stairstep interpolation to the narrow pulse train of amplitude-modulated samples. Exact analysis is straightforward but messy [4, 30].

Finally, applications of the switched filter version of the N-path filter are covered in Chapter 11.

10.4 APPROXIMATE ANALYSIS AND CHARACTERIZATION

Steady State Analysis of Linear Variable Networks

It will be helpful in tying together the several methods for the approximate analysis of switched networks to recall a few of the very simple results on linear variable networks [33]. Zadeh, among others, has made useful contributions here.

The response $y(t)$ of a linear time varying system to an input $x(t)$ with Fourier transform $X(\omega)$ is defined by

$$y(t) = \frac{1}{2\pi}\int_{-\infty}^{\infty} H(\omega, t)X(\omega)e^{j\omega t}\, d\omega \tag{10.26}$$

where $H(\omega, t)$ is the transfer function of the system. In particular, it may be determined by the response to the periodic probe $Ie^{j\omega t}$. With this input, $H(\omega, t)$ is readily seen to be the ratio of the network response, $y(t)$, to $Ie^{j\omega t}$ divided by $Ie^{j\omega t}$, namely

$$y(t) = IH(\omega, t)e^{j\omega t}. \tag{10.27}$$

If we specialize further to linear periodically time varying networks of period T, it should be clear that the *steady-state* behavior of the network remains unchanged as we change t to $t + T$. This means that the response to $e^{j\omega(t+T)}$ must be $y(t + T)$.

$$y(t + T) = IH(\omega, t + T)e^{j\omega(t+T)}. \tag{10.28}$$

If we multiply both sides of (10.27) by $e^{j\omega T}$, then

$$y(t)e^{j\omega T} = IH(\omega, t)e^{j\omega(t+T)} \tag{10.29}$$

which must be $y(t+T)$ by linearity. Comparison of (10.28) and (10.29) shows immediately that the system function is periodic with period T, and because of linearity, independent of the constant I. That this is true may be shown more rigorously from the theory of differential equations known as Floquet's theory. Since $H(\omega, t)$ is periodic, it may be expanded in a Fourier series to give

$$H(\omega, t) = \sum_{-\infty}^{\infty} H_n(\omega)e^{j(2\pi nt/T)}. \tag{10.30}$$

In the sequel we will make use of this expansion and concentrate primarily on the dc term

$$H_0(\omega) = \frac{1}{T}\int_{-T/2}^{T/2} H(\omega, t)\, dt, \tag{10.31}$$

which may be interpreted as the time varying system function averaged over all possible equally likely starting times. Clearly the remaining Fourier coefficients represent system responses at sidebands about $n\omega_s = n(2\pi/T)$. These responses are of interest where modulators are used for frequency translation. In any case, sufficient attenuation must be provided in filters to keep foldover of the energy from one sideband into another. As noted early in Section 10.2, this is the name of our game.

In the above and throughout the following discussion, it is assumed that the frequency of the input signal (ω) and the sampling frequency (ω_s) are incommensurate. Furthermore, the steady-state response $y(t)$ is *not* periodic in t as may be seen from (10.28) and (10.29); i.e., $y(t+T) = e^{j\omega T}y(t)$. This latter fact has been noted by both Bennett [1] and Belevitch [18].

General Comments

There are several useful approaches to the analysis of switched filters applicable when the switch closure time is short relative to the period. As we have already noted, under these conditions the network can, in effect, be separated into two essentially disconnected pieces. Both Desoer and Cattermole have made use of this approximation in their analysis of switched networks employing resonant transfer. Cattermole has shown that the switch can be replaced by an equivalent current source made up of a modulated sequence of impulses. The effect of this current on the network is determined by both the transfer impedance of the network at the envelope (input) frequency and the *pulse sequence impedance* of the network. This pulse sequence impedance, a term coined by Cattermole, is given by either of the functions

$$\sum_{n=0}^{\infty} z(nT)e^{-j\omega nT} = \frac{1}{T}\sum Z(\omega + m\omega_s) + \frac{z(0+)}{2} \equiv \underline{\underline{Z(\omega,+)}} \qquad (10.32)^*$$

or

$$\sum_{n=1}^{\infty} z(nT)e^{-j\omega nT} = \frac{1}{T}\sum Z(\omega + m\omega_s) - \frac{z(0+)}{2} \equiv \underline{Z(\omega,-)} \qquad (10.33)^*$$

which were derived as equations (10.7) and (10.8) in Section 10.2. Physically, the pulse sequence impedance $\underline{\underline{Z(\omega, +)}}$ represents the response of the one-port $Z(\omega)$ to the periodic impulse train given by (10.1) for all impulses from $-\infty$ up to and including 0. The definition for $\underline{Z(\omega, -)}$ is as above except that the impulse at zero has not yet occurred.

Desoer employs a sampled data approach in which the switch is considered to be an impulse sampler. The impulse sampler excites a network whose impulse response is the half period, sine pulses typifying the resonant transfer action. These shaped pulses are then used to excite the filters surrounding the switch. By using the finite pulse shape, it is possible to include the effects of gate loss neglected in Cattermole's idealized analysis. In addition, Desoer gives a successive approximation procedure to account for the finite closure time of the switch. This higher order approximation has never been required in any of the communications applications with which this writer has been associated. It can be shown that the zeroth approximation given by Desoer, and the results of Cattermole, are for all practical purposes identical. In the limit as the pulse shape assumed by Desoer becomes an impulse, the two results are indeed equal mathematically as we show later.

There is another method originally used by Crowley in unpublished work that is intimately related to Cattermole's technique. This approach takes explicit advantage of the fact that the time varying transfer function of the network is periodic with period equal to the switching (sampling) period. A simple way is found to cater to losses in the switch. Results identical to those obtained by Desoer and Cattermole may be arrived at in this way. We will show that the three results may profitably be combined. Furthermore, we will show how certain switched networks may be reduced to an approximately time invariant filter [11] to permit the use of old fashioned (but useful) image-parameter techniques for purposes of preliminary design [34].

* Note the redundancy in the symbolism for purposes of single error detection. Two bars go with a plus, and one bar with a minus.

Cattermole's Idealized Analysis

In this section we cover Cattermole's technique for determining the transfer function of the ideal resonant transfer, bidirectional (called two-way direct channel by Cattermole) channel shown in Figure 10-7. For convenience, we assume here and in the next sections too that the two lowpass filters are identical. Following Cattermole, the switches are taken to be ideal (zero on resistance and infinite off resistance, and zero transition time to transfer between these states), the resonant transfer coil Q is infinite, and the lowpass filters are lossless. To clarify the concepts involved, we convert the diagram to the equivalent circuit of Figure 10-14. by using the Thevenin equivalent

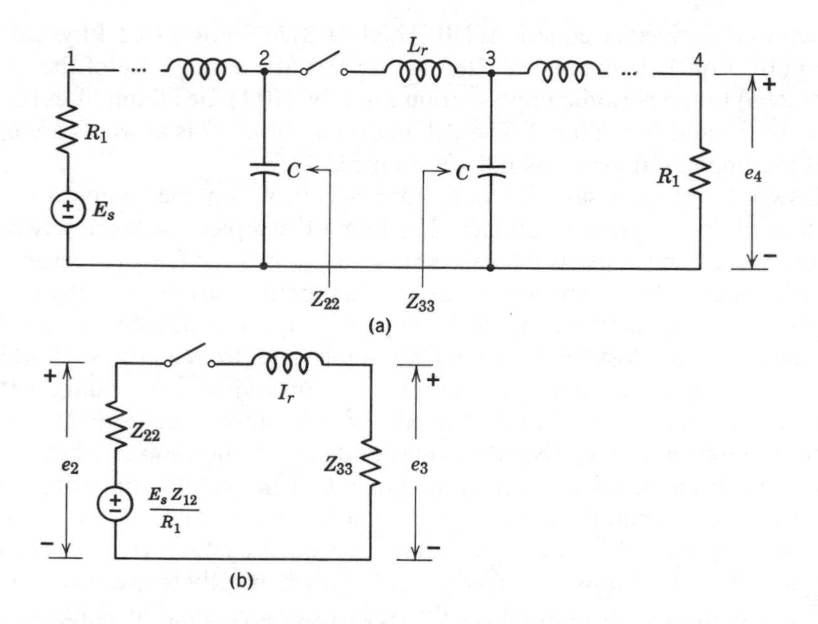

FIGURE 10-14 Equivalent circuit for resonant transfer switching: (a) original circuit, (b) equivalent circuit.

and reciprocity. This will permit us to put all three approaches in the same light. The various symbols are defined as follows:

1. Z_{22} and $Z_{33}(= Z_{22})$ are the impedances seen at ports 2 and 3, respectively;
2. $Z_{12} = Z_{34}$ is the transfer impedance of either of the two identical filters;
3. $T = 2\pi/\omega_s$ is the sampling period;
4. ω_s = sampling frequency.

Cattermole goes one step further by assuming that the switch and resonant transfer inductor may be replaced by the current generator I, where

$$I = CF(\omega)e^{j\omega t} \sum \delta(t - nT) = \frac{CF(\omega)}{T} e^{j\omega t} \sum e^{jm\omega_s t} \tag{10.34}$$

is a modulated sequence of impulses with envelope equal to $CF(\omega)e^{j\omega t}$. The function $F(\omega)$ is to be determined from the boundary conditions, and, as we see later, will turn out to be closely related to the dc term $H_0(\omega)$ in the Fourier series expansion of the system function given in Section 10.4, (10.30). In addition, as in all of the approximate approaches, the time constants of the lowpass filter are very much larger than the short-gate-closure time. Thus, energy stored in these networks remains essentially unchanged during the time the charges are interchanged by the capacitors in the resonant-transfer loop. $F(\omega)$ may be obtained by considering the boundary conditions immediately before and after each impulse has occurred. Two equations result from a consideration of the boundary conditions. The first is

$$\frac{Z_{12}(\omega)}{R_1} - CF(\omega)\underline{\underline{Z_{22}(\omega, +)}} = CF(\omega)\underline{Z_{33}(\omega, -)} \tag{10.35}$$

which states that $e_2(0+) = e_3(0-)$. The second equation

$$\frac{Z_{12}(\omega)}{R_1} - CF(\omega)\underline{Z_{22}(\omega, -)} = CF(\omega)\underline{\underline{Z_{33}(\omega, +)}} \tag{10.36}$$

follows from $e_2(0-) = e_3(0+)$. In the particular case at hand the lowpass filters have been assumed to be identical, thus one of the equations suffices to determine $F(\omega)$ to be

$$F(\omega) = \frac{Z_{12}(\omega)/CR_1}{\underline{\underline{Z_{22}(\omega, +)}} + \underline{Z_{22}(\omega, -)}}. \tag{10.37}$$

Since the component of the current I at frequency ω is $F(\omega)/T$, the component of the output voltage at the input frequency is simply $Z_{34}(\omega)CF(\omega)/T$. This givcs us thc transfer function

$$\frac{E_4(\omega)}{E_1(\omega)} = H_0(\omega) = \frac{[Z_{12}(\omega)]^2}{R_1 T[\underline{\underline{Z_{22}(\omega, +)}} + \underline{Z_{22}(\omega, -)}]}. \tag{10.38}$$

That this expression is indeed $H_0(\omega)$ will be shown later. Toward that end, it will be useful to convert the denominator of (10.38) to another form by means of (10.31) and (10.33). Using these equations we can write

$$\frac{E_4(\omega)}{E_1(\omega)} = H_0(\omega) = \frac{[Z_{12(\omega)}]^2}{2R_1 \sum Z_{22}(\omega + m\omega_s)}. \tag{10.39}$$

Since we have assumed the filters to be lossless, it is well known [34] (and easy to show) that the real part of $Z_{22}(\omega)$ is

$$\mathrm{Re}[Z_{22}(\omega)] = \frac{|Z_{12(\omega)}|^2}{R_1}. \tag{10.40}$$

With this relationship, (10.39) may be written

$$H_0(\omega) = \frac{\mathrm{Re}[Z_{22}(\omega)]}{2\sum Z_{22}(\omega + m\omega_s)}. \tag{10.41}$$

One further idealization will enable us to demonstrate that in the limit, zero insertion loss is attainable for signals over the band up to half the sampling frequency. If each filter is an ideal lowpass filter with a flat passband up to half the sampling frequency and zero transmission beyond this point, then $\mathrm{Re}[Z_{22}(\omega)] = R_1$ and from (10.21) the sum in the denominator of (10.41) is also equal to R_1. Subject to all the above idealizations, it is thus theoretically possible to have zero insertion loss despite the fact that the switch is closed for only a very small fraction of the sampling period. This is indeed a very remarkable result, and as we see later it is approached closely in practice with finite, and in fact rather simple, economically realized networks.

Desoer's Analysis

As noted previously Desoer's zeroth approximation to the response of the resonant-transfer system is also predicated on the large disparity in time constants between the resonant transfer loop and the remaining meshes of the lowpass filters. With this assumption, the currents and voltages external to the central resonant transfer mesh need not be considered during the short period when the switch is closed. This leads directly to the following set of equations describing the initially relaxed resonant-transfer mesh of Figure (10-14) when the switch is closed

$$e_2(t) = v(t) - \int_0^t i_r(\tau)z(t-\tau)\,d\tau \tag{10.42}$$

$$e_3(t) = \int_0^t i_r(\tau)z(t-\tau)\,d\tau \tag{10.43}$$

$$e_2(t) - e_3(t) = v(t) - 2\int_0^t i_r(\tau)z(t-\tau)\,d\tau \tag{10.44}$$

where: $v(t)$ is the open circuit voltage across port 2 and $z(t)$ is the inverse transform of $[Z_{22}(s)]$. As Desoer notes, these equations would also follow from the block diagram of Figure 10-15. In particular, (10.44) describes the

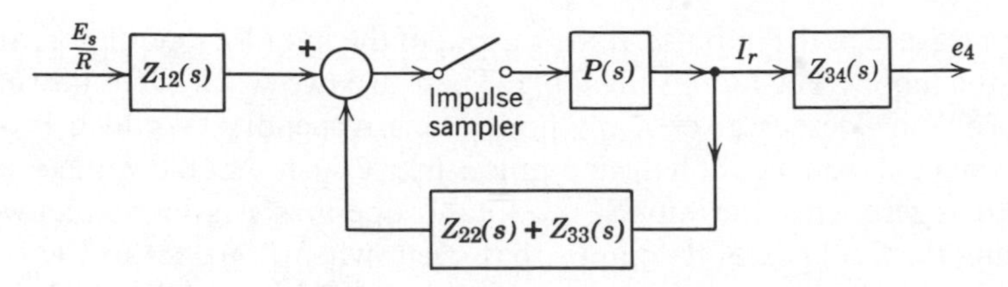

FIGURE 10-15 Sampled data system approximately equivalent to resonant transfer system of Figure 10-14.

input to the impulse modulator of Figure 10-15; namely, $V(s) - 2Z(s)I_r(s)$ which is the Laplace transform of the right hand side of (10.44). From the block diagram, (10.44), and the resonant transfer mesh, it should be clear that the input to the impulse modulator is $e_2(t) - e_3(t)$. For the interval $0 \le t < T$, then the output of the pulse shaping network $CP(s)$ is $C[e_2(0) - e_3(0)]p(t)$, which is the resonant transfer current with

$$p(t) = \begin{cases} \dfrac{\pi}{2\tau}\sin\dfrac{\pi t}{\tau} & 0 \le t < \tau \\ 0 \text{ elsewhere} \end{cases} \tag{10.45}$$

and

$$P(s) = \mathscr{L}[p(t)] = \frac{\omega_0^2}{s^2 + \omega_0^2}(1 + e^{-s\tau}) \tag{10.46}$$

where

$$\tau = \frac{\pi}{\omega_0} = \sqrt{\frac{2}{CL_r}}. \tag{10.47}$$

Use of the block diagram to obtain the zeroth approximation gives (from standard sampled data techniques)

$$E_4(s) = \frac{R_1 \dfrac{1}{T}[\sum Z_{12}(s + jm\omega_s)E_1(s + jm\omega_s)]CP(s)Z_{12}(s)}{1 + 2C\left[\dfrac{1}{T}\sum P(s + jm\omega_s)Z_{22}(s + jm\omega_s)\right]} \tag{10.48}$$

where the initial condition $\lim_{s\to\infty} sP(s)Z_{22}(s)$ is zero. When $\tau \to 0$, $P(s) \to 1$, the initial condition is *not zero*, and indeed the feedback loop is unstable as Desoer has pointed out. This follows from the fact that an impulse from the impulse modulator feeds directly into the impedance Z_{22} in the feedback path and produces a step across its input capacitance. This, in turn, produces a step instantaneously at the input to the impulse modulator, causing a jump in the measure of the impulse at the output of the impulse modulator, and so on to the moon! The loop is unstable. There are two ways of circumventing this problem. Desoer gives a method in his Appendix IV which involves advancing the sequence of half sine-pulses by $\tau/2$ such that the voltage across the gate is zero when the impulse modulator operates. Another direct way of avoiding the feedback of the jumps that occur when $P(s)$ goes to 1 ($\tau \to 0$) is to remove them. This is accomplished by replacing the term in the brackets in the denominator of (10.48) by (10.33).

$$\frac{1}{T}\sum Z_{22}(s + jm\omega_s) - \frac{1}{2C} = \underline{Z(\omega, -)} \tag{10.49}$$

whenever $\tau \to 0$ to obtain:

$$E_0(s) \equiv E_4(s)]_{\tau=0} = \frac{[\sum Z_{12}(s + jm\omega_s)E_1(s + jm\omega_s)]Z_{12}(s)}{2R_1[\sum Z_{22}(s + jm\omega_s)]} \tag{10.50}$$

which is identical with Desoer's result when $\tau \to 0$. To obtain the steady-state response to an input $e_1(t) = Ee^{j\omega t}$, we substitute $E_1(s) = E/(s - j\omega)$ in (10.49) to obtain

$$E_0(s) = \frac{1}{2R_1}\frac{E\left[\sum Z_{12}(s + jm\omega_s)\dfrac{1}{s + jm\omega_s - j\omega}\right]Z_{12}(s)}{\sum Z_{22}(s + jm\omega_s)}. \tag{10.51}$$

The steady-state response is obtained from the imaginary axis poles of the sum in the numerator of (10.51) (since $Z_{12}(s)$ has no such poles). Therefore, the steady-state response has the form

$$\sum_{-\infty}^{\infty} H'_{m0}(\omega)e^{j(\omega + m\omega_s)t} \tag{10.52}$$

where

$$H'_{m0}(\omega) = \frac{EZ_{12}(\omega)Z_{12}(\omega + m\omega_s)}{2R_1 \sum Z_{22}[\omega + (k + m)\omega_s]}. \tag{10.53}$$

That the response should be of the form given by (10.52) follows from the general results for the steady-state response of variable networks covered in Section 10.4—specifically (10.27) and (10.30). Our interest centers around H'_{00}, or in particular

$$\frac{E_0(\omega)}{E_1(\omega)} = \frac{H'_{00}(\omega)}{E_1(\omega)} = \frac{[Z_{12}(\omega)]^2}{2R_1 \sum Z_{22}(\omega + k\omega_s)} \tag{10.54}$$

which is seen to be identical with (10.39) obtained by Cattermole's approach.

The major advantage of Desoer's analysis is the inclusion of the shape factor $P(s)$ when the resonant transfer mesh includes loss due to finite inductor Q and/or switch resistance. Under these conditions it should be clear that complete charge interchange is not possible. The inevitable loss under these conditions may be computed from the more general result (10.48) that includes the pulse shape. In this case, if we denote the coefficients of the steady-state response (as in (10.52)) by $H'_m(\omega)$ (note subscript zero corresponding to $\tau = 0$ has been removed), the reader should be able to start with (10.48) and obtain

$$\frac{2R_1 H'_m(\omega)}{E} = \frac{Z_{12}(\omega)P(\omega + m\omega_s)Z_{12}(\omega + m\omega_s)}{\frac{T}{2C} + [\sum P(\omega + k\omega_s)Z_{22}(\omega + k\omega_s)]} \tag{10.55}$$

by paralleling the steps used in going from (10.50) to (10.53). Later we will indicate the usefulness of the pulse shape factor in examining a nonresonant-transfer-switched filter.

Formal Fourier Series Approach

Cattermole's method is basically a Fourier series approach.* Crowley, in unpublished work in 1956, independently followed a similar path. Our discussion, for the most part, follows Crowley's more formal Fourier series approach. With the exception of the gate characterization, the assumptions are the same as those used previously. The gate is characterized by (See Figure 10-14)

$$e_3(nT + \tau) - e_2(nT + \tau) = G[e_2(nT) - e_3(nT)]. \tag{10.56}$$

Since impulse modulation will be assumed, (10.56) may be written

$$e_3(nT+) - e_2(nT+) = G[e_2(nT-) - e_3(nT-)]. \tag{10.57}$$

When $G = 1$, it is easy to see that (10.57) represents the charge interchange characteristic of the resonant-transfer circuit. On the other hand for $G < 1$, (10.57) represents the charge interchange in a lossy-resonant-transfer mesh. For later reference, we also note that $G = 0$ characterizes circuits where the gate is a low resistance which clamps e_3 to e_2.

Since the current in the resonant transfer mesh is periodic with period T, it may be written explicitly as the Fourier series

$$I = \frac{Q}{T} \sum e^{j(\omega + n\omega_s)t} \tag{10.58}$$

* Fettweis, in his 1959 paper, also took advantage of periodicity in developing an exact analysis procedure for a switched filter.

which is simply (10.58) with $CF(\omega)$ replaced by Q. The quantity Q will be determined from the charge interchange conditions (10.57). With the aid of (10.58) it is easy to see that the voltages $e_2(t)$ and $e_3(t)$ may be expressed as

$$e_2(t) = \frac{Z_{12}(\omega)e^{j\omega t}}{R_1} - \frac{Q}{T}\sum Z_{22}(\omega + n\omega_s)e^{j(\omega+n\omega_s)t} \tag{10.59}$$

and

$$e_3(t) = \frac{Q}{T}\sum Z_{33}(\omega + n\omega_s)e^{j(\omega+n\omega_s)t}. \tag{10.60}$$

Note that by assumption $Z_{22}(\omega) = Z_{33}(\omega)$ and $E_s = e^{j\omega t}$. To solve for Q, we first form $v(t) = e_3(t) - e_2(t)$ and evaluate it at $t+ = NT + \varepsilon$ and $t- = NT - \varepsilon$. Substitution of $v(t)$ in (10.57), followed by taking the limit as $\varepsilon \to 0$ from above, converts (10.57) to

$$\lim_{\varepsilon\to 0} (NT + \varepsilon) = -G \lim_{\varepsilon\to 0} v(NT - \varepsilon) \tag{10.61}$$

where:

$$\lim_{\varepsilon\to 0} v(NT \pm \varepsilon) = e^{j\omega NT}\left[\frac{Z_{12}(\omega)}{R_1} - \lim_{\varepsilon\to 0}\frac{2Q}{T}\sum e^{\pm j(2\pi n\varepsilon/T)}Z_{22}(\omega + n\omega_s)\right]. \tag{10.62}$$

Care must be exercised in evaluating the limit in (10.62). In general, the operations of limit and summation cannot be interchanged unless the sum is uniformly convergent, which is not true for the case at hand. Fortunately we have already paved the way for taking the limit on the right of (10.62) for the positive sign (See (10.6)); namely from (10.5)

$$\frac{1}{T}\sum e^{+j(2\pi n\varepsilon/T)}Z_{22}(\omega + n\omega_s) = \sum_{k=0}^{\infty} z(kT + \varepsilon)e^{-j2\pi f(kT+\varepsilon)} \tag{10.63}$$

which for rational Z may be expressed as a finite sum by means of (10.15).

$$\frac{1}{T}\sum e^{+j(2\pi n\varepsilon/T)}Z_{22}(\omega + n\omega_s) = \sum_{m=1}^{M} \frac{A_m e^{-\gamma_m \varepsilon}}{1 - e^{-(j\omega+\gamma_m)T}}. \tag{10.64}$$

Since γ_m has a positive real part, it is clear that the limit on the right side is finite, consequently we have from (10.6)

$$\lim_{\varepsilon\to 0}\frac{1}{T}\sum e^{+j(2\pi n\varepsilon/T)}Z_{22}(\omega + n\omega_s) = \sum_{k=0}^{\infty} z(kT)e^{-j2\pi f kT} \tag{10.65}$$

which we recognize as our old friend $Z_{22}(\omega, +)$. In a similar manner

$$\lim_{\varepsilon \to 0} \frac{1}{T} \sum e^{-j(2\pi n\varepsilon/T)} Z_{22}(\omega + n\omega_s) = \sum_{k=1}^{\infty} z(kT) e^{-j2\pi fkT}, \qquad (10.66)$$

and this is none other than $Z_{22}(\omega, -)$. Thus (10.62) may be written

$$\frac{Z_{12}(\omega)}{R} - 2QZ_{22}(\omega, +) = G\left[2QZ_{22}(\omega, -) - \frac{Z_{12}(\omega)}{R}\right]. \qquad (10.67)$$

Solving (10.67) for Q yields

$$Q = \frac{\dfrac{Z_{12}(\omega)}{2R_1}[1 + G]}{Z_{22}(\omega, +) + GZ_{22}(\omega, -)}. \qquad (10.68)$$

Comparison of (10.68) and (10.37) reveals that Q is Cattermole's $CF(\omega)$ when $G = 1$, as expected. The component of I at the input frequency ($n = 0$ term of (10.58)) is Q/T. Then, the ratio of the output voltage to the input voltage is simply

$$\frac{E_4(\omega)}{E_1(\omega)} = \frac{Q}{TE} Z_{12}(\omega) = \frac{[Z_{12}(\omega)]^2[1 + G]}{2R_1 T[Z_{22}(\omega, +) + GZ_{22}(\omega, -)]} \qquad (10.69)$$

which becomes (10.38) for $G = 1$.

Before we go on to consider the design of other types of switched filters, it seems appropriate to cap the last three rather detailed sections with a short summary. First, following the work of Cattermole, Desoer, and Crowley, we have seen that it is possible in principle to achieve lossless energy transfer between filters with a switch that is closed for a small fraction of its cycle. All approaches lead to a periodic system function whose dc component $H_0(\omega)$ relates the output voltage of the input frequency to the corresponding component at the input. The pulse sequence impedances appearing in $H_0(\omega)$ show the dependence of the performance at ω on the impedance evaluated at $\omega \pm n\omega_s$ for $n = 0, 1, 2, \ldots$. We saw in Section 10.3 that this behavior was apparent to early workers in the field of switched modulators.

It is difficult to choose between the various approaches; indeed the very reason for covering them is to show their inherent unity. In the following we will generally adhere to the Fourier representation extended to include the finite pulse shape factor. This will lead to a straightforward extension of the concept of pulse sequence impedance. In addition, we will apply this approach to nonresonant transfer filters, determine how the approximate methods compare with exact analysis, and go on to the design problem.

PCM Demultiplex Filter

Consider the switched filter in Figure 10-16(a). This was used in each channel of an experimental PCM terminal to reconstruct an analog signal (voice or data) from its quantized PAM samples. For our purposes, the quantizing noise accompanying the PAM samples is unimportant. The following conditions are assumed to prevail:

1. The gate closes every T seconds for a short interval τ. For a 24-channel PCM system, typical values are: (a) $T = 125$ μsec, corresponding to 8 kHz sampling; (b) $\tau = 2.6$ μsec*.
2. Since $\tau/T \ll 1$, during the time the switch is closed the shape of the switch current is determined by the pulse response of the circuit in

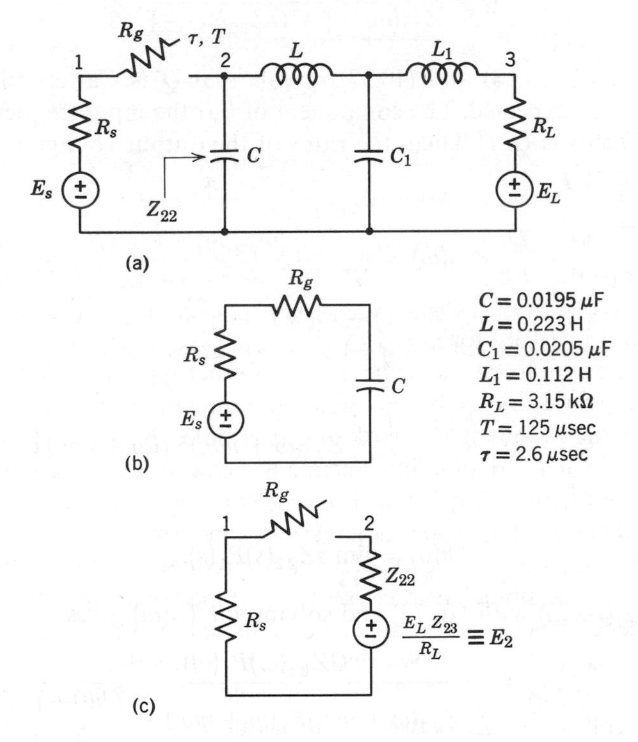

FIGURE 10-16 Early experimental demultiplex filter: (a) demultiplex filter, (b) approximate equivalent circuit during charging ($E_L = 0$), (c) equivalent circuit for $E_s = 0$, voltage at (3) to measure equivalent impedance.

* The total time allotted per channel is 125 μsec/24 or approximately 5.2 μsec. Of this interval, 2.6 μsec is used for signal reconstruction and a 2.6 μsec guard band.

Figure 10-16(b). The source resistance R_s and the switch "on" resistance R_g are lumped together in $R = R_s + R_g$. When the switch is open its impedance is assumed to be infinite. Thus the unit pulse shape is

$$p_1(t) = ae^{-at} \qquad 0 \le t \le \tau \tag{10.70}$$

where: $a = 1/RC$
with Laplace transform $P_1(s)$ given by

$$P_1(s) = \frac{a}{s+a}[1 - e^{-(s+a)\tau}] \tag{10.71}$$

3. The switch is characterized by

$$Ge_1(NT+) - e_2(NT+) = 0. \tag{10.72}$$

Since it is assumed that the impulses are shaped by $P_1(s)$,

$$I = \frac{Q}{T}\sum P_1(\omega + n\omega_s)e^{j(\omega + n\omega_s)t} \tag{10.73}$$

where Q will be determined from (10.72). From (10.73) and Figure 10-16 it may be seen that

$$e_1(t) - e_2(t) = Ee^{j\omega t} - \frac{Q}{T}\sum P_1(\omega + n\omega_s)Z_{22}(\omega + n\omega_s)e^{j(\omega + n\omega_s)t}. \tag{10.74}$$

Forming $V(t+) = [Ge_1(NT + \varepsilon) - e_2(NT + \varepsilon)]$ with $\varepsilon > 0$, substituting in (10.72) and taking the limit $V(t+)$ as $\varepsilon \to 0$ gives the following equation for Q

$$GE_1 = Q\left[\frac{h(0)}{2} + \frac{1}{T}\sum Z_{22}(\omega + n\omega_s)P_1(\omega + n\omega_s)\right] \tag{10.75}$$

where:

$$h(0) = \lim_{s\to\infty} sZ_{22}(s)P_1(s). \tag{10.76}$$

Combining (10.75) with (10.73) and solving for $E_3(\omega)$ gives

$$\frac{E_3(\omega)}{E_1(\omega)} = \frac{GZ_{23}(\omega)P_1(\omega)}{\sum Z_{22}(\omega + n\omega_s)P_1(\omega + n\omega_s) + \dfrac{Th(0+)}{2}} \tag{10.77}$$

where the denominator of (10.77) is the generalized pulse sequence impedance. If we allow $P_1(\cdot)$ to go to 1, then $h(0+) = 1/C$ and (10.77) becomes

$$\left[\frac{E_3(\omega)}{E_1(\omega)}\right]_0 = K_0(\omega) = \frac{Z_{23}(\omega)G}{\dfrac{T}{2C} + \sum Z_{22}(\omega + n\omega_s)}. \tag{10.78}$$

The value to be used for G is just the value assumed by $P(\omega)$ at $\omega = 0$; namely, $1 - e^{-a\tau}$. This quantity is seen to be nothing more than the fraction of the input voltage on the capacitor C at the end of the charging interval. Therefore, an equivalent way of obtaining (10.78) from (10.77) is to allow $P_1(\cdot)$ to go to 1 in the denominator and to $P_1(0)$ in the numerator of (10.77) for $G = 1$.

Equivalent Impedance Substitution for the Switch

To avoid problems associated with the initial conditions, it is advisable to carry $h(0+)$ along even though it is 0 when the pulse shape is used. Thus returning to (10.77), we rearrange it to the alternate form*

$$\frac{E_3(\omega)}{E_1(\omega)} = K(\omega) = \frac{GZ_{23}(\omega)}{\dfrac{Th(0+)/2 + \sum' P_1(\omega + n\omega_s)Z_{22}(\omega + n\omega_s)}{P_1(\omega)} + Z_{22}(\omega)}. \tag{10.79}$$

We have purposely separated out the $n = 0$ term in the sum in the denominator to define an equivalent impedance to replace the switch [11] in Figure 10-16. From (10.79) the equivalent impedance $Z_e(\omega)$ is easily seen to be

$$Z_e(\omega) = \frac{Th(0+)/2 + \sum' P_1(\omega + n\omega_s)Z_{22}(\omega + n\omega_s)}{P_1(\omega)} \tag{10.80}$$

and the circuit with the switch is replaced by the equivalent circuit shown in Figure 10-17. Clearly when $P_1(\omega) \to 1$ (as $\tau \to 0$), we have

$$Z_{e0}(\omega) = \frac{T}{2C} + \sum' Z_{22}(\omega + n\omega_s). \tag{10.81}$$

Since we have eliminated the switch and replaced it with the impedance (10.80) or (10.81), it might appear that standard design techniques can be

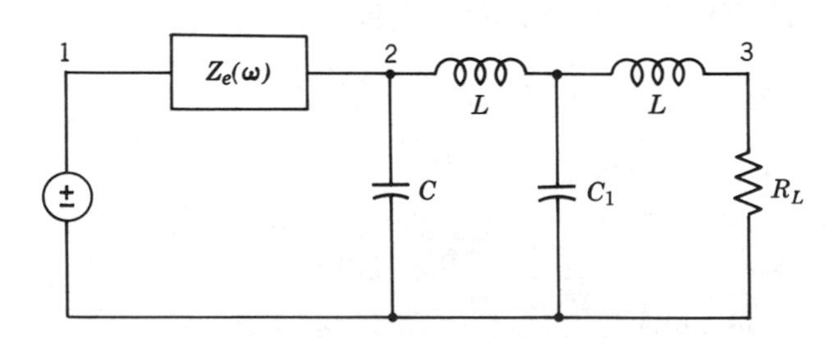

FIGURE 10.17 Equivalent time-invariant network.

*The prime on the sum in (10.79) denotes omission of the $n = 0$ term.

used. Difficulties in design still remain due to the dependence of the equivalent impedance on the out-of-band translates of Z_{22}. Soon we will show how this remaining obstacle can be overcome.

To illustrate that care must be exercised in defining impedance and handling limits, we derive an equivalent impedance by another approach. Assume that the voltage source is moved to the output. Then via Thevenin's theorem and reciprocity, we get the circuit of Figure 10-16(c). The current I is given as before by (10.73) and the voltage $e_2(t)$ by the right hand side of (10.74) with E replaced by E_2. Now, Q will be determined from the charge that appears on the filter capacitor C just prior to the N-th operation of the gate; namely

$$Q_N = Qe^{i\omega(NT-\varepsilon)} = \lim_{\varepsilon \to 0+} Ce_2(NT - \varepsilon) \tag{10.82}$$

From (10.74) and (10.82) we get

$$Q = \lim_{\varepsilon \to 0} \left[CE_2 e^{-i\omega\varepsilon} - \frac{CQ}{T} \sum e^{-i\omega_s \varepsilon} P_1(\omega + n\omega_s) Z_{22}(\omega + n\omega_s) \right]. \tag{10.83}$$

Solving for Q yields

$$Q = \frac{CE_2}{1 + \frac{C}{T} \lim_{\varepsilon \to 0} \sum e^{-i\omega_s \varepsilon} P_1(\omega + n\omega_s) Z_{22}(\omega + n\omega_s)}. \tag{10.84}$$

With the aid of (10.66) the limit is seen to be the generalized pulse sequence impedance defined by

$$\lim_{\varepsilon \to 0} \frac{1}{T} \sum e^{-i\omega_s \varepsilon} P_1(\omega + n\omega_s) Z_{22}(\omega + n\omega_s) = \sum_{k=1}^{\infty} h(kT) e^{-j\omega kT}, \tag{10.85}$$

which by comparison with (10.33) is the same as

$$[PZ(\omega, -)] = \frac{1}{T} \sum P_1(\omega + n\omega_s) Z_{22}(\omega + n\omega_s) - \frac{h(0+)}{2}. \tag{10.86}$$

If we take the ratio of the ω component of $e_2(t)$ from (10.74) (with E replaced by E_2) to the ω component of I, we get the impedance

$$Z(\omega) = \frac{E_2 - \frac{Q}{T} P_1(\omega) Z_{22}(\omega)}{\frac{Q}{T} P_1(\omega)}. \tag{10.87}$$

Using Q as given by (10.84) and (10.86), in (10.87) we get

$$Z(\omega) = \frac{\dfrac{T}{C} - \dfrac{Th(0+)}{2} + \sum' P_1(\omega + n\omega_s) Z_{22}(\omega + n\omega_s)}{P_1(\omega)}. \tag{10.88}$$

This is not identical with $Z_e(\omega)$ given by (10.80). When we allow $P_1(\omega)$ to go to unity, $h(0+)$ becomes $1/C$ and (10.88) and (10.81) become the same. For narrow pulses, $P_1(\omega)$ is a much broader band than $Z_{22}(\omega)$, and the principal effect of the pulse shape is specified by $P_1(0)$. Stated another way, the charging time constant $1/\alpha$ must be much less than the charging time τ to avoid considerable loss, and to avoid loss *variation* if τ varies due to sampling jitter. For these reasons, we concentrate on the equivalent impedance given by equation (10.81) and the transfer function given by (10.78).

Comparison of Approximate and Exact Transfer Functions

By this time the reader certainly must be asking at least four questions:

1. How does the approximate transfer function given by (10.78) compare with exact analysis?
2. How well does it predict measured results?
3. How is it used in design?
4. How about some examples?

Here we address the first question. The next part and all of Section 10.5 is aimed at the remaining questions.

Consider the switched demultiplex filter of Figure 10-18, used in the $D2$ PCM channel bank [29]. The design approach will be covered later. S. Yang has written a computer program to calculate the magnitude of the transfer function of this filter as specified by two approaches—namely (10.78) as it stands and as modified by an approximation to the equivalent impedance that we derive in the next section, (10.99). In addition using a computer

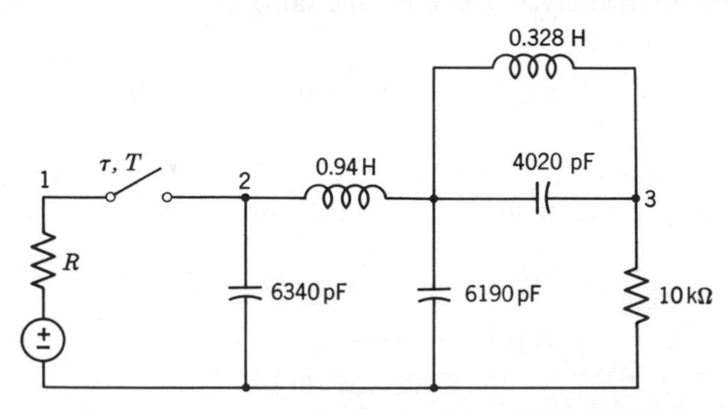

FIGURE 10-18 Demultiplex filter for D2 channel bank.

program based on a state variable formulation developed by M. L. Liou and Mrs. P. H. McDonald, S. Yang has also computed the exact transfer function for the demultiplex filter. Negligible switch loss has been assumed (0.1 milliohm!) as has switch closure time of 0.1 μsec (1/1250 of the 125 μsec sampling period). The two approximate transfer functions are listed below for ready reference (for $G = 1$)

$$K_0(\omega) = \frac{Z_{12}(\omega)}{\dfrac{T}{2C} + \sum Z_{22}(\omega + n\omega_s)} \tag{10.89}$$

$$K_1(\omega) = \frac{Z_{12}(\omega)}{\dfrac{T}{2C} + \sum_{-1}^{1} Z_{22}(\omega + n\omega_s) + j\dfrac{1.29\omega}{\omega_s^2 C}\left[1 + .127\left(\dfrac{\omega}{\omega_s}\right)^2\right]}. \tag{10.90}$$

Results for the magnitude of the transfer function in decibels for the three cases [(10.78), (10.99), and exact]* are given in Table 10-1. It may be seen that the three methods agree to within 0.005 dB in the passband and 0.03 dB in the stopband. This agreement between (10.78) and the state variable approach is expected, especially when the switch closure time and on-resistance are purposely made extremely small. However, the excellent results obtained with the approximation (10.99) given in the next section serve

TABLE 10-1 Effect of Approximations Attenuation of Switched Filter of Figure 10-18

Freq. (kHz)	Attenuation (dB)			Magnitude of
	Exact ($\tau = 0.1\ \mu$sec, $R = 10^{-4}$ ohms)	Approx 1 (10.78)	Approx 2 (10.99)	|Max. Diff.| (uses 5 place data)
0.5	5.97	5.97	5.97	0.005
1.0	6.00	6.00	6.00	0.004
1.5	6.03	6.04	6.04	0.004
2.0	6.08	6.08	6.08	0.005
2.5	6.07	6.07	6.07	0.003
3.0	5.97	5.97	5.97	0.001
3.5	7.74	7.73	7.74	0.012
4.0	20.80	20.78	20.78	0.019
4.5	44.52	44.51	44.49	0.029
5.0	35.94	35.94	35.94	0.003
5.5	35.23	35.23	35.22	0.008
6.0	36.98	36.98	36.97	0.008

* 1001 terms were used to evaluate the nominally infinite sum in the denominator of (10.78).

to give confidence in its use. It considerably simplifies both analysis and design. An even simpler approximation derived in the next section provides a useful starting point for design.

Let us consider other comparisons where the switch resistance is more realistic and the on-time is a larger fraction of the sample period. This will enable us to determine the region of validity of (10.78) or (10.99), at least for the present example. First, we retain the gate on-resistance at the low value of 0.1 ohm and let the on-time of the gate vary from 0.08% to 8% of the 125 μsec sampling period. Computed attenuation for $\tau = 0.1$, 1.0, 5.0, and 10.0 μsec is given in Table 10-2. Increasing the on-time from 0.1 to 1.0 μsec

TABLE 10-2 Effect of On-Time (τ) Attenuation of Switched Filter of Figure 10-18

	Attenuation (dB) ($R = 0.1\ \Omega$)			
Freq. (kHz)	$\tau = 0.1\ \mu$sec	$\tau = 1.0\ \mu$sec	$\tau = 5\ \mu$sec	$\tau = 10\ \mu$sec
0.5	5.96	5.92	5.71	5.43
1.0	5.99	5.95	5.76	5.50
1.5	6.03	6.00	5.85	5.64
2.0	6.08	6.05	5.95	5.80
2.5	6.07	6.06	6.01	5.91
3.0	5.97	5.98	5.97	5.92
3.5	7.73	7.81	8.11	8.34
4.0	20.80	20.97	21.64	22.31
4.5	44.52	44.59	44.77	44.67
5.0	35.94	35.94	35.77	35.27
5.5	35.23	35.21	34.99	34.41
6.0	36.98	36.95	36.63	35.88

results in a slight decrease in the in-band attenuation by as much as 0.04 dB at the low end of the band. A further increase in the gate on-time to 10 μsec gives a 0.5 dB rising attenuation slope across the passband. If the on-time is maintained constant at the low value of 0.1 μsec and the gate resistance varied, we obtain the curves shown in Figure 10-19. Examination of these responses shows that the in-band attenuation varies very closely with $1 - e^{-\alpha\tau} = P(0) = G$; namely the voltage loss in the charging circuit. Thus, the use of the factor G with any of the approximate expressions serves to bring them in very close agreement with exact analysis. For the larger values of on-time (1, 5, and 10 μsec), the variation of attenuation with gate resistance from 0 through 25 ohms has been computed to be virtually zero (actually a maximum of 0.02 dB across the 0 to 6 kHz band). This should

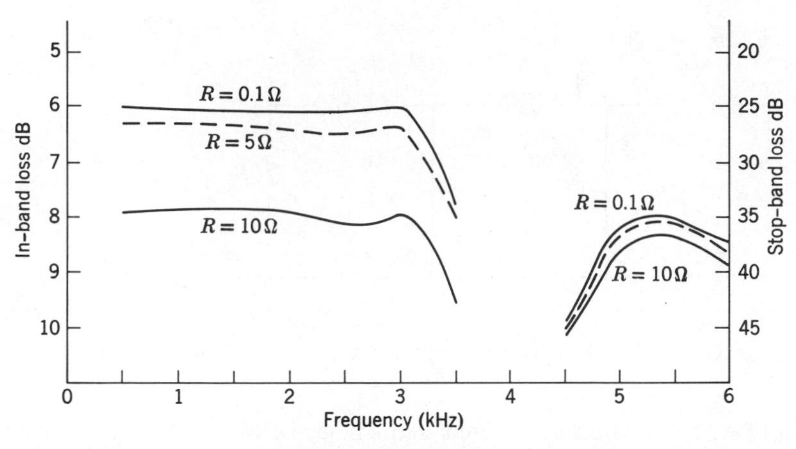

FIGURE 10-19 Variation of performance with switch resistance.

not be surprising, because the capacitor is virtually fully charged at the end of even the minimum charging interval of 1.0 μsec for the largest on-resistance. Thus, we conclude, for gate resistances achievable in communications applications, the simple approximation (10.99) is excellent.

Measurement of the Equivalent Impedance

Since we are dealing with the somewhat unfamiliar concept of replacing a time-varying network with a stationary network containing the impedance function (10.81), it behooves us to indicate how good the approximation is experimentally.* We have already seen that the loss characteristic computed from the approximate-transfer function is virtually indistinguishable from the exact result computed from the state-variable formulation. Here we compare measurements of the equivalent impedance (10.81) with computed results and develop another interesting approximation to (10.78) as a prelude to design. In the design section we make further comparisons with experiment.

Care must be exercised in making measurements on time varying networks. This follows from the fact that inputs at frequency ω produce outputs at frequencies $n\omega_s + \omega$, as is obvious from (10.30). This means that tuned detectors must be used and that measurements in the neighborhood of $\omega = \omega_s/2$ are particularly difficult due to the aliasing with the $\omega_s - \omega$ component. In addition, there is a loss in sensitivity beyond the filter cut-off frequency of about 3 kHz. Therefore, we concentrate in the impedance measurements on frequencies below $\omega_s/2$. The approach used to make the measurements was based on a substitution technique. First, the output impedance

* Most of the results of this section and those of the previous three sections were given by the author in an unpublished report in 1959.

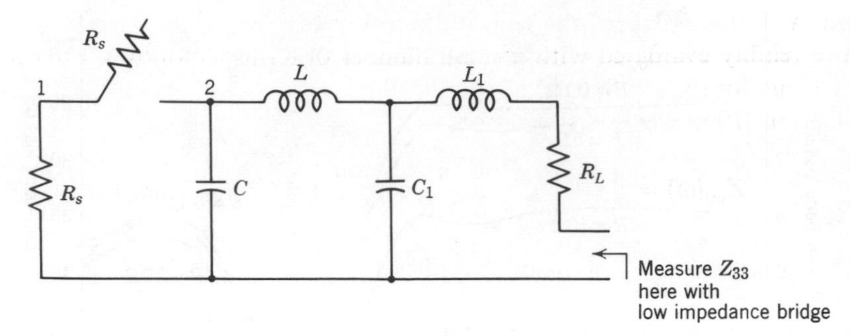

FIGURE 10-20 Measurements to determine equivalent impedance.

$Z_{33}(\omega)$ was measured at selected frequencies below $\omega_s/2$ by an impedance bridge using a tuned detector (about 50 Hz bandwidth). Under this condition, the input voltage at the switch side was shorted, the switch continued to operate. and the termination R_L remained connected (See Figure 10-20). At each frequency, the switch is replaced by a resistance and a reactance (inductance), and each is adjusted until $Z_{33}(\omega)$ reads the same as when the switch was connected. In this manner, we obtain the impedance replacement for the switch. Measured results* for the filter of Figure 10-16 compared favorably with computations. Over most of the passband the resistance component is essentially $T/2C = 3.2$ K ohms in excellent agreement with measurements. Agreement between the measurements and computations of the equivalent inductance are within 6% over the entire frequency range—well within the accuracy range of the bridge used at that time.

Computations on $Z_e(\omega)$ are facilitated by using either of two approaches to handle the infinite sum in (10.81). From (10.17) it is apparent that we can write

$$Z_{e0}(\omega) = \frac{T}{2C} - Z_{22}(\omega) + \frac{T}{2}\sum_{m=1}^{M} A_m\left[\frac{1 + e^{-(j\omega+\gamma_m)T}}{1 - e^{-(j\omega+\gamma_m)T}}\right]. \tag{10.91}$$

Therefore, if the singularities (γ_m) and the residues (A_m) are determined, (10.91) may be evaluated without difficulty. It is more appealing physically to deal with the element values. Indeed, this will permit us to establish a procedure for getting a first cut at switched filter design.

If we remove the asymptotic capacitance from Z_{22} to obtain

$$Z_{22}(\omega) = \frac{1}{j\omega C} + Z_{22}(\omega) - \frac{1}{j\omega C} = \frac{1}{j\omega C} + Z_a(\omega), \tag{10.92}$$

*These measurements were made in 1959 by the author and C. L. Maddox.

then, with this breakup, the sum in the equivalent impedance expression is more readily evaluated with a small number of terms as follows. First, the exact sum for the asymptotic term in (10.92) is carried out. Using (10.92) and (10.19) in (10.81), it is easy to show that

$$Z_{e0}(\omega) = \frac{T}{2C} + \frac{j}{\omega C} - j\frac{\pi}{\omega_s C}\cot\frac{\pi\omega}{\omega_s} + \sum{}' Z_a(\omega + n\omega_s). \qquad (10.93)$$

For the case at hand (and in most applications), Z_{22} is well approximated by its asymptotic value for values of $|n| > 2$. This means the last sum in (10.93) need only be carried out for at most the terms $n = \pm 1, \pm 2$. A further approximation is useful for the region of interest here below the sampling frequency. In this region, expansion of the cot in (10.93) and combination with the term $j/\omega C$ (valid for $|x| = |\omega/\omega_s| < 1$) gives

$$\frac{j}{\omega C} - j\frac{\pi}{\omega_s C}\cot\frac{\pi\omega}{\omega_s} = j\frac{2x}{\omega_s C}\left[\frac{\pi^2}{6} + \frac{\pi^4}{90}x^2 + \cdots + \frac{(-1)^k(2\pi)^{2k}}{2\cdot(2k)!}B_{2k}x^2\cdots\right]$$

$$(10.94)$$

where the B_{2k} are the Bernoulli numbers. An alternate representation for the coefficient of x^k in the brackets in (10.94) is

$$\sum_{n=1}^{\infty}\frac{1}{n^{2k}} = \frac{(-1)^k(2\pi)^{2k}B_{2k}}{2\cdot(2k)!}. \qquad (10.95)$$

The first few coefficients in the series in the brackets in (10.94) are $\pi^2/6 = 1.645$, $\pi^4/90 = 1.082$, $\pi^6/9450 = 1.017$, and so on toward unity. Using this fact, the series in (10.94) may be written

$$j\frac{2x}{\omega_s C}[.645 + .082x^2 + .017x^4 + \cdots] + j\frac{2x}{\omega_s C}\left(\frac{1}{1-x^2}\right). \qquad (10.96)$$

Combining (10.94) and (10.95) with (10.93), we finally get

$$Z_{e0}(\omega) = \frac{T}{2C} + j\frac{1.29x}{\omega_s C}[1 + .127x^2 + .027x^4 + \cdots]$$

$$+ \sum_{-2}^{2}{}' Z_a(\omega + n\omega_s) + j\frac{2x}{\omega_s C}\frac{1}{1-x^2} \qquad (10.97)$$

In most cases, the terms in the sum in (10.97) involving $n = \pm 2$ can be neglected. Under this assumption we arrive at the more compact approximation

$$Z_{e0}(\omega) = \frac{T}{2C} + j\frac{1.29x}{\omega_s C}[1 + .127x^2] + \sum_{-1}^{1}{}' Z_{22}(\omega + n\omega_s), \qquad (10.98)$$

which is an excellent approximation for $|x| < 1$. This has been demonstrated in Table 10.1 in connection with the comparison of transfer functions. With (10.98), the transfer function becomes

$$K_1(\omega) = \frac{Z_{12}(\omega)}{\dfrac{T}{2C} + \sum_{-1}^{1} Z_{22}(\omega + n\omega_s) + \dfrac{1.29jx}{\omega_s C}(1 + .127x^2)}. \tag{10.99}$$

An even rougher approximation suitable for purposes of design is to neglect the sum in (10.97) to obtain

$$Z_{e02}(\omega) = \frac{T}{2C} + j\frac{1.29x}{\omega_s C}\frac{1}{(1 - 0.127x^2)} + j\frac{2x}{\omega_s C}\frac{1}{(1 - x^2)}. \tag{10.100}$$

The reader should not find it difficult to show that (10.100) corresponds to the physical impedance given in Figure 10-21(a). Note that this impedance only depends upon the input capacitance C, and the sampling frequency ω_s. The input capacitance is generally chosen on the basis of impedance level and power handling capabilities of the gate. Once C is chosen, we can use

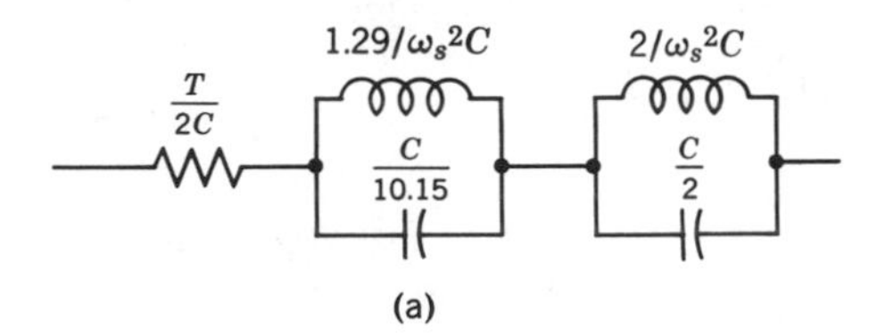

FIGURE 10-21 Approximations to equivalent impedance.

(10.99) as a vehicle for design. This will be explored in the next section. We complete the present section by comparing the transfer functions using (10.100) or the cruder approximation

$$Z_{e03}(\omega) = \frac{T}{2C} + j\frac{3.29x}{\omega_s C} = \frac{T}{2C}\left(1 + j\frac{3.29}{\pi}x\right) \tag{10.101}$$

with (10.98) whose validity was established in Table I. If we use (10.100) or (10.101), we get the approximate transfer functions

$$K_2(\omega) = \frac{Z_{12}(\omega)}{\dfrac{T}{2C} + Z_{22}(\omega) + j\dfrac{1.29x}{\omega_s C}\dfrac{1}{(1 - 0.127x^2)} + j\dfrac{2x}{\omega_s C}\dfrac{1}{(1 - x^2)}} \tag{10.102}$$

and

$$K_3(\omega) = \frac{Z_{12}(\omega)}{\dfrac{T}{2C} + Z_{22}(\omega) + j\dfrac{3.29x}{\omega_s C}}, \tag{10.103}$$

whose attenuation characteristics are given in decibels in Table 10-3. For purposes of ready comparison, the attenuation values obtained from (10.99) ($K_1(\omega)$) are repeated there. Agreement up to 3 kHz is clearly very good (less than 0.1 dB discrepancy). Even in the transition region between 3 kHz and 4 kHz, the simplest function (10.103) is good enough for many purposes. If the reactive components are dropped and $Z_e = T/2C$, then one obtains the results given in the last column of Table 10-3. Clearly this is a very poor approximation.

TABLE 10-3 Effect of Equivalent Impedance Simplifications Attenuation of Switched Filter of Figure 10-18

	Attenuation (dB)			
Freq (kHz)	From 10.99 (repeated from Col 3 of Table 10.1)	From 10.102 [Z_{eq} of Figure 10.21(a)]	From 10.103 [(Z_{eq} of Figure 10.21(b)]	Crudest Equivalent $Z_e = T/2C$
0.5	5.97	5.97	5.97	5.98
1.0	6.00	5.99	5.99	6.09
1.5	6.04	6.04	6.04	6.43
2.0	6.08	6.11	6.11	7.12
2.5	6.07	6.13	6.15	8.13
3.0	5.97	6.01	6.03	9.06
3.5	7.74	8.23	8.41	8.84
4.0	20.78	21.34	21.71	8.50
4.5	44.49	38.57	38.71	32.10
5.0	35.94	30.17	29.83	25.00
5.5	35.22	30.58	29.45	25.00
6.0	36.97	32.82	30.47	25.92

10.5 DESIGN OF SWITCHED FILTERS

General

Two approaches to the design of switched filters suggest themselves. We will emphasize one based on the equivalent impedance concept. This may be labeled a frequency-domain approach. An alternate time-domain design technique has been used successfully, and we simply note it briefly in passing. The work of May and Stump [38], Thomas [39], and Gibbs [40] is representative of this work. The underlying unifying theme follows from equations (10.12), (10.20), and (10.21) in which, roughly speaking, the objective is to make the impulse response of the impedance function close to zero at all sample times except at the initiation of the impulse. In other words, the capacitance in the resonant-transfer mesh discharges to zero before the next charging interval. Exact satisfaction of this condition requires an infinite network (See Problem 10.5). Gibbs [40] presents an interesting method for obtaining a finite equi-ripple in-band approximation, and he compares his technique to that given by Thomas [39] (See Problem 10.5 for Thomas' method). Thomas has designed several filters that are used in electronic-time-division switching systems. Space precludes a detailed coverage. It should be apparent that this method can be used as a prelude to the computer-aided design method discussed later.

Design Based on Equivalent Impedance

The impedance looking back into the capacitor C, and the simple R, L approximation for the gate* is shown in Figure 10-22(a). Examination of the plot of the real part as a function of frequency reveals that it approximates the mid-series impedance [34] of a constant k lowpass filter (with cut-off frequency about 3.8 kHz) very closely over most of the passband. The imaginary component, which is negative, may be cancelled over most of the signal band by the addition of a series inductance. The remaining sections include a constant K half section followed by a m-derived half section with an $m = 0.6$ as shown in Figure 10-22(b). The m-derived section is required to obtain a good match to the terminating impedance. The composite filter in 10-22(c) maintains a better than 30 dB return loss up to 3 kHz. Fortunately, this approach resulted in a filter whose attenuation characteristic also satisfied the requirements for the D1 Channel Bank of the T1 PCM Carrier system [28].

In general it will be difficult, with the design constraints imposed, to meet the most stringent requirements by this approach, but it is a useful vehicle for getting into the right ballpark. Let us summarize the approach.

1. Use the simple R, L gate approximation—$R = T/2C$, $L = 3.29/\omega_s^2 C$. This sets the impedance for the filter and the cut-off frequency close to 4 kHz.

* This design procedure was evolved in 1960.

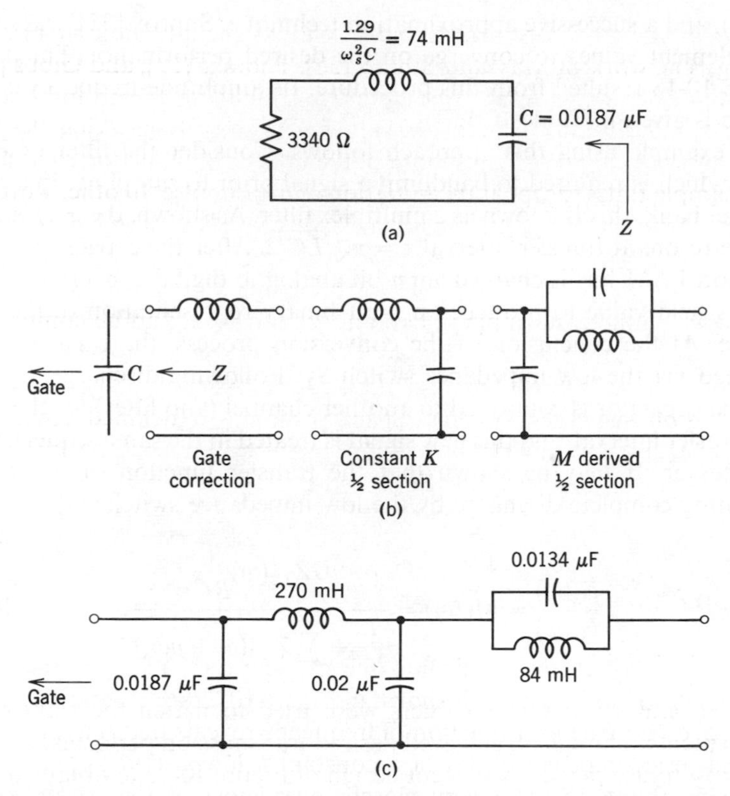

FIGURE 10-22 Demultiplex filter design for D1: (a) approximate impedance Z looking back into C and gate, (b) filter sections, (c) composite filter.

2. Use a series L to cancel the reactive component of the impedance viewed looking back at the capacitor C toward the gate. This gives a mid-series match.
3. Add constant k and m-derived sections to satisfy cut-off and out-of-band loss requirements as well as impedance matching.

Computer-Aided Design—Passive Filters

When the need arose to design a demultiplex filter for an application requiring a more confined passband, it became necessary to develop a computer-aided approach to design.* A subroutine was written to evaluate

* P. E. Fleischer developed the approach discussed in this section in 1966.

(10.98), and a successive approximation technique, Suprox [35], was used to vary element values to converge on the desired performance. The filter of Figure 10-18 resulted from this procedure. Its amplitude-frequency characteristic is given in Table 10-1.

An example using this approach follows. Consider the filter of Figure 10-10 which is required to bandlimit a signal prior to sampling. In a digital channel bank, this is known as a multiplex filter. As shown, the gate is closed for the resonant transfer interval $\tau = \pi\sqrt{LC/2}$. After the capacitor C of the common PAM bus is charged, an n-bit analog to digital converter operates on this held value to produce an n-bit binary representation of the PAM sample. At the conclusion of the conversion process, the capacitor is discharged via the low impedance switch S_2. Following discharge, S_2 opens and the capacitor is connected to another channel (and filter) for the resonant transfer interval, and this new signal is treated in the same sequence as its predecessor. It may be shown that the transfer function of this filter is (assuming complete discharge by the low impedance switch, S_2)

$$\frac{E_2(\omega)}{E_1(\omega)} = M_0(\omega) = \frac{GZ_{12}(\omega)\dfrac{T}{RC}}{\dfrac{T}{2C} + \sum Z_{22}(\omega + n\omega_s)}. \tag{10.104}$$

Five, six, and seven element filters were tried to match the transmission requirements. Though all achieved desired transmission performance, it was necessary to use the seven element design of Figure 10-23 to obtain an input

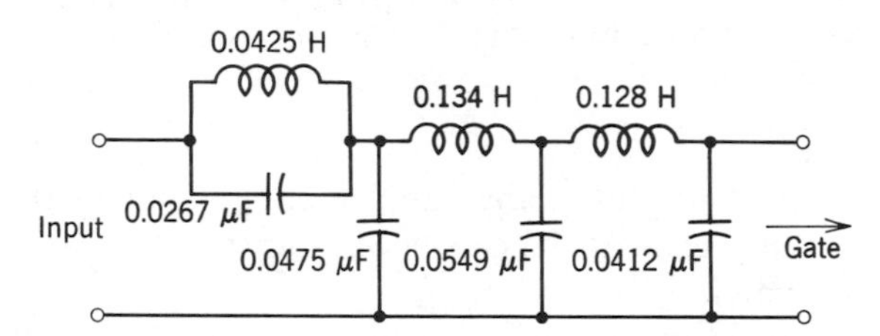

FIGURE 10-23 Multiplex filter designed for D2.

impedance match to the source. About 0.01 hr of IBM 7094 time was required for design.

Other filters, including the demultiplex filter for the D2 Channel Bank [29] and filters for TASI-B [37], have been designed by Fleischer's technique.

Computer-Aided Design—Active Filters

Recently (1970), Liou, McDonald, and Thelen [13] described a computer-aided design approach that follows Fleischer's basic philosophy. It differs in detail in a couple of quite interesting respects. It continues to use the Suprox successive approximation approach, but it uses an exact analysis routine based on state-space techniques [6]. Second, it is applicable to both active and passive filters, while the approach of Section 10.5 was limited to passive ladder networks. For communications applications, the second modification may be more significant than the first. This follows for two reasons. First, the approximate analysis techniques are usually more than adequate.* More importantly, active filters have become cost competitive with passive filters for the applications considered herein. One such active filter designed by Liou, McDonald, and Thelen is given in their paper and will not be repeated here.

The basic idea behind this approach is summarized either by the old adage, "If at first you don't succeed try and try again," or by its more modern version, called variously a "design strategy," an algorithm, or flow chart. In the first cycle of operation, requirements are compared with the calculated performance of a specific circuit (a first try). The difference between desired and actual performance, called the error, is evaluated, and the circuit is altered via the optimization strategy to reduce some measure of this error. Various measures can be implemented, but perhaps the simplest and most common measure is a weighted sum of the squares of the errors in the amplitude frequency characteristic at a selected set of frequencies. Steepest descent and/or least squares [14] is used in Suprox [35], but in principle any of the common optimization strategies [36] can be programmed.

It should be noted that for multiple op-amp realization of active switched filters, a natural partitioning occurs. For example, in the demultiplex filter, the first op-amp isolates the switch from the rest of the filter. Thus, a portion of the design utilizes the approaches given here, and the remaining design follows conventional practice for time-invariant networks. This technique has been used by Aaron and Mitra [41] and Liou, McDonald, and Thelen [13].

Other Factors

The reader should not be deluded into thinking that switched filter design is merely a matter of exercising a computer program. Realistic design encompasses considerably more than what we have covered. In the design of high performance sample and hold circuits [30], it is mandatory to consider the nonlinearities of the switch. Effects of jitter in the switch control waveform must be understood. Choice of the switch circuit (diode bridge, transistor, JFET, MOSFET, ...) is based on many factors including speed, power requirements, balance of the control circuit, and control feedthrough, offset,

*Of course, the most demanding performance objectives may require exact analysis routines [42].

and so on. With active-switched filters, interaction of control feedthrough with op-amp overload may be important, and the stability of amplifiers working into time-varying loads must be considered. In thin-film circuits, total substrate capacitance and resistance have to be taken into account. Of course, cost must always receive heavy weight. Detailed discussion of these and other matters is precluded here, but their importance should not be underestimated. See [43] for example.

10.6 SUMMARY

There are several ways of arriving at a useful first approximation to a switched-filter design. We introduced the equivalent impedance approximation for the design of the multiplex and demultiplex filters for the first commercial PCM system in 1962. Literally millions of filters based on this design approach have been produced. As requirements became more stringent the computer was called in as an aid to economical filter realization. Techniques described or referenced herein are sufficient to provide a basis for the switched filter designs required as digital communications and digital switching expand.

ACKNOWLEDGMENT

Special pains have been taken to indicate how today's design techniques and understanding have resulted from the contributions of many workers in the past. Particular credit has been given throughout the text. To this I would like to add my thanks to S. Yang for his help in obtaining the data on Tables 10-1 through 10-3. Over a period of more than ten years, I have had the good fortune of collaboration with C. L. Maddox in this field. P. E. Fleischer has been a helpful ally in illuminating some of the more obscure points.

PROBLEMS

10.1 Derive (10.6).

10.2 Show that as $T \to 0$ (sampling frequency becomes arbitrarily large) that (10.17) reduces to (10.13).

10.3 Show that the initial value theorem is identical to Bode's resistance integral theorem for impedance functions.

10.4 If $Z(\omega)$ is a minimum reactance impedance function whose real part $R(\omega) = R$, a constant, over the band $-\omega_s/2 \leq \omega \leq \omega_s/2$ and zero elsewhere, show that its unit area impulse response is given by (10.20). Show that the corresponding minimum reactance is

$$X(\omega) = -\frac{R}{\pi}\ln\left|\frac{\omega + \dfrac{\pi}{T}}{\omega - \dfrac{\pi}{T}}\right|$$

and that the asymptotic capacitance (large ω) is $T/2R$. Further show that

$$\sum_{m=-\infty}^{\infty} X(\omega + 2\pi m/T) = 0$$

so that (10.21) follows immediately from this result.

10.5 Show that the continued fraction expansion of the reactance function $X(\omega)$ of problem 10.4 yields the unterminated infinite ladder of Figure 10-24(a) with

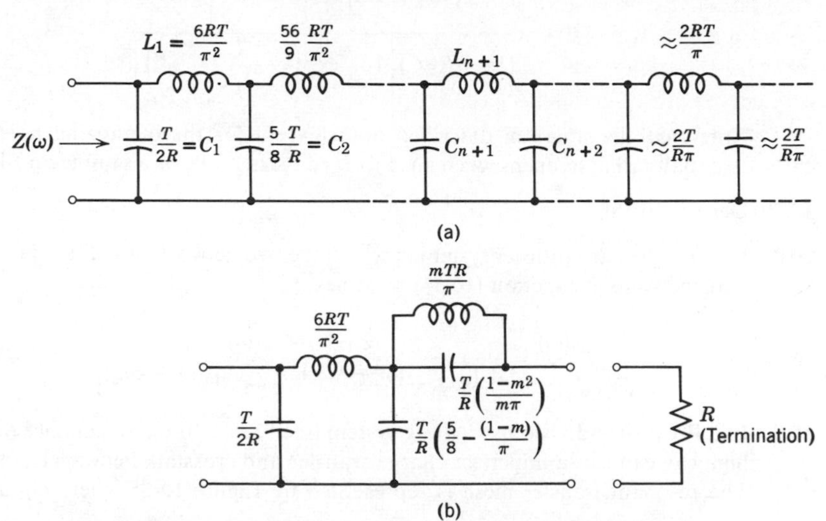

FIGURE 10-24 Approximation of ideal lowpass filter—Problem 10.5: (a) infinite unterminated ladder, (b) 5 element approximation.

element values given by [39]

$$C_1 = \frac{T}{2R}, \qquad L_1 = \frac{6RT}{\pi^2}$$

and for $n = 1, 2, 3 \ldots$

$$C_{n+1} = \frac{T}{2R}\left(\frac{1^2 \cdot 3^2 \cdot 5^2 \cdots (2n-1)^2}{2^2 \cdot 4^2 \cdot 6^2 \cdots (2n)^2}\right)(4n+1)$$

$$L_{n+1} = \frac{2RT}{\pi^2}\left(\frac{2^2 \cdot 4^2 \cdot 6^2 \cdots (2n)^2}{1^2 \cdot 3^2 \cdot 5^2 \cdots (2n+1)^2}\right)(4n+3).$$

For large n show that $C_{n+1} \to 2T/R\pi$, and $L_n = 2RT/\pi$ and that these values are the full shunt and series branches of an iterative constant k ladder. Therefore if the ladder is broken midshunt or mid-series, an m-derived terminating section can be used to match to a resistive load. Show that if the ladder of Figure 10-24(a) is broken at the second capacitor, and an m-derived terminating half section is added, that the filter of figure 10-24(b) results. This is basically the approach used by Thomas [39] and is sufficient for some designs or may be used as a first approximation in an iterative design.

10.6 Let the sample-and-hold circuit of Figure 10-12 be idealized such that $R_2 \to \infty$. Assume that the circuit is followed by an ideal impulse modulator which impulse-samples in the held interval. Show that under these conditions the exact transfer function for the sample-and-hold circuit is $G(s)$ where $G(s)$ is

$$G(s) = \frac{\left[1 - \exp\left(-\frac{T_c}{R_1 C} - sT_c\right)\right] e^{sT_c}}{(1 + sR_1 C)\left[1 - \exp\left(-\frac{T_c}{R_1 C} - sT\right)\right]}.$$

Note that the situation described here (except for the quantizing noise) is essentially what happens when an A to D converter follows a sample-and-hold.

10.7 Derive (10.55).

10.8 For the resonant transfer system in which the two networks are different, show that the transfer function (10.54) becomes

$$\frac{H'_{00}(\omega)}{E} = \frac{Z_{12}(\omega)Z_{34}(\omega)}{R_1[\sum Z_{22}(\omega + k\omega_s) + \sum Z_{33}(\omega + k\omega_s)]}.$$

10.9 In a PAM time-division-switching system (see Figure 10-6), capacitance on the highway can cause imperfect charge transfer and crosstalk between channels. The resonant transfer mesh is represented by Figure 10-25 where C_H is the

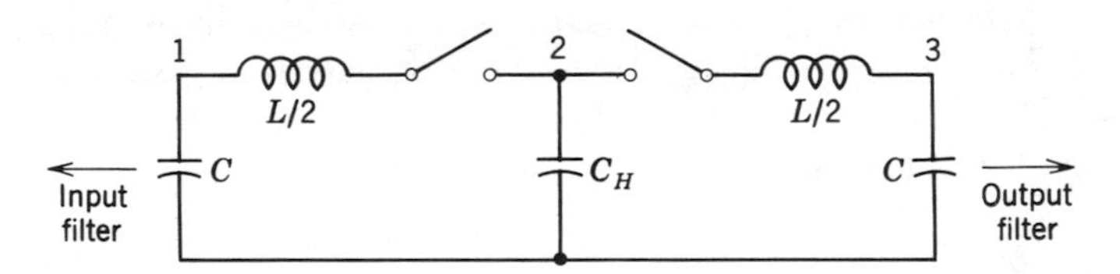

FIGURE 10-25 Highway capacitance in resonant transfer—Problem 10.9.

highway capacitance and the resonant transfer inductance has been split. Relate C_H to C to insure complete charge interchange (See reference [8]). Plot the node voltages and inductor current during the transfer interval when the switches are closed.

10.10 What are the effects of loss in the inductance in the resonant transfer mesh of Figure 10-8(a)? Plot the waveforms for small loss during the transfer time $\pi\sqrt{LC}$. For a high Q inductor, what is the approximate loss in voltage transfer from e_1 to e_2? How can the transfer interval be modified to minimize this loss? What should the transfer interval be to make the current zero when the switch opens?

10.11 Derive the approximate transfer function of the multiplex filter of Figure 10-10, (10.104).

10.12 Consider the multiplex filter of Problem 10.11, Figure 10-10 with the clamping switch S_2 removed. Assuming impulse sampling show that the approximate transfer function is

$$\frac{E_3}{V(\omega)} = \frac{Z_{12}(\omega)\dfrac{T}{RC}}{\dfrac{T}{2C} + \sum Z_{22}(\omega + n\omega_s) - e^{-j\omega T}\left[\sum Z_{22}(\omega + n\omega_s) - \dfrac{T}{2C}\right]}.$$

10.13 Show that the result in Problem 10.12 and (10.104) are identical for the ideal lowpass filter of Problem 10-4.

10.14 Show that (10.94) may be obtained directly from $1/C \sum_{-\infty}^{\infty} 1/j(\omega + k\omega_s)$ without using (10.19). (Hint: Consider pairs of terms (for positive and negative values of k), and use (10.95).)

10.15 It is possible to derive (10.98) directly from (10.88), letting $P_1(\omega)$ *in the denominator* equal unity. It is a messy operation involving use of the asymptotic value of $Z_{22}(\omega)$ in the sum, including the pulse shape in the sum, and taking advantage of the fact that $\alpha\tau \gg 1$ (This is necessary for rapid charging). This is essentially an exercise of manipulatory skills.

REFERENCES

[1] W. R. Bennett, "Steady-state transmission through networks containing periodically operated switches," *IRE Trans. PGCT*, **2**, No. 1, 17–22 (March, 1955).

[2] C. A. Desoer, "Transmission through a linear network containing a periodically operated switch," *Wescon Conv. Record*, 34–41 (1958).

[3] T. R. Bashkow, "The *A*-matrix, new network description," *IRE Trans. PGCT*, **4**, No. 3 (1957).

[4] I. W. Sandberg, "The analysis of networks containing periodically variable piecewise—constant elements," *Proc NEC.*, **17**, 81–97 (1961).

[5] Y. Sun and I. T. Frisch, "Transfer functions and stability for networks with periodically varying switches," *Asilomar Conference Record*, 130–142 (1967).

[6] M. L. Liou and F. R. Mastromonaco, "Exact analysis of linear circuits containing a periodically operated switch using the state space approach," *1968 Int. Symposium on Circuit Theory.*

[7] A. Fettweis, "Steady-state analysis of circuits containing a periodically operated switch," *IRE Trans. Circuit Theory*, **CT-6**, 252 (1959).
[8] K. W. Cattermole, "Efficiency and reciprocity in pulse amplitude modulation, part I—principles," *Proc. IEE*, **105**, Part B, 449–462 (1958).
[9] T. H. Crowley, Unpublished work (1956).
[10] C. A. Desoer, "A network containing a periodically operated switch solved by successive approximations," *BSTJ*, **36**, No. 6, 1403 (1957).
[11] M. R. Aaron, Unpublished work (1959).
[12] P. E. Fleischer, Unpublished work (1966).
[13] M. L. Liou, P. H. McDonald, and W. Thelen, "Computer optimization of active and passive switched low-pass filters for PCM," *Conference Proceedings of the 1970 Electronic Components Conference* (May, 1970).
[14] M. R. Aaron, "The use of least squares in system design," *IRE Trans. Circuit Theory*, **CT-3**, 224–231 (Dec., 1956).
[15] G. Doetsch, *Guide to the Applications of Laplace Transforms*, D. Van Nostrand, New York (1961).
[16] E. Peterson and L. W. Hussey, "Equivalent modulator circuits," *BSTJ*, **18**, 32 (1939).
[17] R. S. Caruthers, "Copper oxide modulators in carrier telephone systems," *BSTJ*, **18**, 315–337 (April, 1939).
[18] V. Belevitch, "Linear theory of bridge and ring modulators," *Elec. Commun.*, **25**, 62 (1948).
[19] D. G. Tucker, "Rectifier modulators with frequency selective terminations," *Proc. IEEE*, **96**, Part III, 422 (1949).
[20] D. G. Tucker, *Modulators and Frequency Changers*, MacDonald, London (1953).
[21] D. G. Tucker, "Input modulation (i.e. interchannel crosstalk) in constant resistance modulators for use in FDM systems," *Proc. IEE*, **114**, No. 10, 1385–1390 (Oct., 1967).
[22] A. Fettweis, "Theory of resonant-transfer circuits," *Network and Switching Theory*, G. Biorci, Ed., Academic Press, New York (1968) pp. 382–446.
[23] C. G. Svala, E. Aro, and S. L. Junker, "A 4-wire solid state switching system: Time division switching network," *IEEE Trans. On Communication Technology*, 197–206 (Dec., 1964).
[24] T. E. Browne, D. J. Wadsworth, and R. K. York, "New time division switch units for no. 101 ESS," *BSTJ*, 443–476 (Feb., 1969).
[25] P. M. Thrasher, "A new method for frequency-division multiplexing, and its integration with time-division switching," *IBM Journal of Res. and Dev.*, 137–140 (March, 1965).
[26] K. M. Roehr, P. M. Thrasher, D. J. McAuliffe, "Filter performance in integrated switching and multiplexing," *IBM Journal of Res. and Dev.*, 282–291 (July, 1965).
[27] W. R. Bennett, "Time division multiplex systems," *BSTJ*, **20**, 199–221 (1941).
[28] C. G. Davis, "Experimental pulse code modulation system for short haul trunks," *BSTJ*, **41**, 1–24 (1962).

[29] H. H. Henning, "A 96-channel PCM channel bank," Conference Record 1969, *IEEE Int. Conf. on Communications*, 34-17–34-22 (1969).
[30] J. R. Gray and S. C. Kitsopoulous, "A precision sample and hold circuit with subnanosecond switching," *IEEE Trans. Circuit Theory*, **CT-11**, No. 3 (Sept., 1964).
[31] Y. Sun and I. T. Frisch, "Analysis and synthesis of ac compensation filters," *IEEE Trans. CT*, **CT-15**, No. 4, 341–349 (Dec., 1968).
[32] K. A. Ivey, "AC carrier control systems," John Wiley, New York, (1964), 349p.
[33] L. A. Zadeh, "Frequency analysis of variable networks," *Proc. IRE*, **38**, 291–299 (March, 1950).
[34] H. W. Bode, "Network analysis and feedback amplifier design," D. Van Nostrand, New York (1945).
[35] P. E. Fleischer, "Optimization techniques in system design," Chapter 6, *System Analysis by Digital Computer*, edited by F. F. Kuo and J. F. Kaiser, John Wiley, New York (1966).
[36] See Chapter 7 of this book.
[37] Robert B. Leopold, "TASI-B, a system for restoration and expansion of overseas circuits," *Bell Labs Record*, 299–306 (Nov., 1970).
[38] P. J. May and T. M. Stump, "Synthesis of a resonant transfer filter as applied to a time division multiplex system," *AIEE Trans. Communications and Electronics*, **79**, 615–620 (Nov., 1960).
[39] G. B. Thomas, "Synthesis of input and output networks for a resonant transfer gate," *IRE International Convention Record*, Part 4, 236–243 (1961).
[40] A. J. Gibbs, "Design of a resonant transfer filter," *IEEE Transactions on Circuit Theory*, **CT-13**, No. 4, 392–398 (Dec., 1966).
[41] M. R. Aaron and S. K. Mitra "Design of active R-C switched filter for PCM applications," *Kyoto International Conference on Circuits and System Theory Digest*, **1**, 14–15 (1970).
[42] M. L. Liou, "Exact analysis of linear circuits containing periodically operated switches—with applications," *IEEE Trans. Circuit Theory*, **CT-19**, No. 2, 146–154 (1972).
[43] R. A. Friedenson, "Active filters make it small in the D3 channel bank," *Bell Labs Record*, 105–110 (Apr., 1973).

11

N-Path Filters

L. E. Franks
University of Massachusetts
Amherst, Massachusetts

The use of circuit elements that vary periodically with time affords some attractive alternatives to the realization of filters using only constant-parameter elements. The basic advantage is that filters, operating in a frequency range where inductors or distributed-parameter elements are too bulky or too expensive to obtain with sufficiently low loss, can be equivalently realized with a combination of time-varying and constant resistors and capacitors. Such circuits can be designed to exhibit time-invariant terminal behavior, despite the presence of internal time-varying elements. Another advantage that often accompanies realizations of this type is that the filter characteristics can be made easily adjustable by electronic means. This is accomplished by controlling the phase, frequency, and harmonic content parameters of the oscillators that provide the periodic signals for the time-variable elements. Using digital circuitry for controlling these parameters, an accuracy and ease of adjustment that far surpasses that obtained with variable inductors or delay lines is provided. In fact, many of the "programmability" features enjoyed by digital filters can be attained with properly designed analog filters using periodically-variable elements.

The time-variable filter is also relatively free from the severe sensitivity problems associated with active RC network realization of high-Q resonator elements. These inherent advantages, taken along with recent advances in the technology of fabricating integrated circuits and thin-film circuits, has given a fresh inpetus to the investigation of realizing filters in microcircuit form by using time-variable elements.

11.1 PRELIMINARY CONSIDERATIONS

The N-path filter [1] is a generalized configuration for realizing a broad variety of filtering characteristics by means of time-variable elements in the form of multipliers or modulators. The basic structure, shown in Figure 11-1, consists of N parallel paths (indexed by n from 0 to $N-1$ for analytical convenience), where each path contains a time-invariant filter with input

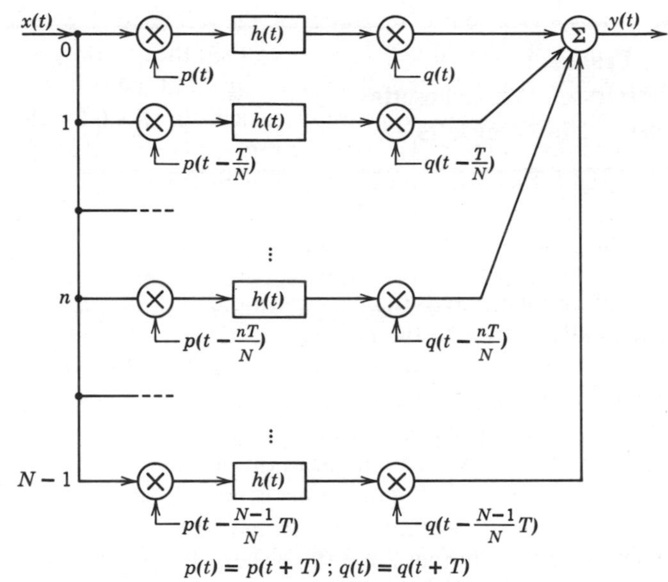

$p(t) = p(t+T)$; $q(t) = q(t+T)$

FIGURE 11.1 The N-path configuration.

and output modulators. The paths are identical except for the uniform staggering of the phases of the modulating functions in consecutive paths. With this particular structure, time-invariant filtering over a bandwidth proportional to the number of paths employed can be obtained. Relatively simple path networks can produce either a highly-selective bandpass filter or a multiple-passband (comb) filter which would normally require high-quality delay lines for realization. For some applications, a large number of modulators will be required, but, on the other hand, the requirements for these applications can often be met with simple "on-off" modulation or sampling switches. Hence the implementation of highly accurate multiplier circuits for modulation is not usually required.

As a preliminary to deriving the response properties of the N-path filter, we present first an analysis of a time-variable network which represents a single path in the N-path configuration. A time-domain approach to deriving the input-output relation corresponding to a general linear operation on a signal is often simpler than the frequency-domain approach commonly used to characterize time-invariant operations. Assuming that the network producing the linear operation has zero output under zero-input conditions (no initial conditions), then the response, $y(t)$, to an input signal, $x(t)$, can be generally expressed as an integral transform on $x(t)$.

$$y(t) = \int_{-\infty}^{\infty} w(t, t_1)x(t_1)\, dt_1 \tag{11.1}$$

where $w(t, t_1)$ is the impulse response of the network. If the impulse is applied at the instant, τ, i.e., if $x(t) = \delta(t - \tau)$, then the response is $y(t) = w(t, \tau)$. For a physically realizable network, the output at the instant t must not be dependent on future values of the input, hence $w(t, \tau) = 0$ for all $t < \tau$. For this reason, the upper limit in (11.1) can be replaced by t. If the network were also time invariant, then the impulse response must be a function only of the difference variable, $t - \tau$.

The operation of interest here is a multiplication followed by a time-invariant operation which is followed by another multiplication as shown in Figure 11-2. The impulse response of this single-path structure is obtained

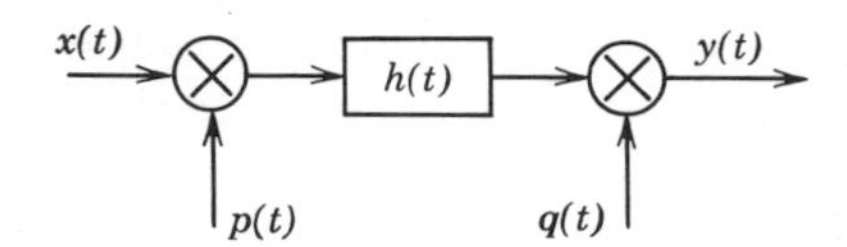

FIGURE 11-2 Single-path time-variable network with input and output multipliers (modulators).

by inspection. A unit impulse applied at the instant τ is simply scaled by the factor $p(\tau)$ at the input to the time-invariant network. The response of the time-invariant network is $p(\tau)h(t - \tau)$, and this gets multiplied by $q(t)$ at the output. Hence the impulse response of the path is

$$w(t, \tau) = p(\tau)h(t - \tau)q(t). \tag{11.2}$$

In the applications to be discussed here, the modulating functions, $p(t)$ and $q(t)$, will both be periodic with period, T. These functions can be represented by their Fourier series

$$p(t) = \sum_{m=-\infty}^{\infty} P_m e^{j(2\pi mt/T)} = p(t + T)$$

$$q(t) = \sum_{l=-\infty}^{\infty} Q_l e^{j(2\pi lt/T)} = q(t + T). \tag{11.3}$$

Some of the significant features of a time-variable network of this type are found by examining the response to an exponential input of the form e^{st} where s is an arbitrary complex frequency. For $x(t) = e^{st}$, and using the change of variable, $\sigma = t - \tau$,

$$y(t) = \int_{-\infty}^{t} p(\tau)h(t-\tau)q(t)e^{s\tau}\,d\tau$$

$$= \sum_l \sum_m Q_l P_m e^{j[2\pi(l+m)t/T]} e^{st} \int_0^{\infty} h(\sigma) e^{-s\sigma - j(2\pi m\sigma/T)}\,d\sigma$$

$$= \left[\sum_l \sum_m P_m Q_l H\left(s + j\frac{2\pi m}{T}\right) e^{j[2\pi(l+m)t/T]}\right] e^{st} \tag{11.4}$$

where $H(s)$ is the Laplace transform of the function, $h(t)$. For a time-invariant network, the response would be proportional to e^{st}, but in this case there are additional components in the response which are the "images" of e^{st}. If $s = j2\pi f$, then there are image components at the frequencies, $f + (l + m)/T$. The strength of any particular image depends on the magnitudes of the Fourier coefficients, P_m and Q_l, and the magnitude of the transfer function $H(s)$ at the frequency, $f + m/T$. The basic operating principle of the N-path filter is that, in combining the responses in the multipath case, the images can be cancelled or "phased-out" over any desired frequency band by making N large enough. The result is that, within this frequency band, the filtering operation is time invariant. Furthermore, the resulting transfer function has a form that affords a unique approach to the realization of many types of filtering operations, in contrast to the methods that must be employed when only constant-parameter elements are used.

11.2 RESPONSE PROPERTIES OF THE *N*-PATH STRUCTURE

Referring again to Figure 11-1, the impulse response for each path is obtained directly from (11.2), and the overall impulse response is obtained by summing over n.

$$w(t, \tau) = h(t - \tau) \sum_{n=0}^{N-1} p\left(\tau - \frac{nT}{N}\right) q\left(t - \frac{nT}{N}\right). \tag{11.5}$$

Evaluation of the factor involving the sum over n is facilitated by using the Fourier series representations in (11.3) for $p(t)$ and $q(t)$.

$$\sum_{n=0}^{N-1} p\left(\tau - \frac{nT}{N}\right) q\left(t - \frac{nT}{N}\right) = \sum_l \sum_m P_m Q_l e^{j(2\pi/T)(m\tau + lt)} \sum_{n=0}^{N-1} e^{-j(2\pi n/N)(l+m)}. \tag{11.6}$$

The sum over n in (11.6) involves a geometric series, hence it can be expressed in closed form.

$$\sum_{n=0}^{N-1} e^{-j(2\pi n/N)(l+m)} = \frac{1 - e^{-j2\pi(l+m)}}{1 - e^{-j(2\pi/N)(l+m)}}$$
$$= N \text{ for } l + m = kN$$
$$= 0 \text{ otherwise} \qquad (11.7)$$

where k is any integer. Substituting $l + m = kN$ for m in (11.6), the impulse response becomes

$$w(t, \tau) = \sum_{k=-\infty}^{\infty} e^{j(2\pi kN/T)\tau} \left[\sum_{l=-\infty}^{\infty} NP_{kN-l} Q_l e^{j(2\pi l/T)(t-\tau)} h(t-\tau) \right]$$
$$= \sum_{k=-\infty}^{\infty} e^{j(2\pi kN/T)\tau} w_k(t-\tau). \qquad (11.8)$$

The factor in braces in (11.8) is a function of $t - \tau$, hence it corresponds to a time-invariant filtering operation. The remaining factors depend only on τ so, by virtue of (11.2), they represent input modulators. By this means we have arrived at a mathematically equivalent structure for the N-path filter as shown in Figure 11-3 [2]. We note that a similar equivalent structure, involving a parallel combination of time-invariant networks with single-frequency output modulators, could be obtained by using $l + m = kN$ to eliminate the index, l, in (11.6).

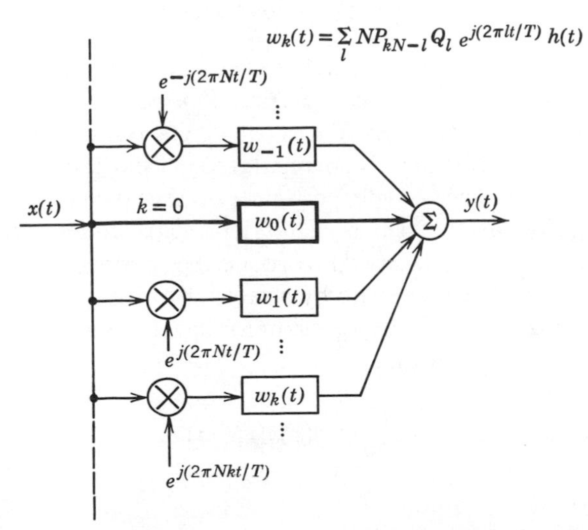

FIGURE 11-3 Equivalent structure involving single-frequency input modulators.

Examining either of these equivalent structures, it is easy to see what is required to make the operation time invariant. Time invariance is obtained if, and only if, transmission through all paths except $k = 0$ in Figure 11-3 can be suppressed. This suppression may be obtained in either of two ways: (1) bandlimited modulating functions, or (2) bandlimited input and output signals. For the first case, if we have $P_l = Q_l = 0$ for all l such that $|l| \geq N/2$, then $P_{kN-l}Q_l = 0$ for any $k \neq 0$. This means that $w_k(t) = 0$ for $k \neq 0$, and signal transmission occurs only through the time-invariant $k = 0$ path. On the other hand, if $x(t)$ contains no frequencies greater than $N/2T$, then the images produced by the $k \neq 0$ paths will all appear at frequencies greater in magnitude than $N/2T$. Thus we see that, if the N-path structure is preceded and followed by ideal lowpass filters with cut-off frequency at $f_c = N/2T$, then the entire structure is equivalent to a time-invariant filter characterized by the $k = 0$ path over the frequency band, $|f| \leq N/2T$, regardless of the harmonic content of the modulating functions.

Assuming that either one or the other of these bandlimiting restrictions is incorporated, we are interested only in the response properties of the $k = 0$ path characterized by the impulse response function

$$w_0(t) = h(t)c(t) \tag{11.9}$$

where $h(t)$ is the impulse response of the filters in each path and $c(t)$ is a periodic function given by

$$c(t) = N \sum_{l=-\infty}^{\infty} P_{-l}Q_l e^{j(2\pi l/T)t}, \tag{11.10}$$

i.e., a function related to the modulating functions in that its Fourier coefficients are $NP_{-l}Q_l$. In the synthesis of a given response characteristic, it is important to note that there is some degree of flexibility in choosing the modulating functions since $c(t)$ depends only on the pairwise products of the corresponding Fourier coefficients of $p(t)$ and $q(t)$. Under certain circumstances, this flexibility can be used to reduce the number of path networks required for a given characteristic by a factor of two [3].

Although (11.10) is a useful expression for characterizing the N-path filter response, it is easily shown that

$$c(t) = \frac{N}{T} \int_0^T p(\sigma)q(t + \sigma)\, d\sigma \tag{11.11}$$

is an equivalent expression for the periodic factor in the impulse response. This alternative expression for $c(t)$ is used to obtain a closed-form relation for the transfer characteristic of the N-path filter.

Transfer Function

Considering input signals of the form, $x(t) = e^{st}$, we get responses of the form, $y(t) = W_0(s)e^{st}$ where the transfer function, $W_0(s)$ is the Laplace transform of $w_0(t)$.

$$W_0(s) = \int_0^{\infty} h(t)c(t)e^{-st}\,dt$$

$$= N \sum_{l=-\infty}^{\infty} P_{-l}Q_l H(s - j(2\pi l/T)). \tag{11.12}$$

The transfer function consists of the sum of frequency-translated versions of the transfer function of the path networks. If the path networks are lowpass filters with bandwidth smaller than $1/2T$, then the N-path transfer function has a multiple passband characteristic with equally-spaced passbands, each modified by a gain factor given by the corresponding Fourier coefficient of $c(t)$. This feature is shown graphically in Figure 11-4.

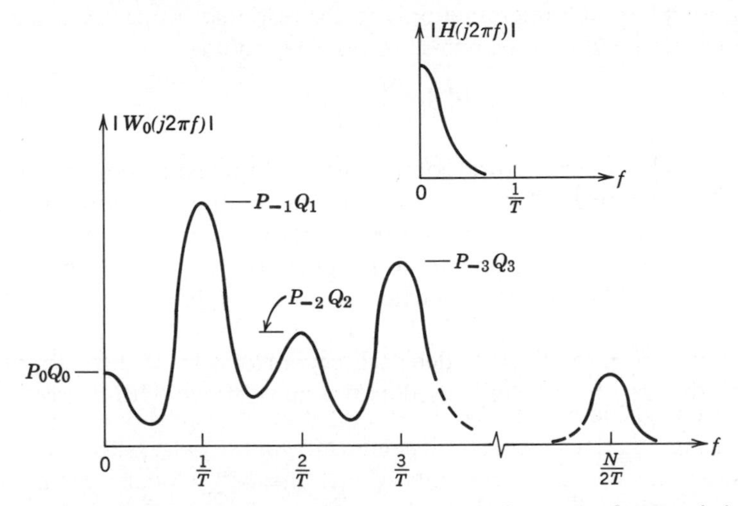

FIGURE 11-4 Multiple passband gain versus frequency characteristic of N-path filter.

Closed-Form Expression for $W_0(s)$

If $H(s)$ is a rational function, corresponding to finite-order path networks, then we can derive an alternative closed-form expression for $W_0(s)$. Consider first the case that $H(s)$ corresponds to a first-order network with a single pole at $s = s_i$. The impulse response has the form

$$h(t) = e^{s_i t} \qquad \text{for} \quad t \geq 0. \tag{11.13}$$

Hence

$$W_0(s) = \int_0^\infty c(t)e^{(s_i - s)t}\, dt$$
$$= \sum_{k=0}^{\infty} \int_0^T c(\tau + kT)e^{(s_i - s)(\tau + kT)}\, d\tau. \tag{11.14}$$

Using the fact that $c(t)$ is periodic, (11.14) can be rewritten as

$$W_0(s) = \int_0^T c(\tau)e^{-(s - s_i)\tau}\, d\tau \sum_{k=0}^{\infty} e^{(s_i - s)kT}, \tag{11.15}$$

and summing the geometric series, we obtain

$$W_0(s) = \frac{D(s - s_i)}{1 - e^{(s_i - s)T}}$$

where

$$D(s) = \int_0^T c(t)e^{-st}\, dt \tag{11.16}$$

which is a closed form expression by virtue of the expression in (11.11) for $c(t)$. The function $D(s)$ is the Laplace transform of a duration-limited time function corresponding to one period of $c(t)$ over the interval $0 \le t < T$. As such, $D(s)$ has no poles, and the only poles of $W_0(s)$ are from the zeros of $1 - e^{(s_i - s)T}$. These poles are located at $s = s_i + j(2\pi l/T)$ where $l = 0, \pm 1, \pm 2, \ldots$.

Now for the case of Mth-order path networks with M distinct poles, we assume that $H(s)$ can be expressed as the ratio of two Mth-degree polynomials, and hence,

$$h(t) = K_0\, \delta(t) + \sum_{i=1}^{M} K_i e^{s_i t} \qquad \text{for} \quad t \ge 0 \tag{11.17}$$

where the s_i are the roots of the denominator polynomial. Now, treating each term separately, we obtain

$$W_0(s) = K_0 c(0) + \sum_{i=1}^{M} \frac{K_i D(s - s_i)}{1 - e^{(s_i - s)T}}. \tag{11.18}$$

Similar expressions can be obtained for the case of multiple-order poles (See Problem 11.3). The pole pattern in the s-plane consists of vertical strings with each s_i repeated in intervals of $2\pi/T$ as illustrated in Figure 11-5. The

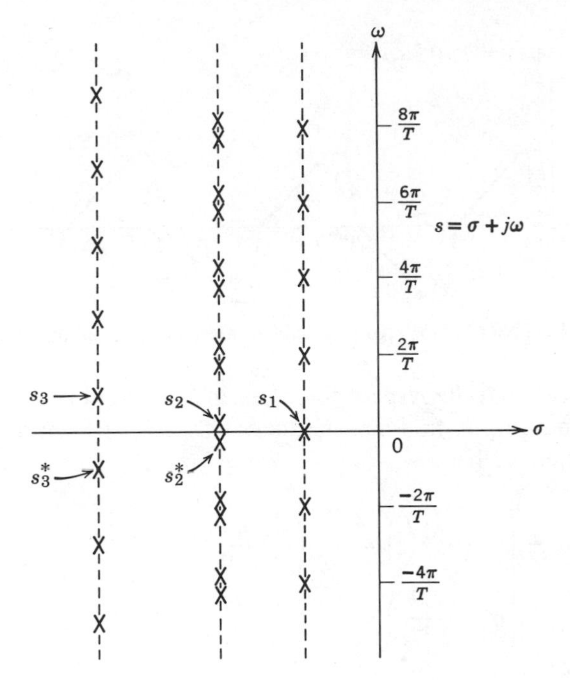

FIGURE 11-5 Pole pattern of $W_0(s)$ for a path network function with poles at s_1, s_2, s_2^*, s_3, s_3^*.

zero pattern of $W_0(s)$ is, in general, not of the same form unless $D(s)$ happens to have a periodic form. Some special cases of periodic $D(s)$ are considered in the following section.

11.3 SPECIFIC TYPES OF MODULATING FUNCTIONS

Alternation Modulation

Certain interesting design features result from the use of modulating functions composed of a repeated waveform with alternating polarity each half period. These types of modulating functions exhibit a kind of symmetry which can be expressed as

$$p(t) = -p\left(t - \frac{T}{2}\right) \tag{11.19}$$

illustrated graphically in Figure 11-6. The Fourier series representation for periodic functions satisfying (11.19) contains only odd harmonics, i.e., $P_m = 0$ for m even.

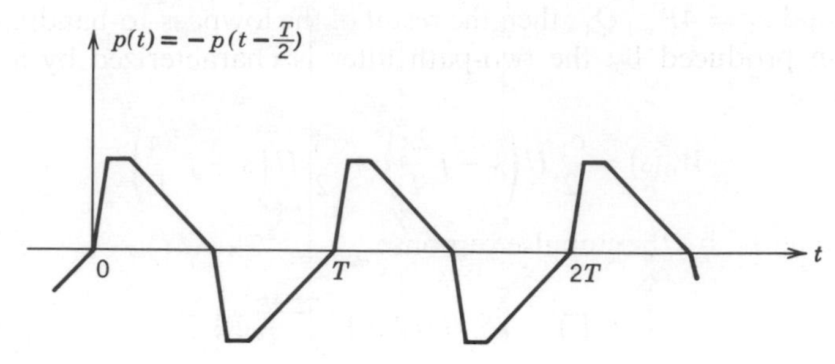

FIGURE 11-6 An example of alternation modulation.

Now suppose that the N-path filter has an even number of paths, and that both input and output modulating functions satisfy the symmetry condition of (11.19). Then we have

$$\left.\begin{aligned} p\left(t - \left[n + \frac{N}{2}\right]\frac{T}{N}\right) &= -p\left(t - \frac{nT}{N}\right) \\ q\left(t - \left[n + \frac{N}{2}\right]\frac{T}{N}\right) &= -q\left(t - \frac{nT}{N}\right) \end{aligned}\right\}, \qquad 0 \le n \le \frac{N}{2} - 1 \tag{11.20}$$

so that transmissions through the nth and $(n + N/2)$th paths are identical. The result is that, except for a gain factor of 2, the lower $N/2$ paths in Figure 11-1 can be eliminated without affecting the transfer function or the frequency band over which it is valid. This significant reduction in the number of paths can be realized only in those cases where the desired transfer characteristic does not call for translated versions of $H(s)$ centered at even multiples of the fundamental modulating frequency. N-path filters of this type are called *quadrature* N*-path filters* by Sun [4].

Sinusoidal Modulation

It was noted in the previous section that if the highest harmonic component in the modulating functions had a frequency less than $N/2T$, then the N-path filter exhibits a time-invariant terminal behavior without the necessity of bandlimiting filters at the input and output terminals. The most frequent application of this result is in effecting a single lowpass-to-bandpass transformation, using single frequency (sinusoidal) modulating functions. This transformation requires $N \ge 3$, but we recognize that sinusoids produce a form of alternation modulation. Hence we can make $N = 4$ and realize the transformation with only two physical paths and a $\pi/2$ phase difference in the modulating functions for each path. Accordingly, if $P_l = Q_l = 0$ for

$|l| \neq 1$ and $c_1 = 4P_{-1}Q_1$, then the result of the lowpass-to-bandpass transformation produced by the two-path filter is characterized by a transfer function

$$W_0(s) = \frac{c_1}{2} H\left(s - j\frac{2\pi}{T}\right) + \frac{c_1^*}{2} H\left(s + j\frac{2\pi}{T}\right) \tag{11.21}$$

or equivalently, by the impulse response

$$w_0(t) = |c_1| h(t) \cos\left(\frac{2\pi t}{T} + \phi\right) \tag{11.22}$$

where

$$c_1 = |c_1| e^{j\phi}.$$

The transformation indicated in (11.21) provides an attractive approach to the realization of highly selective bandpass filters [5–9]. A narrow-bandwidth-lowpass function, $H(s)$, may be accurately realized by RC passive or active networks, and the resulting passband shape can be translated to any desired center frequency simply by using the correct modulation frequency. This also affords a convenient tunability feature since the passband shape is independent of the center frequency, which can be manually or electronically adjusted by controlling the reference oscillator which generates the modulating functions. As a result of this independence of passband shape and center frequency, it is possible to realize extremely narrow-band filter characteristics with a high precision. Equivalent LC realizations would require prohibitively high Q inductors. Similar transfer characteristics can be obtained by inductorless active bandpass filters, but not without severe limitations in regard to stability and sensitivity to variations in element values. Some of the design aspects of the N-path bandpass filter are discussed in the following section. A more detailed discussion is presented by Glaser *et al.* [3].

Staircase Modulation

A practical limitation in the realization of N-path filters is the need for precise multiplier elements which may lead to an uneconomical design. The precision is required, not so much for accurate realization of the transfer function, but rather to provide uniformity in the path characteristics so that the image components will be effectively cancelled out over the desired frequency band. If, on the other hand, it is possible to realize the desired characteristics using modulating functions of a piecewise-constant nature, then realization of good multiplier elements may be quite feasible. For example, if there are not too many modulation levels involved, the multiplier could take the form of a digitally-controlled attenuator.

Modulation involving a waveform with R constant levels and discontinuities appearing at equally-spaced time instants over one period, as shown in Figure 11-7, may be called *staircase modulation.* In this case it is possible to

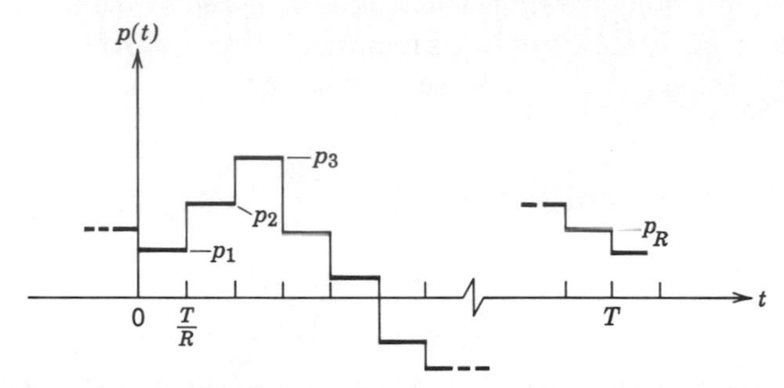

FIGURE 11-7 A typical waveform illustrating staircase modulation.

provide control signals for the multiplier from the states of an R-state shift register clocked at an integral multiple of the fundamental modulation frequency.

A Fourier series expansion of the waveform in Figure 11-7 gives [1]

$$P_m = \sum_{r=1}^{R} \frac{e^{j(m\pi/R)} \sin\frac{m\pi}{R}}{m\pi} e^{-j(2\pi/R)mr} p_r \tag{11.23}$$

which results in the recursion relation,

$$P_{m+kR} = \frac{m}{m+kR} P_m, \qquad k = 0, \pm 1, \pm 2, \ldots . \tag{11.24}$$

For design purposes, it is usually the inverse of the transformation in (11.23) that is desired, i.e., a transformation giving explicitly the appropriate set of levels p_r in terms of a prescribed set of Fourier coefficients. Usually it is the lowest order Fourier coefficients which will be specified since these control the transmission within the passband, $|f| \leq N/2T$, of the N-path filter. If the P_m are specified for $|m| < R/2$ with the constraint that $P_m = P^*_{-m}$ (since $p(t)$ is real) and the additional condition that $P_{R/2} = 0$ if R is even, the desired transformation is given by

$$p_r = \sum_{\substack{m \\ |m| < R/2}} \frac{e^{-j(m\pi/R)} \frac{m\pi}{R}}{\sin\frac{m\pi}{R}} e^{j(2\pi/R)mr} P_m, \qquad r = 1, 2, \ldots R \tag{11.25}$$

The derivation of (11.25) is given in Reference [1]. We note again that there is flexibility in designing the modulating functions since it is only the product of Fourier coefficients, $P_{-l}Q_l$, which determines the transfer characteristics within the passband. One important application for staircase modulation is to obtain wideband tunability of a bandpass filter without the use of bandlimited modulation. This is accomplished by making $P_{-l}Q_l = 0$; $|l| \neq 1$ for as many values of l as is required by the desired tuning range [3]. As expected, the result is a staircase approximation to sinusoidal functions.

Pulse Modulation

A special case of staircase modulation which affords the simplest implementation is one which involves only two-level modulating functions. For convenience we shall assume that the two levels are (1, 0) or (1, -1). The modulation can then be represented by one or more rectangular pulses within each period. Fortunately, many of the useful applications for N-path filters can be realized with this limited form of modulation. The modulation can be interpreted as a gating operation, and implementation can take the form of digitally-controlled solid-state switches or, in low-frequency applications, mechanical switches. In some circuits containing periodically-operated switches, the driving-point properties, in addition to transfer properties, must be taken into account. These considerations are discussed in Section 11.5.

As an example, suppose that the input and output modulating functions contain a single, narrow rectangular pulse in each period as shown in Figure 11-8. In this case the periodic function, $c(t)$, is evaluated directly from (11.11). It is a periodic sequence of pulses wherein each pulse is essentially the convolution of the rectangular pulses in $p(t)$ and $q(t)$. Note that if the gate opening intervals on the input and output modulation were different, then the pulses comprising $c(t)$ would be trapezoidal instead of triangular.

Now with $H(s)$ given in partial-fraction form, we can evaluate the transfer function, $W_0(s)$, for the N-path filter directly from (11.18), once the function, $D(s)$, is evaluated for this case. It is important to distinguish two situations in evaluating $D(s)$ since the two cases can lead to remarkably different results: (Case 1) the gates on any path are not open simultaneously, i.e., $|t_2 - t_1| > d$, and (Case 2) the gates on any path are coincident, i.e., $t_2 = t_1$.

Case 1: $|t_2 - t_1| > d$. In this case, $D(s)$ is the Laplace transform of the single triangular pulse in the $0 \leq t < T$ interval. Assuming $t_2 > t_1$,

$$D(s) = \frac{d^2N}{T}\left[\frac{\sinh\dfrac{sd}{2}}{\dfrac{sd}{2}}\right]^2 e^{-s(t_2-t_1)} \tag{11.26}$$

and $c(0) = 0$.
If $t_2 < t_1$, then $t_2 - t_1$ above is replaced by $T + t_2 - t_1$.

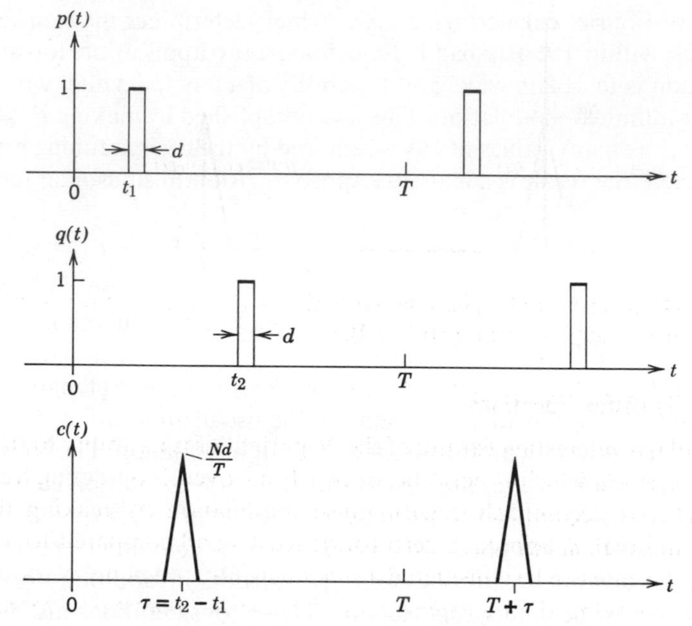

FIGURE 11-8 Rectangular pulse modulating functions and the resulting $c(t)$. The parameter τ indicates the time lag in output gate opening relative to input gate opening.

Case 2: $t_2 = t_1$. In this case, the triangular pulse in $c(t)$ is centered at $t = 0$, hence $D(s)$ is the Laplace transform of two separate halves of the triangular pulse as shown in Figure 11-9. We can write

$$d(t) = f(t) + f(T - t)$$
$$D(s) = F(s) + F(-s)e^{-sT} \qquad (11.27)$$

where

$$F(s) = \frac{Nd}{T}\left[\frac{1}{s} - \frac{1 - e^{-sd}}{ds^2}\right].$$

Also, in contrast to Case 1, we have

$$c(0) = \frac{Nd}{T} \qquad (11.28)$$

which will affect the response if $K_0 \neq 0$ in (11.18). It is apparent that the two cases are substantially different. This is dramatically illustrated for the particular case of a highpass $H(s)$ function in Section 11.4. For partial overlap of the gate openings, the results will be intermediate to the two cases.

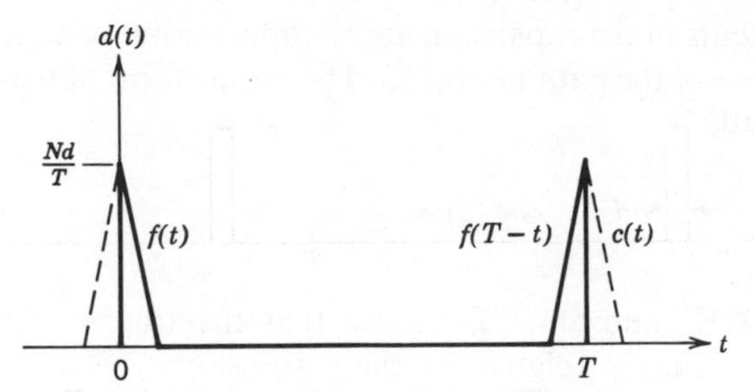

FIGURE 11-9 Pulse pair relevant to the evaluation of $D(s)$ for the case of coincident gate openings (Case 2).

Periodic Transfer Functions

A particularly interesting feature of the N-path filter is its ability to provide a transfer function which is periodic, or nearly so, over its operating frequency range. This is accomplished with pulse modulation by making the gate opening interval, d, approach zero (or at least small compared to T/N). In this case the gate can be considered a *sampling gate*, and many of the analysis techniques developed for *sampled-data filters*, such as the z-*transform*, can be used to advantage. In contrast to the single-path sampled-data filter, however, the N-path filter has a response extended over a frequency range including many periods of the transfer characteristic, in direct proportion to the number of paths employed.

Letting d approach zero in (11.26) we have, for Case 1,

$$D(s) \cong \frac{d^2N}{T} e^{-s(t_2-t_1)} \tag{11.29}$$

and since $c(0) = 0$, the transfer function in (11.18) approaches

$$W_0(s) = \frac{d^2N}{T} e^{-s(t_2-t_1)} \sum_{i=1}^{M} \frac{K_i e^{s_i(t_2-t_1)}}{1 - e^{(s_i-s)T}}$$

$$= \frac{d^2N}{T} e^{-s(t_2-t_1)} W'_0(s) \tag{11.30}$$

which, for $s = j2\pi f$, is periodic in f with a period of $1/T$ provided that the constant delay factor, $e^{-s(t_2-t_1)}$, is considered separately. Expanding the $W'_0(s)$ factor in (11.30) as a power series in e^{-sT}, we get

$$W'_0(s) = \sum_{k=0}^{\infty} h_k e^{-kTs} \tag{11.31}$$

where

$$h_k = h(kT + t_2 - t_1) = \sum_{i=1}^{M} K_i e^{s_i(kT+t_2-t_1)},$$

i.e., the coefficients in the expansion are the appropriate sample values of the impulse response of the path networks. The z-transform of the $\{h_k\}$ sequence is defined as [10]

$$\mathscr{H}(z) = \sum_{k=0}^{\infty} h_k z^{-k}. \tag{11.32}$$

Comparing (11.31) and (11.32), we see that the transfer function of the N-path filter is simply related to the z-transform of the appropriately sampled impulse response, $h(t)$, by means of the change of variable, $z = e^{sT}$.

$$W_0(s) = \frac{d^2N}{T} e^{-s(t_2-t_1)} \mathscr{H}(e^{sT}). \tag{11.33}$$

For Case 2, the corresponding result is

$$D(s) \cong \frac{Nd^2}{2T}[1 + e^{-sT}] \tag{11.34}$$

as d approaches zero. Recalling that $c(0) = Nd/T$ for Case 2, and again expanding in inverse powers of $z = e^{sT}$, we obtain

$$\begin{aligned} W_0(s) &= \frac{K_0 Nd}{T} + \frac{Nd^2}{2T} \sum_{i=1}^{M} K_i \frac{1 + e^{(s_i - s)T}}{1 - e^{(s_i - s)T}} \\ &= \frac{d^2N}{T}\left[\frac{K_0}{d} - \frac{h_0}{2} + \mathscr{H}_0(e^{sT})\right] \end{aligned} \tag{11.35}$$

where

$$\mathscr{H}_0(z) = \sum_{k=0}^{\infty} h_k z^{-k} \tag{11.36}$$

$$h_k = h(kT) \qquad \text{for} \quad k \geq 1$$

$$h_0 = h(0+).$$

Thus, except for the additive constant, the result is similar in form to (11.33). $\mathscr{H}_0(z)$ in (11.36) differs from $\mathscr{H}(z)$ in (11.33) in that the $\{h_k\}$ sequence is obtained by sampling $h(t)$ at different time instants. If $K_0 = 0$ and $h(0+) = 0$, then the two cases are substantially the same.

The practical significance of these results is that a variety of techniques have been developed for obtaining a rational $\mathscr{H}(z)$ to approximate a prescribed periodic transfer characteristic. These techniques have evolved mainly through work relating to the design of digital filters [10] and distributed-element filters (See Chapters 7 and 12).

Passband Design Procedures

If the desired periodic-transfer function has the form of a sequence of very narrow passbands (comb filter), then it may be possible to use conventional lowpass filter design techniques to obtain the desired passband shape. Specifically, if the attenuation of the path networks is sufficiently great at all frequencies above $1/2T$, then the individual terms in (11.12) for $W_0(s)$ can be considered nonoverlapping and each passband is simply a frequency-translated version of the lowpass characteristic. On the other hand, there are many circumstances where this approach is not sufficiently accurate, and it will be necessary to first find a rational $\mathscr{H}(z)$ which leads to a satisfactory passband shape. Once a suitable $\mathscr{H}(z)$ is found, it is a simple matter to find the appropriate $H(s)$ for the path networks. This can be done, for example, by making a partial-fraction expansion of $\mathscr{H}(z)$ in the variable, z^{-1}. Then, from each term in this expansion, a corresponding pole position, s_i, and residue, K_i, for $H(s)$, is found in accordance with (11.30) or with (11.35).

Fortunately, the conventional approaches to the approximation problem in filter design can also be brought to bear on the problem of finding a suitable $\mathscr{H}(z)$. This is usually done by performing a bilinear transformation on the z variable,

$$w = u + jv = \frac{z-1}{z+1} = \tanh \frac{sT}{2}, \tag{11.37}$$

thereby transforming $\mathscr{H}(z)$ into a function, $G(w)$, also rational in w. The transformation indicated in (11.37) maps the horizontal strip, $|\mathrm{Im}(s)| \leq \pi/T$, into the entire w-plane. Furthermore, the left half part of the strip, $\mathrm{Re}(s) < 0$, corresponds to the entire left half of the w-plane, hence left-half-plane poles of $G(w)$ will always result in left-half-plane poles for $H(s)$. In order to make use of conventional approximation theory in the w-plane, we must first "predistort" the desired passband gain-and-phase characteristics to accommodate the frequency-scale warping,

$$v = \tan \pi T f \tag{11.38}$$

produced by the transformation. Note that, in the particular case where it is desired that the passband have an approximately rectangular gain characteristic, this predistortion simply amounts to finding the cut-off point for the $|G(jv)|$ characteristic which corresponds, through (11.38), to the desired passband cut-off frequency.

11.4 PRACTICAL APPLICATIONS

Comb Filter I

One of the useful applications of the periodic response properties of the N-path filter is in selective filtering of periodic signals from a background of nonperiodic noise interference. If the path networks are lowpass filters with bandwidth small compared to $1/2T$, then the N-path filter response is a sequence of equally-spaced passbands forming a comb which introduces little distortion to signals of period T, but substantially reduces the power of the interfering noise.

For example, suppose that the path networks are single-section RC lowpass filters each having a single pole at $s = -\alpha = -1/\text{RC}$.

$$H(s) = \frac{\alpha}{s + \alpha}. \tag{11.39}$$

Then, from (11.18), the transfer function of the N-path filter is

$$W_0(s) = \frac{\alpha D(s + \alpha)}{1 - e^{-(s+\alpha)T}}. \tag{11.40}$$

Suppose also that the modulation involves sampling gates of width d and a time lag $\tau = t_2 - t_1 > d$ in the output gates (Figure 11-8). Then using (11.40) and (11.26), we have

$$W_0(s) = \frac{\alpha d^2 N}{T} \left[\frac{\sinh(s + \alpha)\frac{d}{2}}{(s + \alpha)\frac{d}{2}} \right]^2 \frac{e^{-(s+\alpha)\tau}}{1 - e^{-(s+\alpha)T}}. \tag{11.41}$$

The resulting gain-frequency response is shown in Figure 11-10 for $d = 0.05T$ and $\alpha T = 0.2$. When d is small enough to make the $D(s + \alpha)$ factor in (11.40) essentially constant over $|f| \leq N/2T$, then the transfer function approaches the periodic function,

$$W_0(s) = \frac{\alpha d^2 N}{T} \frac{e^{-(s+\alpha)\tau}}{1 - e^{-(s+\alpha)T}}. \tag{11.42}$$

A simple expression for the performance of the periodic comb filter can be obtained by assuming that the noise has a constant power spectral density of K watts/Hz over the signal band of $|f| \leq N/2T$. Without the comb filter, the signal-to-noise power ratio is

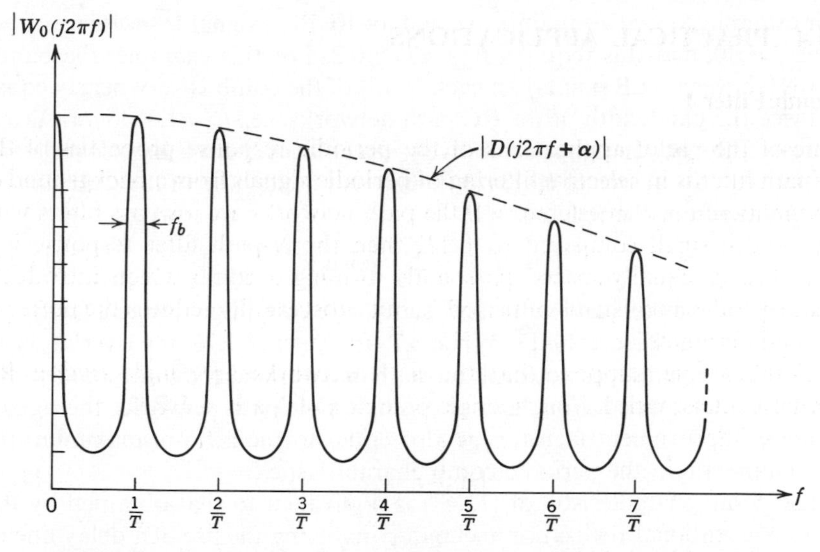

FIGURE 11-10 Frequency response of comb filter with single-pole lowpass path networks and rectangular pulse modulation. The relevant parameters for this graph are $\alpha T = 0.2$ and $d = 0.05T$.

$$SNR_1 = \frac{JT}{KN}$$

where J is the power in the periodic signal. Let $W_0'(s)$ represent the transfer function in (11.42) normalized to unity transmission at multiples of $f = 1/T$, then the signal-to-noise ratio using the comb filter is

$$SNR_2 = J\left[\int_{-N/2T}^{N/2T} |W_0'(j2\pi f)|^2 K \, df\right]^{-1}.$$

Because of the periodic nature of $|W_0'(j2\pi f)|$, we can express the *signal-to-noise improvement factor*, γ, as

$$\frac{1}{\gamma} = \frac{SNR_1}{SNR_2} = T\int_{-1/2T}^{1/2T} |W_0'(j2\pi f)|^2 \, df$$

$$= T\int_{-1/2T}^{1/2T} \frac{(1 - e^{-\alpha T})^2 \, df}{1 + e^{-2\alpha T} - 2e^{-\alpha T}\cos 2\pi T f}.$$

Evaluating this integral, we get

$$\gamma = \coth\frac{\alpha T}{2}. \tag{11.43}$$

For example, to realize an improvement of 10 dB in signal-to-noise ratio, we have $\gamma = 10$, and this requires that $\alpha T \cong 0.2$. For this example, the bandwidth (between 3 dB points) on each tooth of the comb is very nearly equal to twice the bandwidth of the RC path networks, i.e., $f_b = 1.0035\alpha/\pi$. As αT increases, the bandwidth f_b is relatively larger since there is more overlap of the translated versions of $H(j2\pi f)$. The actual bandwidth of the teeth can be determined from the relation

$$\cos \pi f_b T = 2 - \cosh \alpha T. \tag{11.44}$$

The dependence of bandwidth and signal-to-noise improvement factor on αT is shown in Figure 11-11. When αT approaches 1.76, the overlap is so great that the gain does not drop more than 3 dB, and the tooth bandwidth cannot be determined. For the case of single-pole path networks, the signal-to-noise improvement factor, γ, is also equal to the ratio of maximum-to-minimum gain in the periodic comb characteristic.

The comb characteristic of (11.42) is equivalent to that obtained by the more conventional realization technique involving the use of a delay line in the feedback loop of a wideband amplifier [11]. The network shown in

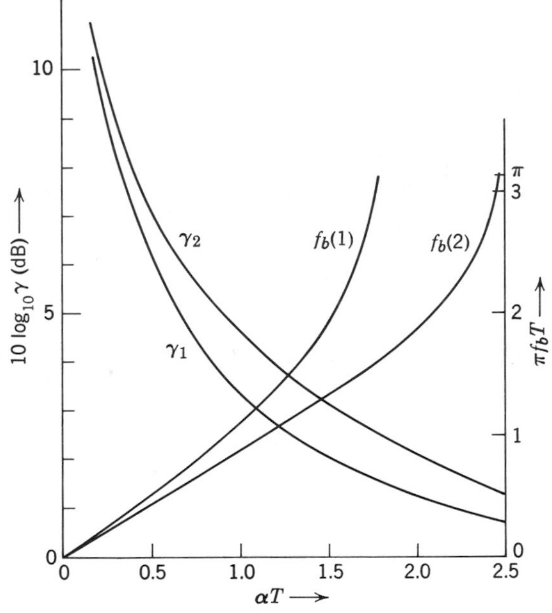

FIGURE 11-11 Signal-to-noise ratio improvement factor, γ, and tooth bandwidth, f_b, for comb filter as a function of pole position in the path network function. The graphs compare the results for nonoverlapping gate openings (Case 1) and coincident gate openings (Case 2).

Figure 11-12 exhibits the same transfer function as the periodic factor in the N-path comb-filter-transfer function.

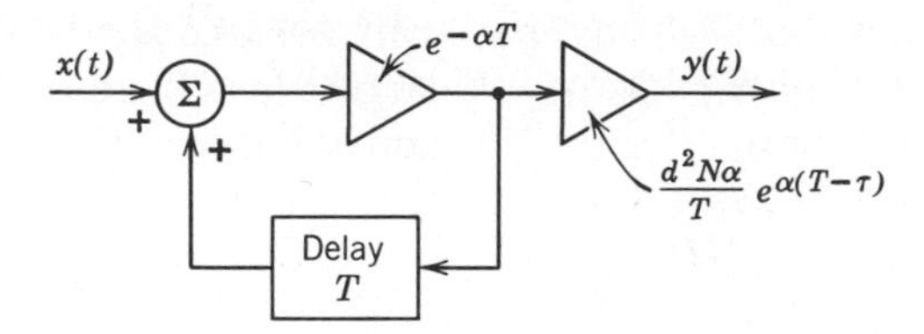

FIGURE 11-12 Equivalent circuit for the comb filter using a delay line (Case 1).

For the Case 2 situation (coincident gate openings on each path) and small d, we have a somewhat different comb characteristic since, from (11.34),

$$D(s) \cong \frac{Nd^2}{2T}(1 + e^{-sT}),$$

and hence,

$$W_0(s) = \frac{\alpha N d^2}{2T} \frac{1 + e^{-(s+\alpha)T}}{1 - e^{-(s+\alpha)T}}. \tag{11.45}$$

This characteristic provides substantially more attenuation in the region between the teeth than provided by (11.42) with the same value of α. The corresponding signal-to-noise improvement factor and tooth bandwidth is also shown graphically in Figure 11-11 for comparison with the Case 1 results.

Comb Filter II

Another practical application of periodic filtering characteristics is the *comb-notch filter*. This filter exhibits a gain characteristic which is essentially constant over the band except for a sequence of narrow notches centered at multiples of some specified frequency. The application is inverse to the one considered previously. In situations where a nonperiodic signal is corrupted with a periodic interference, the appropriate comb-notch filter can completely eliminate the interference without excessive distortion on the signal, provided that the notches are narrow enough. A specific application is in moving-target-indicator (MTI) radar, where the comb-notch filter suppresses the time-periodic returns from stationary targets [11]. In this case, the notches are placed at frequency multiples of the rate of revolution of the scanning antenna.

One of the N-path realizations of the comb-notch filter involves narrow sampling gates at input and output, as in the previous example, and single-pole RC *highpass* filters as path networks. The desired notch characteristic is critically dependent on the time lag, τ, by which the output gates lag the input gates for this particular method of realization.

We let

$$H(s) = \frac{s}{s+\alpha} = 1 - \frac{\alpha}{s+\alpha} \tag{11.46}$$

so that $K_0 = 1$ and $K_1 = -\alpha$ in (11.18) for this case. The periodic $c(t)$ has the shape shown in Figure 11.8 where it is assumed that d is small enough so that essentially periodic gain-frequency characteristics are obtained. Then we can write

$$D(s) \cong A_1(\tau) + A_2(\tau)e^{-sT} \tag{11.47}$$

where A_1 and A_2 represent the areas under the shaded regions in Figure 11-13. Since $c(0) = (N/T)(d - |\tau|)$ for $|\tau| \leq d$, we have, from (11.18),

$$\begin{aligned} W_0(s) &= \frac{N}{T}(d - |\tau|) - \frac{\alpha A_1(\tau) + \alpha A_2(\tau)e^{-(s+\alpha)T}}{1 - e^{-(s+\alpha)T}} \\ &= \frac{\frac{N}{T}(d - |\tau|) - \alpha A_1(\tau) - \left[\frac{N}{T}(d - |\tau|) + \alpha A_2(\tau)\right] e^{-(s+\alpha)T}}{1 - e^{-(s+\alpha)T}}. \end{aligned} \tag{11.48}$$

Now if τ is adjusted so that

$$\frac{\frac{N}{T}(d - |\tau|) + \alpha A_2(\tau)}{\frac{N}{T}(d - |\tau|) - \alpha A_1(\tau)} = e^{\alpha T}, \tag{11.49}$$

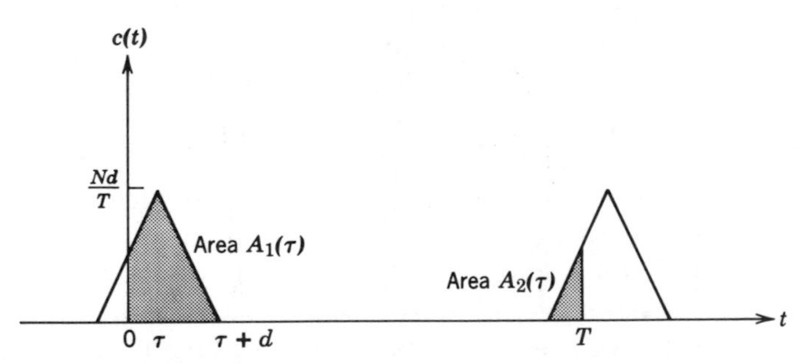

FIGURE 11-13 Illustration of the dependence of $D(s)$ and $c(0)$ on the lag time, τ.

then, using a proportionality constant, A, we have,

$$W_0(s) = A\frac{1 - e^{-sT}}{1 - e^{-(s+\alpha)T}}. \tag{11.50}$$

This transfer function exhibits the desired nulls in transmission to provide the comb-notch characteristic with complete suppression at the center of the notches; i.e., $W_0(j2\pi f) = 0$ for $f = m/T$, m any integer. The pole-zero pattern and the gain-frequency characteristic is illustrated in Figure 11-14. For narrow notches, $\alpha \ll 2\pi/T$, the notch bandwidth (between 3 dB points) is approximately α/π (Hz). Note that it is also possible to obtain this result with the correct negative value of τ, i.e., with output gates slightly leading the input gates. An obvious disadvantage of this method of realization of the comb-notch filter is the high precision in the value of τ that is required to maintain a suitable notch characteristic. In fact, if τ changes by the slight amount necessary to make $|\tau| > d$, the notches disappear entirely and the

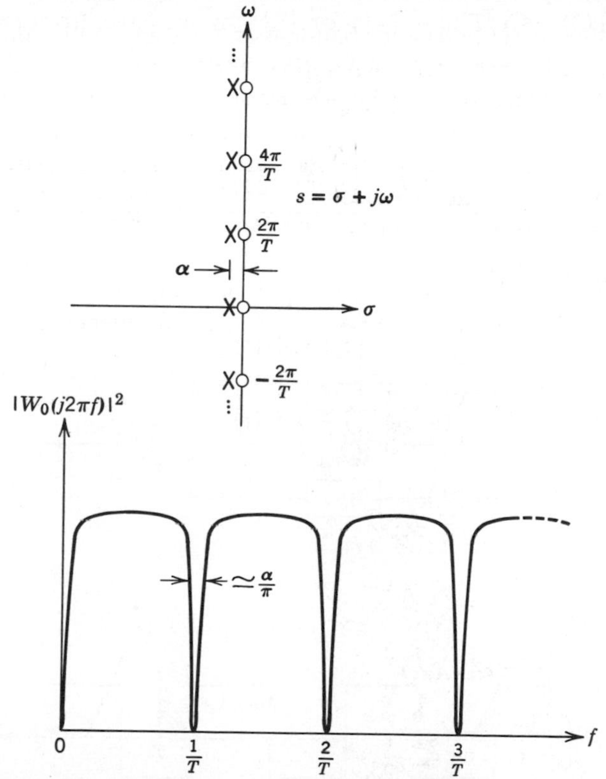

FIGURE 11-14 Pole-zero pattern and frequency response of the comb-notch filter.

result is a comb filter of the first type, obtained with the use of lowpass path filters. This is because with $|\tau| > d$, $c(0) = 0$ and $D(s) \cong (Nd^2/T)e^{-s\tau}$ so, even with highpass path filters, we have

$$W_0(s) = -\frac{\alpha d^2 N}{T}\frac{e^{-(s+\alpha)\tau}}{1 - e^{-(s+\alpha)T}} \tag{11.51}$$

which, except for the reversal in polarity, is the same as the result indicated in (11.42). This example is included primarily to illustrate the existence of situations where the phase of the modulating functions has a pronounced effect on the response characteristic of the N-path filter. A similar effect has been noted by Saraga [12] for N-path filters employing sinusoidal modulating functions. A more practical realization for the comb-notch filter, which is not critically dependent on the relative phase of input and output modulation, is presented in Section 11.5.

Bandpass Filter

Realization of a single passband, bandpass filter using sinusoidal modulation was discussed in Section 11.3. A simpler implementation using two-level staircase modulation, is discussed here. In the circuit shown in Figure 11-15, square-wave input and output modulation is employed. The modulators can be implemented as polarity reversing switches or as a conventional balanced modulator involving a ring of four diodes. Since the square wave is a form of alternation modulation, we can let $N = 4$ but realize the filter with only two paths and a quarter-period phase difference in the modulation on each path. We assume that the two-path structure is accompanied with input and output bandlimiting filters having a cut-off at the frequency, $2/T$. Since the

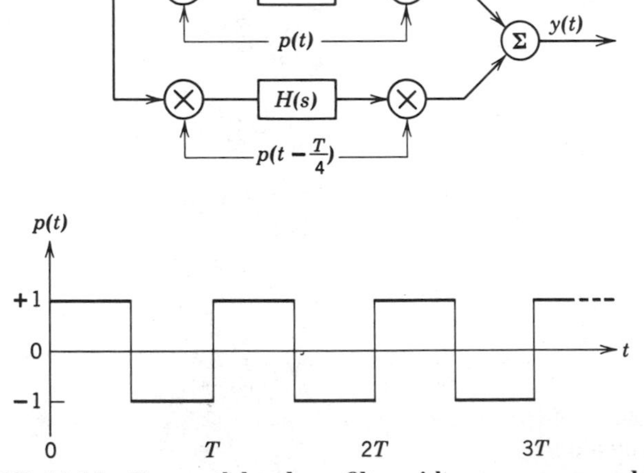

FIGURE 11-15 Two-path bandpass filter with square-wave modulation.

square wave has no dc or second harmonic components, there is only one passband (centered at $f = 1/T$) in the frequency range, $0 \leq f < 2/T$.

Evaluating the fundamental Fourier component of the square wave, we obtain the voltage-transfer ratio

$$W_0(s) = \frac{8}{\pi^2}\left[H\left(s - j\frac{2\pi}{T}\right) + H\left(s + j\frac{2\pi}{T}\right)\right] \tag{11.52}$$

where $H(s)$ is the voltage-transfer ratio of the path filters. An experimental realization of this type using second-order lowpass active path networks is reported by Glaser *et al.* [3]. The measured image components were found to be approximately 37 dB below the signal level. This reference also presents an analysis of the image component level in terms of the amount of mismatch in the transfer functions of the two paths. To minimize the effects of path mismatch for high-order path networks, it has been suggested that the overall characteristic be realized as a parallel combination [3, 12], or a cascade combination [13], of two-path filters, each involving only first- or second-order path networks and possibly different modulating frequencies in each two-path filter.

The bandpass filter described here is tunable over a limited frequency range by varying the frequency of the square-wave modulating functions. Suppose that the input and output bandlimiting filters have a sharp cut-off at f_0 Hz., then a tuning range of approximately 2 : 1 for a narrow passband could be obtained ($f_0/2 < 1/T < f_0$) with image rejection provided by the bandlimiting filters. If the passband is sufficiently narrow so that the path networks themselves provide image rejection, then the tuning range can be extended upward towards 3 to 1 ($f_0/3 < 1/T < f_0$). To obtain a larger tuning range with fixed bandlimiting filters, more levels in the staircase modulation can be used to make the higher order Fourier coefficients vanish [3]. Another approach, which has been the subject of more recent investigation, is to retain two-level modulation but to increase the number of transitions per period, and to position the transition instants in such a manner that the modulating functions exhibit the desired harmonic content. A general method for determining the appropriate set of transition instants (over a single period) has been devised, and the specific results for implementing a 10-to-1 tuning range have been reported [14].

11.5 IMPLEMENTATION WITH PERIODICALLY OPERATED SWITCHES

Commutated Networks

The simplest implementation of pulse modulation (using levels 0 and 1) is provided by periodically operated switches connected in series with the input and output terminals of the path networks. The N-path filter then takes the form of a sequence of N identical two-port networks with N-pole rotary switches at input and output as shown in Figure 11-16. The switch rotors have a constant angular velocity at $1/T$ rps. For this reason, the structure is called a *commutated network*. It is normally assumed that the rotor connects only one path at a time. For high-frequency applications, of course, the rotary switch would be implemented with electronically controlled solid-state switches. Except for some early versions of the two-path bandpass filter, the N-path commutated network using single-capacitor elements for each path was the first application of N-path filter principles [15, 16].

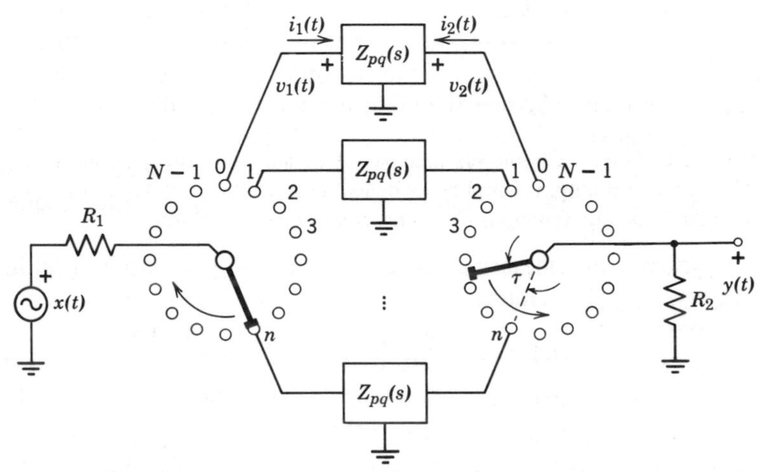

FIGURE 11-16 Structure for commutated version of the N-path filter. The $Z_{pq}(s)$ functions are the open-circuit impedance parameters of the two-port path networks.

If active elements are used in each path to provide the appropriate isolation, then the switches provide pulse modulation on the input and output signals, and the transfer function analysis discussed previously can be applied without difficulty. Without the isolating elements, however, the driving-point properties of the path networks, as well as their transfer properties, must be taken into account, and the analysis becomes considerably more complicated. Consider a single path in Figure 11-16. The external circuit imposes the following voltage-current constraints.

$$\begin{aligned} R_1 i_1(t) &= [x(t) - v_1(t)]p(t) \\ R_2 i_2(t) &= -v_2(t)q(t) \end{aligned} \tag{11.53}$$

where $p(t)$ and $q(t)$ are the periodic repetitions of the rectangular pulses, $\gamma_1(t)$

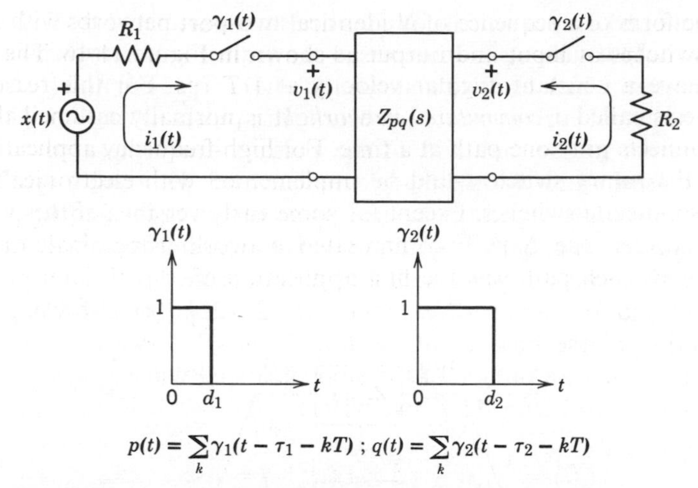

FIGURE 11-17 Circuit for analysis of response of two-port network with input and output sampling switches. The functions $p(t)$ and $q(t)$ are periodic indicator functions characterizing the switch closures.

and $\gamma_2(t)$ respectively, shown in Figure 11-17. The path networks impose the additional terminal constraints,

$$\begin{aligned} v_1(t) &= [z_{11} \otimes i_1](t) + [z_{12} \otimes i_2](t) \\ v_2(t) &= [z_{21} \otimes i_1](t) + [z_{22} \otimes i_2](t) \end{aligned} \tag{11.54}$$

where the $z_{pq}(t)$ functions in the convolution products indicated in (11.54) are the open-circuit driving-point and transfer-voltage responses to applied unit-impulse currents at the two ports. The Laplace transforms of these functions are the conventional open circuit impedance parameters of the two-port network.

To obtain the relationship between output current in R_2 and the source voltage, $x(t)$, we need the simultaneous solution of (11.53) and (11.54). Bennett [17] and others [18, 19] have presented analysis techniques for networks which are time variable by virtue of containing a periodically operated switch. More recently, the approach has been simplified by using a state-variable formulation of the problem. Sun and Frisch [20] use this approach with particular reference to the analysis of the N-path commutated network.

The Narrow-Pulse Approximation

In special circumstances, there are valid approximate solutions which lead to simplified analysis and design of commutated networks. The situation we shall investigate here is the one that results when the dwell times, d_1 and d_2, of the input and output switch contacts are small compared to T/N, so that input and output currents are comprised of a sequence of narrow pulses. The approximation which leads to a simple solution of (11.53) and (11.54) is to replace the actual current pulses by rectangular pulses having a height equal to the current at the midpoint of the pulse. Clearly the approximation improves as d_1 and d_2 are made smaller, and the results are usually sufficiently accurate, especially in the lower portion of the frequency range. The switches are essentially providing a signal sampling operation, and these filters are often called *sampled-data filters*. Similar methods of analysis have been employed for *resonant-transfer circuits* [21, 22]. (See also Chapter 10).

Derivation of the final results, although straightforward, is somewhat tedious, and only an outline of the steps involved will be presented here. Referring again to Figure 11-17, we write

$$i_q(t) = \frac{1}{T}\gamma_q(t - \tau_q) \otimes j_q(t); \quad q = 1, 2 \tag{11.55}$$

where

$$j_q(t) = i_q(t + \tfrac{1}{2}d_q + \tau_q)\left[T \sum_{k=-\infty}^{\infty} \delta(t - kT)\right].$$

The Laplace transform of (11.55) is

$$I_q(s) = \frac{1}{T}\Gamma_q(s)e^{-s\tau_q}J_q(s) \tag{11.56}$$

where

$$J_q(s) = \sum_{l=-\infty}^{\infty} I_q\left(s - j\frac{2\pi l}{T}\right)e^{[s - j(2\pi l/T)](d_q/2 + \tau_q)}$$

Now Equation (11.54) can be expressed as

$$V_p(s) = \frac{1}{T}\sum_{q=1}^{2} Z'_{pq}(s)J_q(s)e^{-s\tau_q}; \quad p = 1, 2 \tag{11.57}$$

where

$$Z'_{pq}(s) = \Gamma_q(s)Z_{pq}(s),$$

and these primed open-circuit impedance parameters simply represent the two-port responses to unit rectangular current pulses rather than to current impulses. The next step is to take the Laplace transform of (11.53) and substitute (11.57) into it. The result is considerably simplified by noting the periodic nature of $J_q(s)$; i.e., $J_q[s - j(2\pi l/T)] = J_q(s)$. The resulting pair of simultaneous equations involves the unknown quantities $J_1(s)$ and $J_2(s)$. Eliminating $J_1(s)$ gives the equation for $J_2(s)$, and hence $I_2(s)$.

$$I_2(s) = \frac{-\dfrac{1}{d_1}\Gamma_2(s)\tilde{Z}_{21}(s) \sum_{l=-\infty}^{\infty} X\left(s - j\dfrac{2\pi l}{T}\right) e^{sd_1/2} e^{-j(2\pi l/T)(d_1/2+\tau_1)}}{\left[\dfrac{R_1 T}{d_1} + \tilde{Z}_{11}(s)\right]\left[\dfrac{R_2 T}{d_2} + \tilde{Z}_{22}(s)\right] - [\tilde{Z}_{12}(s)][\tilde{Z}_{21}(s)]} \tag{11.58}$$

where still another set of open-circuit impedance parameters have been defined.

$$\tilde{Z}_{pq}(s) = \frac{e^{sd_p/2}}{d_p} \sum_{l=-\infty}^{\infty} Z'_{pq}\left(s - j\frac{2\pi l}{T}\right) e^{j(2\pi l/T)(\tau_q - \tau_p - d_p/2)}. \tag{11.59}$$

These functions are very similar to the *pulse impedance* functions employed by Fettweis [22] in the analysis of resonant transfer circuits. For analysis of the N-path filter, we characterize the nth path by

$$\tau_1 = \frac{nT}{N}, \tau_2 = \frac{nT}{N} + \tau; \qquad n = 0, 1, 2, \ldots, N-1 \tag{11.60}$$

so that the parameter, τ, characterizes the lag in the output commutator relative to the input commutator. It is important to note that the $\tilde{Z}_{pq}(s)$ parameters will depend on τ, but they are independent of the path number, n. The total current in R_2 in Figure 11-16 flows in nonoverlapping pulses, given for each value of n by an expression in the form of (11.58). The output voltage is obtained by summing over n and retaining only the $l = 0$ term in the result, because, as before, we shall apply bandlimiting restrictions to $x(t)$ and $y(t)$. The final result is a transfer function given by

$$W_0(s) = \frac{\left[\dfrac{Nd_2}{Td_1}\Gamma_2(s)e^{sd_1/2}\right][\tilde{Z}_{21}(s)]\left[\dfrac{R_2 T}{d_2}\right]}{\left[\dfrac{R_1 T}{d_1} + \tilde{Z}_{11}(s)\right]\left[\dfrac{R_2 T}{d_2} + \tilde{Z}_{22}(s)\right] - [\tilde{Z}_{12}(s)][\tilde{Z}_{21}(s)]} \tag{11.61}$$

over the frequency band, $|f| \le N/2T$.

The transfer function is intentionally expressed in the particular form that appears in (11.61) because we recognize this form as one that gives the overall voltage-transfer ratio of a two-port network with source and load resistances taken into account. Thus, except for the first factor in the numerator of (11.61), the voltage-transfer ratio is equivalent to that of a time-invariant two-port network with open-circuit impedance parameters given by $\tilde{Z}_{pq}(s)$, and operating between source and load resistances of $R_1\,T/d_1$ and $R_2\,T/d_2$, respectively. The equivalent circuit is shown in Figure 11-18. The resistance multiplication factors, T/d_1 and T/d_2, are typically very large, and this is often an advantage in microcircuit fabrications [23]. From (11.59), it is interesting to note that even with path networks that are reciprocal $[Z_{12}(s) = Z_{21}(s)]$, e.g., passive RC networks, the equivalent network characterized by $\tilde{Z}_{pq}(s)$ will be nonreciprocal unless $d_1 = d_2$ and $\tau = 0$. In other words, the equivalent circuit in Figure 11-18 is reciprocal only if the input and output switches on each path make simultaneous contact.

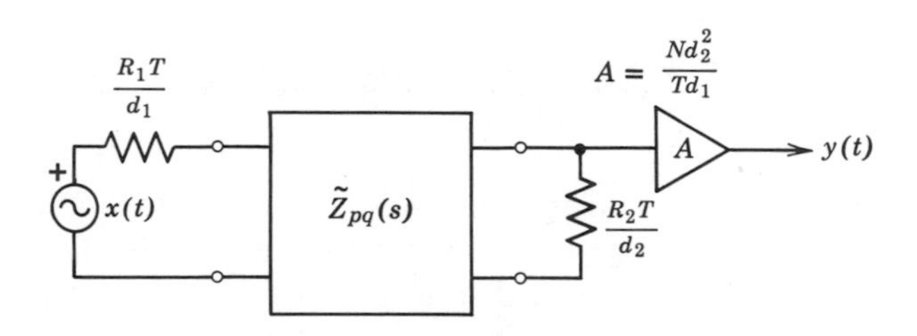

$$\tilde{Z}_{pq}(s) \triangleq \frac{e^{sd_p/2}}{d_p} \sum_{l} Z'_{pq}\left(s - j\,\frac{2\pi l}{T}\right) e^{-j(2\pi l/T)\,(d_p/2 + \tau_p - \tau_q)}$$

and $Z'_{pq}(s) = Z_{pq}(s)\,\Gamma_q(s)$ for $p, q = 1, 2$

FIGURE 11-18 Equivalent time-invariant circuit for *N*-path commutated network.

As the dwell times, d_1 and d_2, approach zero, the transfer function of (11.61) exhibits the expected periodic-gain characteristic with frequency. If the open-circuit impedance parameters, $Z_{pq}(s)$, of the path networks are expressed in partial-fraction form, then simple closed-form expressions for the parameters can be obtained by essentially the same technique used to arrive at (11.18). The expressions can be obtained directly from the general relationship,

$$\begin{aligned}\tilde{Z}(s) &= \frac{e^{-st_0}}{T} \sum_{l=-\infty}^{\infty} Z\left(s - j\,\frac{2\pi l}{T}\right) e^{[s - j(2\pi l/T)]t_1} \\ &= \sum_{k=0}^{\infty} z(kT + t_1) e^{-s(t_0 + kT)}. \end{aligned} \tag{11.62}$$

Using the appropriate values for t_0 and t_1 and a $z(t)$ corresponding to one term in a partial fraction expansion of $Z(s)$, the resulting geometric series can be summed giving $\tilde{Z}(s)$ in closed form.

Commutated Capacitor Bank

The comb filter characteristics described in Section 11.4 can be obtained in the commutated network form by using path networks consisting of single capacitors either in shunt or series connection [15, 16, 24]. In certain cases, only one rotary switch is required. For example, using shunt capacitors in each path and letting $d_1 = d_2 = d$ and $\tau = 0$, the commutated network can be equivalently implemented with one rotary switch and a single-pole output switch operated every T/N sec during the contact intervals of the rotary switch. This is shown schematically in Figure 11-19. In this case, the parameters for the path network are

$$Z_{pq}(s) = \frac{1}{Cs}, \qquad p, q = 1, 2. \tag{11.63}$$

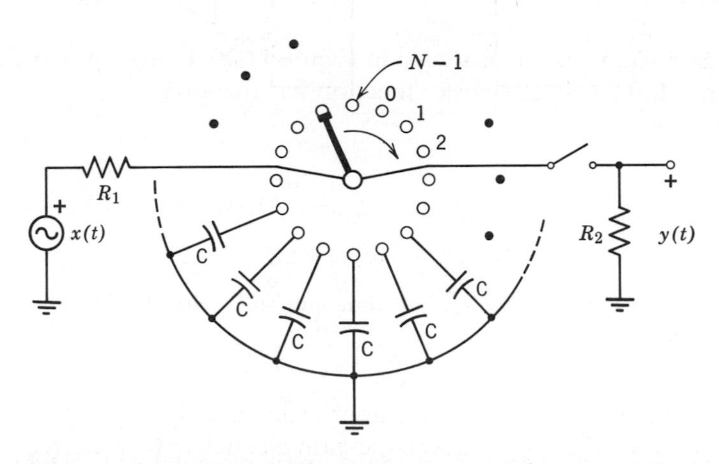

FIGURE 11-19 Commutated-capacitor-bank comb filter. The output switch operates at a rate N times the revolution rate of the rotary switch.

Using (11.62) and letting d approach zero, the $\tilde{Z}_{pq}(s)$ parameters become

$$\tilde{Z}_{pq}(s) = \frac{T}{2C}\coth\frac{sT}{2}, \qquad p, q = 1, 2. \tag{11.64}$$

We recognize the expression in (11.64) as the driving-point impedance of an ideal delay line with length corresponding to a delay of $T/2$, a characteristic impedance of $R_0 = T/2C$, and an open-circuit termination. Thus, for small d, the equivalent circuit for the configuration in Figure 11-19 is simply a shunt-connected delay line with the appropriate source and load resistances

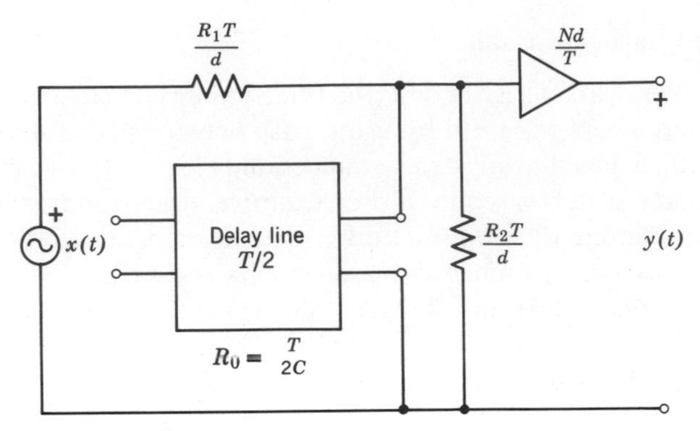

FIGURE 11-20 Equivalent circuit for the comb filter using an open-circuit terminated delay line. R_0 is the characteristic impedance of the line.

and a gain-scaling factor as shown in Figure 11-20. Using a proportionality constant, A, the voltage transfer function has the form

$$A \frac{1 + e^{-sT}}{1 - e^{-(s+\alpha)T}}$$

where

$$e^{-\alpha T} = \frac{1 - \dfrac{(R_1 + R_2)d}{2R_1 R_2 C}}{1 + \dfrac{(R_1 + R_2)d}{2R_1 R_2 C}}. \tag{11.65}$$

This characteristic is very similar to the one corresponding to (11.45) for the Type I comb filter except that transmission nulls occur at odd multiples of $\frac{1}{2}T$ in the gain-frequency characteristic. When α is small, the tooth bandwidth is given approximately by

$$f_b \cong \frac{\alpha}{\pi} \cong \frac{d}{\pi T} \cdot \frac{R_1 + R_2}{R_1 R_2 C}. \tag{11.66}$$

The effects of capacitor losses represented by shunt resistances can be easily incorporated [1] (See Problem 11.7). Implementations of this circuit with solid-state switches (usually with $N = 4$) to obtain single-passband bandpass filters have also been described [25–27]. In these cases, the passband at $f = 0$ is eliminated by using bandlimiting filters of a broad bandpass type.

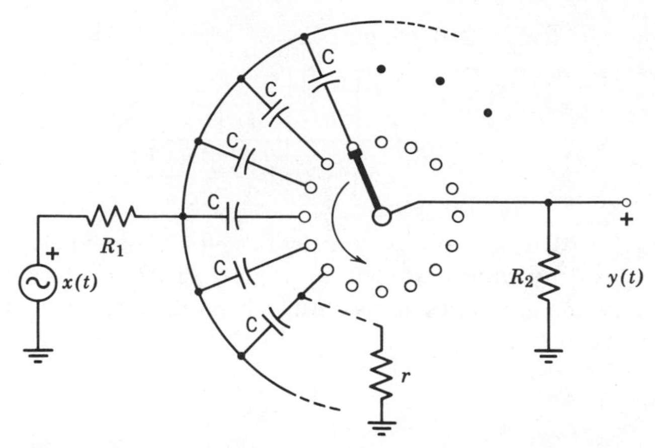

FIGURE 11-21 Commutated-capacitor-bank comb-notch filter.

The comb-notch filter described in Section 11.4 can be realized with series-capacitor path networks in the commutated form. In this case, only a single rotary switch is required as shown in Figure 11-21. Again, this corresponds to the situation where $d_1 = d_2 = d$ and $\tau = 0$. For analysis purposes, we let the path network include a shunt resistance, r, so that the open-circuit impedance parameters are finite. In the final expression we can let $r \to \infty$.

With

$$Z_{11}(s) = r + \frac{1}{sC}$$

$$Z_{12}(s) = Z_{21}(s) = Z_{22}(s) = r, \tag{11.67}$$

and letting d approach zero, we obtain

$$\tilde{Z}_{11}(s) = \frac{rT}{d} + \frac{T}{2C}\coth\frac{sT}{2}$$

$$\tilde{Z}_{12}(s) = \tilde{Z}_{21}(s) = \tilde{Z}_{22}(s) = \frac{rT}{d} \tag{11.68}$$

so the voltage transfer function of (11.61) becomes

$$W_0(s) = \frac{\left[\dfrac{Nd}{T}\right]\left[\dfrac{rT}{d}\right]\left[\dfrac{R_2 T}{d}\right]}{\left[\dfrac{R_1 T}{d} + \dfrac{rT}{d} + \dfrac{T}{2C}\coth\dfrac{sT}{2}\right]\left[\dfrac{R_2 T}{d} + \dfrac{rT}{d}\right] - \left[\dfrac{rT}{d}\right]^2}. \tag{11.69}$$

Now letting $r \to \infty$ in (11.69), we get

$$W_0(s) = \frac{\left[\frac{Nd}{T}\right]\left[\frac{R_2 T}{d}\right]}{\frac{R_1 T}{d} + \frac{T}{2C}\coth\frac{sT}{2} + \frac{R_2 T}{d}}. \tag{11.70}$$

Interpreting (11.70) as the voltage-transfer ratio for a ladder network, we see that the equivalent circuit, shown in Figure 11-22 involves the open-circuited delay line in a series connection. The voltage transfer can also be expressed as

$$W_0(s) = A\frac{1 - e^{-sT}}{1 - e^{-(s+\alpha)T}}$$

where

$$e^{-\alpha T} = \frac{R_1 + R_2 - \frac{d}{2C}}{R_1 + R_2 + \frac{d}{2C}}. \tag{11.71}$$

For small α, this is the desired comb-notch characteristic with narrow notches at multiples of $1/T$. The notch width is approximately

$$f_b \cong \frac{\alpha}{\pi} \cong \frac{d}{\pi T} \cdot \frac{1}{(R_1 + R_2)C}. \tag{11.72}$$

Because of the single commutator configuration, the notch characteristic is not critically dependent on the phasing of input and output modulation as it was in the case of the Type II comb filter described in Section 11.4.

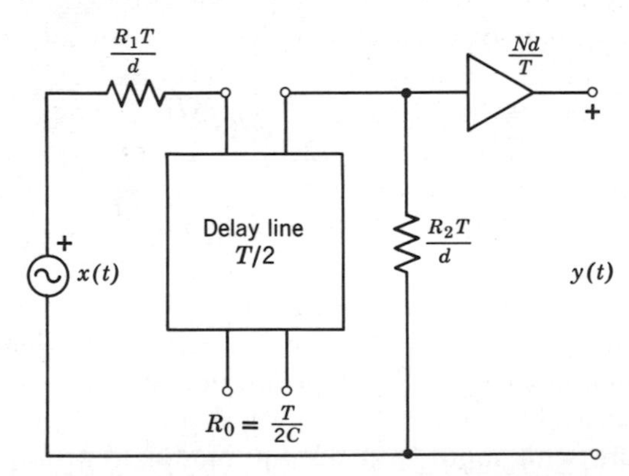

FIGURE 11-22 Equivalent circuit for the comb-notch filter using open-circuit terminated delay line.

11.6 SUMMARY

The simplest way to summarize the response properties of the N-path filter is to point out that the impulse response of the system is given by a bandlimited version of

$$w_0(t) = c(t)h(t)$$

valid over the frequency band, $|f| \leq N/2T$. The factor, $c(t)$, is a periodic function of time, with period T, determined entirely by the periodic input and output modulating functions. The second factor, $h(t)$, is the impulse response of a time-invariant network which must be capable of accurate replication in each of the N paths. If the path network is a finite order system, then there is a simple closed-form expression for the transfer function, $W_0(s)$, involving a partial-fraction expansion of $H(s)$ and the function, $D(s)$, which is the Laplace transform of one period of the $c(t)$ function. This expression is especially useful for determining the pole-zero pattern and residues for the overall transfer function.

Because of the periodicity of $c(t)$, the transfer function, $W_0(s)$, consists of a sequence of frequency-translated versions of $H(s)$, centered about the harmonic frequencies in $c(t)$. This feature provides the attractive alternative to realization of filtering characteristics which would otherwise require high-quality inductors or delay lines. A corollary advantage of the N-path filter realization is the suitability for implementation in microcircuit form and also the possibility for accurate electronic adjustability of the filter response. This latter feature is obtained by changing the parameters of the modulating waveforms rather than by changing circuit element values.

Recent studies of the N-path filter have concentrated on the possibilities for realization of single-passband bandpass filters. It is expected that future investigations will be directed more toward the unique capability for providing multiple passband filtering using the N-path configuration. For some signal processing applications, where optimum estimation of signal parameters calls for periodically time-variable filtering operations, it is likely that the N-path configuration can be usefully employed. It has been noted that a somewhat generalized version of the N-path configuration serves as a canonical representation for any finite-order periodically varying linear system [28]. Some particular applications of periodically varying processing to provide, simultaneously, bandpass filtering and frequency translation have been extensively examined [29, 30]. These applications involve the N-path structure with input and output modulating functions having a different period, with particular reference to producing single-sideband amplitude modulation signals in frequency-division multiplex systems.

Another area which deserves further investigation is the use of digital filtering techniques in conjunction with the N-path configuration. Glaser [31] has considered a structure wherein the path filtering operations are all provided by a single digital filter operating in a time-division multiplexing mode. A significant advantage of this approach is the elimination of the problem of dissimilar path network responses afforded by the use of a single digital filter.

PROBLEMS

11.1 Derive the equivalent structure, similar to that shown in Figure 11-3, which involves a parallel combination of paths with time-invariant networks followed by single-frequency modulators.

11.2 Determine the pair of staircase modulating functions, $p(t)$ and $q(t)$, involving the fewest number of jumps in the interval, $0 \leq t < T$, which will make $|P_{-1}Q_1| = 1$ and $P_{-l}Q_l = 0$ for $l = 0, 2, 3, 4, 5, 6$.

11.3 In obtaining the closed-form expression of (11.18) for $W_0(s)$, suppose that $H(s)$ contained a term of the form,

$$\frac{1}{(s+\alpha)^2}.$$

Show that the corresponding term in $W_0(s)$ is given by

$$\frac{TD(s+\alpha)e^{-(s+\alpha)T}}{[1-e^{-(s+\alpha)T}]^2} - \frac{D'(s+\alpha)}{1-e^{-(s+\alpha)T}}$$

where $D'(s)$ denotes the derivative of $D(s)$ with respect to s. (*Hint:* Use the relation

$$\sum_{k=0}^{\infty} k\beta^k = \frac{\beta}{(1-\beta)^2} \qquad \text{for} \quad |\beta| < 1.)$$

11.4 Provide the missing steps in the derivation of (11.35).

11.5 Find a simple pair of modulating functions, $p(t)$ and $q(t)$, which will provide a comb filtering characteristic which has teeth only at odd multiples of $1/T$.

11.6 For the transfer function, $W_0(s)$, given in (11.48), plot the locus of transmission zeros in the s-plane as τ varies in the range $|\tau| \leq T/2$.

11.7 In the commutated-capacitor comb filter shown in Figure 11-19, suppose that capacitor loss is represented by a resistance, R, in parallel with each capacitor. Show that the equivalent delay-line circuit in Figure 11-20 is modified by terminating the delay line with the appropriate resistance.

11.8 In the N-path commutated network shown in Figure 11-16, suppose that the two-port path networks are single-section RC lowpass filters as shown below.

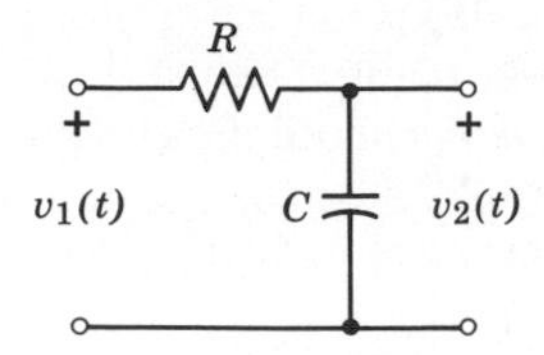

Let $d_1 = d_2 = d \to 0$, and let the output commutator lag the input commutator by $\tau = T/2$. Derive the corresponding $\tilde{Z}_{pq}(s)$ parameters, the transfer function, $W_0(s)$, and show an equivalent network using delay lines which provides the same voltage transfer ratio.

11.9 Show that the integral expression for $c(t)$ in (11.11) yields a periodic function with Fourier coefficients given by $NP_{-l}Q_l$.

11.10 It is desired to realize a filter whose transfer function, within an arbitrary delay factor, is proportional to

$$\frac{\cos 6\pi x + \cos 2\pi x + j(\sin 6\pi x + \sin 2\pi x)}{\cos 6\pi x - 1.5 \cos 4\pi x + \cos 2\pi x - 0.25 + j(\sin 6\pi x - 1.5 \sin 4\pi x + \sin 2\pi x)}$$

where $x = f/f_0$. Design an N-path filter using pulse modulation which will match this characteristic closely out to a frequency of $10f_0$. Specify all the modulation parameters and the transfer function, $H(s)$, for the path networks.

REFERENCES

[1] L. E. Franks and I. W. Sandberg, "An alternative approach to the realization of network transfer functions: the N-path filter," Bell System Tech. J., **39**, No. 5, 1321–1350 (Sept., 1960).

[2] V. Cizek, "Analysis of a filter with N channels," *Proceedings of the Third Colloquim on Microwave Communication, Budapest* (April, 1966).

[3] A. B. Glaser, C. C. Halkias, and H. E. Meadows, "A tuneable, bandwidth-adjustable solid-state filter," *J. Franklin Inst.*, **288**, No. 2, 83–98 (Aug., 1969).

[4] Y. Sun, "Network functions of quadrature N-path filters," *IEEE Trans. Circuit Theory*, **CT-17**, No. 4, 594–600 (Nov., 1970).

[5] N. F. Barber, "Narrow band-pass filter using modulation," *Wireless Engineer*, **24**, No. 5, 132–134 (May, 1947).

[6] G. B. Madella, "Single-phase and polyphase filtering devices using modulation," *Wireless Engineer*, **24**, No. 10, 310–311 (Oct., 1947).

[7] I. F. MacDiarmid and D. G. Tucker, "Polyphase modulation as a solution of certain filtration problems in telecommunication," *Proc. IEE*, Pt. III, **97**, 349–358 (Sept., 1950).

[8] H. B. Paris, Jr. "Utilization of the quadrature functions as a unique approach to electronic filter design," *IRE Intern. Conv. Record*, Pt. 9, 204–216 (1960).
[9] G. Rigby, "An integrated selective amplifier using frequency translation.," *IEEE-Journal on Solid-State Circuits*, **SS-1**, 39–44 (Sept., 1966).
[10] B. Gold and C. M. Rader, *Digital Processing of Signals*, McGraw-Hill, New York, 1969.
[11] H. Urkowitz, "Analysis and synthesis of delay line periodic filters," *IRE Trans. Circuit Theory*, **CT-4**, No. 2, 41–53 (June, 1957).
[12] W. Saraga, "Realizability conditions and associated invariance relations for *N*-path networks with sinusoidal carrier oscillations," *1969 IEEE International Symposium on Circuit Theory*, (Dec., 1969).
[13] G. Szentirmai, "Active switched bandpass filters," *1969 IEEE International Symposium on Circuit Theory*, San Francisco, Calif. (Dec., 1969).
[14] L. E. Franks and L. D. Dann, "Implementation of *N*-path filters using polarity-reversal modulation," *1970 Canadian Symposium on Communications*, Montreal, Canada (Nov., 1970).
[15] W. R. LePage, C. R. Cahn, and J. S. Brown, "Analysis of a comb filter using synchronously commutated capacitors," *AIEE Trans. Comm. and Elect.*, **72**, 63–68 (March, 1953).
[16] B. D. Smith, "Analysis of commutated networks," *IRE Trans. Aeronautical Electronics*, **AE-10**, 21–26 (Dec., 1953).
[17] W. R. Bennett, "Steady-state transmission through networks containing periodically-operated switches," *IRE Trans. Circuit Theory*, **CT-2**, No. 1, 17–21 (March, 1955).
[18] C. A. Desoer, "Transmission through a network containing periodically-operated switches," *Wescon Convention Record*, Pt. 2, 34–41 (1958).
[19] A. Fettweis, "Steady-state analysis of circuits containing a periodically-operated switch," *IRE Trans. Circuit Theory*, **CT-6**, No. 3, 252–260 (Sept., 1959).
[20] Y. Sun and I. T. Frisch, "A general theory of commutated networks," *IEEE Trans. Circuit Theory*, **CT-16**, No. 4, 502–508 (Nov., 1969).
[21] C. A. Desoer, "A network containing a periodically operated switch solved by successive approximations," *Bell System Tech. J.*, **36**, 1403 (Nov., 1957).
[22] A. Fettweis, "Theory of resonant-transfer circuits," *Network and Switching Theory*, G. Biorci, Ed., Academic Press, New York, 1968.
[23] Y. Sun and I. T. Frisch, "Resistance multiplication in integrated circuits by means of switching," *IEEE Trans. Circuit Theory*, **CT-15**, 184–192 (Sept., 1968).
[24] R. Fischl, "Analysis of a commutated network," *IEEE Trans. Aerospace and Navigational Electronics*, **ANE-10**, 114–123 (June, 1963).
[25] L. E. Franks and F. J. Witt, "Solid-state sampled-data bandpass filters," *Proceedings 1960 Solid State Circuits Conference*, Philadelphia, Pa.
[26] J. Thompson, "RC digital filters for micro-circuit bandpass amplifiers," *Electronic Equipment Engineering*, **12**, 44–49 (March, 1964).
[27] W. R. Harden, "Digital filters with *IC*'s boost *Q* without inductors," *Electronics* (July, 1967).

[28] H. E. Meadows, L. M. Silverman, and L. E. Franks, "A canonical network for periodically variable linear systems," *Fourth Allerton Conference on Circuit and System Theory*, 647–657 (1966).

[29] D. K. Weaver, "A third method of generation and detection of single-sideband signals," *Proc. IRE*, **44**, 1703–1705 (Dec., 1956).

[30] S. Darlington, "Some circuits for single-sideband modulation," *Proceedings Third Princeton Conference on Information Sciences and Systems* (1969).

[31] A. B. Glaser, "Digital time-division multiplexed *N*-path system," *IEEE Trans. Circuit Theory*, **CT-17**, No. 4, 600–604 (Nov., 1970).

12
Digital Filters

Roger M. Golden
Technology Service Corporation
Santa Monica, California

The theory, design, and synthesis techniques presented in the preceding chapters have as their goal the eventual realization of filters using analog circuit elements. Among the most familiar of these components are resistors, capacitors, inductors, and special devices such as operational amplifiers, piezoelectric components, and mechanical transducers. In complex design procedures, frequent use is made of the digital computer to perform the design steps (and iteration where necessary) rapidly and accurately. However, the digital computer itself can be used for the filtering of digitized continuous waveforms. For many applications, special purpose digital hardware would be more economical for performing such filtering operations. This means that it is possible to build signal-wave filters from the same building blocks as are used in the construction of the digital computer: adders, multipliers, shift registers, memory components, etc.

The purpose of this chapter is to show how the design techniques thus far presented are readily made use of in the design and synthesis of digital filters.

Digital filter design as considered here will concern the realization of digital processing structures which approximate the frequency or time characteristics of continuous filters. Once the design parameters are determined, the filter may be implemented by software (i.e., a digital computer algorithm or program), or it may be realized in hardware form. The way in which a digital filter is implemented depends on the system in which it is to be incorporated. Generally, two distinct methods of implementation present themselves: frequency-domain weighting, and time-domain weighting. The first method would be used in systems which perform spectral analysis of time waveforms such as discrete (fast) Fourier transform processors. For these systems, the filter function might consist of a stored table of complex weights. These weights represent the complex frequency response of the filter evaluated at those frequency values for which the spectrum analyzer yields a corresponding value. These weights can be determined from either a continuous or digital filter transfer function.

The second method of implementation is used for processing a sequence of numbers which represent a sampled version of a time waveform. The filter yields an output sequence which is related to the input sequence by the digital transfer function of the filter. The output sequence results from convolving the input sequence with the system function sequence of the filter.

In the design techniques to be considered, it is assumed that the sequence of input samples is uniformly spaced in time. In other words, the input sequence is synchronously sampled (e.g., a sequence derived by uniformly sampling a continuous signal every 100 μsec). The mathematics of the sampling process and z-transform, along with the consequences resulting therefrom, are considered in the following sections.

12.1 THE SAMPLING PROCESS AND THE z-TRANSFORM

The usual approach taken by texts on sampled-data theory [1] is to introduce the concept of sampling a continuous signal by impulse or delta function modulation. Thus, the output signal, $o(t)$, produced by modulating a continuous signal, $g(t)$, by a sampler, $m(t)$, is just,

$$o(t) = m(t)g(t). \tag{12.1}$$

Mathematically, the impulse modulation function, $m(t)$ is represented by

$$m(t) = T\sum_{n=0}^{\infty} \delta(t - nT) \tag{12.2}$$

where T denotes the unit sampling interval.

The factor of T is introduced in front of the summation so that the integrated signal between sample points will be approximately equal to the integrated continuous signal over an interval T, thus preserving the power in the signal. Properties of $m(t)$ useful in the discussion to follow are considered next.

Sampling Process

Modulation of a continuous function by $m(t)$ yields samples of the continuous function at the sampling intervals, nT. This result follows from the fact that $\delta(x) = 0$ for $x \neq 0$. Thus

$$m(t)g(t) = T\sum_{n=0}^{\infty} g(t)\,\delta(t - nT) = T\sum_{n=0}^{\infty} g(nT)\,\delta(t - nT) = g^*(t), \quad (12.3)$$

the * denotes that $g(t)$ is a sampled function.

Laplace Transform of $m(t)$

The Laplace transform of (12.2) is just

$$\mathscr{L}[m(t)] = M(s) = T\sum_{n=0}^{\infty} \exp(-snT) = T\sum_{n=0}^{\infty} [\exp(-sT)]^n. \quad (12.4)$$

Equation (12.4) can be written in closed form as

$$M(s) = \frac{T}{1 - \exp(-sT)}. \quad (12.5)$$

The variable $\exp(-sT)$ is identified as the unit sample interval delay. This function defines a new variable, z, as

$$z \equiv \exp(sT).$$

Using the above properties, it is possible to proceed to a discussion of the standard z-transform.

Digitizing a Continuous Filter Transfer Function

The digitization of a continuous filter transfer function is accomplished by sampling the impulse response of the filter by $m(t)$. Thus the sampled-impulse response using (12.3) is

$$h^*(t) = h(t)m(t) = T\sum_{n=0}^{\infty} h(nT)\,\delta(t - nT). \quad (12.6)$$

It is possible to think of this process as one in which a set of synchronous sampling switches are placed at the input and output of a continuous filter as depicted in Figure 12-1. Hence, the standard z-transform [2] results from finding the Laplace transform of $h^*(t)$ which is

$$\mathscr{L}[h^*(t)] = H^*(s) = T\sum_{=0}^{\infty} h(nT)\exp(-snT) \quad (12.7a)$$

$i(t)$ $i^*(t)$ $I^*(s)$ $h(t)$ $H(s)$ $o(t)$ $o^*(t)$ $O^*(s)$

$$\frac{O^*(s)}{I^*(s)} \equiv H^*(s)$$

FIGURE 12-1 Digitization of a continuous filter transfer function.

or

$$H(z^{-1}) = T \sum_{n=0}^{\infty} h(nT) z^{-n} = \sum_{n=0}^{\infty} A_n z^{-n} \tag{12.7b}$$

where

$$A_n = Th(nT).$$

The digital filter defined by (12.7(b)) represents an infinitely long tapped delay line with tap weights equal to the sampled values of the impulse response. For practical realizations of continuous filters, $h(t)$ may be assumed to be of finite duration so that there are only a finite number of taps, say $N + 1$. Thus (12.7(b)) becomes

$$H(z^{-1}) = \sum_{n=0}^{N} A_n z^{-n} \tag{12.8}$$

and the implementation thereof is as shown in Figure 12-2.

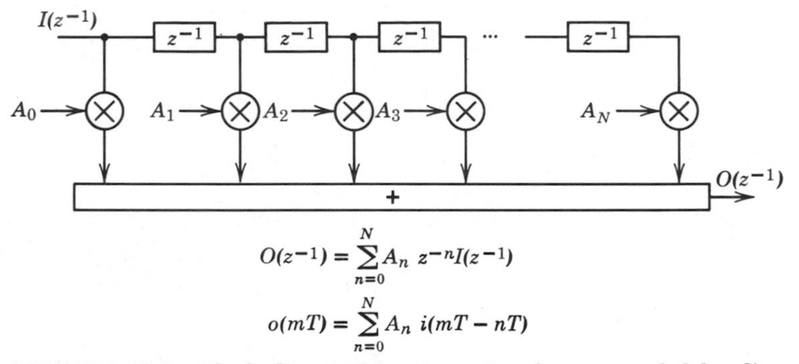

$$O(z^{-1}) = \sum_{n=0}^{N} A_n \, z^{-n} I(z^{-1})$$

$$o(mT) = \sum_{n=0}^{N} A_n \, i(mT - nT)$$

FIGURE 12-2 Block diagram representation for a tapped delay line digital filter approximation to a continuous filter.

Figure 12-3 shows pictorially the relationship between the impulse response of the continuous filter and that of the digital filter. Because the sampled values of the continuous response are the same as the values of the impulse response of the digital filter, this type of transformation is said to be impulse invariant.

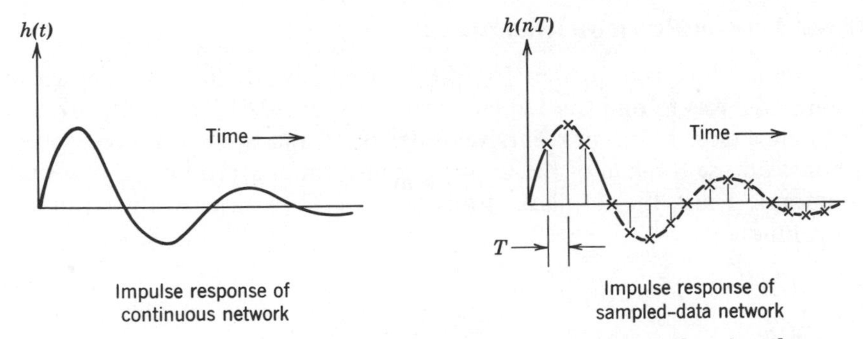

FIGURE 12-3 Comparison of continuous and digital filter impulse responses.

Sampled signal values propagate through the filter or delay line at the sampling rate. At each sample or clock pulse time, the samples stored in the delays at the previous clock pulse time are weighted by the appropriate coefficient and added together to produce an output sample. The transfer function relation between input and output is just

$$O(z^{-1}) = \sum_{n=0}^{N} A_n z^{-n} I(z^{-1}). \tag{12.9}$$

In terms of actual input and output samples, one may write the output at the mth sampling interval as,

$$o(mT) = \sum_{n=0}^{N} A_n i(mT - nT), \tag{12.10}$$

which is a difference equation relating input and output. The z-transform is often used to change a linear difference equation with constant coefficients into an algebraic equation. Thus, the z-transform performs the same type of operations on difference equations as does the Laplace transform on differential equations.

This type of filter is often called a transversal or nonrecursive design; transversal because of its transverse nature with respect to an input signal, nonrecursive because the output is not recursively related to past values of the output, but only to the last N values of the input. Furthermore, this type of realization exhibits finite memory of the input. A finite recursive digital filter which exhibits infinite memory of the input also can be derived from (12.6). This type represents a closed-form expression for the infinite summation given in the equation.

Closed Form or Recursive Realizations

The method followed for deriving the nonrecursive digital filter shown in Figure 12-2 was to find the Laplace transform of (12.6). This same method will yield a closed-form or recursive design if $h(t)$ and $m(t)$ are not combined, but instead use is made of the Laplace transform of a product of two time functions. Thus, the Laplace transform of (12.6) yields the complex convolution

$$\mathscr{L}[h^*(t)] = H^*(s) = \mathscr{L}[h(t)m(t)]$$

$$= \frac{1}{2\pi j}\int_{c_2-j\infty}^{c_2+j\infty} H(s-\omega)M(\omega)\,d\omega = \frac{1}{2\pi j}\int_{c_2-j\infty}^{c_2+j\infty} H(\omega)M(s-\omega)\,d\omega. \quad (12.11)$$

Substitution of (12.5) in the above yields

$$H^*(s) = \frac{T}{2\pi j}\int_{c_2-j\infty}^{c_2+j\infty} \frac{H(s-\omega)\,d\omega}{1-\exp(-sT)}$$

$$= \frac{T}{2\pi j}\int_{c_2-j\infty}^{c_2+j\infty} \frac{H(\omega)\,d\omega}{1-\exp(-sT+\omega T)}. \quad (12.12)$$

The integral in (12.12) represents a line integral in the complex plane. Evaluation of the integral is found by contour integration with the contour being closed to the left or right. The integral is easily evaluated if the filter transfer function, $H(\omega)$, is expressed in partial fraction expansion form as

$$H(\omega) = \sum_{n=1}^{NP} \frac{R_n}{\omega + \alpha_n} \quad (12.13)$$

where

$$\mathscr{R}_n = \text{residue of } H(-\alpha_n).$$

Substituting (12.13) into (12.12), and closing the contour to the left thereby enclosing the NP poles of $H(\omega)$, yields

$$H^*(s) = T\sum_{n=1}^{NP} \frac{\mathscr{R}_n}{1-\exp(-\alpha_n T)\exp(-sT)} \quad (12.14a)$$

or, substituting z^{-1} for $\exp(-sT)$,

$$H(z^{-1}) = T\sum_{n=1}^{NP} \frac{\mathscr{R}_n}{1-z^{-1}\exp(-\alpha_n T)}. \quad (12.14b)$$

Equation (12.14)(b) establishes a z-transform relationship between first-order filter transfer functions in the continuous domain or s-plane and the discrete (sampled-data) domain or z-plane. The transform pair for first-order terms is,

$$\frac{\mathcal{R}_r}{s+\alpha} \Leftrightarrow \frac{A_0}{B_1 z^{-1} + 1}$$

where

$$A_0 = \mathcal{R}_r T$$
$$B_1 = -\exp(-\alpha T).$$

Complex conjugate roots are handled in the same manner except that the conjugate terms resulting from complex conjugate pairs are combined to yield second-order functions. Thus, a complex conjugate pair of poles in the s-plane transforms according to the relationship

$$\frac{\mathcal{R}_r + j\mathcal{R}_i}{s+\alpha-j\beta} + \frac{\mathcal{R}_r - j\mathcal{R}_i}{s+\alpha+j\beta} \Leftrightarrow \frac{A_1 z^{-1} + A_0}{B_2 z^{-2} + B_1 z^{-1} + 1}$$

where

$$A_1 = -2T\exp(-\alpha T)[\mathcal{R}_r \cos(\beta T) + \mathcal{R}_i \sin(\beta T)],$$
$$A_0 = 2T\mathcal{R}_r,$$
$$B_2 = \exp(-2\alpha T),$$
$$B_1 = -2\exp(-\alpha T)\cos(\beta T).$$

Hence, higher-order continuous filter functions are readily digitized by means of the z-transform if the original function is first expressed as a partial fraction expansion. The individual terms in the expansion are then digitized according to whether they give rise to first-order or second-order terms. The complete digital filter function is then made up of first-order and second-order terms, each of which exhibit a recursion relationship between output and past samples of the output. The recursion relationship is easily demonstrated by finding the inverse z-transform of either the first- or the second-order transfer function above. From this function the relationship between input and output is

$$(1 + B_1 z^{-1} + B_2 z^{-2})O(z^{-1}) = (A_0 + A_1 z^{-1})I(z^{-1}). \tag{12.15}$$

Taking the inverse, z-transform yields

$$o(mT) + B_1 o[(m-1)T] + B_2 o[(m-2)T] = A_0 i(mT) + A_1 i[(m-1)T]. \tag{12.16}$$

Hence, the general recursion relationship for a digital filter transfer function expressed as a ratio of an NAth degree polynomial in z^{-1} to an NBth degree polynomial in z^{-1} is

$$o(mT) = \sum_{n=0}^{NA} A_n i[(m-n)T] - \sum_{n=1}^{NB} B_n o[(m-n)T]. \tag{12.17}$$

Equation (12.17) is a general representation for the output samples of a digital filter in terms of the input and past values of input and output. Equation (12.17) reduces to (12.10) when the B coefficients are all zero. Implementation of first- and second-order terms is considered next.

Digital Filter Building Blocks

The general expression for first-order and second-order terms in the transfer function for a digital filter is

$$\frac{A_0}{1 + B_1 z^{-1}}, \quad \text{and} \quad \frac{A_0 + A_1 z^{-1}}{1 + B_1 z^{-1} + B_2 z^{-2}} \quad \text{respectively.}$$

Two possible implementations [3] of the second-order function are shown in Figure 12-4. First-order functions are realized by omitting the signal paths

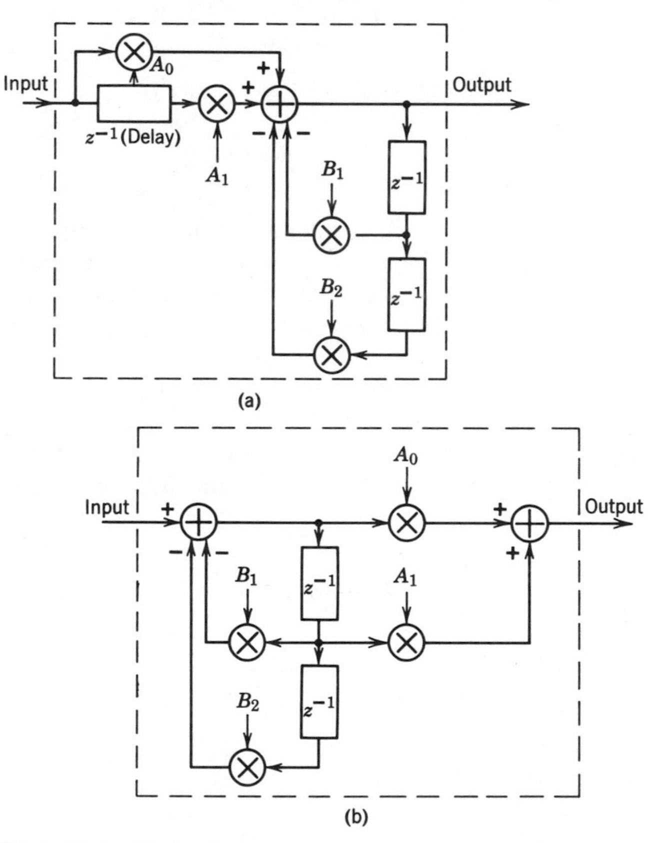

FIGURE 12-4 Block diagram realizations for the digital transfer function $(A_0 + A_1 z^{-1})/(1 + B_1 z^{-1} + B_2 z^{-2})$.

containing A_1 and B_2. Combining the appropriate number of first-order and second-order functions in parallel, as shown in Figure 12-5, yields the desired digital filter transfer function and is the synthesis desired.

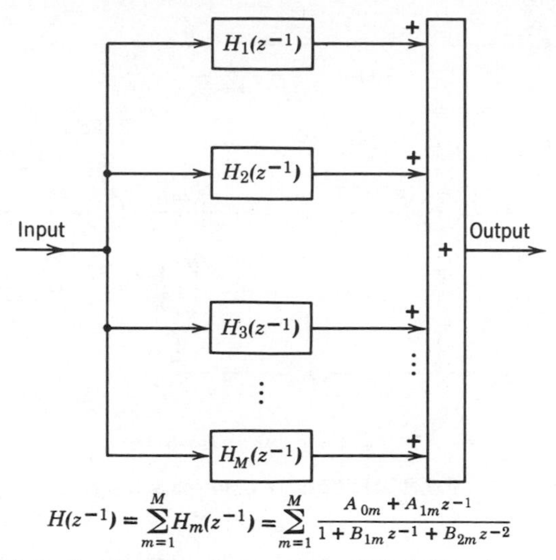

FIGURE 12-5 Realization of a recursive digital filter in parallel form.

Example 12-1. A simple example will show how the z-transform is applied. Consider a simple RC filter function as shown in Figure 12-6(a). The transfer function for this filter is

$$\frac{O(s)}{I(s)} = H(s) = \frac{1/CS}{1/CS + R} = \frac{1}{1 + RCs}. \tag{12.18}$$

The z-transform of this function is

$$\mathscr{Z}\left[\frac{1}{1 + RCs}\right] = H(z^{-1}) = \mathscr{Z}\left[\frac{1/RC}{1/RC + s}\right] = \frac{T/RC}{1 - z^{-1}\exp(-T/RC)}. \tag{12.19}$$

Thus, the constants are

$$A = T/RC$$
$$B = -\exp(-T/RC),$$

and the realization is that of a first order recursive section shown in Figure 12-6(b). For this function to be sufficiently bandlimited at $T/2$, $1/RC$ must be small compared to π/T so that,

$$\frac{1}{RC} \ll \frac{\pi}{T} \quad \text{or} \quad \frac{T}{RC} \ll \pi.$$

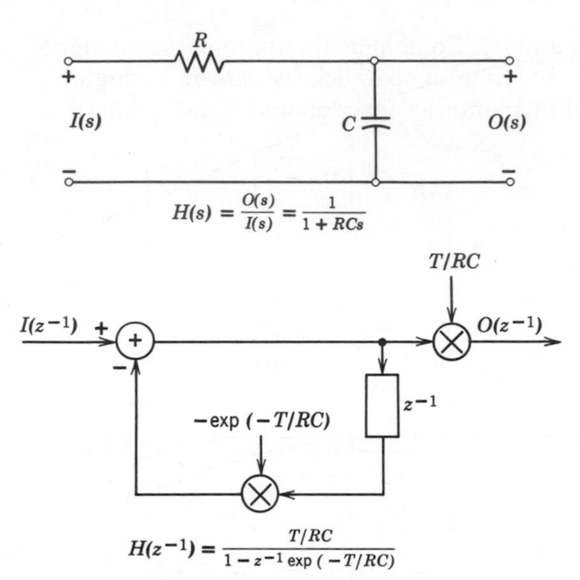

FIGURE 12-6 RC lowpass filter.

As a practical matter, if $|H(j\omega)| \leq .01$ for $\omega \geq \pi/T$, then the filter can be assumed to be sufficiently bandlimited so as to yield a useful standard z-transform design. A comparison of the magnitude response of three z-transformed RC filters is shown in Figures 12-7 (a), (b), and (c) for cut-off frequencies ($1/2\pi RC$) of 1, 10, and 100 Hz respectively. The sampling rate used was 2 kHz. The coefficients for these three filters are:

1 Hz	10 Hz	100 Hz
$A_0 = 3.1367 \times 10^{-3}$	$A_0 = 3.09283 \times 10^{-2}$	$A_0 = 2.69597 \times 10^{-1}$
$B_1 = -0.99686$	$B_1 = -0.96907$	$B_1 = -0.73040.$

The problems of input impedance levels and component tolerances are not pertinent to digital filters. However, the accuracy with which the filter coefficients are implemented, and the number of bits used to perform the arithmetic operations in the digital system of which the filter is part, are important [4, 5]. The effect of coefficient quantization on the response of the digital filter is treated in Section 12.4.

It must be recognized that the digital filter found by means of applying the standard z-transform to a continuous-filter transfer function results in a digital approximation to the continuous filter. Although the time responses are essentially the same (due to the impulse invariance of the transform), the frequency responses will be different. This may be shown by considering

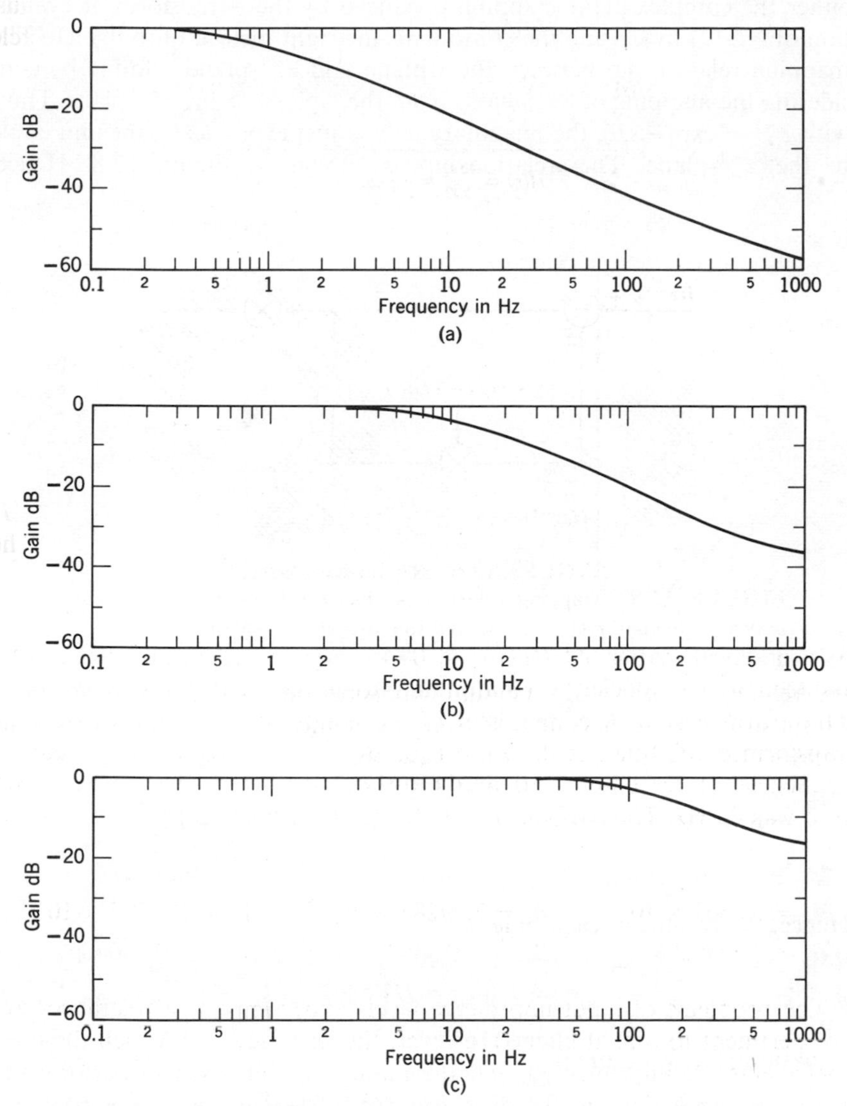

FIGURE 12-7 Magnitude-frequency responses of three digital RC filters designed for a sampling frequency of 2 kHz and cut-off frequencies of (a) 1 Hz, (b) 10 Hz, and (c) 100 Hz.

either the complex plane mapping produced by the z-transform or evaluation of (12.12) by closing the contour to the right instead of to the left. The mapping relationship between the s-plane and z^{-1}-plane is found by considering the mapping of the $j\omega$-axis from the s-plane to the z^{-1}-plane. Thus, with $z^{-1} = \exp(-sT)$, the imaginary axis is just $\exp(-j\omega T)$, the unit circle in the z^{-1}-plane. This relationship is shown in Figure 12-8. Hence,

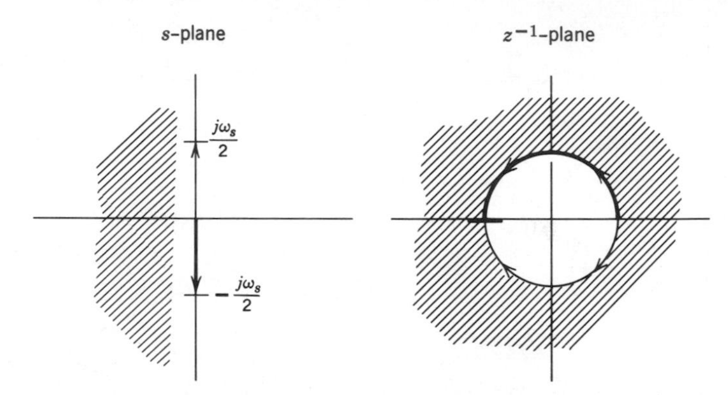

FIGURE 12-8 Mapping relation produced by standard z-transform between the continuous s-plane and the discrete z^{-1}-plane.

the frequency response of the digital filter is periodic with period equal $2\pi/T$. This also follows from (12.12). Closing the contour to the right encloses the poles of $1/[1 - \exp(-sT + \omega T)]$. These poles are located at

$$\exp(\pm j2\pi n) = \exp(-sT + \omega T) \qquad \text{or} \qquad \omega_n = s \pm j\frac{2\pi n}{T} \qquad \text{for}$$

$$n = 0, 1, \ldots.$$

Hence, the residue at each pole is just

$$\mathscr{R}_n = H(\omega_n) = H\left(s \pm j\frac{2\pi n}{T}\right).$$

Equation (12.13) becomes

$$H^*(s) = T \sum_{n=-\infty}^{\infty} H\left(s + j\frac{2\pi n}{T}\right). \tag{12.20}$$

Thus, the frequency response of the sampled-filter function is equal to the frequency response of the continuous function plus contributions of the response displaced by multiples of $2\pi/T$. The addition or folding in of these extra terms is known as aliasing of the frequency response [6]. If $H(s) \simeq 0$ for $|s| \geq j\pi/T$, then the difference between $H(s)$ and $H^*(s)$ in the region $-j\pi/T \leq s \leq j\pi/T$ is negligible, and the frequency responses of the digital filter and continuous filter are essentially equal. This result is depicted in Figure 12-9. (When discussing digital-filter frequency characteristics, the frequency band of primary interest is $-\pi/T \le \omega \le \pi/T$—the fact that the frequency response is periodic is assumed.)

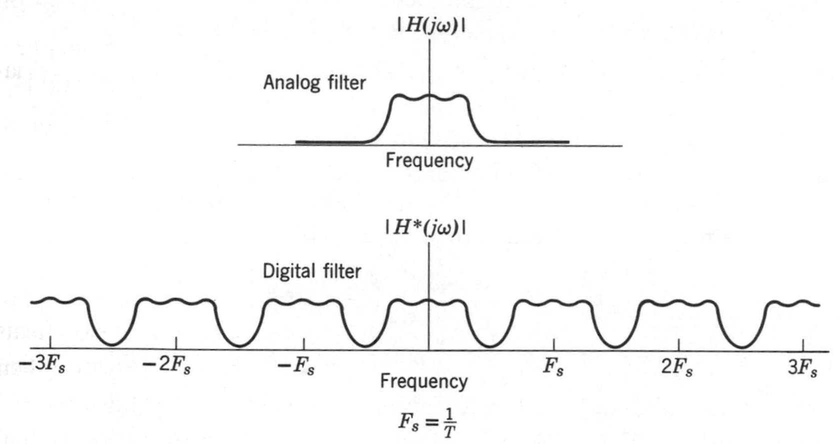

FIGURE 12-9 Comparison of the frequency responses of a continuous bandlimited lowpass filter and a digital filter found by means of the standard *z*-transform.

Guard Band Filter

A technique that is sometimes useful for increasing the attenuation of $H(j\omega)$ for $\omega \ge \pi/T$ consists of multiplying the filter function $H(s)$ by a so-called [7] guard band filter $G(s)$. The guard band filter is chosen so that the product, $G(j\omega)H(j\omega) \to 0$ for $\omega \ge \pi/T$. The disadvantage of this technique is that it yields digital filter designs of higher order than the original continuous design. It does, however, provide one means of realizing useful standard z-transform functions of filters which inherently are not bandlimited. The following example will illustrate the technique as applied to a bandstop design.

Example 12-2. The design of a digital bandstop filter by means of the standard z-transform cannot be accomplished without bandlimiting the bandstop function to be transformed. A digital 12th-order Butterworth bandstop design with stopband between 400 and 600 Hz is desired for operation at a sampling rate of 2 kHz. The continuous filter was designed from a 6th-order lowpass prototype by using the lowpass-to-bandstop transformation

$$s \to \frac{s(\omega_u - \omega_l)}{s^2 + \omega_u \omega_l}$$

where $\omega_l = 2\pi \times 400$ and $\omega_u = 2\pi \times 600$. The 6th-order lowpass prototype has complex poles at

$$s_1 = -\alpha_1 \pm j\beta_1 = -0.9659 \pm j0.2588$$
$$s_2 = -\alpha_2 \pm j\beta_2 = -0.7071 \pm j0.7071$$
$$s_3 = -\alpha_3 \pm j\beta_3 = -0.2588 \pm j0.9659.$$

The order of the guard band filter is determined by two considerations. The first is the highest frequency that the bandstop filter must pass. For this example, let it be 800 Hz. The second consideration is the attenuation required to prevent folding into the periodic response at 2 kHz–600 Hz. Requiring an attenuation of about 40 dB at this frequency allows determination of the order, N, of a Butterworth guard band filter from

$$N = \frac{40}{20 \log\left(\frac{1400}{800}\right)} = 8.25.$$

Hence, a first-trial design requires combining the 12th-order bandstop function with an 8th-order lowpass function. The standard z-transform of this combination yields a 20th-order digital filter! The filter coefficients are given in Table 12-1.

Figure 12-10(a) shows the magnitude frequency response of the original continuous filter. The response of the 20th-order digital approximation is shown in Figure 12-10(b). To achieve greater rejection in the stopband of the

TABLE 12-1 z-Transform Coefficients for a Digital Filter Approximation of a 12th-Order Continuous Bandstop Filter Combined with an 8th-Order Guard Band Filter.*

Section	A_0	A_1	B_1	B_2
1	3.0574	0.1263	−0.2109	0.5631
2	3.4686	−1.1815	0.0314	0.5275
3	0.6434	0.7462	−0.4148	0.6837
4	1.1387	−1.2040	0.2941	0.6015
5	−0.1158	0.2201	−0.5671	0.8771
6	−0.0329	−0.3526	0.5352	0.8236
7**	39.6822	−0.8972	−0.1500	0.0072
8	−42.3199	3.2615	−0.0430	0.0153
9	−7.9140	−5.1976	0.2455	0.0613
10	2.3924	0.3037	0.9550	0.3751

* Bandstop 400–600 Hz. Guard band cutoff at 800 Hz. Sampling frequency is 2 kHz.
** Terms 7 through 10 are due to the guard band filter.

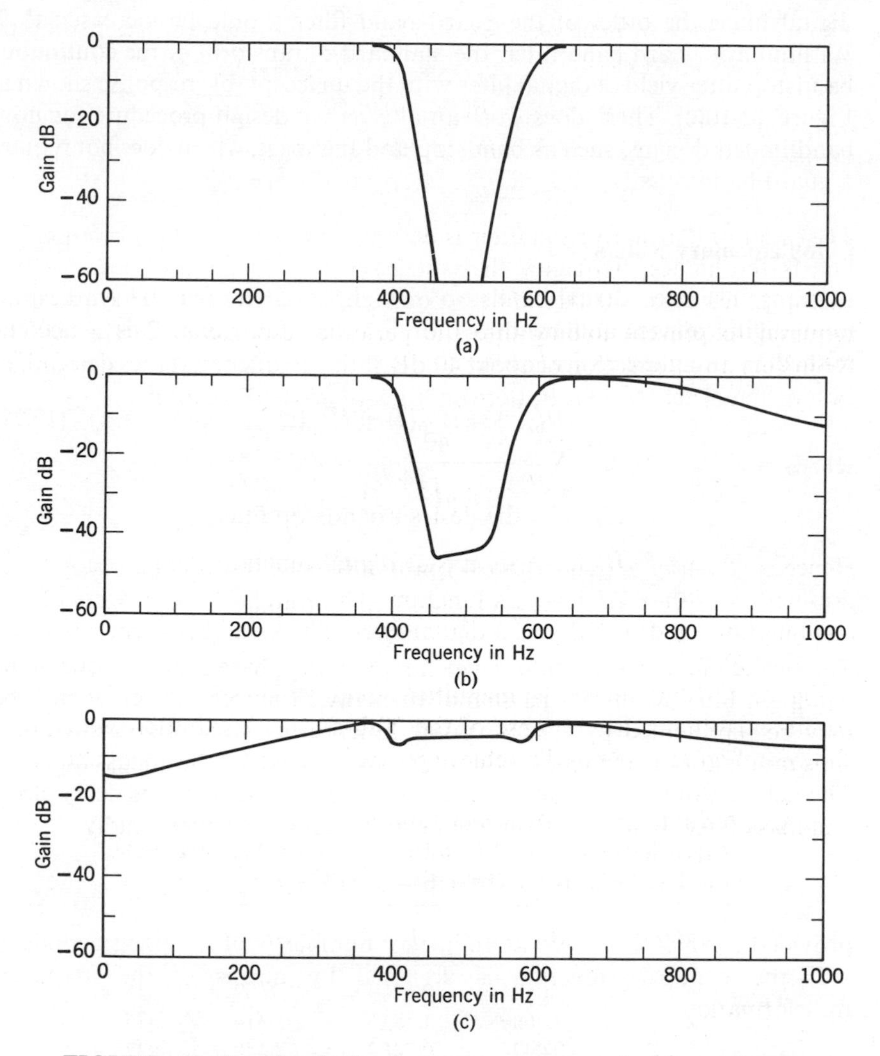

FIGURE 12-10 Magnitude frequency response of: (a) continuous 12th-order Butterworth bandstop filter with cut-off frequencies of 400 and 600 Hz, (b) standard *z*-transform of 12th-order bandstop combined with 8th-order lowpass guard filter with cut-off at 800 Hz, and (c) standard *z*-transform of bandstop filter alone.

digital filter, the order of the guard band filter should be increased to 9. Without any guard band filter, the standard z-transform of the continuous bandstop filter yields a digital filter with the unacceptable response shown in Figure 12-10(c). There does exist an alternative design procedure for non-bandlimited designs, such as bandstop and highpass, which does not require a guard band filter.

Complementary Filters

In many instances, digital bandstop or highpass filters may be synthesized from digital allpass and bandpass, or lowpass, functions. This is accomplished by using the relationship

$$H_{\text{bs}}(z) = H_{\text{ap}}(z) \pm H_{\text{bp}}(z) \tag{12.21}$$

where

$$H_{\text{bs}}(z) = \text{the desired bandstop filter}$$

$$H_{\text{ap}}(z) = \text{an allpass digital function}$$

$$H_{\text{bp}}(z) = \text{a bandpass digital filter.}$$

The choice of allpass function is dependent on the phase characteristic of the bandpass filter within its passband. In many instances, this characteristic may be represented by a linear phase function as a first approximation.* This means that the function $\exp(-s\tau)$ may be used for the allpass function. The delay, τ, of the function is chosen so as to approximate the delay of the bandpass filter. In digital form, the delay function becomes simply

$$\exp(-s\tau) = \exp(-sNT) = z^{-N},$$

provided τ can be expressed as an integer number N of sampling periods T.

If the bandpass function is designed by means of the frequency transformation

$$s \to \frac{s^2 + \omega_u \omega_l}{s(\omega_u - \omega_l)} = \frac{s^2 + (2\pi)^2 f_u f_l}{s 2\pi (f_u - f_l)},$$

then the filter will have zero phase at the geometric center frequency $f_0 = \sqrt{f_u f_l}$. Thus, in order to be able to combine the delay function and bandpass to give a bandstop filter, the phase of the delay function should be a multiple of π at f_0. (Odd multiples would require use of the plus sign in

* The design of allpass functions requires a comprehensive treatment by itself which cannot be considered here.

(12.21), while even multiples require the use of the minus sign.) Applying this condition to the phase of the delay function z^{-N} at f_0 gives

$$2\pi f_0 NT = M\pi$$

or

$$\frac{f_s}{f_0} = \frac{2N}{M}.$$

Hence, the ratio of sampling frequency to bandcenter frequency should be in the ratio of two integers. The following example will illustrate the above concepts.

Example 12-3. A bandstop filter similar to the one specified in Example 12-2 is desired using the complementary-filter technique. The major requirements are that the filter have maximum attenuation at 500 Hz and have a nominal stop bandwidth of 200 Hz. These requirements determine the following relationship between the upper and lower cut-off frequencies.

$$f_u f_l = (500)^2 \qquad f_u - f_l = 200$$

so that

$$f_l = 409.9 \qquad \text{and} \qquad f_u = 609.9.$$

The delay [8] of a 200 Hz bandwidth filter at bandcenter is approximately 6.14 msec. (Sometimes it is necessary to adjust the bandwidth of the filter so that $f_s\tau$ at bandcenter is a particular value. This adjustment is readily accomplished by noting that τ is inversely proportional to the bandwidth of the filter. Once the bandwidth is set, the lower and upper cut-off frequencies, f_l and f_u respectively, can be recomputed. For this example, the above frequencies will be used as originally computed.) The required value of N is

$$N = 6.14 \times 10^{-3} \times 2000 = 12.28 \simeq 12.$$

Checking that M is an integer gives

$$M = \frac{f_0}{f_s} \times 2N = \frac{500}{2000} \times 2 \times 12 = 6,$$

and, therefore, the negative sign in (12.21) should be used. The design then requires two steps:

1. Transform the 6th-order normalized lowpass function (See Example 12-2) into a bandpass function using the frequency band transformation and cut-off frequencies given above, and
2. Find the standard z-transform of the continuous bandpass function.

TABLE 12-2 ***z*-Transform Coefficients for a Digital Filter Approximation to a Continuous 12th-Order Bandpass Filter.***

Section	A_0	A_1	B_1	B_2
1	0.5022	−1.1743	−0.1627	0.5627
2	1.1559	1.0262	0.0784	0.5279
3	−0.8193	0.2558	−0.3636	0.6829
4	−1.0953	−0.0872	0.3424	0.6022
5	0.1147	0.1270	−0.5110	0.8766
6	0.1418	−0.1907	0.5892	0.8241

* Cutoff frequencies are 409.9 and 609.9 Hz.

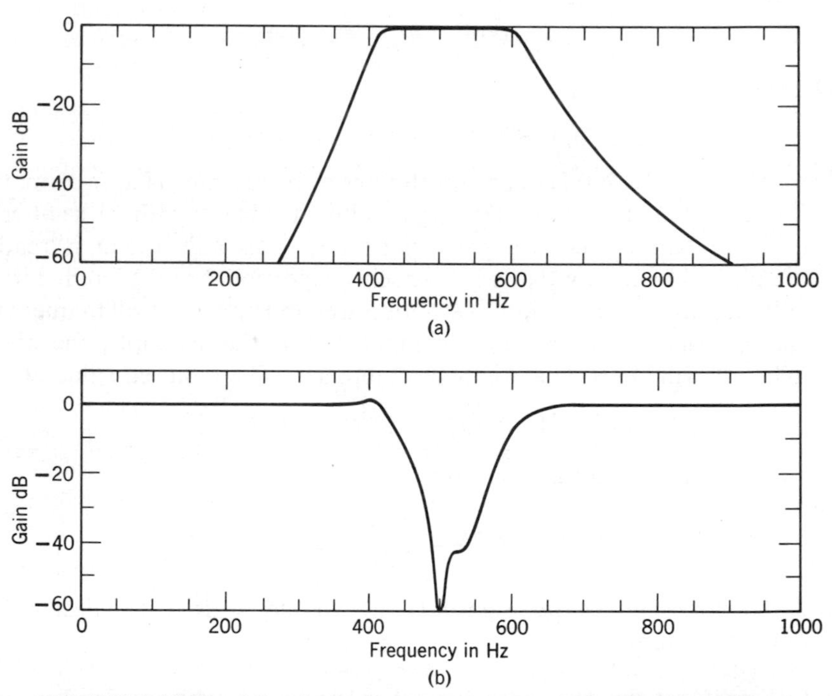

FIGURE 12-11 Magnitude frequency responses of a digital bandpass and complementary bandstop filter: (a) 12th-order Butterworth bandpass with cut-off frequencies 409.9 and 609.9 Hz, (b) complementary bandstop formed by subtracting (a) from z^{-12}.

The result of these two steps is a digital bandpass filter which is to be combined with a delay of 12 units. The coefficients for the digital filter are given in Table 12-2. Figures 12-11(a) and (b) show the magnitude frequency responses of the digital bandpass and complementary bandstop filter. It is seen that the filter does indeed achieve maximum attenuation (minimum gain) at 500 Hz. In Section 12.6, it will be seen how the use of a complementary filter yields exceptionally good results.

Realistic Considerations

Practically, the standard z-transform is best suited for finding digital representations of lowpass and bandpass filters. It is particularly applicable to continuous designs derived from all-pole lowpass prototypes such as Butterworth, Bessel, transitional, and Chebyshev Type I designs. For bandstop and highpass filters, and without resorting to either guard band filters or complementary filters, digital filters with desired frequency characteristics may be designed by means of the bilinear z-transform or matched z-transform [9, 10].

12.2 THE BILINEAR z-TRANSFORM

The design specification of many filter functions requires the realization of a given response characteristic in the frequency domain, but it does not demand any particular impulse or step response characteristic. Specification of frequency response only permits the use of a different digitizing transformation—the bilinear z-transform. This transform is used to realize digital filters that have relatively constant-magnitude passband and stopband requirements.

The bilinear z-transform given by

$$s = \frac{2}{T}\tanh\left(\frac{s_1 T}{2}\right) = \frac{2}{T}\left(\frac{1 - \exp(-s_1 T)}{1 + \exp(-s_1 T)}\right) = \frac{2}{T}\left(\frac{1 - z^{-1}}{1 + z^{-1}}\right) \tag{12.22}$$

maps the entire complex s-plane into an infinite set of horizontal strips in the s_1-plane bounded by the lines

$$s_{1l} = \frac{j\pi(2n - 1)}{T} \quad \text{and} \quad s_{1u} = \frac{j\pi(2n + 1)}{T}.$$

This mapping transformation is illustrated in Figure 12-12. Hence, the bilinear z-transform may be looked upon as a bandlimiting transformation. This algebraic transformation uniquely maps the left half s-plane into the exterior of the unit circle in the z^{-1}-plane. The transformation may be applied to either the rational fraction form or the partial fraction expansion

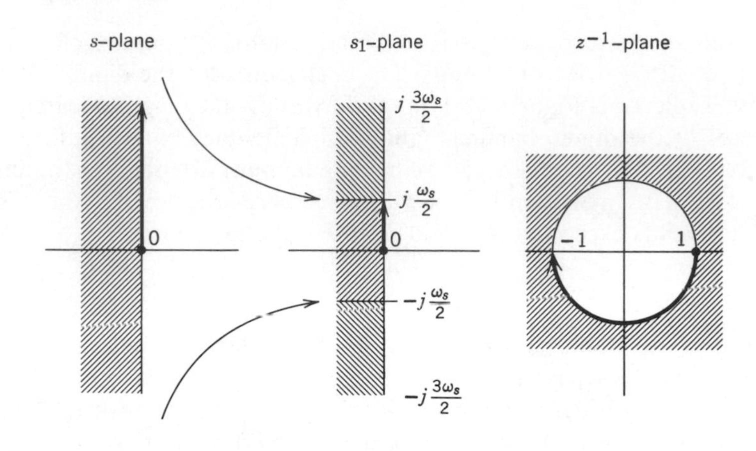

FIGURE 12-12 Mapping relation produced by bilinear z-transform between the continuous s-, s_1-plane and the discrete z^{-1}-plane.

of $H(s)$. When applied to the partial fraction form, the resulting terms for first-order and second-order poles are respectively,

$$\frac{R_r}{s+\alpha} \Leftrightarrow \frac{(1+z^{-1})A_0}{1+B_1 z^{-1}} \tag{12.23}$$

where

$$A_0 = \frac{R_r T}{2(1+\alpha T/2)} \qquad B_1 = -\frac{(1-\alpha T/2)}{(1+\alpha T/2)}$$

and

$$\frac{R_r + jR_i}{s+\alpha-j\beta} + \frac{R_r - jR_i}{s+\alpha+j\beta} \Leftrightarrow \frac{(1+z^{-1})(A_0 + A_1 z^{-1})}{1+B_1 z^{-1} + B_2 z^{-2}} \tag{12.24}$$

where

$$A_0 = \frac{T[R_r(1+\alpha T/2) - R_i(\beta T/2)]}{d} \qquad B_1 = \frac{-2[1-(\alpha T/2)^2 - (\beta T/2)^2]}{d}$$

$$A_1 = \frac{-T[R_r(1-\alpha T/2) + R_i(\beta T/2)]}{d} \qquad B_2 = \frac{[(1-\alpha T/2)^2 + (\beta T/2)^2]}{d}$$

$$d = (1+\alpha T/2)^2 + (\beta T/2)^2.$$

Because the bilinear z-form is algebraic in character, it may be applied directly to the individual terms in the rational fraction representation of a transfer function. Thus real factors transform according to

$$s + \alpha \Leftrightarrow C_0 \frac{(1 + B_1 z^{-1})}{(1 + z^{-1})} \tag{12.25}$$

where

$$C_0 = (2/T)(1 + \alpha T/2) \qquad B_1 = -\left(\frac{1 - \alpha T/2}{1 + \alpha T/2}\right)$$

while complex factors become

$$(s + \alpha)^2 + \beta^2 \Leftrightarrow C_1 \frac{(1 + B_1 z^{-1} + B_2 z^{-2})}{(1 + z^{-1})^2} \tag{12.26}$$

where

$$C_1 = (2/T)^2\, d$$

and B_1, B_2, and d are as defined for (12.24).

Using the above relations derived from the rational fraction and partial fraction expansion respectively yields either the series form,

$$\mathscr{H}(z^{-1}) = \frac{\prod_{m=1}^{M} (A_{0m} + A_{1m} z^{-1} + A_{2m} z^{-2})}{\prod_{n=1}^{N} (1 + B_{1n} z^{-1} + B_{2n} z^{-2})}, \tag{12.27}$$

or the parallel form,

$$\mathscr{H}(z^{-1}) = (1 + z^{-1}) \sum_{n=1}^{N} \frac{A_{0n} + A_{1n} z^{-1}}{1 + B_{1n} z^{-1} + B_{2n} z^{-2}}. \tag{12.28}$$

The realization of the parallel form given by (12.28) is essentially the same as shown in Figures 12-4 and 12-5 with the addition of the term $(1 + z^{-1})$ in series with the input or output line shown in Figure 12-5. Implementation of the terms given by (12.27) is shown in Figure 12-13(a). Numerator and denominator terms are paired so as to reduce total storage or memory requirements. Cascading the required number of sections results in a series realization as shown in Figure 12-13(b).

The bilinear z-transform is particularly suited for filters which exhibit relatively flat magnitude characteristics in contiguous passbands and stopbands. However, care must be used in designing filters with critical (cut-off) frequencies greater than 10% of the sampling frequency. This is due to the nonlinear warping of the frequency scale caused by the bilinear z-transform.

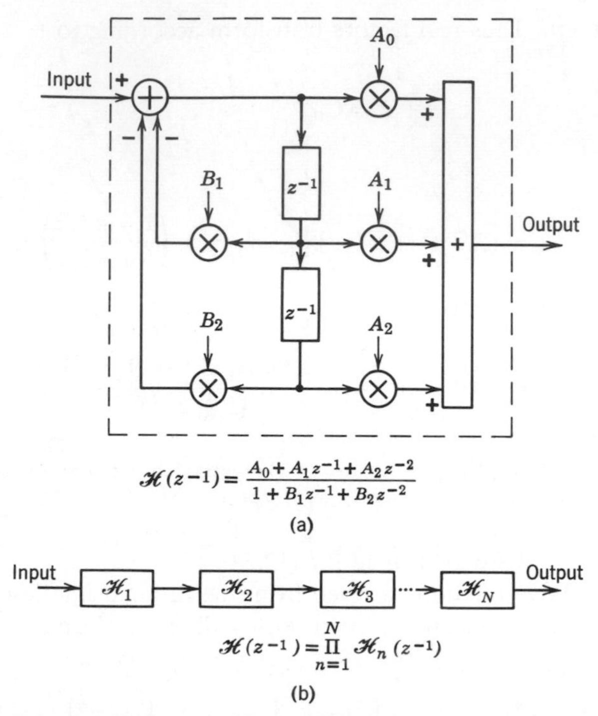

FIGURE 12-13 Block diagram of series realizations for a digital filter.

Frequencies in the z^{-1} plane are related to frequencies in the s-plane by

$$f_s T = \frac{1}{\pi} \tan \pi f_z T. \tag{12.29}$$

This relationship is shown sketched in Figure 12-14. Correction for frequency scale warping may be accomplished by prewarping the critical frequencies of the desired digital filter before applying one of the frequency band transformations. The resulting prewarped continuous filter function will transform to the desired digital function by means of the bilinear z-transform. The following example will demonstrate the use of prewarping when using the bilinear z-transform.

Example 12-4. The bandstop filter described in Example 12-2 can be realized by means of the bilinear z-transformation. However, the critical frequencies of the continuous bandstop filter to which the bilinear z-transformation is applied must be determined from the cut-off frequencies desired of the digital filter. Application of (12.29) gives

$$f_l = \frac{2000}{\pi} \tan(\pi 400/2000) = 462.53$$

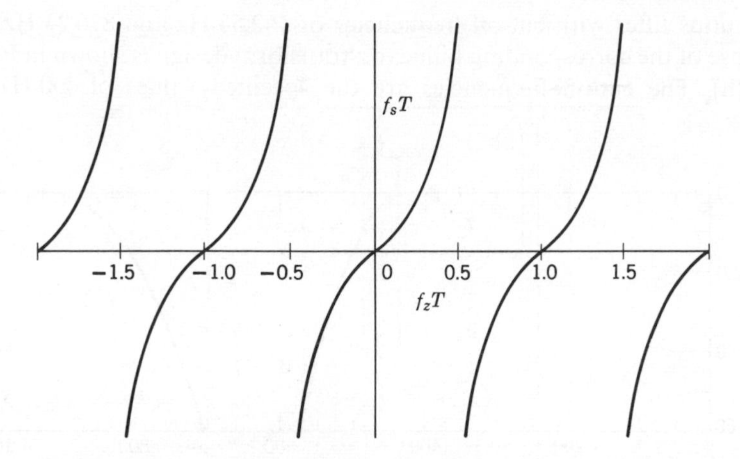

FIGURE 12-14 Frequency scale correspondence between s-plane and z^{-1}-plane produced by the bilinear z-form.

and

$$f_u = \frac{2000}{\pi} \tan(\pi 600/2000) = 876.23.$$

These frequencies are used in the stopband frequency transformation to determine the required continuous-filter transfer function. Applying the transformations given by the transform pairs of (12.23) and (12.24) yields the desired digital bandstop design in parallel form. This form, as given by (12.28), has been reduced to proper fraction form plus a constant (See Problem 12.6). The coefficients for the digital bandstop filter are given in Table 12-3. Figure 12-15(a) shows the magnitude response of the prewarped

TABLE 12-3 Bilinear z-Transform Coefficients for a Parallel Form Digital Bandstop Filter 400 Hz to 600 Hz at a Sampling Rate of 2 kHz.*

Section	A_0	A_1	B_1	B_2
1	−1.6837	−0.4971	−0.1345	0.5251
2	−1.6837	0.4971	0.1345	0.5251
3	−0.1034	−0.7434	−0.3768	0.6425
4	−0.1034	−0.7434	0.3768	0.6425
5	0.2042	−0.1730	−0.5580	0.8579
6	0.2042	0.1730	0.5580	0.8579

* Constant term in parallel with filter = 3.4553.

continuous filter with cut-off frequencies of 462.53 Hz and 876.23 Hz. The response of the corresponding bilinear z-transform design is shown in Figure 12-15(b). The cut-off frequencies are the specified values of 400 Hz and 600 Hz.

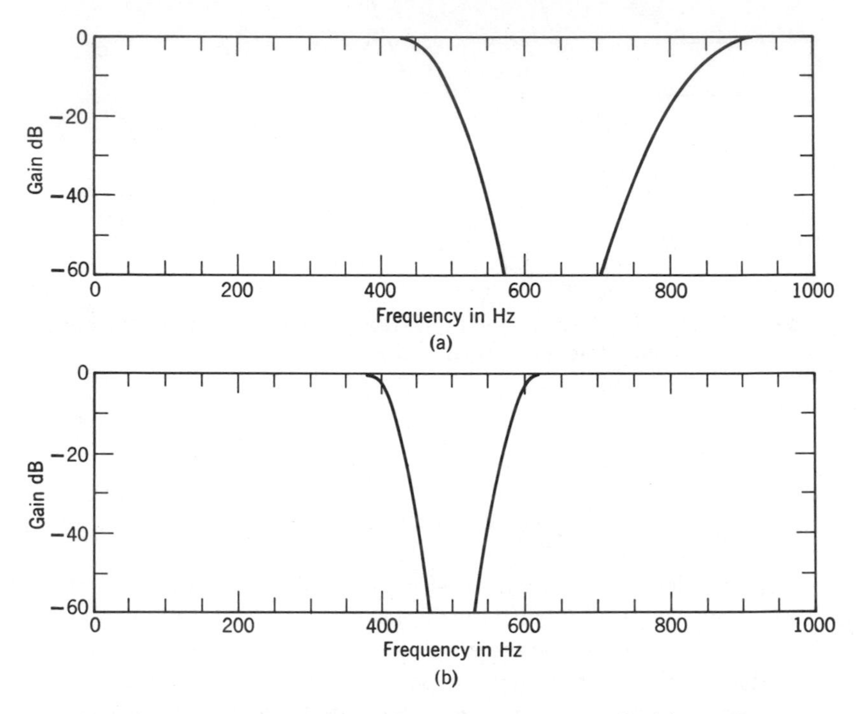

FIGURE 12-15 Magnitude frequency response of: (a) continuous bandstop filter with cut-off frequencies of 462.53 Hz and 876.23 Hz, and (b) digital bandstop filter with cut-off frequencies of 400 Hz and 600 Hz derived by applying bilinear z-transform to (a).

The frequency scale warping introduced by the bilinear z-transform can produce a digital filter with frequency response characteristics considerably different from the corresponding continuous filter desired. This is especially true for wide bandwidth designs. Furthermore, the time response characteristic of such digital filters is not impulse invariant under the bilinear z-transform as it is under the standard z-transform. (This can be disastrous in digital radar and sonar systems where waveform preservation is essential to the satisfactory operation of the entire system.) Fortunately, the matched z-transform is available as an alternative to the standard z- and bilinear z-transform for realizing satisfactory digital designs.

12.3 THE MATCHED z-TRANSFORM

Another technique for digitizing a continuous filter function consists of constructing a digital transfer function with poles and zeros corresponding or matched to the poles and zeros of the continuous filter. The transform relation is

$$s \to \exp(sT) = z. \tag{12.30}$$

Thus, a pole at $s = \alpha_n$ would map to a pole at $\exp(-\alpha_n T)$, thereby yielding a *match* between the continuous function poles and zeros and the digital function poles and zeros. The factor in the digital filter function corresponding to $s + \alpha_n$ is $z - \exp(-\alpha_n T)$, or in terms of z^{-1} is $1 - z^{-1}\exp(-\alpha_n T)$. Thus real poles and zeros transform according to

$$s + \alpha \leftrightarrow 1 - z^{-1}\exp(-\alpha T) = 1 + B_1 z^{-1} \tag{12.31}$$

while complex poles and zeros give,

$$(s+\alpha)^2 + \beta^2 \leftrightarrow 1 - 2z^{-1}\exp(-\alpha T)\cos\beta T + z^{-2}\exp(-2\alpha T) = 1 + B_1 z^1 + B_2 z^{-2}. \tag{12.32}$$

Applying the matched z-transform to a continuous function expressed in rational fraction form gives the digital filter function,

$$H(z^{-1}) = K\frac{\prod_{m=1}^{M}[1 - z^{-1}\exp(-x_m T)]}{\prod_{n=1}^{N}[1 - z^{-1}\exp(-\alpha_n T)]} = K\frac{\prod_{m=1}^{M}[1 + z^{-1}A_m]}{\prod_{n=1}^{N}[1 + z^{-1}B_n]}, \tag{12.33}$$

where K is chosen to adjust the passband insertion gain.

The poles of this function are the same as those derived from the standard z-transform whereas the zeros are usually different. The matched z-transform leads directly to a series (or cascade) realization. However, a parallel representation can be found by finding the partial fraction expansion of the above $H(z^{-1})$.

Because of the difference in zero locations between matched z- and standard z-derived functions, the matched z-transform may be used on nonband-limited functions to obtain satisfactory digital filter designs.* This will be illustrated in the following example.

Example 12-5. The matched z-transform will be used to design the same bandstop filter discussed in Examples 12-2, 12-3 and 12-4. Because the poles of the matched z-transform filter are the same as the poles determined for the standard z-transform, the first six B_1 and B_2 coefficients given in Table 12-1

* In the matched z-transform case, if the continuous filter had a zero on the $j\omega$- (imaginary) axis of the s-plane, this is transformed onto the unit circle in the z-plane. Hence, the transmission zeros are preserved. This is not the case with the standard z-transform.

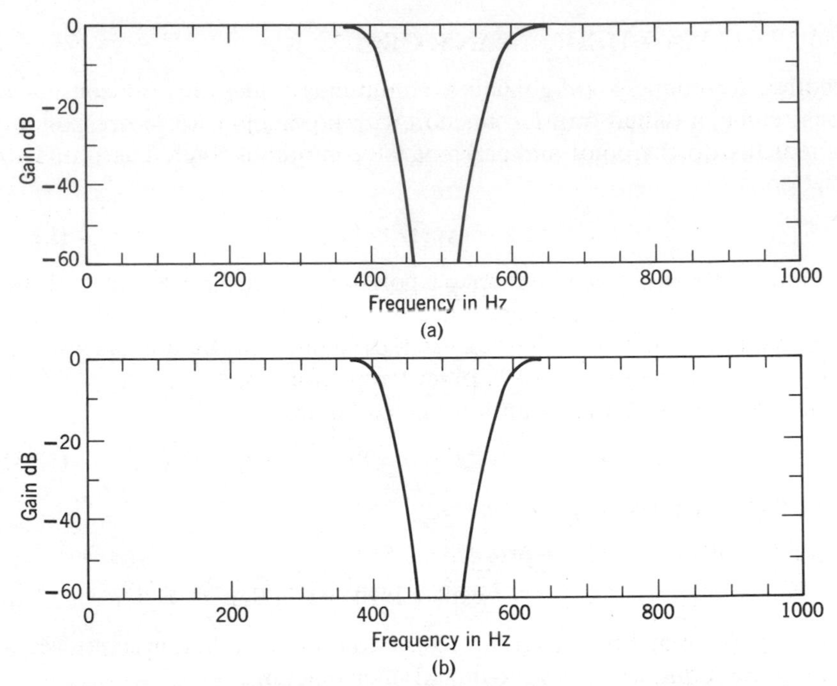

FIGURE 12-16 Magnitude frequency response of a 12th-order digital bandstop filter (a) designed by using the matched *z*-transform applied to a continuous filter with response shown in (b). Cut-off frequencies are 400 and 600 Hz with sampling frequency of 2 kHz.

are also the B_1 and B_2 coefficients for this design. The numerator of the continuous bandstop filter is of the form $(s^2 + \beta^2)^6$. According to (12.32), this term transforms to $(1 - 2z^{-1} \cos \beta T + z^{-2})^6$. For this filter β is 3.0780×10^3 so $-2 \cos \beta T$ is -0.06346. Thus, the numerators for each of the six cascaded digital sections consist of the polynomial $(z^{-2} - 0.06346z^{-1} + 1)$. Mechanization of the filter is shown in Figure 12-13. (The coefficients A_2 and A_0 are unity, and, therefore, the corresponding signal paths would not require multipliers.) The normalizing gain K for 0 dB insertion loss at bandcenter is 0.29706. Figure 12-16(a) shows the magnitude frequency response of the matched *z*-transform digital filter and Figure 12-16(b) shows the response of the original continuous bandstop filter for comparison. (This is the same as Figure 12-10(a).)

Need for Additional Zeros

For all-pole continuous functions, such as the Butterworth, Bessel, transitional, and Chebyshev I designs, it is sometimes necessary to introduce additional zeros at the origin or half-sampling frequency in order to achieve a satisfactory design.* This is accomplished by multiplying $H(z^{-1})$ by $(1 + z^{-1})^N$ with N equal to the order or degree of the zero desired at the half-sampling frequency $1/2T$.

For elementary designs such as the simple RC lowpass filter, the matched z yields the same results as the standard z-transform. The matched z is a useful alternate to the bilinear z-transform for obtaining satisfactory digital transfer functions for bandstop and highpass filters.

12.4 QUANTIZATION OF RECURSIVE DIGITAL FILTER COEFFICIENTS

In the preceding discussions, it has been assumed that the actual computation of the digital filter coefficients would be of sufficient accuracy to achieve the desired digital-filter response characteristics. Computation carried out with the aid of a digital computer will readily yield decimal accuracies of 10 to 16 digits. Generally, coefficient accuracy will not be an important consideration when the digital filter is to be simulated on a general purpose computer. (Accuracy might become a consideration when designing very narrow bandwidth filters at high sampling rates.) Coefficient accuracy is, however, a major consideration when implementing a digital filter in hardware. Practical hardware and economic considerations usually set the upper bound on the number of bits (word size) that may be used to represent the digital-filter coefficients. The result of this constraint is that the positions of the poles and zeros in the desired digital-filter function will be shifted due to coefficient quantization. Depending on filter bandwidth, sampling frequency used, and word size, the "quantized" coefficients may result in a filter that is unstable. Instability results when the poles of the filter move across the stability boundary of the unit circle.

The effect of quantization on the poles of $H(z^{-1})$ (and, therefore, on stability) may be investigated by examining the first- and second-order denominator terms, $1 + B_1 z^{-1}$ and $1 + B_1 z^{-1} + B_2 z^{-2}$, respectively, in the z-plane. For the first-order term, the real root in the z-plane is at

$$z_r = -B_1, \tag{12.34}$$

and for the second-order term, the root, which is usually complex, is

$$z_c = \frac{-B_1}{2} \pm j\sqrt{B_2 - \left(\frac{B_1}{2}\right)^2}. \tag{12.35}$$

Thus, stability is assured for first-order terms if $|B_1| < 1$ and for second-order terms if $|B_2|^{1/2} < 1$ or $B_2 < 1$. Quantization of (12.34) means choosing that quantized value of B_k, either $[B_k]$ or $[B_k] + \Delta$, *truncated* to N bits.

* Additional zeros are needed primarily for bandpass and bandstop designs that have complex poles near the real axis and unit circle. The effect is similar to the folding error introduced when a function is sampled at too low a frequency.

The brackets signify a truncated value, and Δ represents the minimum quantization increment of 2^{-N}. Quantization of (12.35) results in a choice of four possible pole locations for a complex pole depending on the choice for the quantized coefficients, B_1 and B_2. The four possible locations, shown in Figure 12-17, are

$$p_1 = \frac{-[B_1]}{2} + j\sqrt{[B_2] - \frac{[B_1]^2}{2}}$$

$$p_2 = -\left(\frac{[B_1] + \Delta}{2}\right) + j\sqrt{[B_2] - \left(\frac{[B_1] + \Delta}{2}\right)^2}$$

$$p_3 = \frac{-[B_1]}{2} + j\sqrt{[B_2] + \Delta - \left(\frac{[B_1]^2}{2}\right)}$$

$$p_4 = -\left(\frac{[B_1] + \Delta}{2} + j\sqrt{[B_2] + \Delta - \left(\frac{[B_1] + \Delta}{2}\right)^2}\right). \qquad (12.36)$$

In most cases, the quantized pole position chosen would be that value of p_k which is closest to the actual pole. However, situations may arise where a quantized approximation can be improved by choosing one of the other pole

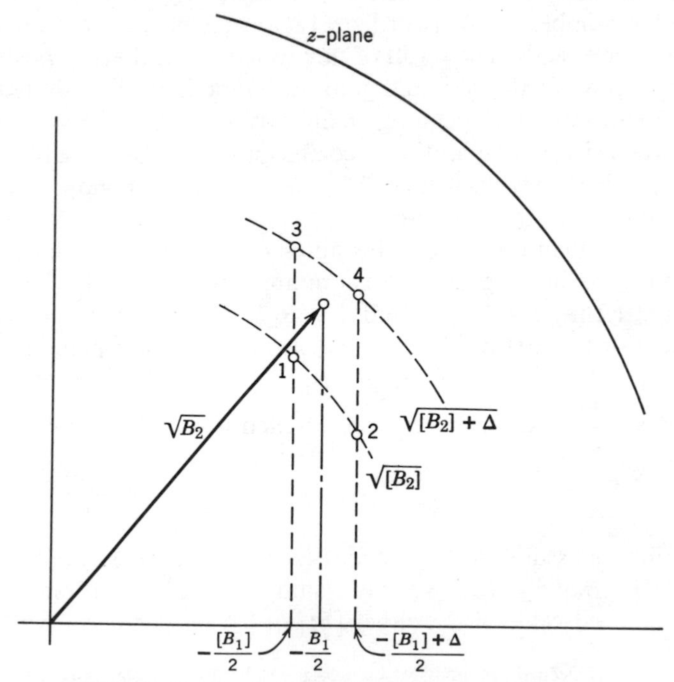

FIGURE 12-17 Quantized root locations in z-plane.

positions. For example, quantization may yield a pole on the unit circle as being closest to the actual pole. This condition arises from continuous-function poles that have small real parts with respect to the sampling period used in the digitizing transformation. Namely, the product αT is very much less than one. The relation between the sampling rate, number of bits used to mechanize a digital filter, and pole position α, is considered next.

Interrelations Between Pole Position, Sampling Rate, and Number of Bits

The transfer function considered is of the form

$$H(s) = \frac{\alpha}{s + \alpha}. \tag{12.37}$$

The standard z-transform of this function yields

$$H(z^{-1}) = \frac{\alpha T}{1 - \exp(-\alpha T) z^{-1}}, \tag{12.38}$$

while the bilinear z-transform gives

$$H(z^{-1}) = \frac{1 + z^{-1}}{1 - z^{-1}\left(\dfrac{1 - \alpha T/2}{1 + \alpha T/2}\right)} \left(\frac{\alpha T/2}{1 + \alpha T/2}\right). \tag{12.39}$$

Both functions are of the form

$$H(z^{-1}) = \frac{A_0 + A_1 z^{-1}}{1 + B_1 z^{-1}}. \tag{12.40}$$

When $\alpha T \ll 1$, B_1 for either z-form representation becomes

$$B_1 \approx -(1 - \alpha T). \tag{12.41}$$

If only N-bits are available for representing B_1 (excluding the sign bit), then the smallest αT that may be accurately represented is just 2^{-N}, the smallest quantization interval. Thus the largest value of B_1 less than one that may be represented with N bits is

$$B_1 = -(1 - 2^{-N}). \tag{12.42}$$

Knowing α and T determines the minimum number of bits required to represent B_1. This is

$$N = -\frac{\log_{10}(1 + B_1)}{\log_{10} 2} \simeq -\frac{\log_{10} \alpha T}{\log_{10} 2}. \tag{12.43}$$

Hence, specification of any pair of the parameters α, T, and N determines the value required for the third. (For second-order poles, B_2 determines the stability, and, therefore, B_2 should be substituted for B_1 in (12.42).) Several examples may help to clarify the consequences of choosing different combinations of variables.

Example 12-6. If $|\alpha| = 1$ rad/sec and $N = 8$ bits, then the smallest T possible (highest sampling frequency) is $\simeq (2^{-N})/\alpha = 2^{-8} = (1/256)$ sec. However, if the system sampling interval is already fixed, then the number of bits required may have to be adjusted. If $T = 1/1024$ sec and $|\alpha| = 1$ rad/sec, then the minimum number of bits required (excluding the sign bit) is given by (12.43) as $(\log_{10} 1024)/(\log_{10} 2)$ which is just 10. Increasing the sampling rate to 4096 Hz requires that the number of bits be increased to 12. On the other hand, if only 8 bits are available and the system sampling rate is 2048 Hz, then the lowest pole frequency that can be realized is $\alpha = 8.00$ rad/sec. It is seen, therefore, that the number of bits used to represent the quantized coefficients of the filter absolutely limits the lowest pole frequency that can be represented at a given sampling rate. It is also important to note that for a fixed word length, N, increasing the sampling rate increases the frequency at which the lowest-frequency pole can be located.

The number of quantization bits actually used in a hardware implementation of a digital filter is usually dictated by economic considerations. However, the data rate or sampling interval is determined by the highest-frequency component present in the signal, or by the highest-frequency pole to be represented in the system function. Hence, the required sampling rate may dictate the need for a coefficient word size (i.e., number of bits) which is not economically practical or is not compatible with existing hardware. It is, therefore, desirable to implement lightly-damped poles (small α) at high data rates and with fewer bits than dictated by the preceding analysis. Such implementation can be accomplished by designing the function for a submultiple of the actual system sampling frequency. The result will be to reduce the number of bits needed to represent the filter coefficients in exchange for additional memory or storage for each filter section.

Lower Data Rates Using Functions with Periodic Zeros

Consider a transfer function with a lightly-damped pole such that the combination of pole position and number of bits used permits a maximum sampling rate of $f_1 = 1/T_1$. The standard z-transform, as given by (12.38) is

$$H(z_1^{-1}) = \frac{A}{1 + B_1 z_1^{-1}} \tag{12.44}$$

where

$$z_1^{-1} \equiv \exp(-sT_1)$$
$$A = \alpha T_1$$
$$B_1 = -\exp(-\alpha T_1).$$

However, suppose that the bandwidth of the incoming data requires a sampling frequency $f_2 = 1/T_2$, such that

$$Mf_1 = f_2 \qquad \text{or} \qquad MT_2 = T_1.$$

The problem then is to find a way of representing $H(z_1)$ in terms of z_2 where

$$z_1 = z_2^M \qquad \text{or} \qquad z_2^{-M} = z_1^{-1}$$

with $z_2^{-1} \equiv \exp(-sT_2)$.

Direct substitution of z_2^{-M} for z_1^{-1} is by itself unsatisfactory since the resulting function,

$$H(z_2) = \frac{A}{1 + B_1 z_2^{-M}} \tag{12.45}$$

exhibits a periodic frequency response with period equal to the low-frequency sampling rate, $1/T_1$.

This may be seen by observing that the function given in (12.45) has M poles corresponding to the M roots of $(-B_1)$. These are

$$z_2 = |B_1|^{1/M} \exp\left[j\pi\left(\frac{2n+1}{M}\right)\right], \qquad n = 0, 1, \ldots, M-1. \tag{12.46}$$

The root at $z_2 = -B_1^{1/M}$ is the value that would have been used in (12.44) with z_1 replaced by z_2. However, the number of bits available prevented accurate representation of $B_1^{1/M}$.

The unwanted poles in the z_2-plane may be effectively cancelled by a function that has zeros on the unit circle opposite the undesired poles. Locating the zeros on the unit circle avoids the problem associated with coefficient quantization because the root magnitudes are just unity. The transfer function of a unity gain zero-order hold function of duration T_1 is

$$H_0(z_2) = \frac{(1 - z_2^{-M})}{M(1 - z_2^{-1})} \tag{12.47}$$

and has the desired zeros at $\exp(j2\pi n/M)$, $n = 1, 2, \ldots, M-1$.

Combining this function with that given in (12.45) yields

$$H_2(z_2) = \frac{(1 - z_2^{-M})}{M(1 - z_2^{-1})} \frac{A}{1 + B_1 z_2^{-M}}, \tag{12.48}$$

which is the desired transfer function. The location of the poles and zeros of this function are shown in Figure 12-18.

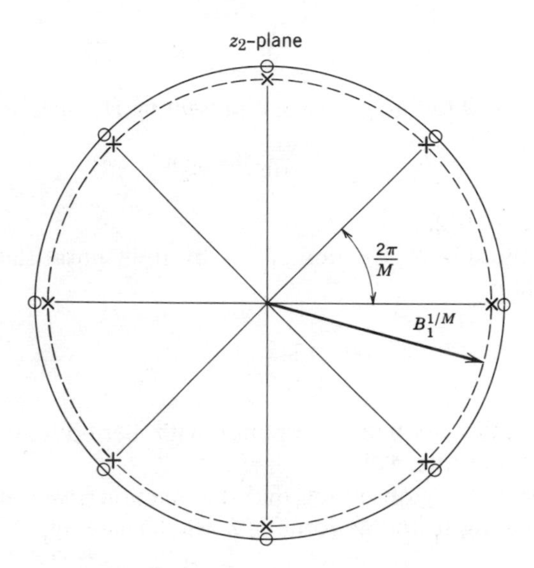

FIGURE 12-18 Pole position (×) for the transfer function $H(z_2) = A/(1 + B_1 z_2^{-M})$ and zero positions (○) for the function $H_0(z_2) = (1/M)(1 - z_2^{-M})/(1 - z_2^{-1})$.

Example 12-7. As an example, if $\alpha = 1$ rad/sec and $N = 8$ bits (excluding sign), the maximum sampling rate that may be used is 256 Hz. Figures 12-19 and 12-20 show, respectively, the frequency and time responses of the function realized with 48-bit coefficient accuracy and 8-bit accuracy (plus 1 bit for the sign). Within the accuracy of the plotter, these response curves are coincidental. If, however, the complete system must operate at 2560 Hz, either an additional 4 bits or the use of a hold with $m = 10$ would be required. Figure 12-21 shows the response of the function when realized by (12.45) alone, and Figures 12-22 and 12-23 show the responses when the hold function is introduced as per (12.48). It is seen that the resulting function achieves the desired characteristics.

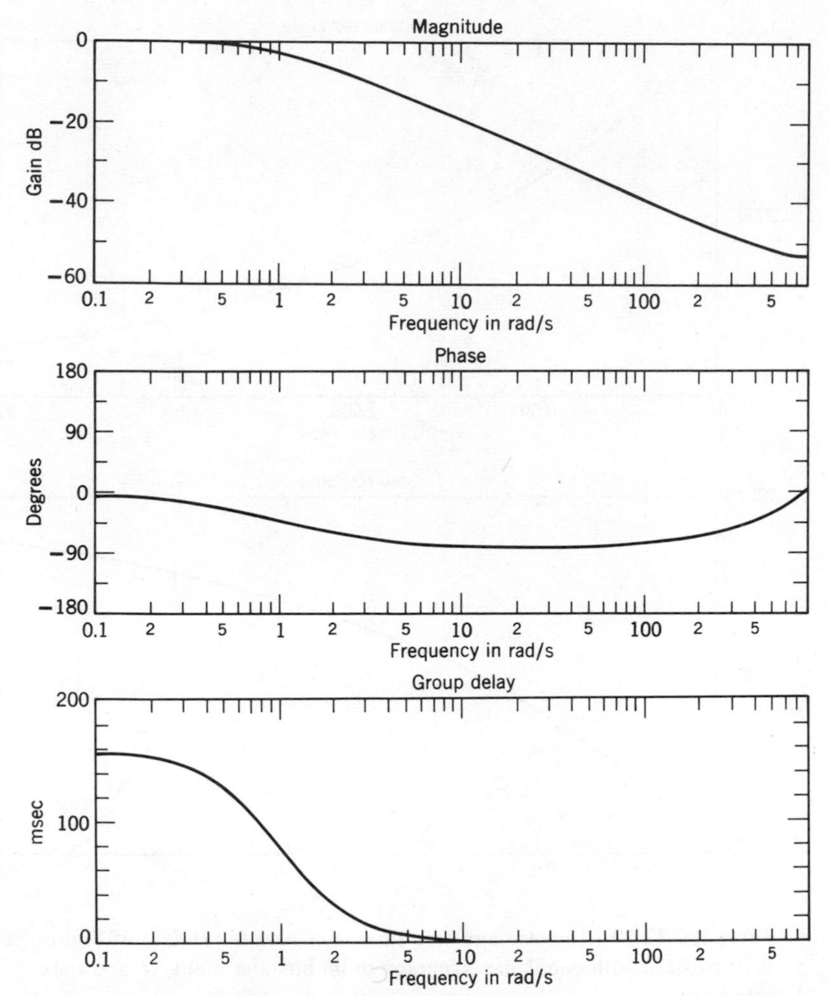

FIGURE 12-19 Frequency response of $H_1(z_1) = \alpha T_1/[1 - z_1^{-1}\exp(-\alpha T_1)]$, realized with coefficient accuracy of 48 bits and 8 bits. $\alpha = 1$, $T_1 = 1/256$. **(Curves are coincident.)**

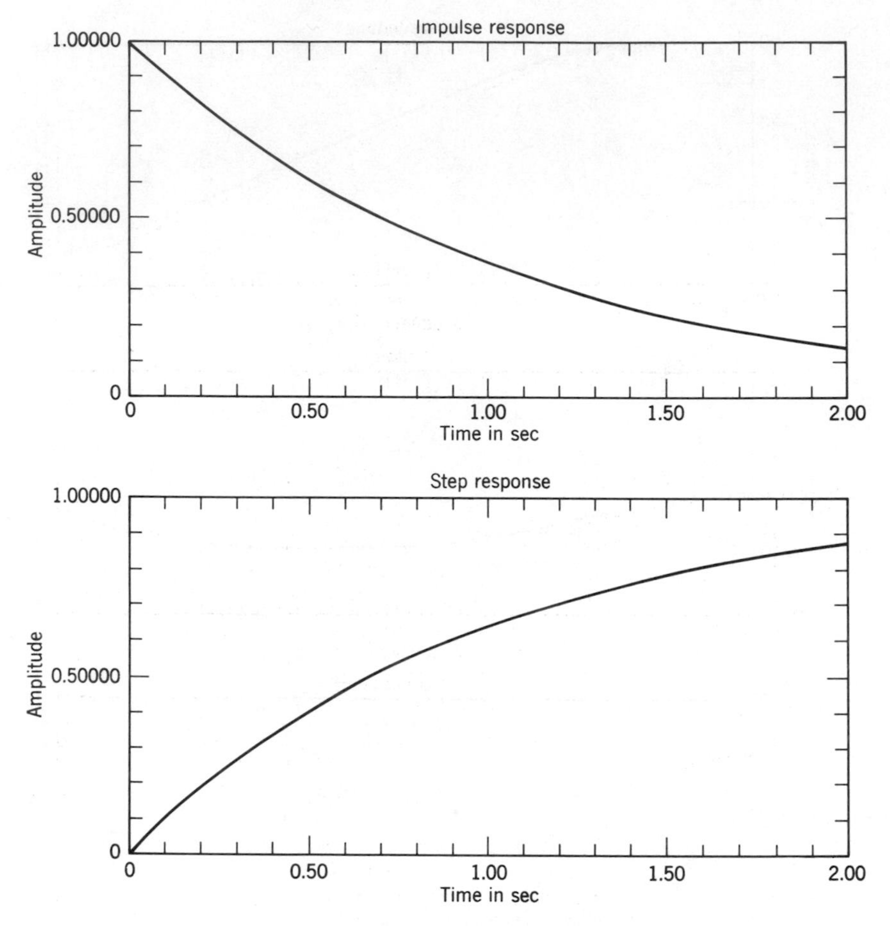

FIGURE 12-20 Impulse and step-time responses for $H_1(z_1)$ of Figure 5-19 realized with coefficient accuracy of 48 bits and 8 bits. (Curves are coincident.).

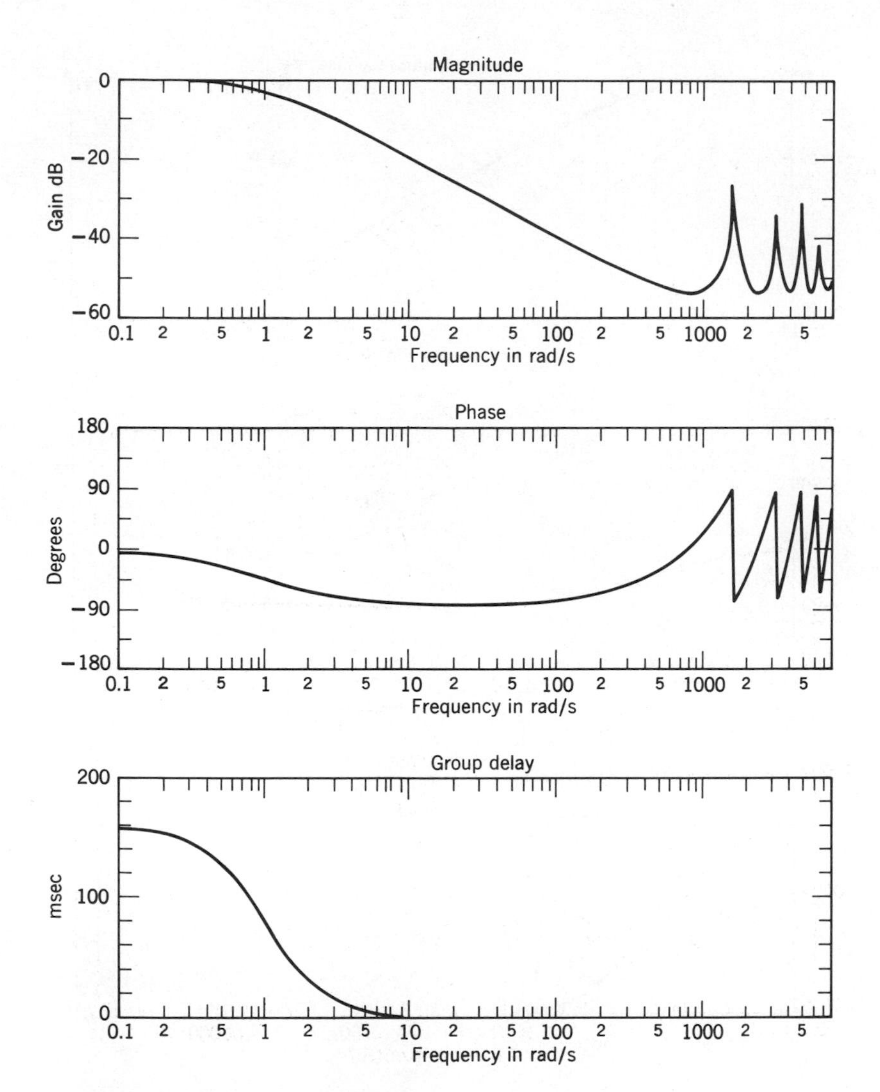

FIGURE 12-21 Frequency response of

$$H_1(z_2) = \alpha M T_2/[1 - z_2^{-M} \exp(-\alpha M T_2)]$$

with coefficient accuracy of 8 bits. $\alpha = 1$, $T_2 = 1/2560$, $M = 10$.

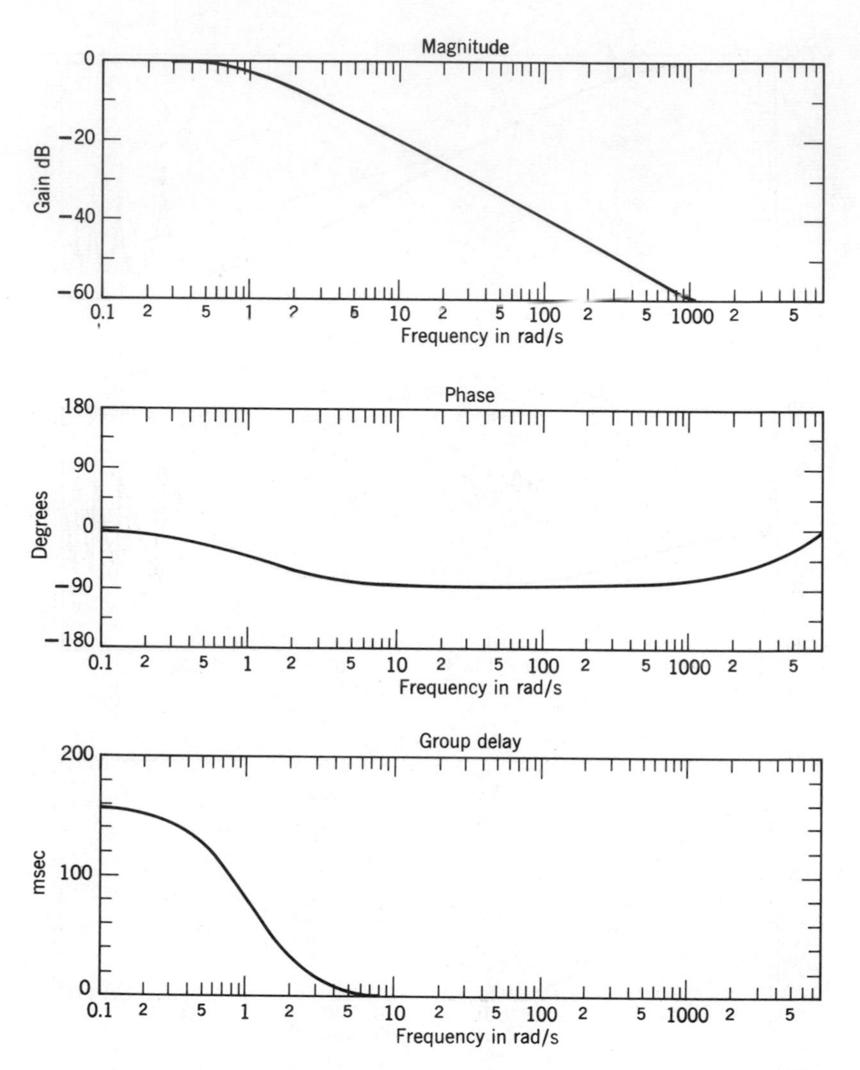

FIGURE 12-22 Frequency response of $H_{10}(z_2) = (1 - z_2^{-M})/(1 - z_2^{-1})$ $\alpha T_2/[1 - z_2^{-M} \exp(-\alpha M T_2)]$ realized with coefficient accuracy of 8 bits. $\alpha = 1$, $T_2 = 1/2560$, $M = 10$.

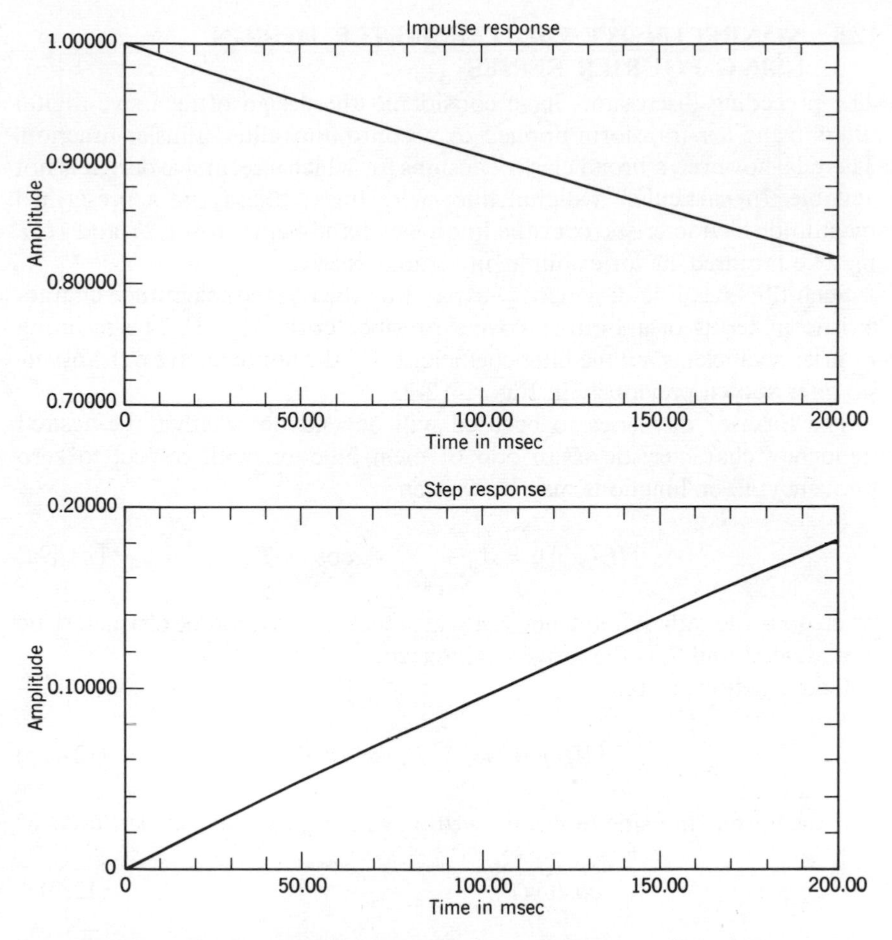

FIGURE 12-23 Impulse and step-time responses of $H_{10}(z_2)$ of Figure 12-22 realized with coefficient accuracy of 8 bits.

12.5 NONRECURSIVE DIGITAL FILTER DESIGN USING FOURIER SERIES

The preceding discussions have considered the design of recursive digital filters using a z-transform applied to a continuous-filter transfer function. There is, however, a broad class of designs for which a recursive design is not suitable. In particular, a digital filter with linear phase and a prescribed magnitude characteristic over the frequency band between $-1/2T$ and $1/2T$ may be required, as for example, in a radar receiver.

Such filters can be designed by expanding the desired magnitude characteristic in terms of a Fourier cosine or sine series [12, 13]. The resulting Fourier coefficients are the filter coefficients for the nonrecursive or transversal filter shown previously in Figure 12-2.

The form of the series to be used will depend on whether the desired frequency characteristic is an odd or even function with respect to zero frequency. Even functions may be written

$$|H_e(j\omega)| = A_0 + \sum_{n=1}^{\infty} A_n \cos \omega nT, \tag{12.49a}$$

where ω is the radian frequency $(2\pi f)$ at which the magnitude characteristic is evaluated and T is the sampling interval.

Odd functions result in

$$|H_o(j\omega)| = \sum_{n=1}^{\infty} B_n \sin \omega nT. \tag{12.49b}$$

If the cosine and sine terms are written in complex exponential terms as

$$\cos(\omega nT) = \frac{e^{j\omega nT} + e^{-j\omega nT}}{2} \tag{12.50a}$$

$$\sin(\omega nT) = \frac{e^{j\omega nT} - e^{-j\omega nT}}{2j} \tag{12.50b}$$

then, from the definition of z,

$$\cos(\omega nT) = \frac{z^n + z^{-n}}{2} \tag{12.50c}$$

$$\sin(\omega nT) = \frac{z^n - z^{-n}}{2j}. \tag{12.50d}$$

Therefore,

$$|H_e(j\omega)| = A_0 + \sum_{n=1}^{\infty} \frac{A_n}{2}(z^n + z^{-n}) \tag{12.51a}$$

$$|H_o(j\omega)| = \sum_{n=1}^{\infty} \frac{B_n}{2j}(z^n - z^{-n}). \tag{12.51b}$$

To obtain filters with real coefficients, the j is dropped from the summation. The resulting filter will have a phase shift of 90° from the theoretical function defined by the sine series, but, of course, the magnitude function will be unaffected.

The two representations given above are not physically realizable due to the presence of the z^n terms (unit advance or prediction operator) and the infinite sum. In practical cases, the sum may be truncated after N terms. Hence, the infinite sums reduce to

$$|H_e(j\omega)|_N = A_0 + \sum_{n=1}^{N} \frac{A_n}{2}(z^n + z^{-n}) \tag{12.52a}$$

$$|H_o(j\omega)|_N = \sum_{n=1}^{N} \frac{B_n}{2}(z^n - z^{-n}). \tag{12.52b}$$

It is now possible to factor out z^N from the above expressions so that

$$|H_e(j\omega)|_N = z^N\left[A_0 z^{-N} + \sum_{n=1}^{N} \frac{A_n}{2}(z^{n-N} + z^{-(n+N)})\right], \tag{12.53}$$

and similarly for $|H_o(j\omega)|$. The term z^N is a unity magnitude, linear phase term and does not affect the shape of the desired magnitude characteristic. The effect of discarding this term is to introduce a linear phase shift of $\exp(-jN\omega T)$, or, equivalently, a time delay of NT sec or N samples. Note that coefficients for the digital filter are symmetric about the center point of the filter, and that the total number of delay storage units is $2N$. This symmetry permits reduction of the actual number of multiplier coefficients required by a factor of two. The reduction is accomplished by folding the filter about its center. Figure 12-24 shows the folded mechanization of a cosine design. A similar configuration results for a sine series.

Use of a finite number of terms in any physical mechanization will introduce truncation errors in the frequency response in the form of frequency ripple. Furthermore, responses with sudden transitions between pass- and stopbands will exhibit a pronounced Gibbs phenomenon which may prove troublesome. In order to reduce the above undesired effects, additional weighting should be applied to the series coefficients. In filter terms, this amounts to lowpass *filtering* or smoothing *of the magnitude response.* Several functions are available for use as smoothing functions;

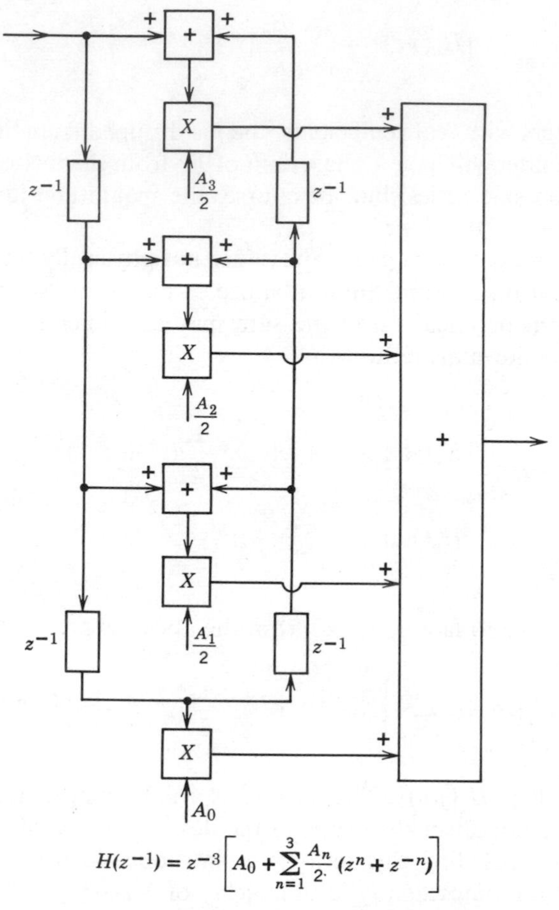

FIGURE 12-24 Block diagram of a nonrecursive filter representation of a finite Fourier cosine series of three terms plus constant.

however, a simple *and* effective one is the normalized Hamming Window which is

$$w_H(n) = 1.0 + (46/54)\cos\frac{2\pi n}{2N}, \qquad n = 0, 1, \ldots, N. \tag{12.54}$$

Therefore, the coefficient A_n should be weighted by $(1.0 + 46/54 \cos \pi n/N)$.

Example 12-8. As an example of the use of this technique, consider a cosine series approximation for an ideal lowpass filter. The function and its series approximation are

$$M(\omega) = U(\omega + \omega_u) - U(\omega - \omega_u) = A_0 + \sum_{n=1}^{N} A_n \cos\frac{2\pi n\omega}{\omega_s} \quad (12.55)$$

$$A_0 = \frac{2}{\omega_s}\int_0^{\omega_u} d\omega = \frac{2\omega_u}{\omega_s}$$

$$A_n = \frac{4}{\omega_s}\int_0^{\omega_u} \cos\frac{2\pi n\omega}{\omega_s}\, d\omega$$

$$= \frac{4}{\omega_s}\frac{\omega_s}{2\pi n}\sin\left(\frac{2\pi n\omega_u}{\omega_s}\right) = \frac{2}{\pi n}\sin\left(\frac{2\pi n\omega_u}{\omega_s}\right). \quad (12.56)$$

Figure 12-25 shows the magnitude response of this function for $n = 10$ and 40 without coefficient weighting. The effect on the response of weighting the coefficients given by (12.56) by $w_H(n)$ (as defined in (12.54)) is shown in Figure 12-26.

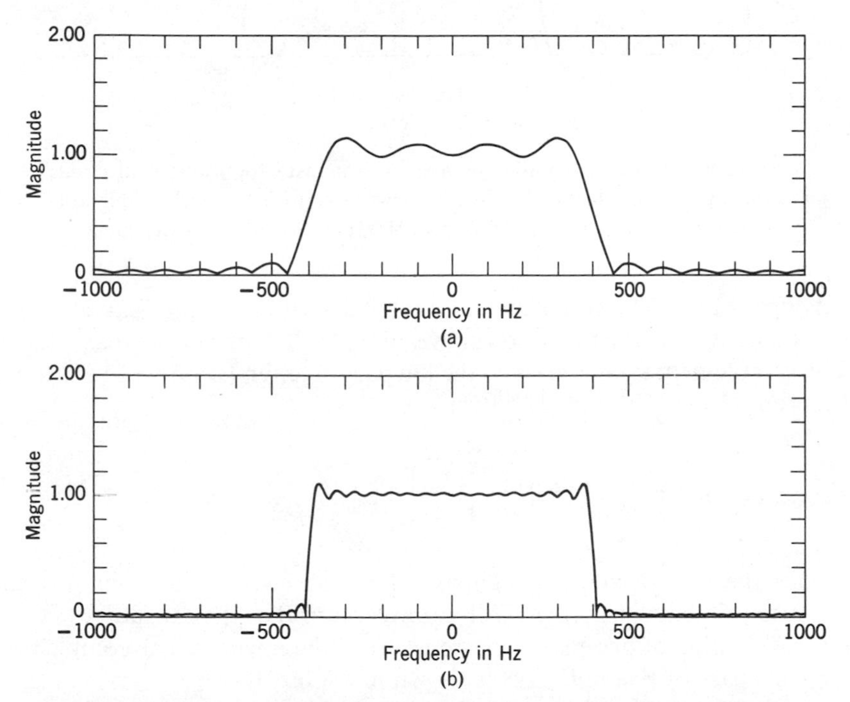

FIGURE 12-25 Magnitude-frequency responses for transversal cosine filter approximation to an ideal lowpass filter with cut-off frequency of 400 Hz and sampling frequency of 2 kHz: (a) 11 terms (21 taps), (b) 41 terms (81 taps).

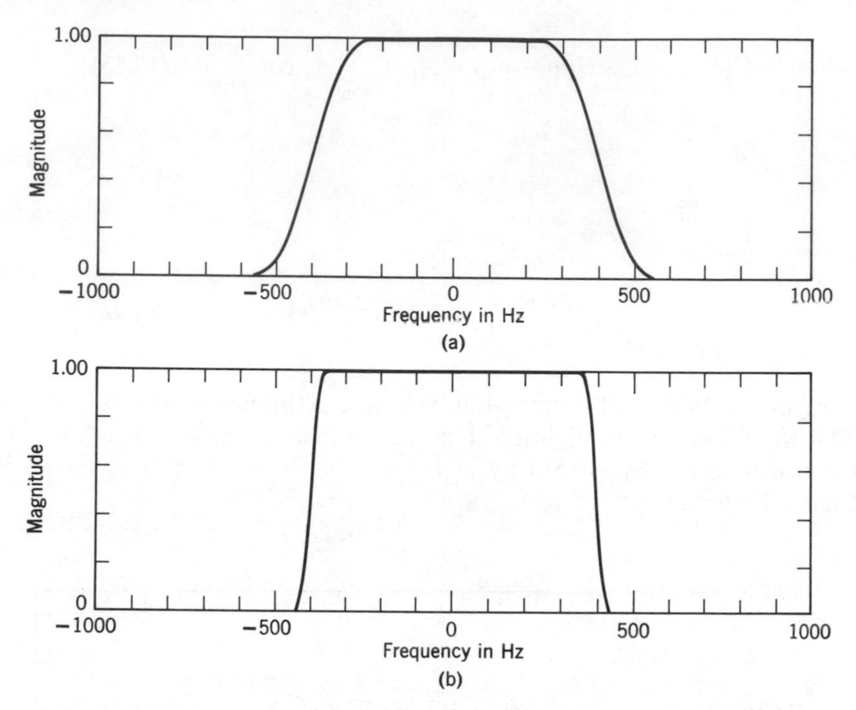

FIGURE 12-26 Magnitude-frequency responses for transversal cosine filter approximations to an ideal lowpass filter with Hamming weighting applied to coefficients: (a) 11 terms (21 taps), (b) 41 terms (81 taps).

Example 12-9. The complementary filter design technique is an ideal way of obtaining a highpass-nonrecursive digital filter from a lowpass design. The complementary highpass to the lowpass given in Example 12-8 is found according to (12.21) and (12.53) as

$$H_{\text{highpass}}(z^{-1}) = 1 - |H_e(j\omega)| = \left\{1 - \left[A_0 + \sum_{n=1}^{N} \frac{A_n}{2}(z^n + z^{-n})\right]\right\} z^{-N}. \quad (12.57)$$

Hence, the complementary highpass filter has coefficients which are the negative of the lowpass filter, with the exception that the constant or center tap has a value of one minus the constant of the lowpass. The complementary highpass of Example 12-8 is shown in Figure 12-27.

Synthesis of a digital filter with arbitrary magnitude response characteristic is possible by utilizing the discrete or fast Fourier transform (FFT) computational algorithms [14, 15]. Thus, the magnitude shape desired

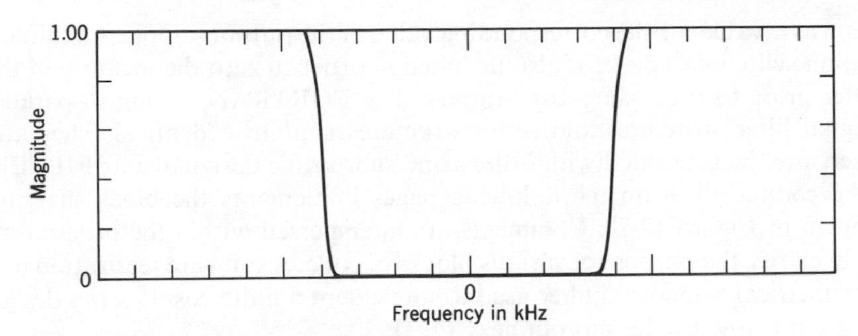

FIGURE 12-27 Magnitude response of complementary highpass filter derived from lowpass design shown in Figure 12-26b.

would be the input to an FFT program, and the output would consist of the filter tap weights. Window function weighting (such as the Hamming window given by (12.54)) could then be applied to provide a smooth response characteristic.

The Fourier series design technique just discussed can be extended to a general technique of filter function expansion in terms of orthogonal trigonometric functions and polynomials. It is only necessary to find a suitable series expansion in terms of $\exp(\pm j\omega T) = z^{\pm n}$ for the nonrecursive filter function desired.

12.6 DIGITAL COMPUTER SIMULATION OF RECURSIVE AND NONRECURSIVE DIGITAL FILTERS AND FREQUENCY RESPONSE EVALUATION

Hardware mechanization of the block diagrams given for the recursive and nonrecursive digital filters consists of implementing the multiplier and delay storage for a particular design with suitable digital circuitry. Hence, this subject is better treated in a course on digital and logical circuit design. However, there often arises the need for implementing a digital design, as part of computer software, to perform signal analysis and processing. Because computer programming is as essential a tool to the engineer and scientist as is a slide rule, this section will show how recursive and nonrecursive designs are programmed in the scientific programming language—FORTRAN. The filters are programmed in general form as subroutines* to which must be supplied arrays containing the coefficient values and delay storage as well as the current value of the input sample. The subroutine

* This is just one of many possible software implementations. Filter software also can be realized as FUNCTION subprograms or coded in some other programming language [3, 16].

returns a value for the corresponding value of the output sample. An initialization switch parameter is also included in order to zero the memory of the filter prior to processing any samples. The FORTRAN coding of parallel digital filter structures and series structures is almost identical. They are, therefore, incorporated within the same subroutine designated as FILTER. The coding given on the following pages implements the block diagram shown in Figure 12-28. Comments are interspersed within the program to help clarify the purpose of various blocks of code. A software realization of a symmetrical transversal filter used to implement a finite cosine series design is given below by the subroutine, CØFIL.

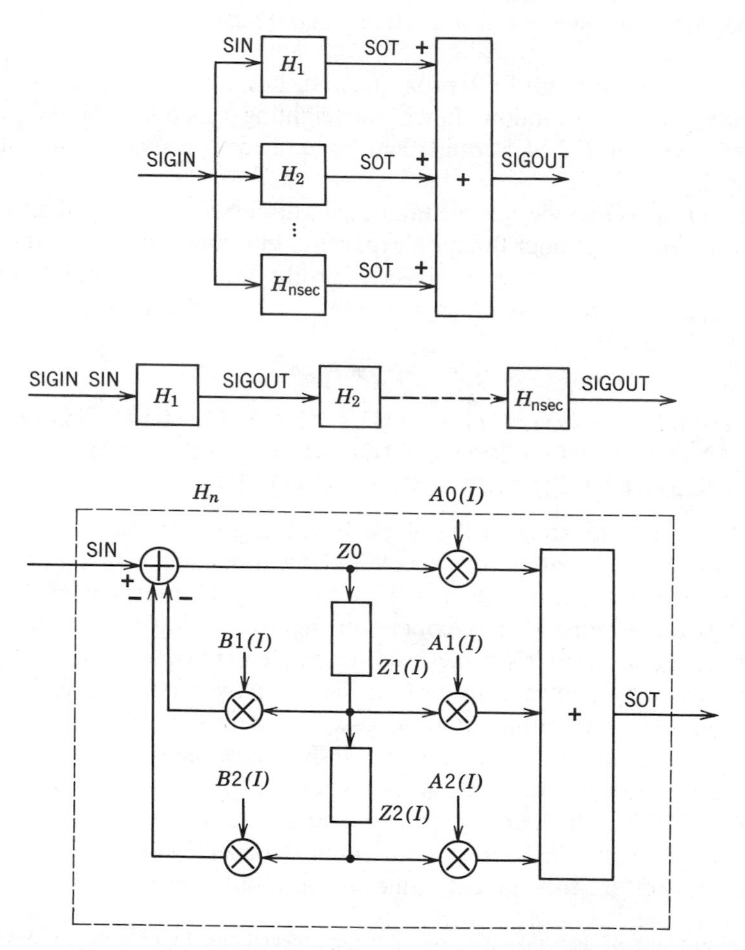

FIGURE 12-28 Block diagrams used for preparing software representation of parallel or series recursive digital filter designs.

The preceding two illustrations show how rapidly a digital filter design can be implemented, and, therefore, utilized for digital signal processing. The frequency response of either recursive or nonrecursive filters is found by substituting $\exp(-i\omega T)$ for z^{-1}. The general second-order function becomes,

$$\frac{A_2 z^{-2} + A_1 z^{-1} + A_0}{B_2 z^{-2} + B_1 z^{-1} + 1} \rightarrow \frac{A_2 e^{-j\omega T} + A_1 + A_0 e^{j\omega T}}{B_2 e^{-j\omega T} + B_1 + e^{j\omega T}}.$$

Making use of complex arithmetic, the complex response of either parallel or series forms is respectively,

$$H(W) = \sum_{n=1}^{N} \left(\frac{A_{2n} W^{-1} + A_{1n} + A_{0n} W}{B_{2n} W^{-1} + B_{1n} + W} \right), \tag{12.58}$$

where

$$W = \exp(j\omega T) = \cos \omega T + j \sin \omega T$$

or

$$H(W) = \prod_{m=1}^{M} \left(\frac{A_{2m} W^{-1} + A_{1m} + A_{0m} W}{B_{2m} W^{-1} + B_{1m} + W} \right). \tag{12.59}$$

Evaluation of the frequency response of either the symmetrical cosine filter or the antisymmetrical sine filter is simply an evaluation of (12.52)(a) and (12.52)(b) respectively. These are,

$$|H_e(j\omega)|_N = |W^{-N}| \left| \left[A_0 + \sum_{n=1}^{N} A_n \cos(\omega n T) \right] \right| \tag{12.60}$$

and

$$|H_o(j\omega)|_n = |jW^{-N}| \left| \sum_{n=1}^{N} B_n \sin(\omega n T) \right|. \tag{12.61}$$

For nonrecursive filters of the general form,

$$H(z) = \sum_{n=0}^{N} C_n z^{-n}, \tag{12.62}$$

the response is best evaluated by nesting the factors as,

$$H(W) = C_0 + W^{-1}(C_1 + W^{-1}(C_2 + W^{-1}(C_3 + \cdots + W^{-1} C_N))) \cdots). \tag{12.63}$$

```
 1:        SUBROUTINE FILTER(SIGIN,NSEC,A2,A1,A0,B2,B1,Z2,Z1,ISOP,ISW,SIGOUT)
 2: *
 3: *            THIS SUBROUTINE REALIZES RECURSIVE FILTER FUNCTIONS IN
 4: *            EITHER PARALLEL OR SERIES FORM.  THE ARGUMENTS OF THE
 5: *            SUBROUTINE ARE...
 6: *
 7: *                  SIGIN        THE CURRENT VALUE OF THE INPUT WAVEFORM
 8: *
 9: *                  NSEC         THE NUMBER OF FILTER SECTIONS
10: *
11: *                  A2,A1,A0,B2,B1
12: *                               ARRAYS OF LENGTH NSEC CONTAINING THE
13: *                               FILTER COEFFICIENT VALUES FOR EACH SECTION
14: *
15: *                  Z1,Z2        ARRAYS OF LENGTH NSEC WHICH REPRESENT THE
16: *                               DELAY STORAGE REQUIRED BY EACH SECTION
17: *
18: *                  ISOP         SERIES OR PARALLEL CONFIGURATION OPTION
19: *                               ISOP=1 DENOTES SERIES REALIZATION
20: *                               ISOP=2 DENOTES PARALLEL REALIZATION
21: *
22: *                  ISW          INITIALIZATION SWITCH.  IF ISW=0, FILTER
23: *                               DELAYS ARE SET EQUAL TO ZERO BEFORE
24: *                               PROCESSING BEGINS.  IF NOT= 0, PROCESSING
25: *                               CONTINUES USING LAST VALUES STORED IN Z2
26: *                               AND Z1
27: *
28: *                  SIGOUT       NEWLY COMPUTED VALUE FOR OUTPUT SIGNAL
29: *
30:        DIMENSION A2(1),A1(1),A0(1),B2(1),B1(1),Z2(1),Z1(1)
31: *
32:        IF (ISW.NE.0) GO TO 10
33: *            ZERO DELAY MEMORY
```

```
34:          DO 11 I=1,NSEC
35:          Z2(I) = 0.
36:          Z1(I) = 0.
37:   11     CONTINUE
38: *
39: *               PROCESS CURRENT INPUT
40:   10     CONTINUE
41:          SIN = SIGIN
42:          SIGOUT = 0.
43:          DO 12 I=1,NSEC
44:          Z0  = SIN - B1(I)*Z1(I) - B2(I)*Z2(I)
45:          SOT = A0(I)*Z0  + A1(I)*Z1(I) + A2(I)*Z2(I)
46:          Z2(I) = Z1(I)
47:          Z1(I) = Z0
48:          IF (ISOP.GT.1) GO TO 14
49: *
50: *               SERIES MODE
51:          SIN = SOT
52:          GO TO 12
53:   14     CONTINUE
54: *
55: *               PARALLEL MODE
56:          SIGOUT = SIGOUT + SOT
57:   12     CONTINUE
58: *
59: *               STORE OUTPUT IN SIGOUT IF SERIES MODE
60:          IF (ISOP.EQ.1) SIGOUT = SOT
61:          RETURN
62:          END
```

```
 1:        SUBROUTINE COFIL(SIGIN,NTRMS,A,Z,ISW,SIGOUT)
 2: *
 3: *            THIS SUBROUTINE PERFORMS THE FUNCTION OF A SYMMETRICAL
 4: *            TRANSVERSAL FILTER WITH TRANSFER FUNCTION
 5: *                  Z**(-N)(A(0) + SUM(.5*A(I)*(Z**(N) + Z**(-N))))
 6: *
 7: *            THE ARGUMENTS OF THE SUBROUTINE ARE...
 8: *
 9: *                  SIGIN THE CURRENT VALUE OF THE INPUT SIGNAL
10: *
11: *                  NTRMS THE NUMBER OF TERMS IN THE SUM EXCLUDING A(0)
12: *
13: *                  A     AN ARRAY OF (NTRMS + 1) CONTAINING THE
14: *                        COEFFICIENTS OR FILTER TAP WEIGHTS
15: *
16: *                  Z     AN ARRAY OF (2*NTRMS) WHICH PROVIDES THE STORAGE
17: *                        FOR THE DELAY LINE
18: *
19: *                  ISW   INITIALIZATION SWITCH FOR DELAY LINE.  IF ISW =0
20: *                        THE ARRAY Z IS SET = 0 PRIOR TO SIGNAL
21: *                        PROCESSING.  IF ISW NOT= 0, THE PROCESSING
22: *                        CONTINUES USING THE LAST VALUES STORED IN Z.
23: *                        GENERALLY ISW = 0 ONLY ON THE FIRST CALL TO THE
24: *                        SUBROUTINE.
25: *
26: *                  SIGOUT      NEWLY COMPUTED OUTPUT SIGNAL
27: *
28:        DIMENSION A(1),Z(1)
29: *
30:        NVAL = 2*NTRMS
31:        IF (ISW.NE.0) GO TO 10
32: *            ZERO OUT DELAY LINE
33:        DO 11 I=1,NVAL
```

```
34:          Z(I) = 0.
35:  11      CONTINUE
36: *
37: *              ENTER HERE FOR SIGNAL PROCESSING
38:  10      CONTINUE
39:          M = NTRMS + 1
40: *
41: *              FIND CONTRIBUTION OF A(0) AND TAP WEIGHTS AT END OF LINE
42:          SIGOUT = A(1)*Z(NTRMS) + .5*A(M)*(SIGIN + Z(NVAL))
43: *
44: *                   FIND CONTRIBUTION OF REMAINING TAP WEIGHTS
45:          NTM1 = NTRMS - 1
46:          K = NVAL
47:          DO 12 I=1,NTM1
48:          M = M - 1
49:          K = K - 1
50:          SIGOUT = SIGOUT + .5*A(M)*(Z(I) + Z(K))
51:  12      CONTINUE
52: *
53: *              UPDATE DELAY LINE
54:          K = NVAL + 1
55:          DO 14 I=2,NVAL
56:          K = K - 1
57:          J = K - 1
58:          Z(K) = Z(J)
59:  14      CONTINUE
60:          Z(1) = SIGIN
61:          RETURN
62:          END
```

TABLE 12-4 A Comparison of Digital Filter Design Techniques

Standard z-Transform	Bilinear z-Transform	Matched z-Transform	Fourier Series
Yields a parallel realization	Yields parallel or series realization	Yields a series realization*	Yields a nonrecursive or transversal realization
Requires partial fraction expansion of filter transfer function	Requires pre-warped filter transfer function	Requires factored form of filter transfer function	Requires a mathematical or actual representation of the frequency response
Preserves shape of impulse-time response	Preserves flat magnitude gain-frequency response characteristics	Preserves shape of frequency response characteristics	Finite memory and therefore finite time response. Magnitude is a finite Fourier series approximation to the actual response and therefore linear in phase
Suitable for band-limited functions (lowpass and bandpass) only	Suitable for all filter types especially wide bandwidth filters	Suitable for all types but may require insertion of additional zeros at the half-sampling frequency	Suitable for all types but may require a large number of terms to achieve either bandpass or bandstop filters
Exponential transform	Algebraic transform	Exponential transform	Requires Fourier coefficients
$\dfrac{R}{s+\alpha} \to \dfrac{RT}{1 - z^{-1}\exp(-\alpha T)}$	$s + \alpha \to \dfrac{2}{T}\dfrac{(1 - z^{-1})}{(1 + z^{-1})} + \alpha$	$s + \alpha \to 1 - z^{-1}\exp(-\alpha T)$	$a_n = \dfrac{4}{\omega_s}\int_0^{\omega_s/2} H(\omega)\cos\dfrac{2\pi n\omega}{\omega_s}\,d\omega$ or $b_n = \dfrac{4}{\omega_s}\int_0^{\omega_s/2} H(\omega)\sin\dfrac{2\pi n\omega}{\omega_s}\,d\omega$

* Parallel form may be obtained by finding the partial fraction expansion of the matched z-transform function.

12.7 SUMMARY

Three mathematical transformations have been presented for designing recursive digital filters from continuous filter transfer functions, namely: the standard z-transform, the bilinear z-transform, and the matched z-transform. When the sampling frequency is high compared to the upper passband cut-off frequency of the filter, all three z-transforms yield similar results. In general, each design requirement must be evaluated separately and a suitable transformation chosen accordingly.

In addition to the three transformations, attention was focused on the use of guard band and complementary filters as well as a consideration of quantization effects in recursive designs. A design technique for nonrecursive filters was also presented which utilized a Fourier series representation. Generally, nonrecursive designs are less sensitive to coefficient rounding than are recursive designs. However, nonrecursive designs usually require more storage and multiplication operations (i.e., more hardware) than do recursive designs.

Finally, examples were given of software implementation for the recursive and nonrecursive designs discussed. The evaluation of the frequency response of both types of designs was also presented. A brief comparison of the design techniques presented is given in Table 12-4.

PROBLEMS

12.1 Show that the highpass prototype filter for any Butterworth design is related to the lowpass prototype by the relationship:

$$H_{\text{HI}}(s) = s^N H_{\text{LO}}\,(s),$$

where N is the order of the lowpass design. (The Butterworth prototype is given by

$$H_N(s) = 1/\prod_{n=1}^{N} (s + \alpha_n) \text{ with } \alpha_n = \text{a complex root of unity.})$$

12.2 Find lowpass, bandpass, bandstop, and highpass transfer functions of $1/(s + 1)$. The nominal lower radian cut-off frequency for the bandpass, bandstop, and highpass filter is ω_1, while the upper radian cut-off for the lowpass, bandpass, and bandstop is ω_2.

12.3 Given the continuous transfer function

$$H(s) = \frac{s + \alpha}{(s + \alpha)^2 + \beta^2},$$

(a) Find the standard z-transform of $H(s)$ and draw a block diagram of $H(z^{-1})$.
(b) Find the bilinear z-transform of $H(s)$. (What approximate prewarped value should be used for β?) Assume $\alpha \ll \beta$.
(c) Find the matched z-transform of $H(s)$ and indicate how to normalize the digital function to have unity gain at $\omega = \beta$.

12.4 With reference to (12.19) find the maximum value of RC for which $H(z^{-1}) \leq 0.1$ at $\omega = \pi/T$.

12.5 Show that the standard z-transform and the bilinear z-transform of $H(s) = a/(s + a)$ are approximately equal for small values of ω when $\omega_s \gg 1$.

12.6 The bilinear z-transform pairs given by (12.23) and (12.24) are improper fractions which can be reduced to proper fractions plus a constant. Perform this reduction.

$$\left(\frac{(1 + z^{-1})A_0}{1 + B_1 z^{-1}} = C_0 + \frac{D_0}{1 + z^{-1}} \quad \text{and} \right.$$

$$\left. \frac{(1 + z^{-1})(A_0 + A_1 z^{-1})}{1 + B_1 z^{-1} + B_2 z^{-2}} = E_1 + \frac{F_0 + F_1 z^{-1}}{1 + B_1 z^{-1} + B_2 z^{-2}}\right)$$

$$C_0 = ? \qquad E_1 = ?$$
$$D_0 = ? \qquad F_0 = ?$$
$$F_1 = ?$$

12.7 With reference to (12.44) and (12.46), show that $B_1^{1/M}$ is essentially equal to $(1 - \alpha T_2)$.

12.8 The Hamming window function given by (12.54) may also be used as a lowpass digital filter function.
(a) Find the Laplace transform of the continuous Hamming window function, and thereby determine a recursive digital filter representation (with the aid of the standard z-transform) for this function.
(b) What is the gain of this function at $\omega = 0$?
(c) Draw the block diagram of the function using only 3 multipliers and any number of power of 2 scalars (e.g., 1/4, 1/2, 2, 4, etc.) in place of multiplication by 2^N.

12.9 Draw a block diagram for a nonrecursive digital filter realization of a Fourier sine series of 4 terms.

12.10 The frequency magnitude characteristic of the "ideal" bandpass filter, shown below,

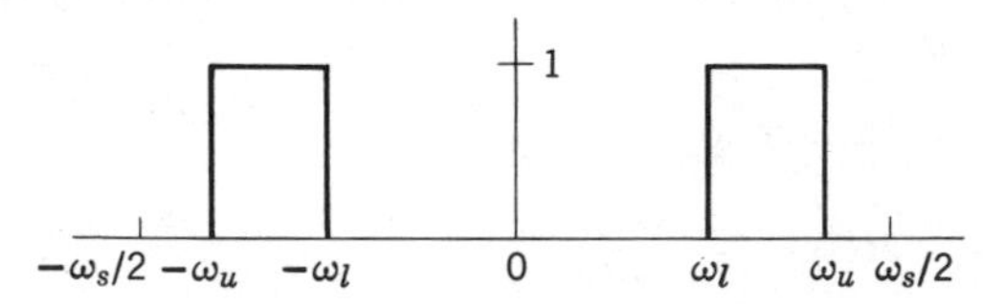

may be written as

$$M(\omega) = U(\omega + \omega_u) - U(\omega + \omega_l) + U(\omega - \omega_l) - U(\omega - \omega_u).$$

Find the tap weights (i.e., coefficients) for a nonrecursive (transversal) digital filter that will approximate this response.

REFERENCES

[1] E. I. Jury, *Sampled-Data Control Systems*, John Wiley, New York, 1958.

[2] E. I. Jury, *Theory and Application of the z-Transform Method*, John Wiley, New York, 1964.

[3] R. M. Golden, "Digital computer simulation of a sampled-data voice-excited vocoder," *J. Acoust. Soc. Am.*, **35**, 1358–1366 (Sept., 1963).

[4] I. J. Sandberg, "Floating point roundoff accumulation in digital-filter realizations," *Bell System Tech. J.*, **46**, No. 8, 1775–1779 (Oct., 1967).

[5] C. Weinstein and A. V. Oppenheim, "A comparison of roundoff noise in floating point and fixed point digital filter realizations," *Proc. IEEE*, **57**, 1181–1183 (June, 1969).

[6] R. B. Blackman and J. W. Tukey, *The Measurement of Power Spectra*, Dover Publications, New York, 1958.

[7] J. F. Kaiser, "Digital filters," *System Analysis by Digital Computer*, F. F. Kuo and J. F. Kaiser, Ed., John Wiley, New York, 1966, 218–285.

[8] R. M. Golden and J. F. Kaiser, "Root and delay parameters for normalized Bessel and Butterworth low--pass transfer functions," *IEEE Trans. Audio Electroacoustics*, **AU-19**, No. 1, 64–71 (March, 1971).

[9] R. M. Golden and J. F. Kaiser, "Design of wideband sampled-data filters," *Bell System Tech. J.*, **43**, Part 2, 1533–1546 (July, 1964).

[10] R. M. Golden, "Digital filter synthesis by sampled-data transformation," *IEEE Trans. Audio Electroacoustics*, **AU-16**, No. 3, 321–329 (Sept., 1968).

[11] R. M. Golden and S. A. White, "A holding technique to reduce number of bits in digital transfer functions," *IEEE Trans. Audio Electroacoustics*, **AU-16**, No. 3, 433–436 (Sept., 1968).

[12] H. Wayland, *Differential Equations*, D. Van Nostrand, Princeton, New Jersey, 1957, Chapters VI and VIII.

[13] A. Papoulis, *The Fourier Integral*, McGraw-Hill, New York, 1963.

[14] J. W. Cooley and J. W. Tukey, "An algorithm for the machine calculation of complex Fourier series," *Math. Comput.*, **19**, 297–301 (April, 1965).

[15] G. D. Bergland, "A guided tour of the fast Fourier transform," *IEEE Spectrum*, **6**, 41–52 (July, 1969).

[16] R. M. Golden, "Digital computer simulation of sampled-data communication systems using the block diagram compiler: BLODIB", *Bell System Tech. J.*, **45**, 345–358 (March, 1966).

Index